HARLOW & HARRAR'S
TEXTBOOK OF OF DENDROLOGY

McGraw-Hill Series in Forestry

HARLOW & HARRAR'S
TEXTBOOK OF DENDROLOGY

NINTH EDITION

JAMES W. HARDIN
**Emeritus Professor of Botany,
North Carolina State University, Raleigh**

DONALD J. LEOPOLD
**Distinguished Teaching Professor,
Environmental and Forest Biology,
State University of New York College of
Environmental Science and Forestry, Syracuse**

FRED M. WHITE
**Chief Forester,
The Forestland Group LLC,
Chapel Hill, North Carolina**

Boston   Burr Ridge, IL   Dubuque, IA   Madison, WI   New York   San Francisco   St. Louis
Bangkok   Bogotá   Caracas   Lisbon   London   Madrid
Mexico City   Milan   New Delhi   Seoul   Singapore   Sydney   Taipei   Toronto

McGraw-Hill Higher Education

A Division of The McGraw-Hill Companies

HARLOW & HARRAR'S TEXTBOOK OF DENDROLOGY, NINTH EDITION

Published by McGraw-Hill, an imprint of The McGraw-Hill Companies, Inc., 1221 Avenue of the Americas, New York, NY 10020. Copyright © 2001, 1996, 1991, 1979, 1968, 1958, 1950, 1941, 1937 by The McGraw-Hill Companies, Inc. All rights reserved. Copyright renewed 1969, 1965 by William M. Harlow and Ellwood S. Harrar. All rights reserved. Copyright renewed 1978 by William M. Harlow. All rights reserved. No part of this publication may be reproduced or distributed in any form or by any means, or stored in a database or retrieval system, without the prior written consent of The McGraw-Hill Companies, Inc., including, but not limited to, in any network or other electronic storage or transmission, or broadcast for distance learning.

Some ancillaries, including electronic and print components, may not be available to customers outside the United States.

 This book is printed on recycled, acid-free paper containing 10% postconsumer waste.

5 6 7 8 9 0 QPF/QPF 0 9 8 7 6 5

ISBN 0–07–366171–6

Vice president and editor-in-chief: *Kevin T. Kane*
Publisher: *Edward E. Bartell*
Developmental editor: *Pat Forrest*
Marketing manager: *Debra A. Besler*
Associate producer: *Judi David*
Project manager: *Vicki Krug*
Production supervisor: *Sandy Ludovissy*
Coordinator of freelance design: *Rick D. Noel*
Cover designer: *Annis Wei Leung*
Cover photography by: *Donald J. Leopold*
Senior photo research coordinator: *Carrie K. Burger*
Compositor: *Precision Graphics*
Typeface: *10/12 Times Roman*
Printer: *Quebecor Printing Book Group/Fairfield, PA*

On the cover: Large ponderosa pine in foreground; incense-cedars and ponderosa pines in background. Taken at Yosemite National Park, California. Photo by Dr. Donald J. Leopold with a Canon camera using Ektachrome 100.

Library of Congress Cataloging-in-Publication Data

Hardin, James W. (James Walker), 1929–
    Harlow & Harrar's textbook of dendrology / James W. Hardin, Donald J. Leopold,
Fred M. White. — 9th ed.
      p. cm.
    Rev. ed. of: Textbook of dendrology / William M. Harlow . . . [et al.]. 8th ed. © 1996.
    Includes bibliographical references (p.).
    ISBN 0–07–366171–6
     1. Trees—United States.   2. Trees—Canada.   3. Forest plants—United States.
4. Forest plants—Canada.   I. Leopold, Donald Joseph, 1956– .   II. White, Fred M.
III. Textbooks of dendrology.   IV. Title.

QK110.T48   2001
582.16'0973—dc21
                                     00–024601
                                        CIP

www.mhhe.com

CONTENTS

ABOUT THE AUTHORS

JAMES W. HARDIN received his Ph.D. in systematic botany from the University of Michigan. After teaching dendrology for several years, he went on to teach systematic botany and plant morphology for 35 years before his retirement from North Carolina State University in 1996. He twice received an Outstanding Teaching Award at NCSU, and in 1991, received the Meritorious Teaching Award from the Association of Southeastern Biologists. He is the author of two books, several laboratory manuals and agricultural bulletins, and many journal articles, most dealing with woody plants. Dr. Hardin has served as President of the Southern Appalachian Botanical Society, the Association of Southeastern Biologists, and the American Society of Plant Taxonomists. He has been editor of the *ASB Bulletin* and Editor-in-Chief of the journal *Systematic Botany.*

DONALD J. LEOPOLD received his Ph.D. in forest ecology from Purdue University. He has taught dendrology at the State University of New York College of Environmental Science and Forestry since 1985, as well as various other courses, including wetland ecology. Dr. Leopold has received a number of awards at SUNY-ESF, including the Student Association Distinguished Teaching Award, the College Foundation Award for Exceptional Achievement in Teaching, and the ESF Public Service Award. Dr. Leopold was editor of the *Natural Areas Journal* for six years and is author or coauthor of over 40 refereed publications, six book chapters, two other books, and many other papers.

FRED M. WHITE received his M.F. degree from Duke University and for 23 years was Director of the Duke Forest and Assistant Professor at Duke University School of Forestry and Environmental Studies, where he taught silviculture, dendrology, forest mensuration, and forest fire behavior. For the following 15 years he was Assistant State Forester for Forest Management with the NC Division of Forest Resources. He is at present Chief Forester for the Forestland Group, a timber investment management organization managing eastern U. S. hardwood lands. He is coauthor of several volumes and author of numerous papers relating to silviculture, wetlands, and forest planning.

PREFACE

We appreciate the constructive comments, corrections, and suggestions regarding the eighth edition that have been received from several individuals and the reviewers. These have led to our changes in this ninth edition.

Two major and obvious changes have been made in this edition. They involve the authors and the title. This textbook was begun in 1937 by Dr. William M. Harlow, Professor of Wood Technology, College of Forestry, State University of New York, Syracuse, and Dr. Ellwood S. Harrar, Professor of Wood Technology, School of Forestry, Duke University, Durham, North Carolina. These two men had been graduate students together in the late 1920s at the New York State College of Forestry and by the mid 1930s were teaching dendrology at their respective institutions. They felt the need for a textbook that covered just the species of greatest importance to forestry. It was an immediate success, and they continued to improve it through the fifth edition published in 1969. Professor Harrar died in 1975, prior to the sixth edition published in 1979, and Fred M. White, then Professor of Silviculture at Duke University, was added as an author. White had a major role in the preparation of that edition. Following the death of Professor Harlow in 1986 and in planning for the seventh edition, Fred White asked Dr. James Hardin, Professor of Botany, N.C. State University, to be coauthor and to assume the major responsibility for its publication in 1991. These two published the eighth edition in 1996 with the authorship still listed as Harlow, Harrar, Hardin, and White. In preparation for the ninth edition, and following the retirement of Hardin, they invited Dr. Donald Leopold, Professor of Environmental and Forest Biology, State University of New York, Syracuse, to be an author and to be the eventual senior author of this textbook. It is appropriate to find someone who has the enthusiasm and excellence that both Harlow and Harrar had for teaching dendrology. To both commemorate the original authors as well as to acknowledge the present authorship, the title is changed to *Harlow & Harrar's Textbook of Dendrology by Hardin, Leopold, and White.*

The literature on dendrology and allied fields is expanding at a rapid rate. It is necessary, therefore, to include numerous small, but significant, changes throughout the text to keep it up-to-date. We have attempted to update the information where necessary and address an expanded scope of dendrology while trying to not increase the size. Two short discussions have been added, one dealing with old-growth forests and one on dendrochronology. Both topics represent important information for the dendrology student.

We greatly appreciate and are pleased to acknowledge the contributions to this ninth edition from Elizabeth A. Hardin, SAS Institute, Cary, North Carolina, for the original line drawings used in Chapter 7 and for her expertise in technical writing and use of computers; Stephen B. Bambara and Nancy Leidy, N.C. State University, for use of their SEM photos of pollen grains; the U.S. Forest Service Photo Collections, National Agricultural Library, for use of many photographs; Dr. Larry Mellichamp, Biology Department, University of North Carolina, Charlotte, for use of his photograph of *Carpinus* fruit; and Dr. David Stahle, Tree-Ring Laboratory, University of Arkansas, for use of his photograph of cypress tree rings. Maximum tree size information, especially heights, has been revised primarily with the generous assistance from Robert Leverett, Holyoke, Massachusetts, and Dr. Robert Van Pelt, University of Washington. Recent, very precise measurements indicate that many maximum tree heights of "record" trees were substantially in error. Unfortunately, tree heights of many species still need to be verified, so some of the heights reported here may still be in error. Information about tree size and age was used here when numbers could be verified or numbers seemed reasonable.

We also thank the reviewers of this edition for their careful reading of the manuscript and useful suggestions.

James W. Hardin

Donald J. Leopold

Fred M. White

1

INTRODUCTION

A knowledge and understanding of the names of trees, their habits, ranges, habitats, and distinguishing characteristics are basic to all studies in forestry, environmental sciences, and advanced studies of the botany, genetics, and chemistry of tree species. Although *dendrology* literally means "the study of trees," the term has been variously defined and debated over the years, reflecting to a great extent what various instructors think should be included in a dendrology course. The circumscription established in 1937, the first edition of this textbook, was the identification and classification of woody plants including geographical distributions and silvical characteristics. We basically agree, but expand the scope somewhat in saying that *dendrology covers those subdisciplines of botany and forestry that deal with the systematics (including nomenclature), morphology, phenology, ecology, geographic range, and natural history of trees.*

The expected and assumed expertise of the dendrologist has also broadened in recent years and continues to expand to meet the needs of employers and the public in general. The USDA Forest Service has recently established a National Urban and Community Forestry (U&CF) Program. This recognizes the need for comprehensive management of sustainable and productive forests and related natural resources in populated areas. These will serve the environmental, social, economic, physical, and psychological well-being of people in large, medium, and small communities. It is therefore important for dendrologists of the twenty-first century to go beyond just the most important forest trees and to also know the commonly cultivated, ornamental, native, or exotic trees along streets and in lawns, city parks, and greenways. Appropriate references for ornamental trees are Dirr (1997, 1998), Griffiths (1994), and Spongberg (1990).

In addition, the field of wildlife biology places emphasis on the trees that provide cover, nesting sites, and food for various wildlife. Renewed interest in natural history demands some knowledge of edible or poisonous plants and plant parts, as well as their various practical uses by people past and present. Interest in global forestry requires knowledge of important international trees in terms of their economic uses, conservation, and management. The present interest and concern for wetland delineation and management makes it important to identify the tree species that are good indicators of wetlands or that at least usually inhabit such sites. Modern dendrology, therefore, encompasses a broad range of topics concerning trees.

Trees are the largest and some of the most impressive plants in the world. It is not only their size, but their age, shape, dominance in their community, beauty of flowers, fruits, leaves, and potential uses by man that inspire a feeling of wonder and excitement in all of us. The importance of trees to the world (human and the rest of the biota) as sources of food, shelter, shade, fuel, wood products, oxygen from photosynthesis, absorption of CO_2, and erosion control expand our appreciation of the key roles of woody vegetation. To learn their identity, relationships, life history, habitats, distribution, and uses adds to the appreciation of these giants of the plant kingdom. We hope your interest in nature will be enhanced by the study of dendrology. We also hope that your powers of observation, discrimination, conceptualization, and your general curiosity will be greatly expanded by this experience.

The world's population reached just over 6 billion at the beginning of this century. People share over 12 billion acres (over 5 billion hectares) of forests, open woodlands, and shrublands (i.e., about 40% of the earth's land surface). The enormous demands on this area in terms of economic yield, conservation, environmental impacts, sustainability, and management make it imperative that the dendrologist think in terms of "international forestry" (Laarman and Sedjo, 1992). The general principles and techniques of identification learned first as a dendrology student with your local trees could be applied internationally. Additional information and color pictures regarding important trees of the world may be found elsewhere, such as in Hora (1981).

To most of us, a tree is quite distinctive; yet, the line of demarcation between trees and shrubs is by no means clear. A given species may be shrubby near the extremities of its range or at or near timberline and still attain large proportions elsewhere. For example, the Alaska-cedar is ordinarily a moderately large tree, but in exposed situations at timberline it is often reduced to a dwarfed or even heathlike shrub; and white spruce, an important pulpwood tree of eastern Canada, is shrubby in habit in the far north. Poison-sumac is generally a shrub in the bogs of the Great Lakes area, yet it can be a 25-ft tree in the low woods of Georgia. Similarly, the distinction between shrubs and vines is often poorly defined; for example, several native species of woody plants are at first vinelike and then become shrubby as they approach maturity, or, in other cases, some shrubs may become vinelike at the ends of long branches.

For the sake of convenience, a tree may be defined as a woody plant that at maturity is 13 ft (4 m) or more in height, with a single trunk at least 3 inches (7.5 cm) in diameter at breast height ($4\frac{1}{2}$ ft or 1.37 m), unbranched for at least several feet above the ground, and having a more or less definite crown (Little, 1979). Shrubs, in contrast, are smaller and usually exhibit several erect, spreading, or prostrate stems with a general bushy appearance. Lianas are climbing woody or herbaceous vines; plants of this sort are extremely numerous in the rain-drenched forests of the tropics, although a few species such as the wild grape, moonseed, trumpet creeper, crossvine, greenbrier, poison-ivy, and Virginia creeper are indigenous to temperate regions.

The area included in this text is North America, excluding Mexico and extreme southern California, southern Texas, and southern peninsular Florida. These southern areas of the United States have about 120 species of subtropical trees that would greatly expand the scope of this book, and although fascinating, few of these trees are important to forestry in the United States. North America, north of Mexico, has about 750 native and naturalized species of trees. Only about 200 are considered as important forest trees. Over 270 are described in this textbook.

This book is not an identification manual and does not cover all the woody species for any particular portion of the country. Detailed keys to species are omitted. Comparative tables covering family, generic, and specific features are provided in addition to the keys to genera. These generic keys should give you experience with the use of keys as well as the characteristics of genera. More time should be spent with keys to local species.

Although a few nomenclatural changes reflect recent research and literature such as *Flora of North America* vol. 2 (1993) and vol. 3 (1997), the nomenclature generally follows the USDA Forest Service checklist (Little, 1979), as do vernacular or common names. Additional commonly used names have been added for some species. Names of cultivated or other economically important species follow Bailey et al. (1976), Dirr (1998), Griffiths (1994), Mabberley (1997), or Terrell et al. (1986).

USE OF THIS TEXTBOOK

The first seven chapters are intended to provide you with a knowledge and understanding of, and appreciation for, plant classification, nomenclature, identification, tree variation, habitat/range/community, and morphology. These are requisite to all work in dendrology. Your instructor may opt to rearrange, omit, minimize, or expand some of these chapters to best fit the level and objectives of your particular course. The generic keys in Chapter 4 are designed to illustrate the different kinds of keys, give you an opportunity to practice using keys, and help you learn genera of trees and their key characteristics. They are not intended to be all-inclusive field keys to the trees of North America. Chapter 7 on morphology is of primary importance and should be used as an illustrated glossary in addition to the more complete glossary at the end of the book.

Chapters 8 and 9, describing gymnosperms and angiosperms, are arranged into families following a classification system based upon evolutionary relationships. The families are briefly described, and then the distinguishing characteristics of genera and species are given within the families. The most important species, in larger print size, have the most detailed descriptions, often with tables for comparison of the most important characters, discussions, illustrations, and a range map. Less important species, in smaller print, are described more briefly. Geographical area, general habitats, and elevations are given under the heading of "Range." More detailed discussions of tree sizes, habitats, associates, tolerance, economic importance, uses, diseases, and natural history information are given under the heading of "General Description."

Chapter 10 provides a brief discussion of dendrochronology. This is a fascinating study that contributes to our understanding of tree growth and long-term environmental trends.

Chapter 11 gives you an indication of the broad field of dendrology available to you after finishing this basic course. It can be a stimulus to those who may wish to further their investigation into woody plants at the graduate and professional levels.

The glossary should help in understanding the many terms encountered in a course such as this. The reference section serves as a major resource for those interested in the study of trees. It includes recent as well as some older literature.

2

CLASSIFICATION

Classification is the ordering of items into groups having common characteristics and into a hierarchy of successively more inclusive groups. It therefore involves two processes, grouping and ranking. We do this almost instinctively to almost everything in our daily lives, and this organization or compartmentalization aids us in bringing order out of chaos and increases our efficiency of recalling and using information. Think of cars, clothes, furniture, jewelry, tools, or almost anything else. In your own mind, you can think of different kinds of these items and a hierarchical grouping for them. For cars, there are different makes such as GM/Ford/Toyota, then sedans/SUVs/vans/trucks, 4/6/8 cylinder, standard/automatic transmission, 2-wheel/4-wheel drive, etc. This is a classification. Go into any store and you will find that items for sale are grouped by various features in common. You soon learn where to look for certain items, such as kitchen utensils or paints, which makes your shopping much easier and more efficient. You use a classification scheme.

Plant classifications follow the same logic. The system is a hierarchy of categories, each level having the characteristics of those above plus other characteristics of their own. The classification of Arizona cypress and white oak are given as examples in Table 2-1. In this, the Pinophyta are the gymnosperms; the Magnoliophyta are the angiosperms, which are subdivided into two classes, the Magnoliopsida (dicots) and Liliopsida (monocots).

Fagaceae are those flowering dicots including oaks, beech, and chestnuts, and you would know that white oak, *Quercus alba,* would have certain characteristics that place it in each of the more inclusive categories above it. You could also safely predict that another genus of the Fagaceae, such as *Lithocarpus,* would have characteristics of the family and all categories above. Such a classification expresses relationships based on similarities and serves as a powerful data storage and information retrieval system.

TABLE 2-1   SYSTEM OF CLASSIFICATION

Division	Pinophyta	Magnoliophyta
Class	Pinopsida	Magnoliopsida
Subclass		Magnoliidae
Order	Pinales	Fagales
Family	Cupressaceae	Fagaceae
Genus	*Cupressus*	*Quercus*
Species	*Cupressus arizonica*	*Quercus alba*
	(Arizona cypress)	(white oak)

Systematic botany began with classifications based on uses and obvious differences and on predetermined, or *a priori,* decisions about the characteristics to be used. These were *artificial* classifications. For instance, Linnaeus (1753) based his system of classification simply on numbers of stamens and pistils in flowers, so all flowers, for instance, with 10 stamens and 1 pistil, were classified together. This combined redbud and chinaberry—unrelated and in different orders according to modern beliefs. But, it was a simple system to use and served to popularize botany in those times. From the late 1700s to the beginning of the 1900s, classifications were based on character similarities shown by the plants themselves (i.e., an *a posteriori* grouping based on many characters in common). These were the *natural* systems based on observation rather than pure logic. This gave rise to most of our presently recognized families, genera, and species and led to a simple linear sequence of families still used in many floristic manuals and herbaria.

Post-Darwinian thinking led to the realization that these similarities could mean descent from a common ancestor or evolutionary lineage and that groups were "natural" because they were derived from a common ancestor. The evolutionary lineage of a group is its *phylogeny,* and classifications of the early and middle 1900s attempted to be *phylogenetic.* Thus, a taxon at any level should represent one evolutionary line evolved from a common ancestor. Taxonomic methods today utilize many features and techniques (morphology, anatomy, cytology, palynology, analytical chemistry of secondary products, enzymes, ribosomal RNA, and nuclear, mitochondrial, or chloroplast DNA) in an attempt to develop a realistic phylogeny based on as many characteristics as possible.

Although evolutionary thought and phylogenetic classifications gave new meaning to "natural groups," they had very little effect on the recognition of the families, genera, and species described earlier. Phylogenetic classification is a three-dimensional system of branches—an evolutionary tree—so the arrangement of families was changed drastically. Any linear sequence is, of course, artificial.

One of the difficulties of reconstructing plant phylogeny is in recognizing *convergent evolution,* the independent origin of similar features in different evolutionary lineages. The question always arises as to whether similarities among groups or individuals reflect common ancestry (*patristic similarities*) or convergence (*homoplastic similarities*). Homoplastic similarities may be due to similar selective pressures from environmental adaptations or reproductive strategies, or nonadaptive trait fixation. Examples are the similar spiny fruits of Ohio buckeye and horsechestnut of Europe or the serotinous cones of unrelated pines. Misinterpretation of homoplastic for patristic similarities can lead to differences of opinion regarding relationships and therefore different classifications. As techniques of comparative study continue to improve, classifications should change to reflect new information brought to bear on evolutionary relationships. One of the keys to a successful classification is the realization that some plant attributes are more useful than others in determining relationships and establishing a classification. Indeed some variation may be misleading or simply environmental or evolutionary noise. One of the tasks in plant systematics is to determine the most dependable characteristics, and from these, the most realistic phylogenetic classification.

In the early 1960s, an attempt was made to describe the actual character changes and branching patterns of evolutionary lines—known as *cladistics*. Precise analyses of large data sets using computer-generated comparisons were possible, and *cladograms* or phylogenetic trees of many plant groups began to appear in the literature. Many of these have been based on molecular as well as morphological markers. About this same time, there was a revival of phenetic classifications, and phenograms were also computer-generated with the use of several techniques of cluster analysis. For additional reading on classifications, see Cronquist (1988), Jones and Luchsinger (1986), Judd et al. (1999), Radford (1986), Stuessy (1990), Woodland (1997), and Zomlefer (1994).

TAXONOMIC CATEGORIES

Species

(Same word for singular or plural; abbreviations: sp. [sing.], spp. [pl.]. The word *specie* means "money in coin" and cannot be used as the singular of biological species.) In classifications, the species is the focal point—the basic unit of taxonomy and classification. It is the basic group of potentially interbreeding individuals that share many characteristics in common and are more or less distinct from related species in appearance, often ecological requirements, and reproduction. We also think of species as the smallest group whose members share mutual phylogenetic relatedness. To the conservation biologist, species are the basis of biodiversity. In Chapter 5 we consider kinds of within-species variation; however, in defining species we demand a recognizable, consistent, and persistent discontinuity in characters between closely related species. Such discontinuities in characters generally reflect a breeding discontinuity (i.e., lack of interbreeding between species), but this in itself cannot be used as a primary criterion for recognizing plant species. Hybridization between species is fairly common in plants, yet the species maintain their relative distinctness due to a fairly effective barrier to wholesale genetic exchange. A species is a concept that is variously defined, so there can be disagreements among taxonomists about the level of recognition of a taxon. A species by one taxonomist may be a variety or subspecies by another and vice versa. These are taxonomic decisions and some taxonomists maintain a much broader view of the species than others. Also, decisions regarding the level of recognition may change due to a recent and better understanding of the patterns of variation. A flexible species concept is obviously necessary to accommodate the complexity of nature as well as our level of understanding.

Infraspecific Taxa

Some patterns of within-species variation are appropriately recognized by the categories of *subspecies, varietas* (variety), and *forma* (form), in descending order. The classical category below the species was variety, and it is still the most commonly used, although subspecies has become more popular since the 1950s. There are no agreed-upon distinctions between these two categories, and either may be used as the sole infraspecific category representing a genetic and morphological subdivision of a species. They are sufficiently distinct to warrant formal recognition but not enough to be called a species, mainly due to the lack of consistent discontinuities. These infraspecific categories usually reflect a geographical part of the species range, such as the three varieties of American basswood, the coast Douglas-fir and Rocky Mountain Douglas-fir varieties, or the Rocky Mountain bristlecone pine and Great Basin bristlecone pine varieties. Some varieties may represent an ecotype with some reproductive discontinuity, as *Juniperus virginiana* var. *silicicola.* Or they may reflect an early stage in speciation (see Chapter 5). Although seldom done, both subspecies and variety may be used if the pattern of variation warrants this much hierarchical classification.

The category of "form," seldom used now, is of lesser rank than variety, and where used, defines distinct phenotypes of individuals rather than populations. These were often based on some unusual pubescence, flower color, or atypical leaf shape.

The "variety" used in horticulture is now correctly known as *cultivar.* This is a distinctive and uniform morphological or physiological type artificially maintained and propagated. Some have originated in the wild, but the majority have arisen in cultivation either spontaneously or as induced mutants or artificial hybrids. The important point is that they are maintained by cultivation. Examples are the copper beech, *Fagus sylvatica* 'Cuprea', the yellow-berried American holly, *Ilex opaca* 'Canary', and the frequently seen Lombardy poplar, *Populus nigra* 'Italica'.

Superspecific Categories

The categories above the level of species serve to show natural (i.e., evolutionary) groupings of the taxa below them. *Genus* (*genus,* singular; *genera,* plural; there is no such word as *genuses*) functions to group together similar species and indicate the relative differences/similarities among other genera. A genus is *monotypic* if it includes only one species. In like manner, a *family* (family, singular; families, plural) is composed of genera, *order* is composed of families, and *class* is composed of orders. The top level within the green plant kingdom is the *division* (or *phylum*). Depending on the number and complexity of the groupings recognized, there are additional categories available, such as subclass (see the oak classification in Table 2-1), subfamily, subgenus, etc. The hierarchical system is designed to accommodate the range of information available for particular groups.

These various categories are recognizable by the endings of the words. Note the accents for pronunciation (see Chapter 3).

Division	-ó-phy-ta	Order	-à-les
Class	-óp-si-da	Family	-à-ce-ae
Subclass	`-i-dae		

Family names are treated as plural nouns; therefore, it is correct to say "The Pinaceae *are* of great economic importance." The ending -aceae is pronounced as -à-see-ee, the *ae* being a diphthong and pronounced as a long *e*. There are several exceptions to this -aceae ending, sanctioned by long use and permitted as correct alternatives. Two of these, which include trees, are Palmae (Arecaceae, the palms) and Leguminosae (Fagaceae, the legumes).

As previously indicated, plant classification is partly a matter of opinion, and various classifications differ at all levels of the system. Some families recognized here differ from those in earlier editions of this textbook. For example, legumes are treated as three separate families rather than the single Leguminosae or Fabaceae. On the other hand, the Taxodiaceae is here combined with the Cupressaceae based on some recent studies. Again, plant classifications are dynamic rather than static and should reflect the best information available at the time and therefore should change as scientific data dictate.

The assignment of names to organisms is *nomenclature,* and we are concerned here with botanical nomenclature. This is an integral and supporting part of systematics. Historically, there have been two nomenclatural systems, *common names* and *scientific names.* The forester and environmental scientist are well advised to learn both in order to communicate effectively with all people.

Common or Vernacular Names

Common names have a place in everyday speech and writing, but they are not precise enough to be used in very exact or scientific communication. Actually the term *vernacular,* meaning "native to a region," is more correct than the term *common.* The reason is that vernacular names are often locally used, are usable in only a single language, and often differ among regions, and therefore are not at all "common" in the sense of widespread or in general use. *Common name* is used here to mean "familiar" in contrast to the scientific name.

How trees acquire common names is an interesting story not only in dendrology but also in the social development of a people in relation to their use and enjoyment of the forest.

The chief characteristics of trees usually influence the selection of their common names, although the names of many botanists also have been commemorated in this way. A few examples are:

1 *Habitat.* swamp white oak, sandbar willow, subalpine fir, river birch, mountain hemlock

2 *Some distinctive feature.* weeping willow, bigleaf maple, whitebark pine, bitternut hickory, cutleaf birch, quaking aspen, overcup oak

3 *Locality or region.* Pacific yew, Idaho white pine, southern red oak, Ohio buckeye, Virginia pine, Carolina basswood

4 *Use.* canoe birch, sugar maple, tanoak, paper-mulberry

5 *In commemoration.* Nuttall oak, Engelmann spruce, Sargent cypress, Douglas-fir

6 *Adaptation* of names from other languages (e.g., Indian, Latin, Spanish). chinkapin, hickory, arborvitae, tacamahac poplar

The common names in this text follow those given by Little (1979), except in a few cases, and some additional ones are given as secondary alternatives. Commercial names for lumber are also given by Little (1979) and are not repeated here.

Common names are preferably written with lowercase beginning letters except for proper names, which begin with a capital letter (e.g., white oak or Gambel oak, mountain hemlock or Canada hemlock).

Scientific Names

Because of the great profusion of common names, it is necessary to have a standardized, universal system that can be used throughout the world. Such a system has been developed through the use of Latin or Latinized names because the laws governing Latin syntax remain constant relative to modern languages, which tend to evolve over time and with widespread usage. Scientific names have followed standard usage since the middle 1700s when Linnaeus (1753) published his monumental *Species Plantarum.* He gave the generic name; the species designation, consisting then of a short descriptive phrase; and finally in the otherwise blank margin, a single word set in italics. He called this the "trivial" name and used it as an indexing device. His students and followers soon found it convenient to write the generic name followed by the trivial name. This practice led to the *generic name* followed by a *specific epithet* for each plant species, the combination constituting *binomial nomenclature.*

The complete designation of a tree or other plant species consists of three parts: (1) a generic name, (2) a specific epithet, and (3) the full or abbreviated name of the person, or persons, who originally published the name and description or made a later change. Thus in the scientific name *Tilia americana* L., *Tilia* is the generic name, *americana* is the epithet, and L. (for Linnaeus) is the describing author. The name of the species is the binomial, *Tilia americana*; the epithet never stands alone. The author's name is omitted except in scientific writing, where it is used at least once to establish the plant's exact identity. It is customary to underline the binomial in handwritten or typewritten work and to set it in italics in print; the author's name, when it appears, is not underlined or italicized. Generic names begin with a capital letter; epithets begin with a lowercase letter.

Tilia americana is the scientific name of the American basswood, and no other species anywhere in the world may have this name. Other basswoods will, of course, be in the genus *Tilia* but with other epithets. A varietal name, if any, follows the binomial. For example, South Florida slash pine is known as *Pinus elliottii* var. *densa* Little & Dorman.

One frequently encounters a scientific name followed by names of authors first in parentheses and then others following the parentheses. This indicates a transfer between genera or a change in taxonomic level. *Taxodium distichum* (L.) Rich., was originally *Cupressus disticha* L. Linnaeus originally placed baldcypress in the genus with true cypresses. Later, L. C. Richard recognized the differences and made a new genus, *Taxodium,* and transferred this species to the new combination *T. distichum.* Sand post oak was originally described as *Quercus margaretta* Ashe, but later Sargent changed it to *Q. stellata* var. *margaretta* (Ashe) Sarg. In both of these cases, the author's name in parentheses is the original describer, and the second author made the transfer. The original designation is the "name-bringing synonym" or *basionym* of the new combination.

When the scientific name is followed by the names of two or more authors (e.g., *Pinus jeffreyi* Grev. & Balf.), it means that both authors were responsible for the new name. Some binomials are followed by two authors connected by "ex." For example, *Pinus torreyana* Parry ex Carr. In this case, Charles C. Parry, the discoverer, used this name on specimens but not in print, so it was not validly published. A specimen was sent to Carriere, who validly published the name with a description and credited the name to Parry. In other cases, two author names may be connected with "in," which is mainly of bibliographic interest. *Pinus edulis* Engelm. in Wisliz. indicates that Engelmann gave the name and description for the species of pinyon pine, but it was published in the book by Wislizenus. In both of these examples, the describing author's name may be given alone.

Interspecific hybrids are designated by either a formula, such as *Pinus palustris* × *taeda,* or a binomial with a "times" sign between the generic name and epithet, as *P.* × *sondereggeri.* Intergeneric hybrids are designated by the "×" before the generic name (e.g., × *Cupressocyparis* for *Chamaecyparis* × *Cupressus*).

Cultivar names are sometimes designated after an epithet either by "cv." before the name or usually by use of single quotes (e.g., *Gleditsia triacanthos* cv. Moraine or *G. triacanthos* 'Moraine'). The flowering dogwood with rich ruby red bracts is *Cornus florida* 'Cherokee Chief'. These cultivar names are governed by the *International Code of Nomenclature for Cultivated Plants* (Trehane et al., 1995).

Whenever a species is first divided into subspecies or varieties, the part of the species that includes the "type specimen"* of the species is automatically treated as an equivalent taxon to the one or more named. Its epithet is a repeat of the specific epithet and without an authority. For example, when Fernald named *Fraxinus americana* var. *biltmoreana* in 1947 as the pubescent form of white ash, *F. americana* var. *americana* was automatically established for the typical, glabrous individuals of the species. In usual practice, the subspecies or varietal names are omitted unless there is some reason to specify a particular infraspecific taxon.

Sometimes when taxonomic concepts differ among authors or when taxa have been combined, it is important to indicate which concept is being considered. For instance, in splitting the legumes into three families, Fabaceae can refer to the original, all-inclusive one, or just that part minus the Mimosaceae and Caesalpiniaceae. Two Latin terms are

---

*The specimen to which a name is permanently attached. When a new species or infraspecific taxon is described, the author designates a particular specimen, deposited in a herbarium, as the "nomenclatural type." The name of the taxon is tied to this particular specimen, which illustrates what the author had in mind when he/she described it (Greuter et al., 1994).

useful here; *s. lat.* (*sensu lato*) means "in the broad sense"; *s. str.* (*sensu stricto*) means "in the narrow sense." Thus Fabaceae s. lat. would indicate the family including all legumes.

The use of scientific names and the coinage of new ones are governed by the *International Code of Botanical Nomenclature* (Greuter et al., 1994). (See also Jeffrey, 1973; Judd et al., 1999; and Walters and Keil, 1996.) Although taxonomy is unregulated and a matter of opinion, as indicated earlier, the application of names for any group (taxon) below the order level is stringently controlled by 62 articles (the rules), which provide for a method of naming taxonomic groups, each with only one correct name, and a means of putting names of the past in order and providing for names of the future. This may involve avoiding and rejecting names that do not conform to the rules.

Scientific names are generally stable, so when an old familiar name is changed, some people are upset. One must realize, however, that there are two valid reasons for changing a scientific name: nomenclatural and taxonomic. A *nomenclatural reason* is the necessity to bring the name in compliance with the rules of nomenclature in the Code. One rule is that of "priority" (i.e., the earliest valid name must be used). A recent change of this type, for yellow buckeye, was from *Aesculus octandra* Marshall (published in 1785), used in most literature through the 1980s, to *A. flava* Solander (published in 1778) (Meyer and Hardin, 1987). A *taxonomic reason* occurs when the available evidence indicates the need for a change in classification. For example, studies by Grant and Thompson (1975) and O'Connell et al. (1988) indicated that the mountain paper birch is distinct from paper birch so it should be called *Betula cordifolia* Regel rather than *B. papyrifera* var. *cordifolia* (Regel) Fern. Likewise, the study by Gillis (1971) showed a generic difference between the sumacs and their poisonous relatives, so poison-sumac, *Rhus vernix* L., was changed to *Toxicodendron vernix* (L.) Kuntze. When a change is made for a valid reason, it takes less energy to learn the new one than it does to complain about the change. Systematics and nomenclature must keep abreast of new knowledge.

DERIVATIONS OF SCIENTIFIC NAMES

Scientific names come from a variety of sources just as with common names.

There are four main sources for generic names, which follow, with some examples.

1 In commemoration. *Carnegiea, Kalmia, Maclura, Washingtonia.*
2 Descriptive. *Gymnocladus, Liriodendron, Lithocarpus, Oxydendrum.*
3 Fanciful, mythological, or poetic. *Diospyros, Nyssa.*
4 Original common names in their native lands. *Asimina, Catalpa, Ginkgo, Tsuga, Yucca.*

Specific epithets have also been derived from many sources, as indicated by the following examples.

1 Descriptive of appearance or structure. *alba, flava, gigantea, latifolia, rubra, pubescens, spinosa.*
2 Descriptive of habitat. *aquatica, palustris, sylvatica.*

3 Uses. *edulis, tinctoria.*
4 Locality where first found. *alleghaniensis, californica, caroliniana, virginica.*
5 Resembling another plant. *bignonioides, strobiformis, taxifolia.*
6 Commemorative. *ashei, fraseri, menziesii, thomasii.*
7 A noun, rather than adjective, and often an old Latin or Greek common name and sometimes old generic names. *negundo, strobus.*

The endings of epithets follow Latin grammar, so they often differ (e.g., *virginiana, virginiensis, virginica,* or *virginianum*).

The following list includes most of the trees in this text. For more extensive lists and more information, see Borror (1960), Coombes (1985), Fernald (1950), Gledhill (1989), Jones and Luchsinger (1986), Judd et al. (1999), Little (1979), or Radford (1986).

Abies. Tall tree
Acer. Sharp
acerifolia. Leaves like those of maple
acuminata. Acuminate, long-pointed
acutissima. Acutely pointed
Aesculus. Ancient Latin name for "an oak with edible acorns"
Agave. Noble or admired one
agrifolia. Field leaf
Ailanthus. Reaching to heaven
alata. Winged
alba. White
albicaulis. White-barked
Albizia. For F. del Albizzi, Italian who introduced the genus into Europe
Aleurites. Mealy surface
alleghaniensis. From the Alleghany mountains
alnifolia. Leaves like those of alder
Alnus. Ancient Latin name for "alder"
alternifolia. Alternate-leaved
altissima. Tall
amabilis. Pleasing or lovely
Amelanchier. French name for "serviceberry"
americana. From America
angustifolia. Narrow leaf
anomala. Unlike others
aquatica. Growing in water
Aralia. From French-Canadian name *aralie,* for "aralias"
arboreum. Treelike
Arbutus. Ancient Latin name for "European madrone"
aristata. Awned, sharp pointed
arizonica. From Arizona
aromatica. Aromatic
ashei. For W. W. Ashe, pioneer forester of the USDA Forest Service

Asimina. From American Indian name
attenuata. Tapering
australis. Southern
azederach. Persian for "noble tree"

balfouriana. For J. H. Balfour, Scottish botanist
balsamea. Balsamlike
balsamifera. Yielding a fragrant resin (balsam)
banksiana. For Joseph Banks, British botanist and Director of Kew Gardens
barbatum. Bearded
Betula. Shining (bark)
bicolor. Two-colored
biflora. Two-flowered
bifolia. Two-leaved
bignonioides. Like *Bignonia*
biloba. Two-lobed
borbonia. For Gaston Bourbon, French patron of botany
borealis. Northern
bracteata. With bracts
brevifolia. Short-leaved
Bumelia. Ancient Greek name for "European ash"

californica. From California
calleryana. After J. Callery, French missionary
Calocedrus. Beautiful cedar
canadensis. From Canada
Carnegiea. For Andrew Carnegie, philanthropist
carolinae-septentrionalis. From North Carolina
caroliniana. From the Carolinas
Carpinus. Latin name for "hornbeam," or Celtic for "a yoke made of this wood"
Carya. Greek for "nut" or "kernel"
Castanea. Ancient Latin name from Castania, Greece, known for its trees
Castanopsis. Chestnutlike
Casuarina. Malay word describing twigs
Catalpa. Cherokee Indian name
cathartica. Purgative
Cedrus. From the River Cedron in Palestine where cedars grew
Celtis. Ancient Greek name of a tree
cembroides. Like *Pinus cembra*
Cercidium. Latin for a "weaver's comb resembling the pod"
Cercis. Ancient Greek name from *kerkis,* a weaver's shuttle appearing like the fruit
Cereus. Wax candle
cerifera. Wax-bearing
Chamaecyparis. Low-growing cypress

chinensis. From China
Chrysolepis. Golden-scaled
chrysophylla. Golden-leaved
cinerea. Ash gray
Cladrastis. Brittle shoots
clausa. Closed, serotinous
clava-hercules. Hercules' club
coccinea. Scarlet
communis. Common
comutata. Changeable
concolor. One color
contorta. Twisted
copallina. Yielding copal gum
cordifolia. Heart-shaped leaves
cordiformis. Heart-shaped
Cornus. Ancient Latin name for "cornelian cherry"
cornuta. Horn-shaped
Corylus. Ancient Latin name for "hazelnut"
coulteri. For Thomas Coulter, Irish botanist who collected in California and Mexico
crassifolia. Thick-leaved
Crataegus. Greek *kratos* meaning "strength," referring to the hard wood
Cudrania. From a Malayan name, Cudrang
Cupressus. The classical Greek and Latin name of the Italian cypress
Cycas. The classical Greek name for "a palm resembling the cycad habit"

decidua. Deciduous
decurrens. Decurrent leaf bases
deltoides. Delta-like, triangular-shaped
densiflorus. Crowded flowers
dentata. Dentate, toothed
deppeana. For Ferdinand Deppe, a German botanist
dioicus. Dioecious
Diospyros. Divine fruit
distichum. Two-ranked
diversilobum. Variable lobing
douglasii. For David Douglas, Scottish botanist and traveler in North America
drummondii. For Thomas Drummond, Scottish botanist and explorer
dunnii. For G. W. Dunn, collector in California and Mexico

echinata. Prickly
edulis. Edible
Elaeagnus. Ancient Greek name applied to a willow
elliottii. For Stephen Elliott, botanist
ellipsoidalis. Ellipsoidal in shape

emoryi. For W. H. Emory, southwest collector

engelmannii. For George Engelmann, physician and botanist in St. Louis and an authority on conifers

equisetifolia. Horsetail-like "leaves"

excelsior. Lofty, higher

Fagus. Ancient Latin name for "beech"

falcata. Sickle-shaped

Ficus. Ancient Latin name for "fig"

filifera. Bearing threads

flava. Yellowish

flexilis. Flexible

florida. Flowering

floridana. From Florida

fragilis. Fragile

frangula. Fragile or brittle twig

fraseri. For John Fraser, Scottish collector who traveled in North America

Fraxinus. Ancient Latin name for "ash"

fremontii. For J. C. Fremont, explorer of western United States

gale. Old English name for "bog turtle"

gambelii. For William Gambel, American naturalist

garryana. For Nicholas Garry, explorer in the Northwest

geminata. Twinned

giganteus. Gigantic

Ginkgo. From a Japanese name (gin-kyo)

glabra. Glabrous

glandulosa. Glandular

glauca. Glaucous

Gleditsia. For J. G. Gleditsch, German botanist and Director of Berlin Botanical Garden

Gordonia. For James Gordon, British nurseryman

grandidentata. Large-toothed

grandiflora. Showy-flowered

grandifolia. Large-leaved

grandis. Large

Gymnocladus. Bare-branched

Halesia. For Stephen Hales, British botanical writer

Hamamelis. Ancient Greek name for "another tree with pear-shaped fruit"

hemisphaerica. Hemispherical

heterophylla. Diversely leaved

Ilex. Ancient Latin name for "an oak with prickly leaves"

illinoinensis. From Illinois

imbricaria. Overlapping
incana. Gray color

japonica. From Japan
jeffreyi. For John Jeffrey, Scottish plant explorer
Juglans. Jupiter's nut
julibrissin. Native Iranian name
Juniperus. Ancient Latin name for "juniper"

Kalmia. For Peter Kalm, a student of Linnaeus who found the laurel in America
kelloggii. For Albert Kellogg, botanist
kentukea. From Kentucky

laciniosa. Shaggy
laevigata, laevis. Polished, smooth
lambertiana. For A. B. Lambert, English botanist and author of a classic work on pines
lanuginosa. Softly hairy
laricina. Like European larch
Larix. Ancient Latin name for "larch"
lasianthus. Hairy-flowered
lasiocarpa. Hairy-fruited
latifolia. Broad-leaved
laurifolia. Laurel-leaved
lawsoniana. For Peter Lawson, nurseryman of Edinburgh
leiophylla. Smooth-leaved
Leitneria. For E. F. Leitner, German naturalist
lenta. Pliable
Libocedrus. Resinous cedar
Liquidambar. Liquid amber (resin)
Liriodendron. Lily tree
Lithocarpus. Stony seed
lobata. Lobed
lutea. Yellow
lyallii. For David Lyall, Scottish naturalist
lyrata. Lyre-shaped

Maclura. For William Maclure, American geologist
macrocarpa. Large-fruited
macrophyllum. Large-leaved
magnifica. Magnificent
Magnolia. For Pierre Magnol, French botanist
Malus. Ancient Latin name for "apple"
margaretta. For Margaret H. Wilcox, Mrs. W. W. Ash

mariana, marilandica. From Maryland (although black spruce, *Picea mariana,* does not grow in Maryland, it was used here in the broad sense to mean northeastern North America)

maximum. Largest

Melia. From Greek name for "ash"

menziesii. For Archibald Menzies, Scottish naturalist

mertensiana. For Karl H. Mertens, German botanist

Metasequoia. Close (i.e., similar) to Sequoia

michauxii. For F. A. Michaux, French botanist who wrote a classic work on the trees of eastern United States and Canada

microphylla. Small-leaved

monophylla. One-leaved

monosperma. One-seeded

montana. Mountain

monticola. Mountain dwelling

Morus. Ancient Latin name for "mulberry"

muehlenbergii. For G. H. E. Muhlenberg, botanist of Pennsylvania

muricata. Rough with hard, sharp points

Myrica. Fragrant

myristiciformis. Shape of *Myristica* (nutmeg)

myrsinifolia. Myrsine- or myrtle-leaved

neomexicana. From New Mexico

negundo. Ancient Malayan common name for a tree with similar leaves

nigra. Black

nootkatensis. From Nootka Sound, B.C., Canada

nuttallii. For Thomas Nuttall, British-American botanist

Nyssa. Water nymph

oblongifolia. Oblong-leaved

occidentalis. Western

octandra. Eight-stamened

odorata. Fragrant

omorika. Serbian name for the "Serbian spruce"

opaca. Opaque, not glossy

orientalis. Oriental

Osmanthus. Fragrant flower

osteosperma. Bonelike seed or fruit wall

Ostrya. Ancient Greek name meaning "shell," for the nut covering

ovalis, ovata. Ovate-shaped

Oxydendrum. Sour tree

pagoda. Pagoda-shaped

pallida. Pale

palmetto. From Spanish common name, palmito

palustris. Swampy land
papyrifera. Paper-bearing
parryi. For C. C. Parry, botanist and explorer
parviflora. Small-flowered
Paulownia. For Anna Paulowna, daughter of Czar Paul I of Russia
pensylvanica, pennsylvanica. From Pennsylvania
Persea. Ancient Greek name for an unidentified Egyptian tree
persica. Persian
phanerolepis. Conspicuous-scaled
phellos. Corky
Picea. Pitch (resin)
Pinus. Ancient Latin name for "pine"
Platanus. Ancient Latin name for "the plane tree"
plicata. Folded
pomifera. Apple-bearing
ponderosa. Heavy, large
populifolia. Poplar-leaved
Populus. Ancient Latin name for "poplar"
prinus. Ancient Greek name for "an oak"
procera. Very tall
profunda. Deep (swamp)
Prunus. Latin for plum
pseudoacacia. False *Acacia*
Pseudotsuga. False hemlock
Ptelea. Ancient Greek name for "an elm with similar winged fruit"
pubescens. Pubescent
pumila. Dwarf
pungens. Pungent, prickly
purshiana. For F. Pursh, German botanist who collected in North America
pyramidata. Pyramid-like
Pyrus. Ancient Latin name for "pear"

quadrangulata. Four-angled
quadrifolia. Four-leaved
Quercus. Ancient Latin name for "oak"

racemosa. In a raceme
radiata. Radiating outward
radicans. With rooting stems
regia. Royal
resinosa. Resinous
Rhamnus. Ancient Greek name for "spiny shrubs"
Rhododendron. Rose tree
rhombifolia. Diamond-shaped leaves
Rhus. Ancient Greek name for "sumac"

rigida. Rigid, stiff
Robinia. For Jean Robin, French herbalist
robur. Strong, hard; Latin name for "oak"
rubens. Blushed with red
rubra. Red

Sabal. American Indian name
sabiniana. For Joseph Sabine, naturalist
saccharinum, saccharum. Sugary
Salix. Ancient Latin name for "willow"
Sapindus. Indian soap
Sapium. Latin for "a resinous tree"
Sassafras. American Indian name
scopulorum. Of cliffs or rocks
sebiferum. Producing wax
sempervirens. Always living
Sequoia. For Cherokee Indian, Sequoyah, who developed the Cherokee alphabet
Sequoiadendron. Sequoia tree
serotina. Of late season
serrata. Serrate, toothed
serrulata. Serrulate, small teeth
shumardii. For B. F. Shumard, state geologist of Texas
silicicola. Growing in sand
sitchensis. From Sitka Island, Alaska
Sorbus. Ancient Latin name for "mountain-ash of Europe"
speciosa. Showy
spicatum. Spikelike
spinosa. Spiny, prickly
stellata. Starlike (hairs)
strobiformis. Like *Pinus strobus*
strobus. Ancient Latin name for "gum-yielding tree"
styraciflua. Resin (styrax)-flowing
sylvatica, sylvestris. Of the forest
Symplocos. Connected (stamens)

taeda. Torch of resinous pine wood
Tamarix. Ancient Latin name possibly referring to the Tamaris River in Spain
taxifolia. Taxus (yew)-like leaves
Taxodium. Yew-like
Taxus. Ancient Latin name for "yew"
tetraptera. Four-winged
texana, texensis. From Texas
thomasii. For David Thomas, American horticulturist
Thuja. Ancient Greek name; resinous
thunbergiana. For K. P. Thunberg, Dutch physician, botanist, and student of Linnaeus, who introduced many Japanese plants to Europe

thyoides. Thuja-like
Tilia. Ancient Latin name for "basswood"; wing
tinctoria. Used for dyeing
tomentosa. Tomentose, densely hairy
torreyana. For John Torrey, American botanist of Columbia University
Toxicodendron. Poison tree
tremuloides. Aspen-like, trembling
triacanthos. Three-thorned
trichocarpa. Hairy-fruited
tricuspidata. Three-toothed
trifoliata. Three-leaflet
triloba. Three-lobed
tripetala. Three-petaled
Tsuga. Japanese name for "hemlock"
tulipifera. Tulip-bearing
typhina. Cattail-like

Ulmus. Ancient Latin name for "elm"
umbellata. Umbel-form
Umbellularia. Little-umbeled
unedo. Latin name for "arbutus tree"

velutina. Velvety
vernix. Varnish
virginiana. From Virginia

Washingtonia. For President George Washington
wislizeni. For F. A. Wislizenus, German physician of St. Louis who collected in southwestern United States and northern Mexico
wrightii. For Charles Wright, American collector in southwestern United States

Yucca. Carib Indian name

Zanthoxylum. Yellow wood

PRONUNCIATION OF SCIENTIFIC NAMES

Botanical Latin, distinct from Classical or Reformed Academic Latin, has no standardized, worldwide pronunciation. The sound of a word can vary because of three variables: first the sounds of the individual letters; second the short or long vowel sounds; and third, the syllable being accented. There are a number of rules governing the "correct" sound but also many exceptions to the rules. Many American botanists/dendrologists follow the traditional English system and pronounce scientific names as though they were English words; others prefer the more classical sound. Although there may be a tendency to avoid saying scientific names out loud for fear of mispronunciation, go ahead! You should realize that pronunciations vary markedly even among professionals.

Because oral communication is so necessary, at least an understandable pronunciation of tree names should be considered important along with their correct spelling and identification. Some of the general guidelines are given here.

• Vowel sounds are either "long" and sound more or less like we say the letter, or "short" and modified. For example, the long *A* in *Àcer,* the short *i* in *flóridus,* the long *i* in *bìcolor,* the short *e* in *Oxydéndrum.* In various books (Little, 1979) and in this textbook, long and short vowel sounds are indicated by two different accent marks. The grave accent (`) designates the *long* English sound of the vowel; the acute accent (´) indicates the *short* or modified sound. These accent marks are not to be used when the names are written elsewhere.

• All single vowels are pronounced.

• Double vowels that form a diphthong are pronounced as a single syllable (e.g., *ae* as in see; *au* as in cause; *eu* as in yew). In most other double vowels, both are pronounced (e.g., *-ii* as ee-eye; *-ea* and *-ia* as ee-ah).

• An *a* at the end of a word is always pronounced as a separate syllable and has an "ah" sound (e.g., *vérna* as vér-nah). An *e* at the end of a word is also always pronounced and has a long "ee" sound (e.g., *arvénse* as are-vén-see; *occidentàle* as oc-see-den-tày-lee; *vulgàre* as vul-gày-ree).

• A final *es* sounds like "ees" (e.g., *clava-hércules* as cla-vah-hér-cue-lees).

• *ch* within a word is generally pronounced as a *k* (e.g., *echinàta* as ek-i-này-ta); if at the beginning of a word, it sounds like the *ch* in "choice" (e.g., *chi-nén-sis*). *ph* has the sound of *f* (e.g., *phéllos* as fél-los).

• When *Ts, Ps,* or *Pt* is at the beginning of a word, the first letter is silent (e.g., *Tsùga* as sù-ga; *Pseudolárix* as su-do-lá-rix; *Ptèlea* as teè-lee-ah).

• The *-oi-* in many specific epithets, meaning "like," or "having the form of," is pronounced in two different ways. Some consider it as two separate vowels (following Latin rules) and say *platanoides* as plat-a-no-ì-dees. Others consider it a diphthong and pronounce it as in oil (e.g., plat-a-noí-dees; *deltoides* is either del-to-ì-dees or del-toí-dees).

• *y* within a name is always a vowel and either with the sound of a short *i* (e.g., *diphýllus* as di-fíll-us) or long *i* (e.g., *Mỳrica* as mì-ri-ca).

As for the syllable accented, there are some rules, but there are also many exceptions—some of which preserve the sound of the related English word or commemorative name. Again, some guidelines:

• The accent is *never* on the last syllable.

• Words of two syllables are *always* accented on the first syllable (e.g., *Quércus, Àcer*).

• Words of more than two syllables are accented on the next to the last syllable (*penult*) if that syllable is *long.* It is long if the penult ends in a long vowel, a diphthong, or a consonant (which makes the vowel short but the syllable long) (e.g., *banksiàna* as bank-si-à-nah; *formòsus* as for-mò-sus; *amoènus* as a-moè-nus; *decúmbens* as de-cúm-bens; *cruéntus* as cru-én-tus).

• Words of more than two syllables are accented on the third syllable from the end (*antepenult*) if the *penult* is *short.* It is therefore the penult that governs the position of the accent (e.g., *serótina* as se-ró-ti-nah; *flóridus* as fló-ri-dus; *sylvática* as syl-vá-ti-ca).

Some exceptions preserve the sound of commemorative names (e.g., *thómasii* as tómas-ee-eye; *jamesii* as jàmes-ee-eye; but yet we say *Halesia* as Ha-lè-see-ah rather than Hàle-zee-ah although it is for Steven Hales).

• The accent is never farther from the end than the antepenult, and usually only one syllable is accented.

• Words derived from a single classical root are pronounced like the related English word (e.g., *rígida* = rigid; *ovàta* = ovate).

• Words from two roots should be accented to preserve the sounds of both related English words (e.g., Rhodo-dén-dron; macro-phỳllum rather than ma-cró-phyl-lum; sem-per-vì-rens rather than sem-pér-vi-rens [although the latter is correct according to a different rule). Yet there are many exceptions, such as Zan-thóx-y-lum rather than Zantho-xỳlum (the *x* in the latter is pronounced as a *z*) and Gymnóc-cla-dus rather than Gymno-clà-dus.

For more discussion of this, and more rules for pronunciation, see Radford (1986) and Stearn (1992).

4

IDENTIFICATION

Identification of trees is one of the primary objectives of a course in dendrology and certainly a necessity for those in various disciplines dealing with vegetation. The process of identifying an unknown tree or any plant is one of mental multivariate analysis because there are numerous characters and character states that are potentially analyzed by our minds to reach a decision. (A *character* is a kind of trait [e.g., bark or leaf shape]; *character state* is the particular form of that trait [e.g., shaggy bark or ovate leaf].) Obviously, the more characters available and known, the more dependable the answer. Yet in many cases, tree species can be identified easily and correctly by analysis of less than a half dozen characters. Tree identification takes into account geographic area, habitat, form, bark, twig, leaves, taste or aroma, flowers, and fruits or cones (i.e., the whole plant in its natural setting). One consciously, or subconsciously and almost automatically after some experience, considers many characters and character states before an identification is made.

There are some shortcuts to identification. The easiest method is to ask someone who knows. Another popular method is to use illustrations in state or regional tree guides. Several are probably available in your local book store. This method can be slow, undependable, and discouraging but still appropriate for the novice. Often a better approach is to use identification keys, available to genera in this text and to species in some field guides and more technical manuals of the plants of a state or region. One must understand considerable morphological terminology to use these manuals effectively. (For a guide to floristic manuals, see Frodin, 1984, or Little and Honkala, 1976.)

When attempting to identify an unknown tree, one should analyze the character states available, possibly overall form, bark, habitat, and locality, then leaf and/or twig, odor or taste, fruit or cone, then flower and any other details. The order of analysis is not important and is generally governed by conspicuousness and availability of the various parts.

To confirm a tentative identification, one could ask someone more knowledgeable, carefully compare the unknown with available illustrations and descriptions of the suspected species, check labeled trees in a botanical garden or *arboretum* (a garden of trees), or check labeled specimens in a *herbarium*. A herbarium is a collection of dried, pressed, and mounted specimens carefully arranged in one of several sequences. Most colleges and universities throughout the world maintain a herbarium, often with several hundred thousand specimens and generally in a biology or botany department. The largest herbaria in the United States and Canada are at museums, botanical gardens, and universities. Some examples are the New York Botanical Garden (with over 5 million specimens), Harvard University Herbaria (nearly 5 million), U.S. National Herbarium at the Smithsonian Institution (ca. 4.5 million), Missouri Botanical Garden (5 million), the Field Museum in Chicago (ca. 2.5 million), and the Canadian Agricultural Herbaria in Ottawa (over 1 million). Most college or university herbaria have between 25,000 and 500,000 specimens (Holmgren et al., 1990). Information on various herbaria is available by searching by institution, city, and state or country at http://www.nybg.org/bsci/ih.html.

Identification, in the field, laboratory, or herbarium, is sometimes routine and simple, but at other times it is a real challenge—a job of detective work demanding ingenuity in pulling together all the available evidence to solve the riddle. This keeps tree identification exciting.

KEYS TO GENERA

The *genus* is the common level of recognition, even in "folk botany" or "ethnobiological classifications" throughout the world (Berlin, 1992). It is the usual group of plants, or animals, most easily recognized as distinct, based on general appearance. Most people, therefore, have a name for pines, oaks, hickories, cherries, maples, etc. These are genera. Distinguishing the family and species often requires detailed examination of more technical characters. Therefore, the genus is the most useful and practical unit for recognition and the reason to begin your study of tree identification with an artificial key to genera.

Using keys is one of the fastest and most popular methods of identifying unknowns. A key is an analytical device that usually has dichotomies or a series of two brief contradictory descriptions. To use a key, you compare the characteristics of the unknown plant to both descriptions. The one matching the plant either leads to a name for the plant or to another set of contrasting statements. You continue this pattern until, if all the correct choices have been made, you arrive at the correct name for the unknown. Making "correct" choices hinges upon your ability to understand the descriptive terminology used in the key as well as your ability to accurately interpret those features of the plant in question. (See Chapter 7 and the Glossary for this terminology.) Another possible stumbling block to using keys correctly is the nature of the key itself. You will find that some keys are ambiguous and more difficult to use than others. When you key something out, always check a description, illustration, or herbarium specimen to verify your identification. If you have made a wrong choice, you may reach a place in the key where neither of the choices fits the characteristics of your unknown. In that case, go back to a place in the key where you were uncertain and try the other direction. Or, you may need to start over.

There are two basic types of keys. The "synoptic key" presents a condensed description of the more important technical characters that form the basis of their phylogenetic arrangement. This type is seldom used for ordinary identification. The "diagnostic key" contains only the conspicuous characters that are most usable for identification. These keys are usually "artificial" in that the sequence of the plants is based upon obvious similarities or "key characters" that do not necessarily indicate evolutionary relationships. This will be obvious in the keys here because some combine both conifers and flowering plants.

The most acceptable key is *dichotomous,* which presents only two contrasting statements. Each pair of choices is called a *couplet,* and each choice is called a *lead.* Keys may be written in either of two formats based on how the leads and couplets are arranged. In the *indented key,* one or more couplets are indented (and included) under the lead having the characters that are correct for those couplets. The contrasting lead is presented only after the indented couplet or couplets. (See the Summer Keys A–F for this type of format.) The advantage of the indented key is that it is readily apparent which plants share certain characteristics because they are all indented under those characteristics. Also, the couplets do not have to be numbered (see the first Summer Key to Keys) because you follow the key by the degree of indentation (i.e., if a lead is correct, it will either give you the name of the plant, or you go to the next couplet indented under it). Numbers are sometimes used to make it easier to recognize the two leads of a couplet. (See the Summer Keys A–F.) It is also easier to work backwards in an indented key to determine the characteristics of a particular plant. The disadvantages of the indented key are that it wastes space when the key is very long and that the two leads of a given couplet may be rather far apart. You recognize the second lead of a couplet by the same number, if used, the same degree of indentation, and the same "lead word."

The second type of format is the *bracketed key.* In this form, the two leads of a couplet are kept together, and each lead will either end in a plant name or the number of the couplet to which you would go next. Alternate couplets are indented for easier recognition. (See the Summer Key G or Winter Key E for this format.) This format is advantageous for very long keys because the left margin is maintained and there is little wasted space. However, the disadvantages are that it is nearly impossible to see similarities among the plants, it is harder to work backwards to obtain the characteristics of a plant, backtracking is more difficult when a mistake has been made, the couplets must be numbered, and there has to be a number at the end of the lead that sends you to another couplet.

The keys that follow are artificial, diagnostic keys to the genera described in this textbook. Each genus is defined only in terms of the species described here. Some genera include species that are so different that they occur in different parts of the key. Such cases are indicated by "in part" (in pt.) placed after the generic name.

The keys are written to give you an opportunity to practice using keys in both formats, to help you learn the genera of trees and their morphological characteristics, and to group them by their obvious features that make them easier for you to remember. The more you use keys, the more proficient you will become.

The abbreviations used in the keys are: **lf.,** leaf; **lvs.,** leaves; **lfts.,** leaflets; **fls.,** flowers; **frt.,** fruit. For the Winter Key, it is advisable to use a 10× hand lens to observe some of the features. It is also a good practice to examine several twigs or several buds, leaf scars, stipular scars, etc., to make certain you are interpreting all the structures correctly. When the character is variable or interpretation is difficult, the genus often appears in more than one place in the key.

SUMMER KEYS TO GENERA (AND WINTER KEYS TO EVERGREENS)

Key to Keys

Plant evergreen.
    Lvs. acicular, linear, subulate, or scalelike; width less than 5 mm.
        Lvs. acicular or narrowly linear . . . . .A.
        Lvs. subulate or scalelike . . . . .B.
    Lvs. of various shapes; width over 5 mm . . . . .C.
Plant deciduous; or, succulent and lvs. lacking.
    Lvs. lacking on succulent plants; or, when present on a nonsucculent
    plant, acicular, linear, scalelike, or subulate; width less than 5 mm . . . . .D.
    Lvs. of various shapes; width over 5 mm.
        Lvs. opposite or whorled . . . . .E.
        Lvs. alternate.
            Lvs. compound . . . . .F.
            Lvs. simple . . . . .G.

Key A. (evergreen; lvs. acicular or narrowly linear)

1 Lvs. acicular and in fascicles of 1–5 in the axils of primary scale lvs . . . .*Pinus*
1 Lvs. narrowly linear and not in fascicles.
    2 Lvs. oriented in 2 ranks; seeds single, axillary, and enclosed
    by a fleshy aril.
        3 Lvs. $\frac{1}{2}$–1 in. long, flexible, the apex soft-pointed;
        seed partially enclosed by a scarlet aril . . . .*Taxus*
        3 Lvs. 1–3$\frac{1}{2}$ in. long, stiff, the apex prickly pointed;
        seed completely enclosed by a green to purplish aril . . . .*Torreya*
    2 Lvs. originating in more than 2 ranks;
    seeds enclosed in a strobilus (cone).
        4 Lf. attached to a peg that persists on the twig
        after lf. dehiscence.
            5 Lf. sessile on a conspicuous, persistent peg;
            lf. 4-angled or flat in cross section . . . .*Picea*
            5 Lf. petiolate on a short, inconspicuous peg;
            lf. flat or rounded in cross section . . . .*Tsuga*
        4 Lf. attached directly on the twig (i.e., without a peg).
            6 Lf. sessile and after dehiscence leaving a round, smooth
            scar on the twig; seed cones erect . . . .*Abies*
            6 Lf. petiolate or with a decurrent base; seed cones pendent.

    7 Lvs. petiolate, diverging in a spiral all around twig;
       seed cones oblong with exserted 3-lobed bracts between
       flattened scales . . . .*Pseudotsuga*
    7 Lvs. slightly narrowed into a long, decurrent base,
       spirally arranged but diverging in 2 ranks; seed cones ovoid
       and smooth with peltate scales . . . .*Sequoia*

Key B. (evergreen; lvs. subulate or scalelike)

1 Lvs. not imbricated, the twig visible in the internodes; fls. present; seeds in
  individual capsules, or samaras in woody conelike aments.
    2 Lvs. alternate; fls. bisexual, white or pinkish;
      capsules splitting lengthwise and exposing
      small seeds with a tuft of hairs at one end . . . .*Tamarix*
    2 Lvs. whorled; fls. unisexual and inconspicuous;
      frt. winged and held by woody bracts forming a conelike ament . . . . . .*Casuarina*
1 Lvs. imbricated, the twig usually hidden by the lvs.;
  fls. absent; seeds in a cone.
    3 Branchlets not flattened into sprays
      (i.e., the twigs arising in radial, 3-dimensional orientations).
      4 Seed cone fleshy and berrylike; seeds wingless;
        both scale and subulate lvs. usually present; plant dioecious . . . .*Juniperus*
      4 Seed cone leathery or woody; seeds winged;
        only scale lvs. present; plant monoecious.
        5 Lvs. alternate; seed cone oblong-ovoid,
          $2–3\frac{1}{2}$ in. long . . . .*Sequoiadendron*
        5 Lvs. opposite or whorled; seed cone subglobose,
          less than 2 in. long . . . .*Cupressus*
    3 Branchlets flattened (i.e., the twigs forming 2-dimensional sprays).
      6 Twigs round in cross section, the facial and lateral lvs. similar;
        seed cones globose with peltate scale/bracts, leathery
        but becoming woody when seeds dispersed . . . .*Chamaecyparis*
      6 Twigs flattened in cross section, the facial and lateral lvs. different;
        seed cones oblong with thin, flattened scale/bracts, woody.
        7 Seed cones $\frac{3}{4}–1\frac{1}{2}$ in. long, with 3 pairs of scales,
          duck-bill-like when closed; lvs. bright yellow-green . . . .*Calocedrus*
        7 Seed cones $\frac{1}{3}–\frac{1}{2}$ in. long, with 6–12 scales; lvs dark green . . . .*Thuja*

Key C. (evergreen; lvs. broad)

1 Lvs. opposite . . . .*Osmanthus*
1 Lvs. alternate.
    2 Lvs. compound or appearing so by mechanical splitting.
      3 Lvs. pinnately compound; seeds borne on the margins of sporophylls . . .*Cycas*

3 Lvs. appearing palmately compound; seeds in a drupe.
    4 Petioles prickly; southwestern U.S. . . . .*Washingtonia*
    4 Petioles smooth; southeastern U.S. . . . .*Sabal*
2 Lvs. simple.
    5 Lvs. with parallel venation.
        6 Stem unbranched below inflorescence;
          lvs. in a basal rosette; ovary inferior . . . . *Agave*
        6 Stem branched; lvs. in terminal clusters; ovary superior . . . .*Yucca*
    5 Lvs. with pinnate venation.
        7 Lvs. aromatic when crushed.
            8 Lvs. covered or dotted with minute,
              yellow glands (use hand lens) . . . .*Myrica*
            8 Lvs. without yellow glands (use hand lens).
                9 Stipular scars encircling the twig;
                  frt. a follicle in a conelike aggregate . . . .*Magnolia* (in pt.)
                9 Stipular scars lacking; frt. a drupe.
                    10 Terminal bud 3–4 scaled; frt. $\frac{1}{2}$–$\frac{3}{4}$ in. in diameter,
                      dark blue-black or black; lvs. spicy when crushed;
                      southeastern U.S. . . . .*Persea*
                    10 Terminal bud naked; frt. $\frac{3}{4}$–1 in. in diameter, yellow-green;
                      lvs. pungent-spicy when crushed; western U.S. . . . . . .*Umbellularia*
        7 Lvs. not aromatic when crushed.
            11 Frt. an acorn in a cup, or nuts enclosed in a spiny bur.
                12 Lvs. without yellow scales below; frt. an acorn in a cup.
                    13 Acorn in a thin, scaly cup . . . .*Quercus* (in pt.)
                  13 Acorn in a prickly cup . . . .*Lithocarpus*
                12 Lvs. covered with yellow scales below;
                  frt. in a spiny bur . . . .*Chrysolepis*
            11 Frt. not an acorn or nut enclosed in a bur.
                14 Lvs. aculeate margined, or at least the apex spiny-tipped . . . .*Ilex*
                14 Lvs. entire (rarely toothed on vigorous shoots)
                or shallowly serrulate.
                  15 Sap milky (check broken petiole).
                    16 Stipular scars encircling twig . . . .*Ficus*
                    16 Stipular scars lacking . . . .*Bumelia*
                  15 Sap clear.
                    17 Frt. a capsule.
                      18 Lf. margin entire, revolute . . . .*Rhododendron*
                      18 Lf. margin serrulate, straight . . . .*Gordonia*
                    17 Frt. a drupe or berry.
                      19 Pith chambered; fls. yellow . . . .*Symplocos*
                      19 Pith solid and homogeneous; fls. white.
                        20 Frt. black; bark gray; twig with
                        pungent odor when broken;
                        southeastern U.S. . . . .*Prunus* (in pt.)

20 Frt. orange; bark red-brown;
twig lacking pungent odor when broken;
western U.S. and Canada . . . .*Arbutus*

Key D. (succulents without lvs.; or, lvs. deciduous and less than 5 mm wide)

1 Lvs. lacking; stem succulent . . . .*Carnegiea*
1 Lvs. present; stem woody.
  2 Lvs. acicular, borne spirally on long and indeterminate short shoots;
seeds in cones with flat scales . . . .*Larix*
  2 Lvs. narrowly linear, subulate, or scalelike; seeds in cones or capsules.
    3 Lvs. narrowly linear or subulate, borne on long
and determinate short-shoots; seeds in cones with peltate scales . . . . .*Taxodium*
    3 Lvs. scalelike on long shoots; seeds in capsules;
flowers white to pink . . . .*Tamarix*

Key E. (deciduous; lvs. broad, opposite or whorled)

1 Lvs. compound.
  2 Lvs. palmately compound; frt. a capsule with large seeds . . . .*Aesculus*
  2 Lvs. pinnately compound or trifoliolate; frt. a samara.
    3 Lfts. coarsely serrate or lobed; twig glaucous;
frt. a double samara . . . .*Acer* (in pt.)
    3 Lfts. entire, serrulate, or serrate; twig not glaucous;
frt. a single samara . . . .*Fraxinus* (in pt.)
1 Lvs. simple.
  4 Lvs. palmately or pinnipalmately veined.
    5 Back of lvs. with erect, branched hairs;
pith hollow in second year; fruit a persistent, ovoid capsule . . . .*Paulownia*
    5 Back of lvs. without branched hairs; pith solid;
frt. an elongated, linear capsule or double samara.
      6 Lf. blades usually all palmately lobed; frt. a double samara . . . .*Acer* (in pt.)
      6 Lf. blades only rarely lobed; frt. an elongated, linear capsule . . . .*Catalpa*
  4 Lvs. pinnately veined.
    7 Lf. venation arcuate; lf. scars connected by a narrow ridge . . . . .*Cornus* (in pt.)
    7 Lf. venation not arcuate; lf. scars not connected by a ridge.
      8 Petiole short, stout, purplish; vascular bundle scars 1;
frt. a drupe; eastern U.S . . . .*Chionanthus*
      8 Petiole long and slender; vascular bundle scars many;
frt. a samara; western U.S . . . .*Fraxinus* (in pt.)

Key F. (deciduous; lvs. broad, alternate, compound)

1 Lf. trifoliolate.
  2 Frt. a thin, round samara; lfts. entire or serrulate; shrub or small tree . . . . . .*Ptelea*

    2 Frt. a drupe; lfts. coarsely toothed or lobed; shrub or vine
      (POISONOUS TO THE TOUCH) . . . .*Toxicodendron* (in pt.)
1 Lf. with 5 or more lfts.
  3 Lvs. 2- or 3-pinnately compound.
    4 Lfts. $\frac{1}{2}$ in. or less long, the main vein of lft. near one margin;
      fls. in globose heads, the stamens long exserted . . . .*Albizia*
    4 Lfts. $\frac{1}{2}$ in. or more long, the main vein of lft. near center;
      fls. in racemes or panicles.
      5 Lvs. $1\frac{1}{2}$–4 ft long.
        6 Lfts. entire; without prickles on stem or lf . . . .*Gymnocladus*
        6 Lfts. serrulate; with prickles on midrib below,
          on petiole, and scattered or in a circle around stem at the node . . . . .*Aralia*
      5 Lvs. $1\frac{1}{2}$ ft or less long.
        7 Lfts. 1–2 in. long; margin serrate or lobed; apex acuminate . . . .*Melia*
        7 Lfts. less than 1 in. long; margin entire
          or slightly crenulate; apex rounded, obtuse, or emarginate . . . . . .*Gleditsia*
  3 Lvs. 1-pinnately compound.
    8 Lvs. and/or twigs with milky sap when broken;
      frts. reddish, hairy, and in an erect, dense panicle . . . .*Rhus*
    8 Lvs. and/or twigs with clear sap (NOTE: If petiole reddish, *do not* break
      petiole; see no. 13); frts. not as above.
      9 Stipular spines or thorns present.
        10 Stipular spines present on vigorous shoots; thorns absent . . . .*Robinia*
        10 Stipular spines absent; thorns long and often branched . . . .*Gleditsia*
      9 Stipular spines and thorns absent.
        11 Margin glandular-toothed only near base;
          crushed lfts. with an unpleasant odor . . . .*Ailanthus*
        11 Margin entire or toothed throughout; crushed lfts. odorless or fragrant
          (NOTE: If petiole reddish, *do not* crush the lfts.; see no. 13).
          12 Lft. margins entire.
            13 Petiole and rachis reddish; frt. a white drupe
              (POISONOUS TO THE TOUCH) . . . .*Toxicodendron* (in pt.)
            13 Petiole and rachis green;
              frt. a legume or berrylike and yellowish.
              14 Lfts. 5–11, each more than 1 in. wide,
                neither lanceolate nor falcate, alternate
                on the rachis; frt. a legume; base of petiole
                nearly enclosing lateral bud; eastern U.S. . . . .*Cladrastis*
              14 Lfts. 11–19, each 1 in. or less wide,
                lanceolate and falcate, opposite or subopposite
                on the rachis; frt. berrylike and yellowish;
                petiole below lateral bud; western U.S. . . . .*Sapindus*
          12 Lft. margins serrate.
            15 Pith chambered; frt. a nut enclosed
              in an indehiscent husk . . . .*Juglans*

     15 Pith solid; frt. a pome or a nut in a dehiscent husk.

        16 Stipules or stipular scars present; buds resinous;
            frt. a cluster of small, red pomes . . . . *Sorbus*

        16 Stipules and stipular scars absent; buds not resinous;
            frt. a nut in a more-or-less dehiscent husk . . . .*Carya*

Key G. (deciduous; lvs. broad, alternate, simple)

1 Lvs. and twigs silvery or brownish lepidote . . . .*Elaeagnus*

1 Lvs. and twigs not lepidote 2

  2 Lvs. typically entire margined; lobed or not 3

  2 Lvs. typically serrate, serrulate, or crenate margined; lobed or not 17

3 Stipules or stipular scars encircling twig 4

3 Stipules or stipular scars lacking, or if present, not encircling twig 5

  4 Lvs. typically 4-lobed with a broadly truncate or notched apex;
     frt. an aggregate of samaras . . . .*Liriodendron*

  4 Lvs. not lobed; apex obtuse to acuminate;
     frt. an aggregate of follicles . . . . *Magnolia* (in pt.)

5 Lf. flabellate (fan-shaped) with dichotomous veins . . . .*Ginkgo*

5 Lf. variously shaped but not flabellate; venation pinnate or palmate 6

  6 Lvs. palmately veined (with 3 or more prominent
     veins arising at or near the base of the blade) 7

  6 Lvs. pinnately veined (with 1 major midrib) 8

7 Lvs. cordate at base, not lobed, not aromatic; frt. a legume . . . .*Cercis*

7 Lvs. cuneate at base, unlobed or 2 or 3 lobed at apex,
  aromatic; frt. a drupe . . . .*Sassafras*

  8 Thorns present; sap milky (observable in broken
     petiole and young frt.) 9

  8 Thorns lacking; sap clear 10

9 Stipules or stipular scars present, frt. a large,
  globose, multiple of dry, greenish drupes . . . .*Maclura*

9 Stipules or stipular scars absent;
  frt. a single, black drupe . . . .*Bumelia*

  10 Lvs. usually oblanceolate, 7–12 in. long,
     malodorous when crushed; frt. a berry 3–5 in. long . . . .*Asimina*

  10 Lvs. and frt. not as above 11

11 Terminal bud lacking, pseudoterminal bud present;
  frt. a globose berry . . . .*Diospyros*

11 Terminal bud present; frt. a drupe, elliptical berry, or acorn 12

  12 Lvs. and twigs spicy-aromatic when crushed . . . .*Sassafras*

  12 Lvs. and twigs not spicy-aromatic when crushed 13

13 Pith chambered . . . .*Symplocos*

13 Pith solid and homogeneous or diaphragmed 14

  14 Pith diaphragmed . . . .*Nyssa*

  14 Pith homogeneous 15

15 Buds with 2 nearly valvate scales; lf. venation arcuate . . . . *Cornus* (in pt.)
15 Buds with imbricate scales; lf. venation typically pinnate 16
    16 Stipules or stipular scars absent; vascular bundle scars 3;
       frt. a dry drupe; plant dioecious . . . .*Leitneria*
    16 Stipules or stipular scars present; vascular bundle scars
       more than 3; frt. an acorn; plant monoecious . . . .*Quercus* (in pt.)
17 Lvs. palmately veined (with 3 or more prominent veins
  arising at or near the base) 18
17 Lvs. pinnately veined (with a single midrib and equal secondary veins)
  or pinnipalmate (with lowest secondary veins larger than the others 21
    18 Lf. lobes 3–7; frt. dry, in a globose multiple 19
    18 Lf. lobes 2 or 3 or not lobed; frt. fleshy, in a multiple or single 20
19 Lvs. finely serrate; frt. a multiple of capsules;
  stipular scars slitlike and eventually inconspicuous . . . .*Liquidambar*
19 Lvs. coarsely serrate; frt. a multiple of achenes;
  stipular scars encircling twig . . . .*Platanus*
    20 Stems often bearing thorns; sap clear; frt a pome . . . .*Crataegus* (in pt.)
    20 Stem lacking thorns; sap of broken petiole milky;
       frt. a multiple of drupelets . . . .*Morus*
21 Lvs. distinctly pinnately lobed 22
21 Lvs. not lobed 23
    22 Twigs lacking thorns; frt. an acorn . . . .*Quercus* (in pt.)
    22 Twigs often bearing thorns; frt. a pome . . . .*Crataegus* (in pt.)
23 Lvs. distichous (2-ranked) 24
23 Lvs. in 3 or more ranks 34
    24 Lvs. with broad, cordate to truncate base;
       frt. nutlike and attached to a long, leafy bract . . . .*Tilia*
    24 Lvs. with rounded, cuneate, or nearly cordate base; frt. not as above 25
25 Lf. margin irregularly crenate, crenate-dentate, or sinuate;
  frt. a woody capsule; buds naked . . . .*Hamamelis*
25 Lf. margin serrate or serrulate;
  frt. a pome, nutlet, nut, drupe, or samara; buds scaly 26
    26 Lf. margin widely serrate with 1 tooth per secondary vein 27
    26 Lf. margin closely serrate, serrulate, or doubly serrate,
       with many teeth per secondary vein 28
27 Stipular scars long and nearly encircling twig; lf. 2× or less longer
  than wide, base broadly cuneate; nuts in a weak-spined husk . . . .*Fagus*
27 Stipular scars short; lf. 3× or more longer than wide,
  base narrowly cuneate or acute; nuts in a stiff-spined bur . . . .*Castanea*
    28 Twigs with conspicuous white lenticels
       lengthening horizontally with age . . . .*Betula*
    28 Twigs with more-or-less circular lenticels, often indistinct 29
29 Terminal bud present; fls. showy, white,
  in crowded racemes; frt. a pome . . . .*Amelanchier*

29 Terminal bud absent, pseudoterminal bud present;
fls. small and in aments or small clusters, individually inconspicuous;
frt. a nut, samara, or drupe 30
   30 Lf. base symmetrical; fls. in aments 31
   30 Lf. base asymmetrical; fls. in small fascicles or cymes 33
31 Lf. with 5–8 pairs of lateral veins; shrubs (eastern U.S. and Canada)
or small trees (western U.S. and Canada) . . . .*Corylus*
31 Lf. with 9 or more pairs of lateral veins; trees (eastern U.S. and Canada) 32
   32 Bark flaking into narrow, vertical shreds; buds not angled;
nutlets enclosed in papery sacs . . . .*Ostrya*
   32 Bark smooth but trunk fluted ("muscular"); buds
4-angled; nutlets subtended by 3-lobed, leafy bracts . . . .*Carpinus*
33 Lf. margin doubly serrate, blade pinnately veined throughout;
frt. a flattened samara . . . .*Ulmus*
33 Lf. margin singly serrate, blade palmately veined at base; frt. a drupe . . . .*Celtis*
   34 Lf. margins deeply crenate, sinuate, or coarsely serrate,
1 tooth per secondary vein; frt. an acorn or nut in a spiny bur 35
   34 Lf. margin serrate, doubly serrate, or serrulate,
more than one tooth per secondary vein; frt. not as above 36
35 Buds clustered at tip of twig,
each with many scales; frt. an acorn . . . .*Quercus* (in pt.)
35 Buds more equally spaced along twig, each with 3 or 4 scales;
frt. a nut enclosed by a spiny bur . . . .*Castanea*
   36 Branches often armed with thorns 37
   36 Branches usually not armed with thorns 39
37 Pomes with stony core (endocarp); fls. solitary or
in compound corymbs; buds small, globose, and bright red . . . .*Crataegus*
37 Pomes with papery cores (endocarp); fls. in simple corymbs;
buds larger, ovoid, and red to brown 38
   38 Buds glabrous; frt. pyriform, with stone cells (grit cells) . . . .*Pyrus*
   38 Buds white pubescent; frt. globose, lacking stone cells . . . .*Malus*
39 Lf. blades long and slender, over 4× longer than wide;
buds with 1 caplike scale . . . .*Salix*
39 Lf. blades only 1–4× longer than wide;
buds with more than 1 scale or naked 40
   40 Pith chambered 41
   40 Pith solid and homogeneous 42
41 Lf. thick and leathery; terminal bud present; frt. an elliptical berry . . . . .*Symplocos*
41 Lf. thin and membranous; terminal bud absent; frt. dry, 4-winged . . . .*Halesia*
   42 Lateral buds stalked; nutlets in a persistent, woody, conelike ament . . . . .*Alnus*
   42 Lateral buds sessile; frt. not as above 43
43 Petiole 1½ in. or more long; blade deltate,
broadly ovate, or ovate-lanceolate 44
43 Petiole 1 in. or less long; blade lanceolate, elliptic, or oval 45

44 Cell sap clear; terminal bud present;
frt. a capsule with comose seeds . . . .*Populus*
44 Cell sap milky (in broken petiole of young lf.);
terminal bud absent; frt. a multiple of drupelets . . . .*Morus*
45 Petiole with 2 glands just below blade or at base of blade . . . .*Prunus* (in pt.)
45 Petiole lacking glands 46
    46 Lf. with stiff, straight hairs along midrib below;
    vascular bundle scars 1; frt. a capsule . . . .*Oxydendrum*
    46 Lf. lacking stiff, straight hairs along midrib below;
    vascular bundle scars usually 3; frt. a pome or drupe 47
47 Buds long and slender, scaly; fls. white; frt. a small pome . . . .*Amelanchier*
47 Buds small, naked and tomentose; fls. greenish-yellow; frt. a drupe . . . . .*Rhamnus*

WINTER KEY TO DECIDUOUS GENERA

Key to Keys or Unique Genera

Lf. scars lacking (the evident round scars are twig scars that lack bundle scars);
    twigs dehiscing with the leaves.
    Twig yellow, red-brown, or purplish; lvs. scalelike
        if persisting on terminal branchlets; pith white or greenish . . . .*Tamarix*
    Twig greenish or brown; lvs. linear or subulate
        if persisting on terminal branches; pith brown . . . .*Taxodium*
Lf. scars present; twigs persisting after leaf dehiscence.
Lf. scars opposite or whorled . . . . .A
    Lf. scars alternate.
        Twigs armed with thorns, spines, or prickles . . . . .B
        Twigs not armed.
            Twigs covered with numerous silvery
            or brownish lepidote scales . . . .*Elaeagnus*
            Twigs lacking lepidote scales.
                Pith diaphragmed, chambered, or chambered
                only at the nodes C
                Pith solid and homogeneous.
                    Buds apparently lacking, or present and naked,
                      indistinctly scaly and appearing naked, or with 1 caplike scale . . . . . .D
                    Buds evident and with 2–many distinct scales .E

Key A. (lf. scars opposite or whorled)

1 Pith chambered or hollow in the internodes . . . .*Paulownia*
1 Pith solid and homogeneous in the internodes.
    2 Lf. scars large and broad; twigs often stout (over $\frac{3}{16}$ in. diameter).
        3 Bundle scars forming a closed circle or ellipse;
        terminal bud absent . . . .*Catalpa*

3 Bundle scars forming an open U or V; terminal bud present.
    4 Bundle scars many, forming a U;
        buds with 2 or 3 pairs of scales ....*Fraxinus*
    4 Bundle scars 3 or in 3 groups in a V;
        buds with about 6 pairs of scales ....*Aesculus*
2 Lf. scars small or narrow; twigs slender or moderate
(usually less than $\frac{3}{16}$ in. diameter).
    5 Bundle scars 1; opposite lf. scars separated ....*Chionanthus*
    5 Bundle scars 3; opposite lf. scars meeting or connected by a ridge.
        6 Buds with more than 2 valvate or imbricate scales ....*Acer* (in pt.)
        6 Buds with 2 valvate scales.
            7 Pith brown ....*Acer* (in pt.)
            7 Pith white or greenish ....*Cornus* (in pt.)

Key B. (lf. scars alternate; twigs with thorns, spines, or prickles)

1 Twig stout, with many prickles; lf. scar nearly encircling twig ....*Aralia*
1 Twig slender, lacking prickles.
    2 Twigs with stipular spines present, at least
    on vigorous branches; thorns lacking ....*Robinia*
    2 Twigs with thorns present; stipular spines lacking.
        3 Thorns 2–3–more branched ....*Gleditsia*
        3 Thorns not branched.
            4 Thorns terminating spur branches.
                5 Sap clear and not sticky.
                    6 Terminal bud lacking ....*Prunus* (in pt.)
                    6 Terminal bud present.
                        7 Buds white pubescent ....*Malus*
                        7 Buds glabrous ....*Pyrus*
                5 Sap milky and sticky ....*Bumelia*
             4 Thorns at nodes in leaf axils.
                8 Twigs with milky sap; bundle scars many in a circle ....*Maclura*
                8 Twigs with clear sap; bundle scars 3 ....*Crataegus*

Key C. (lf. scars alternate; twigs not armed; pith diaphragmed or chambered)

1 Pith chambered, at least at the nodes by the second year.
    2 Lf. scars 3-lobed, each with a U-shaped cluster of bundle scars ....*Juglans*
    2 Lf. scars half-round, narrow, or U-shaped; bundle scars 1–3.
        3 Terminal bud absent.
            4 Lateral buds superposed, with the lower one smaller ....*Halesia*
            4 Lateral buds solitary.
                5 Bundle scars 1; buds black with 2 scales ....*Diospyros*
                5 Bundle scars 3; buds brownish with 4 or more scales ....*Celtis*
        3 Terminal bud present.

   6 Terminal bud long pointed, naked and densely rusty pubescent . . . . .*Asimina*
   6 Terminal bud conical or oval, with scales ciliate on the margin . . .*Symplocos*
1 Pith diaphragmed.
   7 Stipular scar encircling twig.
      8 Terminal bud with a single caplike scale . . . .*Magnolia*
      8 Terminal bud with 2 valvate scales . . . .*Liriodendron*
   7 Stipular scars absent.
      9 Terminal bud naked, rusty pubescent . . . .*Asimina*
      9 Terminal bud scaly and glabrous . . . .*Nyssa*

Key D. (lf. scars alternate; twigs not armed; pith solid and homogeneous; buds apparently lacking, or naked, or with 1 scale)

1 Terminal bud present.
   2 Bud with 1 caplike scale; stipular scar encircling twig . . . .*Magnolia*
   2 Bud indistinctly scaly, naked, or appearing naked;
      stipular scars lacking or not encircling twig.
      3 Bundle scars many scattered or in 3 C-shaped groups;
         buds sessile and scaly . . . .*Carya* (in pt.)
      3 Bundle scars 3 or more but not in 3 groups;
         buds stalked, naked except for 2 small scales . . . .*Hamamelis*
1 Terminal bud lacking.
   4 Twigs with spur branches . . . .*Maclura*
   4 Twigs lacking spur branches.
      5 Buds with 1 caplike scale.
         6 Bundle scars 3 in a narrow U-shaped lf. scar;
            stipular scars not encircling twig . . . .*Salix*
         6 Bundle scars 5–9 in a lf. scar that encircles the bud;
            stipular scar encircling twig . . . .*Platanus*
      5 Buds naked or appearing so, or apparently lacking.
         7 Buds apparently lacking, only barely visible.
            8 Buds covered by leaf scar, but scar often split
               and showing the downy, naked bud . . . .*Robinia*
            8 Buds barely visible, scaly and glabrous . . . .*Gleditsia*
         7 Buds naked or appearing so because of dense covering of hairs.
            9 Stipules or stipular scars conspicuous; lf. scars half-round . . . . . .*Rhamnus*
            9 Stipules lacking or scars inconspicuous;
               lf. scars horseshoe-shaped, U-shaped, heart-shaped, or 3-lobed.
               10 Bundle scars several in 3 groups.
                  11 Lf. scar 3-lobed, each lobe with a C-shaped group
                     of bundle scars; sap clear; pith white . . . .*Melia*
                  11 Lf. scar horseshoe-shaped or U-shaped
                     with 3 irregular groups of bundle scars; sap milky;
                     pith brownish . . . .*Rhus*
               10 Bundle scars 1–6(9), not in 3 groups.

     12 Lf. scar heart-shaped; pith salmon-colored ....*Gymnocladus*
     12 Lf. scar horseshoe-shaped around bud; pith white.
        13 Buds silvery or yellowish silky, partially burried;
           bundle scars 3; lenticels large and brown ....*Ptelea*
        13 Buds bronze silky, protruding from twig;
           bundle scars 3–6(9); lenticels small and white .....*Cladrastis*

Key E. (lf. scars alternate; twig not armed; pith solid and homogeneous; buds with 2 or more scales)

1 Spur shoots conspicuous and at every node after first year 2
1 Spur shoots lacking or less frequent if present 4
    2 Buds elongate-conical; bundle scars 3 ....*Betula*
    2 Buds rounded; bundle scars 1 or 2 3
3 Bundle scars 2; lf. scars without decurrent base ....*Ginkgo*
3 Bundle scar 1; lf. scar raised on a decurrent base ....*Larix*
    4 Buds distinctly stalked; twigs somewhat 3-angled
        with one angle extending down from lf. scar ....*Alnus*
    4 Buds essentially sessile (fl. buds of *Cercis* superposed and stalked);
        twigs essentially terete 5
5 Terminal bud present or apparently so 6
5 Terminal bud replaced by pseudoterminal bud 20
    6 Twigs green with spicy aroma; bundle scar 1 ....*Sassafras*
    6 Twigs brown, gray, or reddish, not spicy; bundle scars 3–many 7
7 Lateral buds mostly $\frac{1}{2}$ in. or more long 8
7 Lateral buds usually less than $\frac{1}{2}$ in. long 10
    8 Lowest bud scale of lateral bud directly over lf. scar;
        lf. scar triangular or 3-lobed ....*Populus* (in pt.)
    8 Lowest bud scale of lateral bud not over lf. scar;
        lf. scar half-round or slender and crescent-shaped 9
9 Stipular scars lacking; lf. scars slender and crescent-shaped;
   bundle scars 3 ....*Amelanchier*
9 Stipular scars present and nearly encircling twig;
   lf. scars half-round; bundle scars usually many ....*Fagus*
    10 Lf. scars somewhat 3-lobed or shield-shaped;
        bundle scars many, mostly in 3 C-shaped groups,
        1 group per lobe ....*Carya* (in pt.)
    10 Lf. scars round, half-round, shield-shaped, or narrow,
        not lobed or only barely so; bundle scars not in 3 C-shaped groups 11
11 Bundle scars 8 or more 12
11 Bundle scars 3–5(7), sometimes divided 13
    12 Buds (terminal and several lateral) clustered at tip of twig;
        bundle scars scattered; pith stellate ....*Quercus*
    12 Buds more equally spaced along twig;
        bundle scars forming a U; pith round ....*Toxicodendron*

13 Stipular scars present 14
13 Stipular scars lacking or inconspicuous even with a hand lens 16
    14 Twig with almond aroma . . . .*Prunus*
    14 Twig lacking almond aroma 15
15 Buds with 1–3 visible scales . . . .*Malus*
15 Buds with 4 or more visible scales . . . .*Populus* (in pt.)
    16 Lf. scars crowded near end of twig, often appearing whorled . . .*Cornus* (in pt.)
    16 Lf. scars more equally spaced along twig 17
17 Terminal bud $\frac{1}{8}$ in. long; pith terete . . . .*Leitneria*
17 Terminal bud over $\frac{1}{4}$ in. long; pith 5-sided 18
    18 Buds resinous with long hairs mainly on inner scales . . . .*Sorbus*
    18 Buds not resinous, the scales glabrous or ciliate margined 19
19 Terminal buds ovoid and glossy; lateral buds small and diverging;
    twigs with or without corky wings . . . .*Liquidambar*
19 Terminal buds elongated and not glossy;
    lateral buds elongated and recurved toward twig;
    twigs without corky wings . . . .*Amelanchier*
    20 Bundle scar 1 21
    20 Bundle scars 2–many 22
21 Lf. scars raised; buds with 2 imbricate scales, nearly black,
    entirely exposed; lenticels orange; twig grayish brown . . . .*Diospyros*
21 Lf. scars sessile; buds with 3–6 imbricate scales, reddish,
    partly embedded; lenticels dark; twig yellow-green to red . . . .*Oxydendrum*
    22 Stipular scars lacking 23
    22 Stipular scars present 26
23 Twigs stout ($\frac{1}{4}$ in. diameter or more) with brown pith;
    bundle scars many . . . .*Ailanthus*
23 Twigs slender with white pith; bundle scars 2 or 3 24
    24 Buds with 4 or more scales, broadly conical;
        lf. scar narrowly elliptical . . . .*Celtis*
    24 Buds with 2 visible scales, often superposed;
        lf. scar broadly cordate or shield-shaped. 25
25 Lf. scars raised, 2-ranked, with a fringe at the top . . . .*Cercis*
25 Lf. scars sessile, more than 2-ranked or sometimes opposite,
    lacking fringe at top . . . .*Sapindus*
    26 Buds long and slender (length 5–7 × width),
        obliquely divergent over lf. scar . . . .*Fagus*
    26 Buds shorter and more ovoid (length 2–4 × width),
        more nearly above lf. scar 27
27 Lf. scars in more than 2 ranks 28
27 Lf. scars in 2 ranks (distichous lf. arrangement) 30
    28 Pith 5-angled; visible bud scales 2 or 3 . . . .*Castanea*
    28 Pith round; visible bud scales 3–many 29
29 Twigs with milky sap; bundle scars many in nearly circular lf. scar . . . .*Morus*
29 Twigs with clear sap; bundle scars 3 in a 3-lobed lf. scar . . . .*Albizia*

5

VARIATION

Natural variation is universal and characteristic of all biological systems. "No two trees are *exactly* alike with respect to all characteristics" would be just as true as saying "no two people are exactly alike." In this chapter, we are concerned with the variation *within* a species (i.e., intraspecific variation).

Intraspecific variation is responsible, at first, for the feeling of frustration and uncertainty in species identification. Unfortunately, such variation may lead to guessing the identification rather than careful consideration of all variable characters. By observing many individuals of a species, one begins to acquire an appreciation for the *range in variation* to be expected in a particular species. As a result, one's concept of a species changes from a narrow, typological one to a broader, populational one. A single tree, therefore, is but one sample of a species population. With a broad concept of a species, differences (discontinuities) among species, or the chance between-species hybrid, can be better understood and recognized.

Although variation may create problems for beginning students, it is the chief reason why one's interest in forest trees is never dulled, even after spending years in the same region. It also adds excitement and may broaden one's species concept and enrich one's appreciation for natural variation to see a well-known species in different habitats and localities and the differences that may exist in various characteristics.

Normal variation between sexually reproducing individuals comes about by mutations, genetic segregation, and recombination. Beyond this, there are distinct patterns of variation that can and should be recognized by the dendrologist when considering the differences observed between individuals and populations of a given species over its entire geographical and ecological range. Following the outline below, variation can be classified initially as either *intrinsic* (originating within the individual or species) or *extrinsic* (originating from outside, i.e., coming from another species). These terms will become clearer after considering the examples.

Types of Variation

   I. Intrinsic
      Phenotypic plasticity
      Developmental plasticity
      Abnormal (mutational)
      Chromosomal
      Nonadaptive
      Ecotypic
      Clinal
      Reproductive
      Speciational
  II. Extrinsic
      Hybrid
      Introgressive

INTRINSIC

Phenotypic Plasticity

Also thought of as *ecophenic variation,* this type of variation is the direct response of plant form to environmental factors (i.e., environmentally induced variation). The term *phenotypic plasticity* means the capacity of organisms with the same genetic makeup (genotype) to vary in its visible characteristics (phenotype) due to varying environmental conditions. Examples of this are seen in both within-tree and between-tree variation. Deeply lobed, smaller, and thicker leaves in the sun contrast sharply with shallowly lobed, larger, and thinner leaves in the shade as seen frequently in many oaks and maples; very hairy leaves in the sun versus glabrous leaves in the shade are found in basswood. The occurrence of two or more leaf types on the same plant, as in these examples, is called *heterophylly.* Also smaller, stunted trees with smaller leaves and nuts growing under poor, dry conditions versus larger trees with large leaves and nuts growing under mesic conditions can be found in the hickories and buckeyes. Such environmental variation affects not only morphological but also anatomical and physiological character states.

Another type of phenotypic plasticity within a tree is often associated with the stress morphology created by extreme midseason drought or defoliation due to a late spring freeze, severe insect damage, or chemical spraying. Leaves of "second flushes" from regular branches or sucker shoots are often atypical and bizarre in shape and size.

Phenotypic plasticity is nongenetic in that the acquired, environmentally induced differences are not inherited. Transplanting trees to other environmental conditions will cause a reversal of form, or different variants from different habitats when brought into a common garden will ultimately result in similar forms. Although these changes are reversible and nongenetic, the *extent* of the plasticity is under genetic control; therefore, different species show different degrees of plasticity and the variation may involve different characters.

Developmental Plasticity

In dendrology, we normally focus our attention on the mature tree and mature leaves and are often unaware of immature stages or the fact that certain well-known leaf types are actually the juvenile form. The developmental or phase change from juvenile to mature form is called a *heteroblastic change* or *phase change.* Two common examples of this are the lobed leaves of sassafras and mulberry found on saplings or vigorous sprouts. The older trees of these two species may not have any lobed leaves. Some other species, such as water oak, characteristically have a remarkable variety of different lobed leaf shapes on seedlings, saplings, or vigorous sprouts in marked contrast with the unlobed leaf of the mature crown. Also, seedlings of trees with pinnately divided leaves may have one or two simple leaves just above the cotyledons as in buckeye, ash, or hickory. In southern red oak, the leaves of the lower part of the crown have a very different shape and lobing than those of the upper part. These patterns of developmental changes are characteristic of certain species and are generally not reversible by environmental conditions. Thus, they are under stronger genetic control than the phenotypic plasticity previously described.

In junipers, false cypress, and arborvitae, the subulate (awl) leaves are juvenile forms, and the scale leaves are mature forms. Pruning or browsing of mature branches causes new growth of juvenile leaf form. This juvenile foliage can be maintained through repeated vegetative propagation and at one time led to the generic name *Retinispora* (or *Retinospora*) for juvenile-like cultivars of *Chamaecyparis* and *Thuja* commonly used in the nursery trade. This type of heteroblastic change is usually not affected by environmental differences, yet when some junipers with scale leaves are placed in the shade, the juvenile awl leaves are formed. This indicates a complex environment-genotype interaction, and the genetics of this phenomenon is not completely understood.

Finally, *seasonal heteromorphism* occurs when the first-formed leaves on a twig of adult individuals normally differ from the later developed leaves on the same twig. This may be seen among the leaves of some deciduous trees such as several species of poplar (Eckenwalder, 1980), sweetgum (Zimmerman and Brown, 1971), and oak (Baranski, 1975). It is not entirely clear whether this is strictly developmental or partly environmentally induced.

Abnormal (Mutational)

Although mutations are the ultimate source of all genetic variation, most effects are rather minor and not obvious. Some examples of obvious mutational effects in trees are genetic dwarfs in conifers (Franklin, 1970). Also the hairless peach (nectarine) arose as a spontaneous mutant in east Asia, although a similar mutant occurred in France in the seventeenth century. Pink-flowered dogwoods were derived as a mutant form of the common "white-flowered" (really white-bracted) type. The navel orange (with a tiny rudimentary second fruit embedded at the apex of the normal one) arose as a mutant; likewise, the original grapefruit was probably a natural mutation of the Asian shaddock (with large fruits weighing up to 11 lb and measuring up to 20 cm in diameter). A number of other mutants have become important in the horticultural trade, such as the thornless forms of honey locust, called the *Moraine* locust.

Chromosomal

The *n* number of chromosomes is the *haploid* number found, following meiosis, in the subsequent gametophyte phase of the life cycle. The 2*n* or *diploid* number is formed at fertilization and is the number in the cells of the sporophyte phase—the tree. The actual chromosome number in trees covers a range from *n* = 8 in hophornbeam or *n* = 20 in buckeye to much higher numbers. Cells in a buckeye leaf, therefore, have 2*n* = 40 chromosomes.

Changes in chromosome number in tree species are of two kinds: *aneuploid* in which a given haploid set is increased or decreased by the addition or deletion of one or more chromosomes; and *polyploid* in which additional full haploid sets become associated in the cells. Examples of aneuploidy are sweetgum (some trees have *n* = 15 and others 16); willow *n* = 19 and 22; beefwood tree *n* = 11, 12, and 13; and hackberry *n* = 10, 11, and 14. In some cases, aneuploid differences are correlated with morphological differences; in others, it is not detectable except by counting the chromosomes.

Polyploidy is rare in gymnosperms (ca. 1.5% of the species) and much more common in angiosperms (50 to 70%). Among gymnosperms, redwood is a good example with 2*n* = 6*x* = 66. (The diploid number of 66 is 6 times the base number of 11, which is 6*x*, a hexaploid. The base number [*x*] is the lowest detectable haploid number within a group of related taxa.) Examples of angiosperm polyploids are the tetraploid cucumbertree (2*n* = 76) and the hexaploid southern magnolia (2*n* = 114) (*x* = 19 in the Magnoliaceae). White ash has three ploidy levels (2*n* = 46, 2*n* = 92, 2*n* = 138; *x* = 23), and paper birch also has three ploidy levels (2*n* = 56, 2*n* = 70, 2*n* = 84; *x* = 14). *Betula* ×*murrayana* (Barnes and Dancik, 1985), of hybrid derivation, is octoploid (2*n* = 112).

In some cases, the differences in chromosome number might be correlated with significant phenotypic differences. In other cases, however, the variation related to aneuploidy or polyploidy might be undetectable by morphological characteristics although they may reflect chemical differences (Black-Shaefer and Beckmann, 1989) or physiological ecotypes. In other cases, particularly if polyploidy has arisen in an interspecific hybrid, there are detectable morphological differences, and often the polyploid shows an adaptive superiority over the ancestral diploids, thus often occupying disturbed or more extreme habitats. Polyploidy may be a mechanism of overcoming hybrid sterility because by chromosome doubling, the hybrid can reproduce and the hybrid character combinations are maintained. An example of this is the *Betula alleghaniensis* × *pumila* derivative, *B.* ×*murrayana,* mentioned above.

Nonadaptive

When a particular variant is not associated with some environmental condition, it may be a nonadaptive trait that occurs scattered throughout the species population. Such is the case in southern magnolia in which some trees have leaves that are green and only slightly hairy below whereas other trees have leaves that are densely rusty-pubescent below. Such trees are often side by side in the same habitat. In some cases, such traits are controlled by single genes. Baranski (1975) found that much of the variation in leaf form (Fig. 5-1), bark, etc., among trees of white oak is explained by nonadaptive, more or less random, genetic variation. In Carolina basswood, different trees in a local population may be very different in leaf pubescence, some glabrous from the beginning, some tomentose at first then becoming glabrate, and others remaining tomentose. This led earlier workers to a multiplication of species and varietal names for what, in reality, is nonadaptive, genetic variation in one species and variety (Hardin, 1990).

FIGURE 5-1   Representative leaf forms of white oak, all drawn to scale. (*From Baranski 1975, N.C. Agric. Exp. Sta. Bul. no. 236, Raleigh.*)

Ecotypic

An *ecotype,* or ecological race, is a distinct morphological or physiological form, or population, resulting from selection by a distinct ecological condition. It is adapted genetically to factors of its local habitat. Ecotypes within a species are able to interbreed freely, but the offspring are naturally selected in each habitat. Ecotypic variation, therefore, is habitat-correlated genetic variation unlike nonadapted variation, and inherited unlike ecophenic variation. Transplant or common garden experiments will distinguish an ecotype (which would maintain its morphological form) from an ecophene (which would change in relation to its new environment).

Many trees with broad ecological ranges have evolved ecotypes adapted to different climatic or edaphic (soil) conditions. Eastern redcedar (*Juniperus virginiana*) has an ecotype (earlier called *J. silicicola*) that is adapted to coastal foredunes and coastal river sandbanks (Adams, 1986). Black cottonwood shows photoperiodic (day length) effects on growth, development, and phenology, which are ecotypic (Howe et al., 1993). Numerous provenance studies of trees reveal such ecotypes and, therefore, it is most important to select the proper source of seeds for any planting to be successful.

On the other hand, some broad-ranging species lack ecotypic or clinal differentiation and instead apparently have a highly plastic "general purpose genotype" such as found in white oak (Baranski, 1975). With such built-in plasticity and broad ecological amplitude or tolerance, there is no selective pressure to establish specific ecotypes for different ecological conditions.

Clinal

A *cline* is a character gradient correlated with a geographical or ecological gradient. Most environmental factors such as temperature, rainfall, and photoperiod vary in a gradual, continuous fashion, and thus one would expect a gradual, continuous variation in a wide-ranging species rather than discontinuous, ecotypic variation. Each population along the gradient is more or less homogeneous and approximately adapted to the environment at that point along the gradient. Mergen (1963) found in white pine a decrease in needle length and number of stomata and an increase in number of resin ducts with increasing latitude in North America. Steiner (1979) described bud-burst timing among populations in several pines. Hardin (1957a) found that red buckeye shows an east-west cline in two characters, the calyx length and pubescence of the lower leaf surface. The calyx becomes shorter and more bell-shaped from the Atlantic coast to Texas, and there is a gradual increase westward in the percent of plants in a population with tomentose or wooly leaves. Barnes (1975) described a north-south cline in leaf shape, size, and tooth number in quaking aspen in western North America. Yellow birch and sweet birch show latitudinal and elevational clines in phenology of growth initiation and cessation (Sharik and Barnes, 1976). Eastern redcedar varies in biochemical constituents as well as vegetative morphology along a northeast-southwest transect (Flake et al., 1969). In a study of the California closed-cone pines, Millar et al. (1988) found that divergence in populations within a species was generally clinal. Clinal variation in morphology, physiology, and chemistry is common in forest trees.

Reproductive

Various reproductive strategies are important determinants of variation at both the population and species level. There are three basic types of breeding systems: outbreeding or outcrossing, inbreeding, and apomixis. Each type has advantages and disadvantages in terms of the long-term survival and evolution of the species. Most species actually maintain a "winning" combination of several strategies.

Outcrossing (*xenogamy,* crossing between individual plants) theoretically leads to greater genetic variation within populations, and, assuming some gene flow among populations, there are fewer differences among populations. Outcrossing is promoted by various floral mechanisms, self-incompatibility, and the monoecious and dioecious conditions. Dioecious plants, such as ginkgo, holly, and ash, are obligate outcrossers. In contrast, inbreeding (*autogamy,* selfing within a single, bisexual flower, and *geitonogamy,* crossing between flowers of one plant) leads to less genetic variation within a population and greater differences among populations of a species. Several floral mechanisms promote such a breeding system.

Apomixis is a specialized mechanism that is a substitution of an asexual process for the normal sexual reproduction. It may give rise to an unusual pattern of within-population and between-population variation. This may occur in two basic ways, the first being *vegetative apomixis* in which the plants reproduce by root sprouts or other means of vegetative propagation in place of sexual seed formation. A good example among trees is bigtooth aspen and quaking aspen, in which, over extensive areas in the Great Lakes region or central Rockies, reproduction is principally by root suckering, despite the potential for regeneration by seeds (Barnes and Wagner, 1981; Mitton and Grant, 1996). The local *clones* (asexually reproduced populations in which all individuals are essentially alike) are extensive.

The other type of apomixis, often called true apomixis, is *agamospermy*—the asexual formation of a seed found in certain tree species such as citrus, alder, mountain-ash, serviceberry, and hawthorn (Campbell and Dickinson, 1990). Many apomicts maintain some degree of sexual reproduction and therefore some genetic variation in their offspring. Other apomictic populations, if only agamospermous, are quite uniform in their characteristics because of the lack of genetic segregation and recombination. If such occurs in interspecific hybrids, it is a way of stabilizing and maintaining particular hybrid forms, with each apomictic population possibly being slightly different. Such a combination of interspecific hybridization and agamospermy leads to the formation of a bewildering array of populations carrying different combinations of the parental characters, but each is different and fairly uniform. This forms what is called the *agamic complex,* which, fortunately for the dendrologist, is very infrequent among forest trees.

Speciational

If ecotypes, portions of a cline, or populations of autogamous plants become *genetically isolated* through spatial or reproductive barriers of some type, the distinct population can continue to diverge genetically. This is the primary mode of species formation (speciation). The degree of differentiation, in terms of morphology, ecology, and reproductive isolation, may be gradual and at first lack distinct discontinuities. Such an early stage of incipient speciation gives rise to patterns of primary intergradation where the divergent population is only slightly distinct from the parental one.

Sometimes these diverging populations are recognized by taxonomists as varieties or subspecies. The coastal and Rocky Mountain forms of Douglas-fir may represent incipient and incomplete speciation, because the two types intergrade in certain areas. The large fruited form of devilwood, found in the dry pine scrub of south peninsular Florida, is probably another good example of incipient speciation and is considered a variety of the widespread *Osmanthus americanus* of more mesic environments (Hardin, 1974). California sycamore of stream banks in California, with 3- to 5-lobed leaves, and Arizona sycamore of canyons in Arizona and New Mexico, with 5-to 7-lobed leaves, are called species, but they are not very distinct and are considered by some as varieties of *P. racemosa* (see Little, 1979).

EXTRINSIC

Hybrid

Chance hybrids between species may occur in nature when the parental types are closely related and not isolated spatially or reproductively. This leads to an occasional F_1 hybrid, generally recognized as intermediate in character states between those of the suspected parents. Such hybrids may survive in a disturbed habitat or in one that is somewhat intermediate between those of the parents. An example is the rather rare hybrid chestnut, which is a cross between the American chestnut and the chinkapin (Johnson, 1988). Other occasional hybrids occur between yellow birch and bog birch, mentioned earlier in connection with polyploidy, and between mountain paper birch and gray birch (DeHond and Campbell, 1989). The hybrid between Coulter pine and Jeffrey pine in California (Zobel, 1951) and the hybrid between white and chestnut oaks in the southern Appalachians (Hardin, 1975) are other good examples, and several others could be given because hybridization is fairly common among forest trees. See Burns and Honkala (1990) and Little (1979) for hybrids listed under the species.

Introgressive

In many cases, hybridization in forest trees goes beyond the F_1 hybrid to the formation of hybrid swarms or introgressive populations when the hybrids backcross repeatedly to one or both of the parental species. Introgression is therefore the transfer of genes (and character states) from one species into another across a strong but incomplete barrier to interbreeding (i.e., secondary intergradation).

Introgression forms not only an intermediate population but also a bridge across the morphological and ecological gaps between the species, and it generally broadens the range of variation in the parental species. When related species overlap in distribution, introgression may occur in local areas of disturbance or intermediate habitats where the introgressants can survive. Much of this is only *localized introgression* where the hybrid population is rather obvious, and there is no widespread influence on the morphology of the parental species (Hardin, 1975). There are some examples of *dispersed introgression* where the genetic influence of one species is detected far from the site of original hybridization (Hardin, 1957b). In all cases, the parental species maintain their general distinctness in the face of this "genetic contamination," and they are still considered to be "good species." Introgression is known to occur in junipers, pines, oaks, basswoods, buckeyes, and other genera. An interesting case in which pollinating hummingbirds apparently play an important role in introgression in buckeyes is given by dePamphilis and Wyatt (1989, 1990).

Hybridization and introgression between species of a genus form a more inclusive unit of interbreeding called a *syngameon* (i.e., a hybridizing group of species). Examples are in pine, juniper, birch, poplar, buckeye, oak, and others. Some involve only a few species, but others may be highly complex. One, for example, includes nearly all the white oaks of eastern North America (Fig. 5-2).

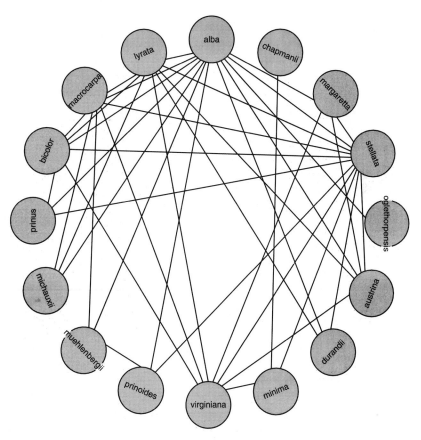

FIGURE 5-2   The white oak syngameon of eastern North America. (*Modified from Journal of the Arnold Arboretum 56:337, Harvard University, 1975, reprinted with permission.*)

For the dendrology student, the "take-home lesson" from this discussion of variation is to realize that species are not always homogenous, static, typological entities easily pigeonholed in a man-made classification system. Rather, they represent variable, dynamic series of populations of individuals that often challenge the plant taxonomist or dendrologist with rather exciting detective work to determine just what an atypical tree really is and how it got that way.

In addition, the recognition, understanding, and use of variation are important in various aspects of forestry. For instance, the success of any tree improvement program depends on a knowledge of variation and its proper utilization. The species that naturally contains the largest genetic variation affords the best opportunity for genetic gains through provenance and trait selection. A knowledge of variation and habitats are also of prime importance in selecting trees for seed orchards or for the most appropriate and successful urban plantings. The understanding and use of variation are becoming increasingly significant in the preservation and protection of natural biodiversity. This assessment of genetic diversity will obviously become more important to the forester in future management planning.

For those who are interested in more extensive discussions of these topics, see Anderson (1949), Briggs and Walters (1984), Grant (1963, 1981), Heiser (1973), Judd et al. (1999), Stace (1989), Stebbins (1950), and Stuessy (1990).

6

HABITAT, RANGE, AND COMMUNITIES

Knowledge of the habitat and range of individual species is of great importance in the practice of forestry. Such ecological knowledge is also helpful, in combination with morphological features, for tree identification. Only brief habitat and community descriptions are given in this text, but more complete information can be found in Barnes et al. (1998), Burns and Honkala (1990), Farrar (1995), Leopold et al. (1998), and Walker (1998).

HABITAT

Factors

Every species of tree is best adapted, by its genetic structure, to a particular combination of environmental factors or conditions. When these conditions are present, the species will grow, develop, successfully compete, and reproduce. The sum of these conditions is known as the habitat of the species. The environmental factors, when combined, produce a particular habitat derived from the climate, soils, physiography, and biota present at the site. *Climate* is the product of the interaction between solar radiation, the atmosphere, and the land and water masses of the earth. Climate is regional, the large canvas upon which are developed a multitude of habitats or forest sites. The climate of the southeastern United States, for instance, differs in many ways from that of the southwestern states. Climate provides the fluctuating temperature, precipitation, atmospheric humidity, wind, lightning, and atmospheric impurities that contribute to the definition of a habitat. *Soils* provide the moisture and the elements required for growth (except for carbon and oxygen) and also provide support for the stem and crown. The physical characteristics of a soil, its texture, structure, and depth determine the availability of moisture and air and often microorganisms to the trees' roots. The chemistry of a soil determines the availability of mineral nutrients. *Physiography* influences habitat through the effects of elevation, slope, aspect, land shape, and the land-water interface. As elevation increases, average temperature and length of growing season and usually soil depth decrease while radiation intensity increases. The direction in which a slope faces is known as its aspect. In the latitudes of North America, south- and west-facing aspects, which are warmer and often drier, differ from north- and east-facing aspects. Depressions, because of cold air drainage, can be much cooler at night than adjacent ridges. Land areas adjacent to large bodies of water are usually slightly warmer in winter, cooler in summer, and more humid than are inland areas. Elevation and aspect will be described more later.

The *biota* is the sum of all living organisms that inhabit an area and exert a positive or negative influence on the suitability of a habitat for a particular species of tree. The relationships between the micro- and macroflora and fauna, played on a background defined by temperature and sunlight and by moisture and soil, create the character of a habitat. From bacteria to birds, from predators to herbivores, from algae and fungi to flowering plants, living organisms both result in and are the result of habitats. This contribution is continued, perhaps even accentuated, as the lifeless and decaying remains of all organisms continue to contribute to the maintenance and modification of habitat. It is from its habitat that a tree obtains its resources for seed germination, establishment, growth, and reproduction. If these resources cease to be present, the tree, ultimately, cannot survive.

Very often, the opportunity for seed germination controls the presence or absence of a given species in a particular habitat. An established seedling/sapling/older tree may be transplanted and grow perfectly well outside of its natural range and usual habitat. Examples are many cultivated urban trees and the plantation culture of Monterey pine in the southern hemisphere.

Elevation and Aspect

The presence of a species is not continuous throughout its range because suitable habitats are not continuous and uniform. The definition of a suitable habitat for a particular species can often be quite sharply defined. The effect of elevation upon habitat is an example of this. Quaking aspen, for example, is rarely found above 3000 feet elevation in the northern part of its range and is abundant only above 10,000 feet in the southern part of its range. In the north, it is more common on warmer southern and western aspects, whereas in the south it occurs on the cooler northern and eastern aspects. In each case, elevation and aspect, through their direct effect upon temperature, are influencing the local presence of this wide-ranging species. Especially in the west, elevation is a limiting factor, and over wide areas certain species are commonly found only in certain elevational belts. The relation between latitude and elevation is also important, and species that occur at sea level in the north are found progressively higher toward the south, until an elevation of 10,000 to 11,500 ft may be reached at the southern limit of distribution.

Tolerance

It has been long known that certain species will grow under varying densities of forest cover, whereas others grow and develop well only in the open. For many years, this was interpreted as a reaction to light; hence, an intolerant tree was one that could bear little or no shade. However, Zon (1907) reported the results of root-cutting experiments by Fricke, in Germany, which seemed to show that the inability of certain species to prosper under old-growth forest was due in large part to excessive root competition. It seems best at present to consider tolerance as the ability of a plant to complete its life history from seedling to adult, under the cover of a dense forest, regardless of the specific factors involved. Or, put more simply, a tolerant tree is one that can survive and grow, if not prosper, under forest competition; when this competition is removed or lessened, the rate of growth and reproduction usually show a marked increase. For a more recent discussion of tolerance, see Decker (1952, 1959).

Tolerance is influenced by site (soil, exposure, climate, etc.) so that the best comparisons are made only between trees growing together under the same set of external conditions. Little is gained in attempting to compare two species of widely separated ranges, growing on different soils and with dissimilar associates. Designations of relative tolerance here follow Burns and Honkala (1990).

Wetland Species

Although some trees are fairly cosmopolitan, most are associated with a characteristic habitat and position along a moisture gradient from wetland to dry upland. For instance, it would be as futile to look for a water tupelo on a dry upland hillside as to expect to find shortleaf pine in a deepwater swamp. Some species cannot survive in saturated soils, whereas others only very rarely can be found elsewhere. In between these extremes, there are a multitude of species that are adapted to varying levels and duration of soil saturation. Site relationships are frequently very complicated, and the absence of a certain species in a given habitat may reflect its inability to compete rather than to simply grow there.

This species-specific habitat characteristic is used in the identification of legal or jurisdictional wetlands. In addition to hydric soils, which are saturated during some part of the growing season, such sites must support a preponderance of plants adapted to such conditions.

Recent attention has been focused on *wetlands* and their delineation, preservation, and management. Wetlands may be permanently or semipermanently flooded or only temporarily or seasonally flooded. Today, it is critical to determine the limits of wetlands because many activities (e.g., draining, filling, or other disturbances) require federal or state permits. Because plant species are a major determinant of wetlands, it has become increasingly important to classify a plant species into one or more of five categories, based on its usual habitat or range of habitats along a soil-moisture gradient. The five categories are (1) obligate wetland, (2) facultative wetland, (3) facultative, (4) facultative upland, and (5) upland. The U.S. Fish and Wildlife Service has compiled a national list of nearly 7000 vascular plants that classifies each species into these categories (Reed,

1988; see also Tiner, 1991). The trees in this text that are designated as "obligate" or "facultative wetland" in the 10 regions of the conterminous United States (Fig. 6-1) are indicated in Table 6-1.

Obligate wetland plants (O) occur *almost always* (probability of 99% or more) in wetlands under natural conditions; facultative wetland plants (F) *usually* occur in wetlands (67 to 99% probability). Certain species may differ in their designations in different parts of their range. For instance, red maple can thrive on a wider range of moisture regimes than any other forest species in North America. It occurs from dry ridges and southwest slopes of uplands to peat bogs and swamps. These kinds of differences may be ecotypes, parts of a cline, or reflect a broad ecological amplitude. Each species has a characteristic pattern of ecological variation (see Chapter 5).

RANGE

Every tree, indeed every land plant, is restricted in nature to a particular geographical area where conditions of climate and soil are suitable and available for its growth and reproduction and where it has been documented to occur. This area is known as its *natural range.* It may be very large, as in the range of quaking aspen that stretches from Newfoundland to Alaska and southward at increasing elevation along the Rocky Mountains to northern Mexico. This represents 111 degrees of longitude and 48 degrees of latitude and is the greatest range of all native North American trees. This vast range is largely defined by growing season, moisture stress, and high summer temperatures. The natural range of Torrey pine, on the other hand, is confined to a very small coastal area of southern California and Santa Rosa Island. Although the dry, sandy bluffs and slopes are found elsewhere, the particular combination of factors allows its persistence in these two areas as relics of a once broader distribution.

The range maps used in this text delineate the larger region in which the species is known to occur naturally. The scale of the maps is too small to show local distribution, but the information given for each species under "Range" does indicate some of the major disjunct areas. Little (1971, 1976, 1977, 1978) and Viereck and Little (1975) show large-scale maps of the various species.

Natural ranges are not fixed but are continually shifting, albeit very very slowly, in response to environmental changes. Although this may not be immediately apparent, it should be remembered that we are in a postglacial period and that some species are still migrating slowly northward following the retreat of the last great ice sheet. Much is known about species distributions and migrations during the last 12,000 years (FNA vol. 1, 1993). Spruce and fir trees were the dominant forest type in coastal North Carolina when the ice began to withdraw. As the climate became too warm and dry for these species, they slowly moved northward following cool growing seasons to be replaced first by an oak-dominated forest, and then, as the weather continued to warm, by forests of the more drought- and fire-resistant southern pines. Today spruce and fir are found in North Carolina only on a few of the highest peaks in the western part of the state where they linger as a tremulous echo of the Ice Age.

Climatic warming due to the greenhouse effect may also create changes in natural ranges of trees. Some believe that the rate of warming is more rapid than that which

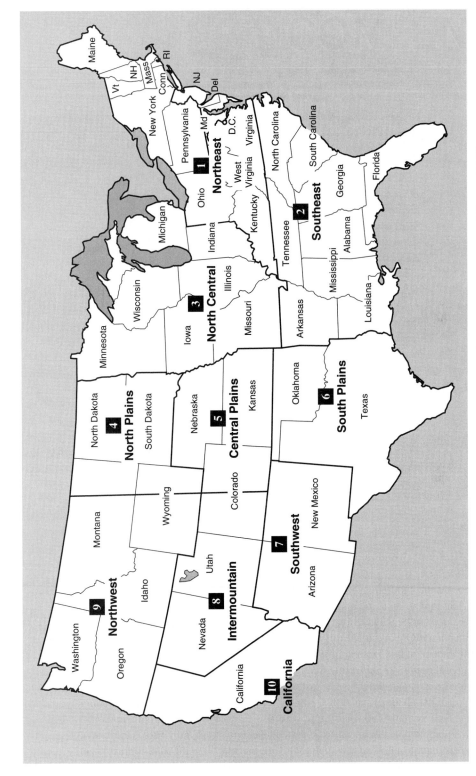

FIGURE 6-1 Regions of distribution for the National List of Scientific Plant Names. (*From Reed, 1988, U.S.D.I. Biol. Rept. 88(24).*)

TABLE 6-1   WETLAND STATUS OF SELECTED TREE SPECIES

Species	Region									
	1	2	3	4	5	6	7	8	9	10
Abies balsamea			F							
Acer negundo		F	F			F	F	F		F
A. rubrum	F	O				F				
A. saccharinum	F	F	F	F	F					
Alnus rhombifolia									F	F
A. rubra										F
Aralia spinosa			F							
Betula nigra	F	F	F		O	F				
Carya aquatica	O	O	O			O				
C. illinoinensis			F							
C. laciniosa		F	F							
C. myristiciformis		F				F				
Celtis laevigata	F	F	F		F					
Fraxinus caroliniana	O	O				O				
F. latifolia									F	F
F. nigra	F		F	F						
F. pennsylvanica	F	F	F		F	F		F		
Gordonia lasianthus		F								
Larix laricina	F		F							
Liquidambar styraciflua			F							
Magnolia virginiana	F	F								
Nyssa aquatica	O	O	O			O				
N. biflora	F	O				O				
Picea mariana	F		F							
Pinus elliottii		F								
P. glabra		F								
P. serotina	O	F								

(continued)

reduced the ice sheets. As vegetation would have less time to adjust, the magnitude of this change could exceed manyfold that caused by the retreating glaciers. Predictions may be made regarding future migrations, which can be expected in response to this climatic change (Critchfield, 1984; Davis, 1976; Delcourt and Delcourt, 1981; Delcourt et al., 1983; Iverson et al., 1999; Jacobson et al., 1987; Wright, 1983).

Mankind has, throughout history, moved trees and plants over long distances bypassing normal barriers to natural range, such as oceans, mountains, and deserts. When the exotic species are introduced to favorable habitats, they can become *naturalized* and reproduce and develop at least as well as in their native area. Scotch pine from Europe has become naturalized in Canada and the northeastern United States; some species of eucalyptus from Australia are now naturalized in California; mimosa and paulownia,

TABLE 6-1   CONTINUED

Species	Region									
	1	2	3	4	5	6	7	8	9	10
Platanus occidentalis	F	F	F							
P. racemosa										F
P. wrightii							F			
Populus angustifolia				F	F	F	F		F	F
P. balsamifera	F		F	F	F			F		F
P. deltoides							F	F		
P. fremontii						F	F	F		F
P. heterophylla	F	O								
Quercus bicolor	F	F	F		O	F				
Q. laurifolia	F	F				F				
Q. lyrata	O	O	O			O				
Q. michauxii	F	F	F			F				
Q. nigra			F							
Q. texana		O	O			O				
Q. pagoda	F									
Q. palustris	F	F	F		F					
Q. phellos		F	F			F				
Q. stellata		F				F				
Salix nigra	F	O	O		O	F		O		
Tamarix chinensis				F	F	F		F	F	F
Taxodium distichum	O	O	O			O				
Thuja occidentalis	F		F							
Toxicodendron radicans							F			
T. vernix	O	O	O							
Ulmus americana	F	F	F							

Regions: see Figure 6-1
Status: F = facultative wet; O = obligate wet

natives of China, have spread throughout much of the southeastern United States; the original distributions of pecan, southern magnolia, and several others were much more limited than their present ranges. In some cases, introduced and naturalized species can become invasive weeds. Punktree and Russian-olive are examples.

COMMUNITIES

North American biotic communities, as we know them today, have been structured by processes operative on both evolutionary and ecological timescales. The forest regions present today have not been stable either in composition or in location through very long periods of time. In fact, they are still responding dynamically to changing climate, geological processes, fire regimes, and the progressively increasing impacts of human populations and needs.

When left unhampered, however, trees and lesser vegetation tend to group themselves on the basis of environmental factors and their ability to compete with their associates into *communities.* Thus the *flora* of specific areas and particularly forest communities are often regarded and designated, especially for practical reasons of description and study, as distinct *forest cover types* or *associations* with one or a few dominant species, such as *pinyon-juniper, sugar maple-beech-yellow birch, oak-hickory, longleaf pine,* or *eastern white pine.* The Society of American Foresters has described 145 forest cover types in the United States and Canada (Eyre, 1980). Each of these cover types occupies a particular geographic region and/or elevational zone. Each is characterized by the species of trees, shrubs, and herbs that grow together. Each may be permanent as a climax type or transitory as a successional stage. In sharply defined and delimited habitats, the forest communities may also be sharply defined and discontinuous. More often, however, community boundaries are rather indistinct, and there is a broad ecotone between two otherwise distinct forest types. An even more gradual change in communities, a *continuum,* may occur along a very gradual environmental gradient, and specific community types may be impossible to define. Regardless of the community designations or delimitations, each species has a characteristic habitat or ecological range in terms of moisture, shade, and other environmental factors.

Other descriptions of forest communities are also in use today. Kuchler (1964) described 116 plant communities in the United States in terms of "potential natural vegetation," meaning what would potentially exist without human influence. Daubenmire and Daubenmire (1968) developed a different system that is widely used in the West.

Elevational zonation of forest communities may be particularly striking in mountainous areas. For instance, in the Sierra Nevada of California (Fig. 6-2), one finds an oak woodland below 1000 ft elevation, then several rather distinct forest associations, and finally the whitebark pines of the subalpine zone at 10,000 ft. In the uplands of the southwest, as another example, pinyon-juniper communities dominate from ca. 4500 to 6500 ft elevation, pine-oak at 6500 to 8000 ft, fir-aspen at 8000 to 9500 ft, and spruce-fir at 9500 to 11,500 ft. These associations are more or less distinct but vary somewhat in their elevational ranges due to latitude and aspect—higher on the drier and warmer south and southwest exposures, and lower on the wetter and cooler north- and northeast-facing slopes and in cool, shaded canyons (Elmore, 1976). Likewise, in the Great Smoky Mountains of eastern Tennessee and western North Carolina, several forest associations occur in rather specific locations with respect to topography, elevation, and moisture (a mixed mesophytic forest in the lower coves to spruce-fir on the peaks and ridges above 6000 ft elevation) (Whittaker, 1956). Even in relatively flat topography, such as in the Green Swamp area of southeastern North Carolina (Kologiski, 1977), vegetation types are sometimes differentiated on the basis of environmental factors such as soil types, relative periods of flooding, and frequency of fire (Fig. 6-3).

Following disturbance by extensive logging, clear-cutting, or fire, each region has a fairly predictable succession of stages eventually culminating in a climax association characteristic of the climate, soils, and other factors. For example, a general successional pattern (from frequent fire to excluded fire) in the Green Swamp area of North Carolina can be interpreted from Figure 6-3 (Kologiski, 1977).

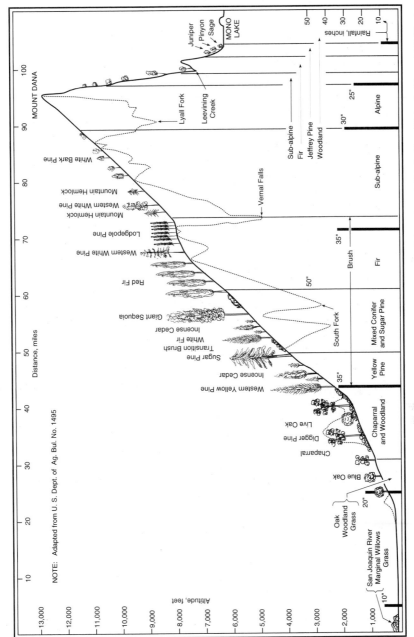

FIGURE 6-2 Profile of the Sierra Nevadas showing the effect of elevation and rainfall on forest distribution. (*Adapted from U.S. Dept. of Ag. Bull. No. 1495.*)

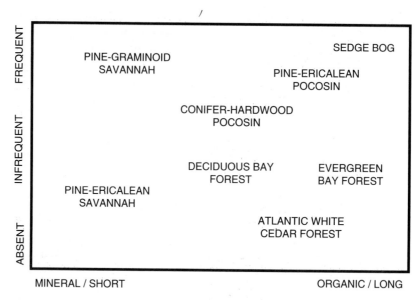

FIGURE 6-3 Relationships among the dominant vegetation types and soil, hydroperiod, and fire in the Green Swamp of North Carolina. (*Modified from Kologiski, 1977, by R. L. Kologiski and T. R. Wentworth.*)

On a much broader scale, one or more plant communities with a definite structure or life form (e.g., deciduous forest, conifer forest, or grassland) and that occupies similar habitats is defined as a *formation*. The distribution of recognizable formations of North America is shown in Figure 6-4.

As indicated earlier, it will be advantageous if you think of habitat and range as important characteristics for understanding and identifying species. It is also useful to think of the communities in which a species occurs and whether the species is a dominant component or lesser associate in the community. When you learn to recognize a particular community by the dominant species present and also by its general aspect (physiognomy), you will learn to predict the presence of certain other species that are often found in that community.

Old-Growth Forests

Photographs of large, old trees (Figs. 7-9, 8-46) have ignited within many of us a life-long interest in trees and forest ecology. Such historical photos depict the nature of our original forest communities. Although most of our original forests have been either harvested two or more times during the past 200 years or destroyed by wildfire or human activities, magnificent glimpses of these forests remain throughout the country. Today, these forests are referred to by most professionals as "old-growth forests," recognizing that truly "virgin" forests (i.e., forests that have never been influenced by humans) probably do not exist.

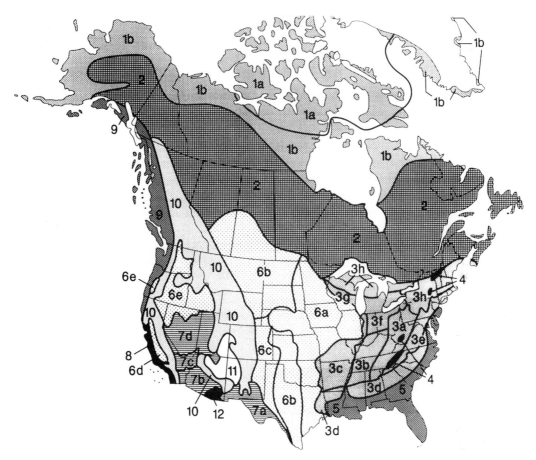

FIGURE 6-4 Generalized vegetation formations of North America north of Mexico. **1a**, high arctic tundra and polar desert. **1b**, low arctic tundra. **2**, boreal forest or taiga. **3a**, mixed mesophytic deciduous forest. **3b**, western mixed mesophytic forest. **3c**, oak–hickory. **3d**, oak–hickory–pine. **3e**, oak–chestnut. **3f**, beech–maple. **3g**, maple–basswood. **3h**, northern hardwoods. **4**, high elevation Appalachians. **5**, Coastal Plain forests, bogs, swamps, strand. **6a**, tallgrass prairie. **6b**, mixed-grass prairie. **6c**, short-grass prairie. **6d**, central California grassland. **6e**, intermountain grassland or shrub steppe. **7a**, Chihuahuan warm desert scrub. **7b**, Sonoran warm desert scrub with saguaro cacti. **7c**, Mojave warm desert scrub with Joshua trees. **7d**, Intermountain, Great Basin, or cold desert scrub. **8**, Mediterranean woodlands and scrub. **9**, Pacific Coast conifer forest. **10**, western montane conifer forests. **11**, mosaic of desert grassland, pinyon-juniper woodland, chaparral, woodlands, and scrub. **12**, mosaic of desert grassland, warm desert scrub, woodlands, and scrub. [*From map developed by W. D. Billings, in M. G. Barbour and W. D. Billings*, North American Terrestrial Vegetation (1988), *with permission of Cambridge University Press, as adapted for* Flora of North America, Vol. 1 (1993), *with permission of Oxford University Press.*]

Some of the most famous examples of old-growth forests anywhere in the world are the redwood forests of coastal California, the mixed conifer forests of the Sierras that are dominated by giant sequoia, and the old-growth Douglas-fir and Sitka spruce forests of the Pacific Northwest. More recently, there has been a growing appreciation for the variety and extent of old-growth forest communities in eastern North America (Davis, 1996). Although these eastern remnants cannot compare in maximum size and age to their western counterparts, they do have numerous common characteristics that collectively define an old-growth forest.

Old-growth forest stands generally have these characteristics: a mosaic of relatively even-aged patches of trees, each patch a different age that resulted from past natural disturbance events, and at some stage of regeneration, maturation, or senescence; presence of, but not dominance by, trees that are near their maximum longevity and often size for that species; substantial number of declining trees and standing dead snags; significant accumulation of coarse woody debris; and pit and mound topography. Furthermore, there should be little if any evidence of timber harvest and/or grazing, and the area should be large enough to allow natural disturbance regimes to operate.

Old-growth forest can educate and inspire us about our natural heritage. Additionally, old-growth forests provide habitat for numerous plants and animals that depend on their unique ecological characteristics.

There is still much to be learned about old-growth forest function and value. A critical understanding of old-growth forests must begin with the study of its key players, the individual tree species that compose these communities.

7

VEGETATIVE AND REPRODUCTIVE MORPHOLOGY

Before attempting to understand descriptions, comparison tables, or identification keys here or in manuals, or to communicate with others about tree identification, one first has to become familiar with the morphology of vegetative and reproductive parts and descriptive terminology for these parts—the necessary jargon of dendrology. The important terms are explained and illustrated here. For more complete illustrated glossaries, see Harris and Harris (1994), Jones and Luchsinger (1986), Judd et al. (1999), Radford (1986), Radford et al. (1974), Smith (1977), Walters and Keil (1996), Woodland (1997), or Zomlefer (1994).

VEGETATIVE MORPHOLOGY

For tree identification, overall form or habit and vegetative parts, including leaves, twigs, and bark, are very important mainly because they are more often present and accessible than reproductive parts.

Habit

Trees growing in the open tend to develop characteristic shapes that may be typical of species or genera; this is well illustrated by the American elm, white oak, Lombardy poplar, or eastern white pine. In general, open-grown specimens have large crowns that may reach nearly to the ground, and the clear trunk is short, with considerable taper. Under forest competition, the form is very different: The bole is long, more cylindrical, often clear of branches for one-half or more of its length, and the crown is small. Certain species, especially those that are extremely tolerant, resist crown restriction, but it may be said that most trees develop typical shapes only in the open; such individuals, of course, are usually less desirable for lumber because of the short clear length and many side branches, which would appear as knots in the finished boards.

Due to variation in various habitats, habit (growth form) is of limited use in identification. Yet, with experience, one can learn the shapes of the crown and pattern of branching of many species. Four primary habits are observable: *excurrent,* with a central dominant trunk and symmetrical, conical, or spirelike crown of many conifers and some hardwoods like yellow-poplar; *deliquescent* or *decurrent,* with repeatedly forked stems giving rise to a spreading form such as in oaks, maples, and most hardwood trees; *palm-like,* with an unbranched trunk and leaves only in a top rosette, in palms and most cycads; *yucca-type,* either with a basal rosette of long stiff leaves and a central tall flowering stalk, or an irregularly branched thick trunk as in the treelike yuccas and cacti.

Within the deliquescent habit, there are a number of distinctive architectural forms depending upon the relative dominance of the primary trunk or the secondary vertical (orthotropic) branches versus horizontal (plagiotropic) branches (i.e., whether height growth or crown spread is favored). The basic growth form within a species is genetically controlled, but the ultimate architecture is also easily modified by habitat (sun, shade, competition), natural or artificial pruning, or branch damage due to lightning, ice, or insects. For a concise treatment of woody plant form and structure, see Barnes et al. (1998).

Sizes of Mature Trees   Sizes are quite variable and influenced by local conditions of site or by geographical location. For example, some species become shrubby or prostrate at the northern limits of their range or at higher altitudes.

Foresters commonly measure tree diameter at $4\frac{1}{2}$ ft (1.37 m) above the ground (diameter breast high, dbh). Diameter at this height is usually above the flaring base of the tree and is a measure of trunk size. On very large trees where butt swell and flaring may extend several feet or more, dbh is measured 18 in. above the butt swell (Avery and Burkhart, 1983). When maximum sizes for a species are given in this text, remember that these may not be common and reflect growth under very optimum conditions.

Leaves

General Features (Fig. 7-1)   Leaves are the primary photosynthetic organs of trees. Although they are often the most variable part, each species has a characteristic range of variation, so leaves are still one of the most important means of identification because of the many macromorphological characters available for observation and comparison. Leaves are formed on stems at regular intervals, and the point on a stem at which one or more leaves arise is called the *node.* The portion of the stem between nodes is the *internode.* The expanded portion of the leaf is the *blade* or *lamina,* and the supporting stalk is the *petiole.* The petiole (considered part of the leaf) may be either short or long, slender or stout, terete, angular, grooved, or laterally flattened. When the leaf is *petiolate,* the petioles are usually attached to the stem just below an axillary bud, but some are swollen at the base and enclose the bud. In some cases, the petiole is lacking and the blade is *sessile* (i.e., the blade is attached directly to the twig). Some leaves are accompanied by a pair of small scalelike or leaflike structures known as *stipules,* which are attached either

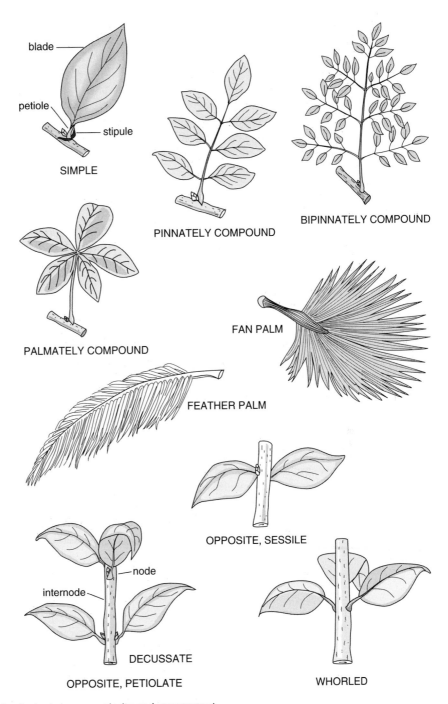

FIGURE 7-1   Leaf morphology, complexity, and arrangement.

to the petiole base or to the twig, one on either side of the petiole. These usually drop off as the leaf expands. Plants with stipules are *stipulate*; those without them are *estipulate*.

A given leaf persists on the tree for one season of growth or two or more years and then will eventually drop. The tree is said to be *deciduous* if all leaves drop after one growing season, or *evergreen* if not all leaves drop at the same season and so appear green all the year round. Some "deciduous" trees keep their leaves in a brown condition into the winter; this is a condition called *marcescent,* particularly noticeable in young beech and some oaks. Some species are deciduous in the northern parts of their range and evergreen southward, such as sweetleaf and laurel oak.

Leaf Complexity (Fig. 7-1)   A leaf with a single blade is *simple*. If the leaf is divided into two or more blades attached to a common stalk, it is *divided* or *compound,* and the individual blades are *leaflets.* The axis supporting the leaflets is the *rachis.* The lateral bud is still in the axil of the petiole of the compound leaf, not in the axil of the leaflet.

When leaflets are attached laterally along the rachis, the leaf is *pinnately* compound. The leaflets may be sessile on the rachis or on a stalklike *petiolule. Odd* or *even* pinnate indicates whether pinnately compound leaves have odd or even numbers of leaflets, respectively. Leaves are twice pinnate or *bipinnate* when the leaflets are attached to a second-order rachis (*rachilla*); an additional division would be *tripinnate.* If the rachis is not elongated, and the leaflets apparently radiate from the top of the petiole, the leaf is *palmately compound.* The "compound" leaves of palms have mechanically split blades either in a pinnate "feather" type, a palmate "fan" type, a somewhat intermediate "costa-palmate" type that looks fanlike but has a short midrib (costa), or a twice-pinnate type.

A *trifoliolate* leaf has three leaflets and may be either pinnate, if the petiolule of the terminal leaflet is attached to an elongated rachis, or palmate, if each of the three leaflets has its petiolule attached at one point.

To distinguish a pinnately compound leaf from a twig with distichous, simple leaves, look for the position of the bud in the axil of the leaf (not leaflet). Also, there is usually a color and/or textural difference between the twig and the petiole/rachis.

In some trees (e.g., *Cercis*), the apparently simple leaf is actually the remaining terminal leaflet of an ancestral pinnately compound leaf. It is called *unifoliolate* although generally treated as *simple* in keys and some descriptions.

Leaf Arrangements (Phyllotaxy) (Fig. 7-1)   On normal twigs, leaves are usually arranged in one of three definite ways: (1) If they are paired at the same height, one on each side of the twig, they are *opposite*; (2) when more than two are found at the same node, they are *whorled,* or *verticillate*; and (3) where only a single leaf is attached at each node, the leaves are *alternate* and also in a spiral up the twig. One other arrangement is *subopposite,* which is when the leaves appear nearly but not quite opposite. This is characteristic of a few trees such as buckthorn, sweetleaf, and cascara.

With the alternate arrangement, the determination of the number of leaves in each complete turn of the spiral is important because it is often the same throughout a genus and sometimes applies to all the members of the same family. In determining spiral phyllotaxy, the twig is held in a vertical position and two leaves are chosen, one of which is directly above the other; twisted, deformed, or very slow growth twigs should not be used, and the inspection should be confined to a single season's growth. Do not count the lower leaf of the two chosen; ascend the spiral and count the number of leaves passed, up to and including the upper leaf, and also note the number of complete turns made around the twig. A fraction may then be formed, using the number of turns as the numerator and the number of leaves as the denominator. If this is done, it will be found that one of the following fractions has resulted; $\frac{1}{2}, \frac{1}{3}, \frac{2}{5}, \frac{3}{8}, \frac{5}{13}, \frac{8}{21}$ .... (Only the first four are common in broad-leaved trees; the higher fractions occur in some of the conifers.)

The $\frac{1}{2}$ phyllotaxy, typical of elm, hackberry, and basswood, is the simplest and most obvious of the spiral arrangements and is known as two-ranked or *distichous*. The $\frac{1}{3}$ type is found in alders, whereas the $\frac{2}{5}$ is found in many trees including the oaks and poplars. Several relationships are immediately evident in this series. The denominator is the number of vertical rows (or ranks) observable by looking down the stem from the tip, and the fraction (of 360 degrees) indicates the angle between successive leaves in the spiral. If the numerators and denominators, respectively, of the first two fractions are added, the result is the next higher fraction of the series, and this applies to all the rest of the members as one ascends the scale; also, the numerator of the third term is the same as the denominator of the first, that of the fourth the same as the denominator of the second, etc. The Italian mathematician, *Leonardo Fibonacci* (ca. 1170–1230) recognized the regularity of this, and it is now called the *Fibonacci series.*

In opposite-leaved plants, a pair of leaves is often oriented at right angles to the pair above and below, so that there are four vertical ranks. This is known as a *decussate* arrangement, common in several conifers with scale leaves or in maples and other hardwoods. If the opposite leaves form just two ranks (i.e., all in one plane), the arrangement is *distichous*; privet has this conspicuous arrangement.

Because lateral buds arise in leaf axils, branch arrangement is the same as leaf arrangement (at least if both branches develop in an opposite arrangement). So, if leaf arrangement is difficult to determine because branches are very high aboveground, look for the branch arrangement.

Leaf Venation (Fig. 7-2)   There are five basic types of venation. (1) *Pinnate* has a single midrib and secondary veins branching off at intervals. The secondaries form a variety of patterns: They may be more or less diffusely branched, angle straight to the margin and end there, bend upward near the margin and sometimes connect to the one above (forming submarginal loops), or the entire secondaries arch upward toward the apex (called *arcuate*). (2) *Palmate* has three or more primary veins arising at or near the base of the blade and spreading out like a fan, with secondaries off of these. (3) *Pinnipalmate* is somewhat intermediate between pinnate and palmate in which the lowermost pair of secondary veins, arising at or near the base of the midrib, is slightly larger than the other secondaries and with large tertiary veins going to the lower margins. (4) *Parallel* or *striate* has many equal veins nearly parallel to each other until they join near the apex. (5) *Dichotomous* has repeated forking or Y-branching, as found in ginkgo. In the cycads and conifers, branch veins are rare; in hardwoods, there are smaller and smaller veins that form a netted pattern, which is often conspicuous and characteristic of genera and species.

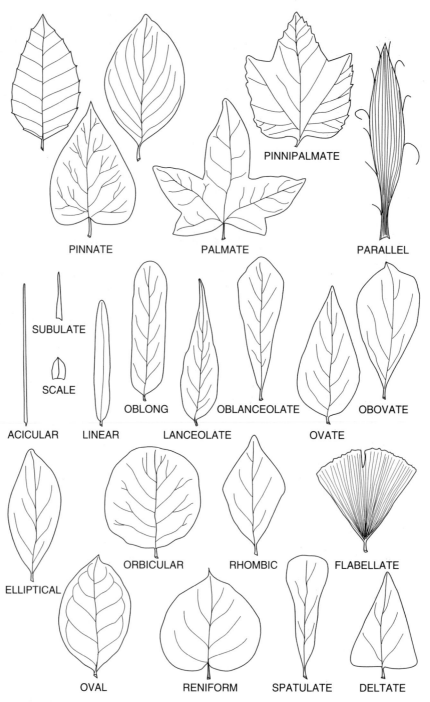

FIGURE 7-2 Leaf venation and shapes.

Leaf Shapes (Fig. 7-2)   The overall shape of a leaf or leaflet is usually characteristic of a species, allowing of course for a range of variation. The shape of compound or lobed leaves conforms to the outline produced by running an imaginary line around the tips of the leaflets or lobes. In addition to overall shape of the blade, the shapes of apices and bases and forms of the margin are important features.

Overall Blade Shapes (Fig. 7-2)
Acicular. Needlelike; very long and slender; variously shaped in cross section
Subulate. Awn- or awl-shaped; narrow lanceolate, flat, stiff, sharp-pointed, usually less than 12 mm long
Scale. Small, sharp-pointed, broadened at the base, usually appressed and imbricated
Linear. Many times longer than broad; narrow, with approximately parallel sides; flat, triangular, or square in cross section
Oblong. Longer than broad, with parallel sides
Lanceolate. Lance-shaped; several times longer than broad, widest at a point about a third or less of the distance from the base, and tapering to the apex
Oblanceolate. The reverse of lanceolate; with the widest point about a third or less of the distance from the apex, and tapering to the base
Ovate. Egg-shaped outline, with the widest point below the middle
Obovate. The reverse of ovate, with the widest point above the middle
Elliptical. Like an ellipse with the widest point at the center
Oval. Broadly elliptical, with the width greater than one-half the length
Orbicular. Circular or nearly so
Reniform. Kidney-shaped; as broad or broader than long with a wide cordate base
Deltate. Delta-shaped; triangular
Rhombic. Diamond-shaped; more or less symmetrical with the widest point at the center and the sides more or less straight to the apex and base
Spatulate. Narrower than obovate; shaped like a spatula with a broad apex and tapering to the base
Flabellate. Fan-shaped

Apices (Fig. 7-3)
Acuminate. Shaped like an acute angle with a long attenuated point
Acute. Shaped like an acute angle but not attenuated
Mucronate. Abruptly tipped with a bristly mucro (midrib extension)
Cuspidate. Tip concavely constricted into an elongated sharp, rigid point
Obtuse. Blunt; the sides forming an angle more than 90 degrees
Rounded. A full sweeping arc
Truncate. As though cut off at right angles to the midrib
Retuse. With a shallow, narrow notch
Emarginate. With a shallow, broad notch

Bases (Fig. 7-3)
Cuneate. Wedge-shaped, acuminate; tapering evenly to a narrow base
Acute. Shaped like an acute angle but not attenuated
Cordate. Heart-shaped; inversely indented

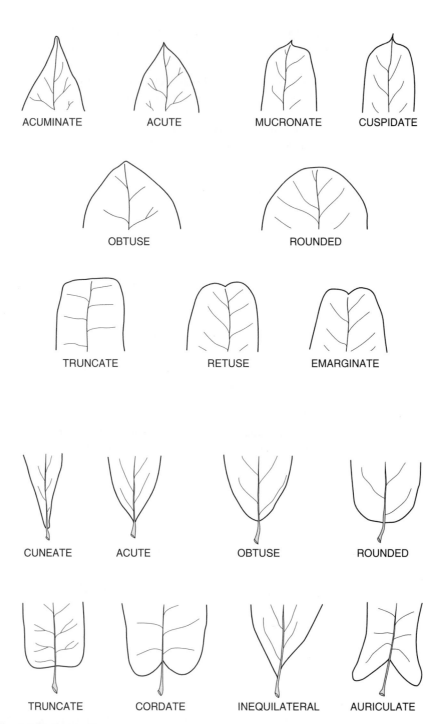

FIGURE 7-3   Leaf apices and bases.

Inequilateral, oblique. Asymmetrical, the two sides different in shape
Obtuse. Blunt, narrowly rounded
Rounded. A full sweeping arc
Truncate. As though cut off at right angles to the midrib
Auriculate. With earlike lobes pointing downward

Leaf Margins (Fig. 7-4)   The edge of the leaf is called the margin.
Revolute. Rolled under just at the margin
Entire. Smooth
Repand. Slightly and irregularly wavy
Sinuate. Shallowly indented; wavy in a horizontal plane
Undulate (or *crisped*). Wavy in a vertical plane
Crenate. With rounded to blunt teeth
Crenulate. Small crenations
Serrate. With sharp teeth pointing toward the apex of the blade
Serrulate. Small serrations
Doubly serrate. With the larger serrations again serrated on their margins
Dentate. With sharp teeth pointing outward
Denticulate. Small dentations
Aculeate. Spiny margined with the teeth long and prickly

Leaf Lobing (Fig. 7-4)   When the blade margins are indented one-quarter to one-half the distance to the midrib or base, it is considered a *lobed* leaf; if indented just over one-half, it is *cleft*; and if cut deeply to near the midrib or base, it is *incised*. In general practice, however, *lobed* is used in a broad sense to include any degree of indentation. Lobing is either *pinnate,* if the indentation is toward the midrib, or *palmate,* if toward the base of the blade. A palmately lobed leaf is usually also palmately veined with a primary vein in each lobe.

Lobing should be considered separately from marginal features or blade shape, as should be obvious from the figures. A lobed leaf may have entire margins such as white oak and sassafras or serrate margins as in red oak, red mulberry, and striped maple.

Conifer Leaves (Fig. 7-5)   There are four types of leaves in the conifers: (1) *acicular* (*needle*), which is long and slender (*Pinus, Larix, Cedrus*); (2) *linear,* which is shorter than the acicular, narrow, and either flat, triangular, or square in cross section (*Abies, Picea, Tsuga, Sequoia, Taxodium*); (3) *subulate* (*awn, awl*), short, narrow, tapered to a sharp point, flat, and stiff (juvenile leaves of *Juniperus* and other Cupressaceae); and (4) *scale,* small, usually appressed and imbricated (*Thuja, Chamaecyparis, Cupressus, Calocedrus, Juniperus*). The distinction between acicular and linear is quite subjective, and various authors may differ in their designations. In FNA, vol. 2 (1993), the authors use "needlelike" to include both acicular and the evergreen linear types.

In some conifers (Pinaceae), linear leaves are either sessile (*Abies*), sessile on a *peg,* known as a *sterigma* (*Picea*), petiolate on a peg (*Tsuga*), or petiolate without a peg (*Pseudotsuga*). When pegs are present, the abscission layer is between the peg and leaf blade so the pegs persist on the twig after leaf dehiscence.

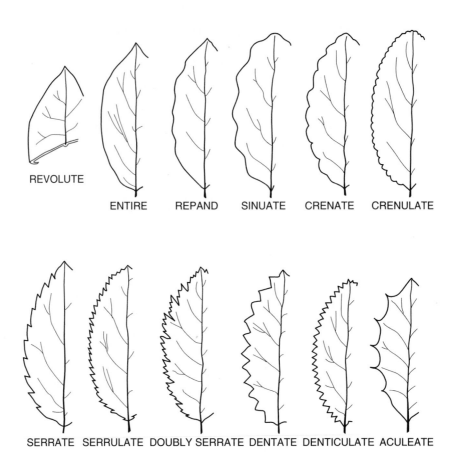

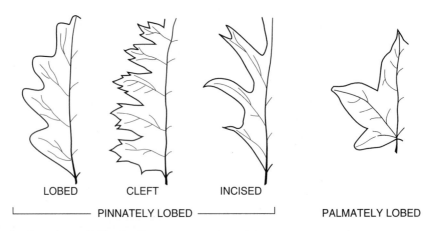

FIGURE 7-4 Leaf margins and lobing.

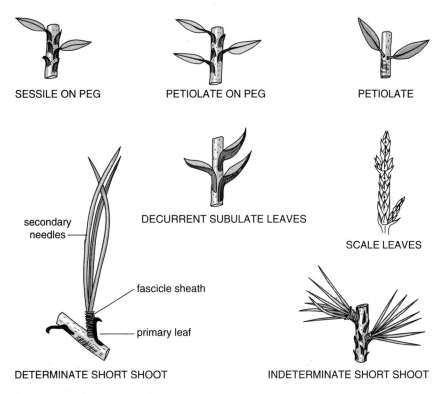

SESSILE ON PEG   PETIOLATE ON PEG   PETIOLATE

DECURRENT SUBULATE LEAVES

SCALE LEAVES

secondary
needles

fascicle sheath

primary leaf

DETERMINATE SHORT SHOOT   INDETERMINATE SHORT SHOOT

FIGURE 7-5   Conifer leaves and short shoots.

In many of the Cupressaceae, juvenile leaves are often subulate (awl or awn), and the mature leaves are scalelike. In *Juniperus virginiana,* for example, leaves of seedlings, vigorous shoots, or spring growth are the stiff subulate ones; those on older or later growth are scales. The subulate leaves are often whorled with three at a node; scale leaves are opposite and decussate, forming four ranks.

The scale leaves of the Cupressaceae are tightly imbricated and completely cover the twig. The twigs (ultimate branchlets) may be terete, angled, or dorsiventrally flattened with two types of scales (and two of each): *lateral,* which cover the sides of the twig and are often keeled (ridged and bent along the center of the leaf), and *facial,* which are flattened or slightly rounded and form the top and bottom of the twig. Scales often have a gland just below the apex on the back (visible) side. The form of these can be important in identification. These branchlets are oriented in either radial (three-dimensional) arrays (*Juniperus, Cupressus*) or are flattened into frondlike *sprays* (*Thuja, Chamaecyparis, Calocedrus*).

In many of the conifers, the linear, subulate, and scale leaves are attached and extend down the stem for some distance below the point of divergence. This is called a *decurrent* leaf base. The pegs of *Picea* and *Tsuga* and the leaves of *Taxus* have decurrent bases.

The pine *fascicle* with one to five needles is a very short stem (determinate short shoot) in the axil of a scalelike, *primary* leaf. The needle leaves are therefore *secondary* because they are on a lateral, axillary branch. The fascicle is wrapped initially at the base by closely overlapping bud scales forming the *fascicle sheath*, which either drops as the needles expand (in soft pines) or persists (in hard pines). The primary leaves in pine may also drop soon (in soft pines) or persist (hard pines).

Surface Features   There are several features of the foliar micromorphology that are important in identification and classification of trees (Hardin, 1992; Meyer and Meola, 1978; Westerkamp and Demmelmeyer, 1997). The surface of blades is usually not completely smooth if examined with high magnification (40× and above). The primary relief is generally formed by the raised or sunken veins (a condition called *rugose* if very pronounced) or by the convexity of the epidermal cell walls. Secondary relief (projections, striations, ridges) may be formed by epidermal cells (papillae) or by a thick and sculptured cuticle on the epidermal walls. Tertiary relief, on top of the cuticle, may be formed by a deposit of flaky, white epicuticular wax. Such micromorphological features are best seen with high magnifications (up to 100 to 500×) using a scanning electron microscope (SEM) (**Fig. 7-6**). A smooth surface usually gives the blade a green, shiny appearance. Added relief, cuticular or epicuticular, increases the reflective surface, and the blade may have a pale color, often found on the lower surface. Epicuticular wax forms a pale or white surface, again usually below, which is called *glaucous* (or *glaucescent* if less). If you can "rub off" this white, it is glaucous or glaucescent.

Hairs (trichomes) on the leaf surface are of numerous types and their structure is best examined with SEM (**Fig. 7-6**). A given species may have a particular type or a complement of types, which is a very stable taxonomic character. The relative density of hairs is much more easily observed with the naked eye or hand lens, but it may be variable in relation to season or habitat. The following terms are a few of the many used to describe the character states (**Fig. 7-7**).

Glabrous. Without hairs of any type; "smooth"
Glabrate. Becoming glabrous or nearly so with age
Puberulent. Minutely hairy as seen with a hand lens or dissecting microscope
Pubescent. With fine, soft, short hairs; often used in a more general way meaning "hairy" without specifying the particular type of vesture
Villous. With long, silky, straight hairs
Tomentose. With curled, matted, woolly hairs
Scabrous. Rough and "sandpapery" with short bristly hairs
Glandular. With numerous glandular hairs, either sessile or stalked
Lepidote. Covered with minute, scalelike hairs
Stellate. With stellate hairs, each having rays parallel with the surface

Finally, leaf blades may be *coriaceous* (thick and leathery) or *membranous* (thin and pliable).

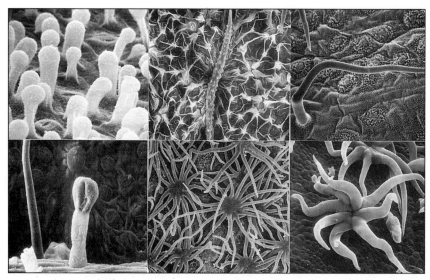

FIGURE 7-6 Scanning electron micrographs of leaf surfaces at different magnifications. Top (left to right): yellow-poplar papillae, dogwood cuticle and trichome, magnolia trichomes and epicuticular wax. Bottom (left to right): trichomes of hickory, basswood, and oak. (*Photos by J. W. Hardin.*)

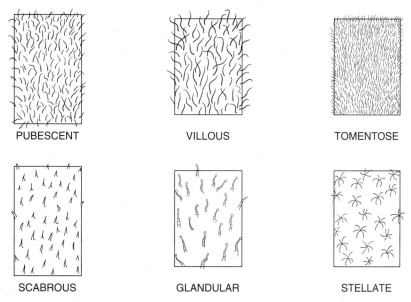

PUBESCENT   VILLOUS   TOMENTOSE

SCABROUS   GLANDULAR   STELLATE

FIGURE 7-7 Vestiture types.

Twigs (Fig. 7-8)

A *twig,* by definition, is the terminal or most recent growth of a branch, often delimited at the base by the terminal bud-scale scars (the position of last year's bud, which developed into the twig). This definition is important because twig characters and twig keys are based on just this part; the second-year and older branches lose some of the twig features. Twigs offer an excellent means of identifying trees and shrubs throughout the year, except for a short time during the spring when the buds formed the previous season are opening and those for the current season have not yet appeared. The most important features of twigs are their buds, leaf scars, stipule scars, vascular bundle scars, and pith, although color, taste, odor, and the presence or absence of cork, spur shoots, spines, thorns, prickles, and pubescence are also valuable characters for purposes of identification.

Buds   These are conspicuous on most twigs and may be of two sorts: Those borne along the twig in the axils of the previous season's leaves are termed *laterals (axillary),* whereas the one at the apex is often larger and is called the *terminal.* In certain trees, however, a true terminal bud is not formed, and growth continues until checked by external factors. When this occurs, the tender leading shoot dies back to the lateral bud below. This lateral bud then assumes the function of the terminal and is called *pseudoterminal.* It is distinguished from the true terminal by its smaller size (it is really a lateral) and the presence of a minute *twig scar* on the side opposite the leaf scar (e.g., *Betula, Tilia*).

If more than one bud appears at a node, the bud directly above the leaf scar is considered to be the true lateral bud and the others are designated as *accessory buds.* If the accessory buds are arranged on either side of a lateral bud, they are said to be *collateral;* if they appear above a lateral bud, they are called *superposed.*

Buds are either *scaly* or *naked.* Bud scales are actually modified leaves or stipules and serve to protect the enclosed embryonic axis and its appendages. In some buds, the scales are often rather numerous and overlap one another in a shinglelike fashion. Bud scales so arranged are *imbricate.* When buds are covered with two or three scales that do not overlap, the scales are described as *valvate.* In a few genera, the buds are covered by a single caplike scale. When the terminal bud opens with the renewal of growth in the spring, the scales slough off, leaving *terminal bud-scale scars. Naked buds* lack scaly coverings and are common to certain trees of tropical climates but are relatively rare elsewhere. *Submerged buds* (buds embedded in the callus layer of leaf scars) are featured by a few species. When two sizes of lateral buds occur on the same twig, the larger usually contains the rudiments of flowers and are called *flower buds. Mixed buds* containing twig, embryonic leaves, and flowers are characteristic of some species, in contrast to those enclosing twig and leaves only (*leaf buds*).

Epicormic buds are either newly formed adventitious buds on roots or latent lateral buds of stems that can give rise, upon cutting or damage of the main stem, to stump sprouts, coppice shoots, or "feathers" along the bole. Coppice is a growth of trees from root or stump sprouts.

Leaf Scars   When a leaf falls from a twig, a *leaf scar* remains at the point of attachment. Leaf scars vary in size and shape with different species and hence are usually diagnostic. On the surface of each leaf scar are found one or more small dots or patches that

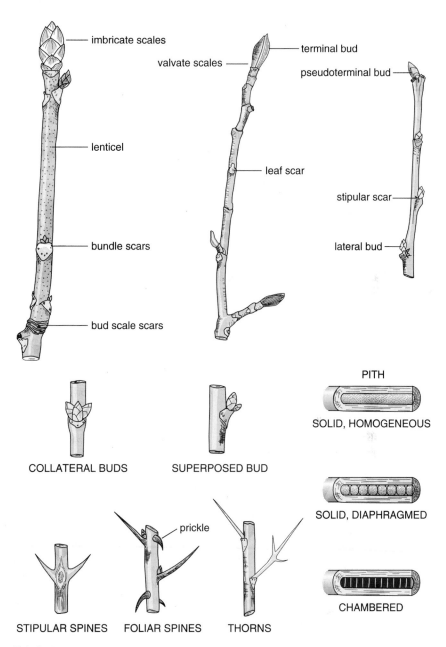

FIGURE 7-8   Twig features.

show where the strands of vascular tissue passed from the twig into the leaf. The size, number, and arrangement of these so-called *vascular bundle scars* are often helpful in identification. It is often necessary to observe these with magnification (10× or more).

Stipular Scars　Stipular scars are not found on all twigs because many species are *estipulate* or the stipules are fused with leaf bases and the scars disappear with the falling leaves. When present, they occur in pairs, one on either side at the top of the leaf scar. As a rule, they are slitlike in shape, and in most instances, they are inconspicuous without a hand lens (10× or more). In a few species, they are fused together and completely encircle the twig.

Lenticels　Lenticels are small, usually lens-shaped patches, sometimes wartlike, consisting of loosely organized tissue, which provide aeration to the tissues beneath. The color, form, orientation, and frequency may be of value.

Pith　The medial or central portion of a twig is composed of *pith.* Pith is usually lighter or darker in color and is of a different texture than the wood that surrounds it; hence, it is readily discernible in cross and longitudinal sections of twigs. The shape of the pith when viewed in cross section usually conforms to one of several patterns. For example, it is *stellate* (star-shaped) or pentagonal in oaks and cottonwoods, *triangular* in alders, and *terete* (round) in ash and elms and most others. Pith often varies in color through shades of pink, yellow, brown, or green, to black or white.

A longitudinal cut should be made through the center of the twig to determine composition. In most native trees, the pith is *solid* (or *continuous*) and *homogeneous.* A distinct modification of this solid type is the *diaphragmed* pith, which has more or less regularly spaced disks of horizontally elongated cells with thickened walls. These appear as bars across the pith and are often readily visible to the naked eye. Yellow-poplar and black tupelo are common examples of this type. In a few trees, such as the walnut, the pith is divided into empty chambers by cross partitions and is termed *chambered.* Chambering in certain species does not take place until the fall of the year, and the season's growth is homogeneous or diaphragmed until that time. A small portion at the end of each season's growth apparently remains unchambered, and therefore the age of a twig with this kind of pith may be estimated by sectioning it and counting the number of yearly "plugs" present. *Hollow* pith and *spongy* pith characterize some species of woody plants.

Lammas Shoots　A new shoot, or sprout, usually consists of a single stem bearing a number of characteristically arranged leaves. Before or soon after twig elongation and leaf enlargement have ceased, minute buds appear in the leaf axils. These buds continue to enlarge, and by midsummer, they have attained full growth. Normally they then remain dormant until the following spring. However, in some species such as yellow-poplar, sweetgum, cascara, beech, sassafras, and water tupelo, these newly formed lateral buds often exhibit renewed growth activity and produce second shoots with green leaves. These second, or midseason, twigs are known as *lammas shoots,* and although rarely diagnostic, they do occur much more frequently on some species than on others. Lammas shoots are most easily recognized in the late summer or early fall when they still may be seen in the axils of leaves.

Shoot Types   The normal twig, with the leaves and buds spread apart by elongation of the internodes, is called a *long shoot.* These may be entirely preformed within the bud or continue to develop during the season. If there is little growth and almost inconspicuous internode length, the stem is called a *short shoot* or "spur" or "dwarf" shoot or branch. The stem axis and embryonic leaves are preformed in the bud. They may bear only leaves, or leaves and flowers (or the reproductive structures in ginkgo). There are two types of short shoots. One is *indeterminate,* in which there is a functional terminal bud, and growth may continue for many years. In this type, the nodes are so close together that otherwise alternate (spiral) leaves may appear whorled. In some cases, the "dwarfness" is released, and subsequent growth in following years may be a normal long shoot. These are commonly found in apple, pear, birch, mountain ash, ginkgo, larch (**Fig. 7-5**), and deodar cedar. The second type of short shoot is *determinate,* and as the term implies, growth length is predetermined in the bud, there is no continued growth, and the short shoot drops as a unit with its leaves. This type forms the pine *fascicle* (**Fig. 7-5**), a short inconspicuous stem (with one to five needles), the deciduous baldcypress twigs that drop in the fall, and the axis of cones and flowers.

Thorns, Spines, and Prickles (Fig. 7-8)   The presence of these pointed structures is often important in identifying certain genera and species. *Spines* are modified leaves or stipules. *Foliar spines* (barberry) are single and have a bud in the axil. *Stipular spines* (black locust) are in the position of stipules; therefore, one is on either side of the leaf scar. *Prickles* (rose, *Aralia*) emerge from the epidermis and cortex and can occur at any location on the twig or leaf. *Thorns* (honeylocust, hawthorn, osage-orange) are modified stems and as such sometimes bear minute leaf scars. They are positioned either above the leaf scar in the position of the bud and may be simple or branched, or terminate a spur branch.

BARK (Fig. 7-9)

Bark features are quite characteristic of tree species, but they vary due to age, growth rate, and habitat and are difficult to describe in words. Yet, bark is still one of the most important features in the identification of large trees, particularly when leaves are absent and twigs inaccessible. Bark characteristics are best learned by actual observation and experience. Beginning students in dendrology often underestimate the value of bark in identification, but with experience, this attitude will change. In trying to learn barks, try not to look up until you have made a tentative identification.

On older trunks, bark characters to consider are texture, thickness, roughness (smooth, flaky, scaly, shaggy, furrowed, etc.), color of the outer and inner layers, depth and width and configuration of furrows (fissures); width, length, and surface of the ridges or plates delimited by the furrows, whether fibrous or not; and horizontal versus

FIGURE7-9   Bark (left to right): Douglas-fir, Sitka spruce, western hemlock, and western redcedar on the Olympic Peninsula near Willapa Harbor, Washington. (*Photo by Asahel Curtis, Seattle.*)

vertical peeling. On younger trunks, before the furrows or scales develop, the form of the lenticels and color are important features.

Bark is also important economically. The high-energy yields obtained from bark as a fuel have prompted most raw wood–conversion plants to convert their processes to utilize fully all the heretofore wasted slabs and bark. Dyes, tannins, alkaloids, and some pharmaceutical compounds are extracted from the bark of certain species, and cordage and paper fibers are obtained from others. Vast quantities of bark from Douglas-fir, Rocky Mountain white fir, and several yellow pines such as longleaf are used for decoration, for mulching, and as a soil conditioner. Mixtures of tupelo, sweetgum, yellow-poplar, and oak barks are also used for the same purposes. Redwood bark is used in walkways and as an underlay in patios. Pine bark has also proved to be an excellent oil scavenger when coarsely ground and spread around oil spills at sea.

Bark is a common habitat for many corticolous green algae, mosses, leafy liverworts, and lichens. In rough-barked trees, foliose and fruticose lichens often inhabit the bark fissures, which stay more moist than the ridges or scales. On smooth-barked trees, such as alder, hornbeam, magnolia, beech, pawpaw, American holly, and young trees of maple, hickory, yellow-poplar, and buckeye, crustose lichens frequently form variously colored blotches on an otherwise gray bark. Some of these are so common, as on American holly, their presence is a dependable characteristic for identification of the tree without even looking at the leaves.

REPRODUCTIVE MORPHOLOGY

Reproductive features offer the most stable characters and thus are the primary basis for classification of seed plants. In trees, they are often unavailable because they are borne in the upper crown and last for a relatively short period of time. When available, however, they may be the most definitive features for identification; therefore, it is important to have a basic knowledge of reproductive morphology.

The primary male reproductive structure is the *pollen grain,* which carries the *sperm*; the corresponding female structure is the *ovule,* which contains the *egg.* Following the processes of *pollination* then *fertilization,* the ovule develops into the *seed* containing an *embryo.*

There are various levels of sexual differentiation. Some flowers contain both male and female structures and are called *bisexual* or *perfect* (found in dogwood, cherry, basswood, black locust, and others). A plant with perfect flowers may be called *synoecious* (*syn* = "together"; *oecious* = "house") or *homoecious* (*homo* = "same"; *oecious* = "house"). If the cones or flowers have only one sex, they are *unisexual* or *imperfect.* If both male and female flowers or cones are on the same tree, then the tree is *monoecious* ("one house") (as in most conifers, oak, hickory, birch, etc.). If the imperfect cones or flowers are on separate trees, then the tree is *dioecious* ("two houses") (e.g., willow, poplar, holly, ginkgo, juniper, ash, boxelder, and a few others). Many trees show developmental or ecological variation in the male/female expression. For instance, a young maple may be male but then switch to female later; or, an ash with male flowers one year may have female flowers and fruits the next, then revert to male again the following year. There may also be a mixture of perfect and imperfect flowers in some trees, which is a *polygamous* condition. For instance, a buckeye inflorescence has a few perfect flowers plus many male flowers, which makes it *polygamo-monoecious*.

Pollen and Pollination

Pollen grains are variable in size, weight, shape, number of pores through which the pollen tube may emerge, and outer-wall sculpturing. Pollen type is characteristic of a family or genus and sometimes species. For some examples of pollen types, as seen by SEM, see **Figure 7-10.** For additional information, see Kapp (1969).

Pollination is the transfer of pollen from the pollen sac to the ovule (in gymnosperms) and anther to stigma (in angiosperms). Among our temperate zone trees, there are three types of pollination mechanisms: wind (*anemophily*), insect (*entomophily*), and bird (*ornithophily*). The usual features of the *anemophilous syndrome* among angiosperm trees are small, lightweight, dry pollen in great overabundance; small, reduced flowers generally lacking a perianth and with exposed anthers/stigmas; many pistillate flowers, each with one to few ovules per ovary; and flowering in early spring before the leaves and therefore with maximum airflow among the flowering branches. Many of these pollens are allergenic, such as oak, hickory, alder, birch, elm, and others, and cause respiratory and nasal discomfort to many people. Some cities, particularly in the southwest, have banned the planting of certain trees due to their allergenic pollen. The conifers are also anemophilous but much less allergenic. Many have air sacs or

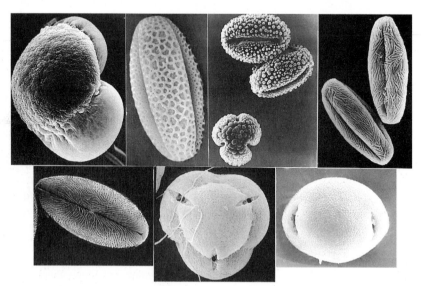

FIGURE 7-10   Scanning electron micrographs of representative tree pollens at different magnifications. Top (left to right): pine with air sacs, willow, holly, maple. Bottom (left to right): buckeye, rhododendron tetrad, basswood. (*Those of top row, courtesy of Mr. Stephen B. Bambara and Mrs. Nancy Leidy, Department of Entomology, N.C. State University, Raleigh; those of bottom row, by J. W. Hardin.*)

"wings" on the pollen grains. The *entomophilous syndrome* involves a floral morphology correlated with the particular pollinator (beetles, bees, bumble bees, wasps, butterflies, moths): showy perianth; small amounts of large, heavy, sticky pollen grains; usually several to many ovules per ovary; often with a nectar supply used for food (although some insects are pollen feeders); and the flowers generally aromatic. Examples are persimmon, magnolia, yellow-poplar, apple, cherry, sourwood, and locust. Among the gymnosperms, the cycads are beetle pollinated. *Ornithophilous* flowers are often red or yellow in color, have a tubular-shaped corolla, and an abundance of nectar (e.g., red buckeye, mimosa tree).

Several insect-pollinated trees are important "bee trees" for honey production. Excellent honey comes from sourwood, tupelo, yellow-poplar, apple, holly, basswood, and many others.

For additional reading on pollination, see Faegri and van der Pijl (1979) and Proctor et al. (1996).

Gymnosperm Parts (Fig. 7-11)

Gymnosperm ovules are exposed (naked) at the time of pollination. These ovules are either borne on modified leaves (in cycads), on scales in the axils of bracts (in conifers),or on stalks (in ginkgo, podocarpus, yew, and some others). The cycad cone (strobilus) is simple (ovule-bearing leaves attached to a simple stem axis), whereas the conifer seed cone (strobilus) is compound; that is, it has a central stem axis bearing highly modified lateral branches (the *scales*), each in the axis of a reduced leaf (the *bract*). Pollen cones of all gymnosperms are simple; that is, they have a stem axis bearing modified leaves, each with two to many pollen sacs.

Pollen cones are deciduous soon after the pollen is released and are usually not important for identification. *Ovulate* or *seed cones,* on the other hand, show a number of differences and are very important in the classification and identification of conifers. The scales may be woody, leathery, or semifleshy, each one with one or more seeds. They may be spirally borne on the axis (Pinaceae) or opposite and decussate (Cupressaceae). They may be separate from the bract (Pinaceae) or fused with it into a scale/bract complex (Cupressaceae) and then either thin or thick, flexible or rigid, or flattened or peltate (shield-shaped). In the latter case, the seed cones are more or less oval or spherical. In most of the Pinaceae, in which the scale and bract are separate, the scale is the larger, obvious structure, and the bract is much smaller, shorter, and inconspicuous. In some firs and Douglas-fir, however, the bract is longer than the scale.

The exposed portion of the cone scale in a mature unopened cone is known as the *apophysis.* The apophysis is usually lighter in color than the remaining portion of the scale, although this is not always the case. In some species, the apophysis is smooth; in others it is wrinkled, grooved, or ridged. In a few species, the scale margins are slightly reflexed or revolute, whereas in others, they appear to be eroded (*erose*).

In pine cones, which mature at the end of two or rarely three seasons, the apophysis terminates in a small protuberance called the *umbo.* The umbo is the visible portion of the scale in the first year of development. When the umbo is found at the tip of the scale, it is *terminal,* but if it appears on the back or raised portion, it is *dorsal.* Umbos are often

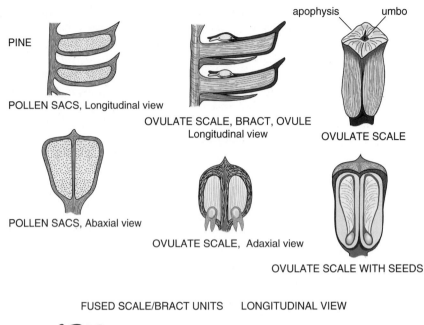

PINE

POLLEN SACS, Longitudinal view

apophysis   umbo

OVULATE SCALE, BRACT, OVULE
Longitudinal view

OVULATE SCALE

POLLEN SACS, Abaxial view

OVULATE SCALE, Adaxial view

OVULATE SCALE WITH SEEDS

FUSED SCALE/BRACT UNITS   LONGITUDINAL VIEW

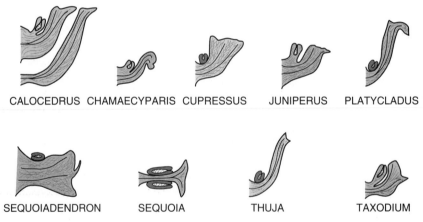

CALOCEDRUS   CHAMAECYPARIS   CUPRESSUS   JUNIPERUS   PLATYCLADUS

SEQUOIADENDRON   SEQUOIA   THUJA   TAXODIUM

FIGURE 7-11   Conifer cone structures.

wartlike or have a raised and rounded projection or boss, or they may terminate in a prickle or claw.

Cones may open and release seeds fairly soon after maturity. If they remain closed for a long time and often require fire or similar level of heat to open, they are considered *serotinous.* These are often associated with fire ecology and lead to rapid seed dispersal and seedling development following a fire. Seed cones, open or closed, may persist on the tree for several years. Most cones remain intact when they open, whereas in *Abies, Cedrus,* and *Taxodium,* the scales and bracts are deciduous. Most conifers complete their life cycle (pollination to seed development) in one year, whereas others (pines) require two or sometimes three years. In pines, therefore, you can often find small, ovuliferous, first-year (pollination) cones, second-year (fertilization) cones, and the mature seed cones on the same branch.

Seeds are borne at the base of the cone scales. In several species, depressions in the scales denote their original position, whereas in others, minute scars are left after seed dispersal. The number of seeds per cone scale varies with the species (usually two in Pinaceae, more in Cupressaceae). Seeds are either terminally winged, laterally winged, or wingless, depending on genus and species.

Angiosperm Parts

Flowers (Fig. 7-12)   A flower is a modified, determinate short shoot with a central stem axis (*receptacle*), highly modified sterile leaves (*perianth*), and reproductive leaves. The perianth includes the outer whorl of *sepals* (collectively called the *calyx*) and the inner whorl of *petals* (collectively the *corolla*). Undifferentiated perianth parts are called *tepals*. The reproductive leaves are the male *stamens* composed of *anther* (pollen sacs) and *filament* (stalk), and female *carpels* composed of *ovary* (containing the *ovules*), *style*, and *stigma*. The term *pistil* is used to describe the visible ovary/style/stigma, which may be a single carpel or the fusion product of several carpels (compound pistil). A flower with all four parts (sepals, petals, stamens, pistils) is *complete*; if any one or more is missing, the flower is *incomplete*. If loss of parts involves the perianth, the flower may be *asepalous* or *apetalous* (*a-* = "without"). If the loss involves functional stamens and pistils, the flower is *imperfect* (rather than perfect) and either *staminate* or *pistillate*.

Several other modifications of the basic floral structure occur, such as variation in the number of parts per whorl (usually three, four, or five, or many), symmetry (radial = *actinomorphic* or bilateral = *zygomorphic*), or fusion of parts. Fusion may be *connation* (i.e., lateral fusion between parts within a whorl) or *adnation* (i.e., fusion between parts of different whorls). In the case of connation, separate parts may fuse to form, for example, a *sympetalous corolla* or *compound pistil*. Adnation between stamens and perianth forms a structure called a *hypanthium*. (In some plants, the hypanthium may be formed by a cup-shaped receptacle rather than by adnation.) This hypanthium may form an open cup-shaped structure, or it may completely envelop and fuse with the ovary wall. Such modifications lead to several different "ovary positions" in the flower. In the simplest case, the ovary is free and above the position of the origin of the perianth and stamens (a *superior ovary*). Even with a cup-shaped hypanthium not fused with the ovary, the ovary is still considered superior. If the hypanthium is fused with the ovary and thus the ovary appears to be below the perianth and stamens, it is called *inferior*. Terms relative to the position of the perianth with respect to the ovary are *hypogynous* (perianth below ovary), *perigynous* (hypanthium around but free from ovary), and *epigynous* (perianth above ovary). (The suffix *-gynous* refers to "gynoecium," the collective term for all the carpels.)

Flowering is the process of floral maturation reaching that point (*anthesis*) when pollination is possible. Because gymnosperms do not have flowers, this term is inappropriate even though frequently heard. *Coning* is correct but less often used.

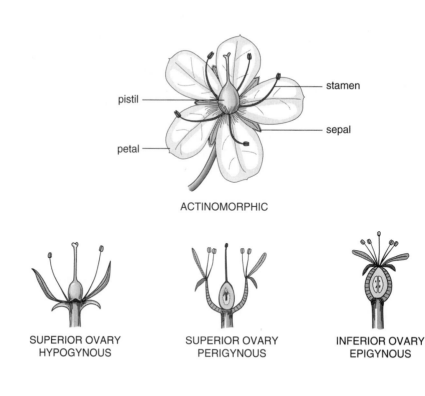

pistil

petal

stamen

sepal

ACTINOMORPHIC

SUPERIOR OVARY
HYPOGYNOUS

SUPERIOR OVARY
PERIGYNOUS

INFERIOR OVARY
EPIGYNOUS

ZYGOMORPHIC

SYMPETALOUS

FIGURE 7-12   Flower parts and variation.

Inflorescence, Infructescence (Fig. 7-13)   Flowers may arise from separate buds or in mixed buds with leaves, and they may be *terminal* on the twig or *axillary* in leaf axils. They may be single or borne in clusters of several to many. An *inflorescence* is the arrangement of several flowers in a branch system; an *infructescence* is the same for fruits. The stalk of a single flower (as in magnolia) is the *peduncle*; the stalk of an inflo-rescence (of many flowers) is also the peduncle; the stalk of an individual flower in an inflorescence is the *pedicel*. Pedicels or primary branches arise from a *rachis,* each from the axil of a bract. Successive branches or pedicels may be in the axils of bractlets, and sometimes the pedicels may bear one to two bracteoles below the flower. There are many types of inflorescences, but the main ones found in trees are illustrated and described briefly.

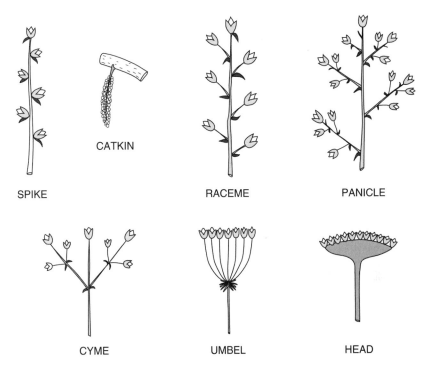

SPIKE   CATKIN   RACEME   PANICLE

CYME   UMBEL   HEAD

FIGURE 7-13   Inflorescence types.

Spike. A central rachis bearing a number of sessile flowers (see *catkin*)

Catkin or *ament.* A dense spike bearing many, apetalous, unisexual flowers; often pendent (staminate flowers in oak, hickory; pistillate and staminate flowers in birch)

Raceme. A central axis bearing a number of pedicelled flowers; all the pedicels are about the same length (black locust)

Panicle. Diffusely branched rachises with many flowers, or the primary branches as racemes or cymes (buckeye, sumac)

Umbel. Usually flat-topped, with many pedicels arising from a common point at the top of the peduncle; sometimes compound (ends of panicle branches in devils-walkingstick)

Cyme. A central rachis terminated early by a flower and with lateral branches, arising below and later, either terminated by a flower or leading to successive cymes (alternate-leaf dogwood, basswood)

Head. A number of sessile flowers clustered on a common receptacle that may be flat-topped or more or less globose (flowering dogwood, sycamore)

Fruits and Seeds (Fig. 7-14)   A *fruit* is a ripened ovary, usually with seeds; a *seed* is a ripened ovule, usually with an embryo. Classification of fruits is on the basis of the ovary structure as well as the form of the mature fruit. If the fruit is from a single ovary in a flower, it is a *simple fruit*; if the fruit is formed from several separate ovaries

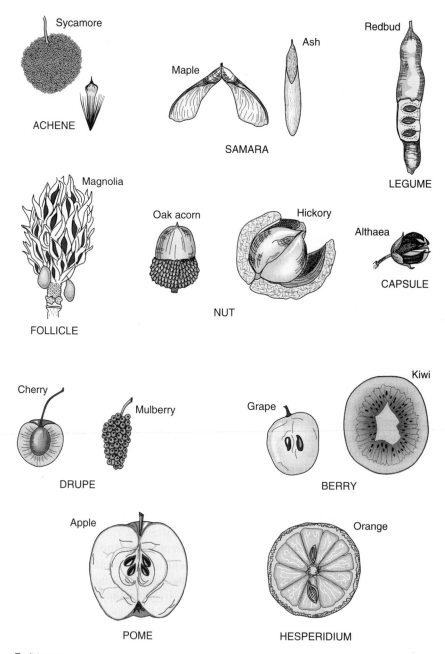

FIGURE 7-14 Fruit types.

that stay together when mature, it is a *compound fruit*. Both types may be either fleshy or dry. Fleshy fruits are classified on the basis of the differentiation of the fruit wall (*pericarp*). Dry fruits are classified on whether they split open (*dehiscent*) and release the seeds or not (*indehiscent*), the number of carpels, and the general form of the mature fruit. The common types are illustrated and described briefly in the following outline.

I. Simple fruits
   A. Dry, indehiscent fruits
       1. *Achene.* Small, one-seeded, unwinged but sometimes plumose (sycamore—multiple of achenes)
       2. *Samara.* Achenelike or nutletlike but winged (alder, birch, elm, maple, ash; yellow-poplar—aggregate of samaras)
       3. *Nut.* Ovary with two or more carpels, but fruit usually one-celled, one-seeded, and with a hard wall (basswood); sometimes partially or entirely encased in a husk or bur (involucre or cupule) (walnut, hickory, oak, beech, chestnut)
   B. Dry, dehiscent fruit
       1. *Legume.* Product of a simple, one-carpellate ovary, splitting along two lines (or occasionally indehiscent) (redbud)
       2. *Follicle.* Product of a simple, one-carpellate ovary, splitting along one line (magnolia—aggregate of follicles)
       3. *Capsule.* Product of a compound ovary, splitting along two or more lines, usually one per carpel (buckeye, poplar, willow; sweetgum—multiple of capsules)
   C. Fleshy fruits
       1. *Pome.* Product of a compound, inferior ovary, the outer hypanthium wall fleshy, the inner pericarp wall (endocarp) papery or woody (apple, pear, hawthorn, serviceberry)
       2. *Drupe.* Product of a simple ovary and usually one-seeded, the outer pericarp walls (*exocarp* and *mesocarp*) fleshy, the inner pericarp wall (*endocarp*) bony; a "stone fruit" (peach, cherry, olive, palm; mulberry—multiple of drupelets)
       3. *Berry.* Usually many-seeded (rarely one seed as in avocado), with the entire pericarp fleshy (persimmon, pawpaw, grape, kiwi)
       4. *Hesperidium.* A type of berry with a thick, leathery outer "rind" (exocarp) and many radial sections (orange, grapefruit, lemon, lime)

II. Compound fruits
   A. *Aggregate.* A cluster of simple fruits from separate pistils of a single flower (magnolia—aggregate of follicles; yellow-poplar—aggregate of samaras)
   B. *Multiple.* A cluster of simple fruits from many separate flowers in a compact inflorescence; a coalesced infructescence (sweetgum—multiple of capsules; mulberry—multiple of drupelets; sycamore—multiple of achenes)

Dispersal Mechanisms.   The dispersed portion (the *diaspore*) of seed plants may be the seed, cone, or fruit. Small, dry, lightweight diaspores (some seeds, achenes, samaras) are generally wind dispersed (*anemochory*). Small, fleshy diaspores of various colors (seeds with fleshy seed coats, berries, drupes, juniper cones) are frequently eaten by birds and small mammals, pass through the digestive tract, and are deposited in the scats (*endozoochory*). Larger, dry diaspores (nuts, large and hard seeds) are often collected by small mammals, and the individuals or cache stored or buried and often forgotten (*synzoochory*). Some diaspores (seeds or fruits) have barbs or spines that get caught in feathers or fur and are carried on the bodies of various animals (*exozoochory*). In a few cases (e.g., witch-hazel), the seeds are projected out of capsules (*ballochory* or *autochory*). Large, heavy diaspores (fruits or serotinous cones) may be limited to dispersal by weight and gravity (*barochory*). Many diaspores are capable of water dispersal (*hydrochory*) along rivers and streams, during floods, or in ocean currents. Actually, many diaspores may have several means of dispersal, often differing between short-range and long-range dispersal. For a more thorough discussion of dispersal mechanisms, see van der Pijl (1982).

Seed Germination and Seedlings   Seed germination is of two types. *Epigeal* (or *epigeous*) is when the cotyledons are brought above ground level by the elongation of the hypocotyl, as in pine (**Fig. 7-15**), beech, basswood, maple, and others; the cotyledons then function as either storage with little photosynthesis, or alternatively, little storage and mainly photosynthesis. *Hypogeal* (or *hypogeous*) is when the cotyledons are the primary storage structures and remain in the seed coat below ground level; the plumule is the first structure to appear aboveground, as in oak or hickory. The form of the cotyledons and/or first leaves can be distinctive, and one can learn to recognize genera and sometimes species from these early reproductive stages. Some examples are shown in **Figures 7-16 and 7-17**. See Young and Young (1992) for more details.

FIGURE 7-15   Three stages (right to left) in the epigean development of a pine seedling, showing cotyledons and first juvenile leaves. (*Photo by B. O. Longyear.*)

FIGURE 7-16   Some coniferous seedlings: (1) red pine, (2) balsam fir, (3) northern white-cedar, (4) white spruce, (5) eastern hemlock, (6) European larch.

FIGURE 7-17 Some broad-leaved (dicot) seedlings: (1) beech $\times\frac{3}{4}$, (2) basswood $\times 1$, (3) aspen $\times 3$, (4) white ash $\times\frac{1}{2}$, (5) white oak $\times\frac{1}{3}$, (6) sugar maple $\times\frac{1}{2}$, (7) yellow birch $\times 1$, (8) American elm $\times 1$.

8

PINOPHYTA (GYMNOSPERMS)

Gymnosperms include many of the world's most interesting and useful trees, and their importance can hardly be overemphasized, especially in the temperate and cold-temperate forests of both Northern and Southern Hemispheres.

Historically, the gymnosperms have received much attention because their lineage extends back to the mid-Paleozoic era—the Devonian period about 350 million years ago. They were among the first seed plants and reached their peak of diversity and dominance during the Mesozoic era between 250 and 100 million years ago. Since that time, there has been a steady decline, leaving at present only five orders, 16 families, some 86 genera, and ca. 840 species worldwide.

Gymnosperms are seed plants with "naked seeds" (i.e., naked ovules at the time of pollination). This is in contrast with the angiosperms, which have ovules (and seeds) enclosed within the carpel. Also, gymnosperms have a prezygotic megagametophyte tissue as the storage tissue in the seed, whereas the angiosperms have a postzygotic endosperm formed by the process of double fertilization.

Commercially and ecologically, conifers are the most important of the gymnosperms, although many others are used extensively as ornamentals.

Cycadales   Cycads

The cycads (four families, 11 genera, ca. 145 species) are tropical and subtropical plants that mostly resemble palms or tree ferns. Their trunk is either generally unbranched and 20 to 60 ft tall or merely a short, tuberous, mostly subterranean stem as in our native *Zàmia* (Zamiaceae). They produce a terminal rosette of large, pinnately divided leaves and seeds borne on modified leaves (megasporophylls) either clustered around the stem

(Cycadaceae) or in a compact cone (Zamiaceae, Boweniaceae, Stangeriaceae). The plants are dioecious. Some have very large pollen cones. Pollination is by beetles, which are attracted to both pollen and ovulate structures by a strong odor. For additional information see Norstog and Nichols (1998).

Cỳcas (Cycadaceae—recognized by the leaflets that have a midrib and no lateral veins) is the most frequently used ornamental, but other genera may be seen in botanical gardens and arboreta in southern Florida, southern Texas, and southern California or in greenhouses. All the cycads of the Western Hemisphere belong to the Zamiaceae (recognized by the leaflets that have many parallel or forked veins). The native species of the United States is the Florida arrowroot or coontie, *Zàmia integrifòlia* L.

Ginkgoales   Ginkgoaceae: The Ginkgo Family

Gínkgo bilòba L.   ginkgo, maidenhair tree

Distinguishing Characteristics (Fig. 8-1). *Leaves* deciduous, alternate, flabellate (fan-shaped), petiole 1 to $3\frac{1}{4}$ in. (2.5 to 8.5 cm) long, blade $\frac{3}{4}$ to $3\frac{3}{4}$ in. (2.0 to 9.5 cm) long, $\frac{3}{4}$ to $4\frac{3}{4}$ in. (2 to 12 cm) wide, with or without one or more narrow apical sinuses, and with small dichotomous veins; borne on long shoots at the tips of branches, but on older growth, they occur only on short spur branches (indeterminate short shoots); fall color a brilliant pale yellow or orange-yellow, and all leaves drop from a tree within a few days. *Twigs* with scaly brown, globose buds, terminal present; leaf scars raised, half round with 2 bundle scars; estipulate; pith homogeneous. *Dioecious*; the pollen sacs in a catkinlike arrangement on spur branches, pollination in spring by wind; the naked ovules in pairs at the end of a long stalk, also from the spur branches. *Seeds* large, ca. 1 in. (2.5 cm)

FIGURE 8-1   *Ginkgo biloba,* ginkgo. (1) Leaves and catkinlike clusters of male pollen sacs on a spur shoot $\times\frac{3}{4}$. (2) Young leaves and ovules $\times\frac{3}{4}$. (3) Seed $\times\frac{3}{4}$. (4) Seed with outer fleshy layer removed $\times\frac{3}{4}$. (5) Leaf $\times\frac{1}{2}$. (6) Twig $\times\frac{1}{2}$.

in diameter; outer, fleshy seed coat, upon ripening, is yellow-orange and has an extremely foul odor, particularly in mass under a tree due to substances (fatty acids: butanoic and hexanoic acid) found in rancid butter and romano cheese; inner seed coat hard and pale in color.

General Description.   This native of China is the only living representative of a once large group that flourished during the late Mesozoic, about 150 million years ago. It was sacred according to the Buddhist religion and has been cultivated for centuries in the temple gardens of China and Japan. It is now used as a "nut" tree in Asia and an ornamental street and yard tree in Europe and North America. It was introduced into the United States in 1784 but has not become naturalized in North America despite over two centuries of cultivation. Ginkgo can reach 100 ft in height and over 4 ft in diameter. The seed, when boiled, fried, or roasted, is considered a delicacy in the Orient, and canned seeds are sold as "silver almonds" or "white nuts." The pollen is suspected of being allergenic. The herbal extract of Ginkgo leaves, said to increase blood flow to the brain, improving alertness, memory, and concentration, is now the source of various over-the-counter "nutriceuticals" in North America. It is a widely prescribed medicine in Germany. For a more thorough discussion, see Del Tredici (1991) and Hori et al. (1997).

Taxales   Taxaceae: The Yew Family

The yew family comprises five or six genera and about 18 species. *Torreya* and *Taxus* are represented in the American flora (Table 8-1). The other genera are mostly in eastern Asia (one in New Caledonia).

Botanical Features of the Family

Leaves persistent, mostly in spirals (rarely opposite), decurrent, linear, with or without resin canals.

Seeds entirely or partly surrounded by a fleshy aril; plants dioecious or monoecious. Pollination by wind.

TORREYA Arn.   torreya

This is a small genus consisting of seven species of trees and shrubs found in North America, China, and Japan. Leaves rigid, spine-tipped; seed entirely enclosed by the aril. Wooden novelties such as chess pieces, rings, and articles of turnery are manufactured from the Japanese kaya, *Tòrreya nucífera* (L.) Sieb. and Zucc.

Two species are found in the United States: *Tòrreya taxifòlia* Arn., Florida torreya or stinking-cedar, is local and federally listed as endangered; it is found along the

TABLE 8-1   COMPARISON OF NATIVE GENERA IN THE TAXACEAE

Genus	Leaves	Seeds
Torreya	Stiff, spine-tipped; with 2 narrow, glaucous bands below; resinous	1–1½ in. (2.5–4 cm) long; completely surrounded by a purplish green aril
Taxus	Flexible, soft-pointed; with 2 broad, glaucous bands below; nonresinous	ca. ¼ in. (6 mm) long; partly surrounded by a scarlet to orange aril

Appalachicola River in northwestern Florida and in southwestern Georgia, but now endangered due to a fungal disease of the stem. *Tòrreya califórnica* Torr., California torreya (nutmeg), is restricted to the west slopes of the Sierra Nevada and central coast region of California. Both are cultivated as ornamental trees or shrubs.

TAXUS L.   yew

This genus includes 6–10 species of trees and shrubs of wide distribution through eastern Asia, Asia Minor, northern Africa, Europe, and North America. *Táxus baccàta* L., the English yew, with many cultivars, is commonly used as an ornamental throughout the northern hemisphere. Japanese yew, *T. cuspidàta* Sieb. and Zucc., and its cultivars, and the hybrid of these two, *T.* × *mèdia,* are also widely cultivated. The foliage and seeds are poisonous if ingested.

Three species of *Taxus* are native to the United States: *T. canadénsis* Marsh., commonly designated as ground-hemlock or Canada yew, is a low sprawling shrub found in moist, shady situations through the northeastern states and Canada; *T. floridàna* Nutt. ex Chapm., the Florida yew, is a shrub or small tree frequenting river banks and ravines in northwestern Florida; *T. brevifòlia* Nutt., the Pacific yew, is a small- to medium-sized tree of the Pacific coast and is the only native species of any commercial importance.

Táxus brevifòlia Nutt.   Pacific yew

Distinguishing Characteristics (Fig. 8-2).   *Leaves* flexible, $\frac{1}{2}$ to $1\frac{1}{8}$ in. (12 to 29 mm) long, apex soft-pointed, linear and decurrent, appearing distichous, dark yellow-green to blue-green above, paler below, petiolate. *Dioecious*; pollen cones and ovules axillary. *Seed* ovoid-oblong $\frac{1}{4}$ in. (6 mm) in length, partially surrounded by a scarlet aril. *Bark* dark reddish purple, thin, and scaly.

Range.   Southeastern Alaska along the coast to Monterey Bay, California; also in the Sierra Nevada Mountains; eastward through British Columbia and Washington to the west slopes of the Rockies in Montana and Idaho. Elevation: sea level northward to 8,000 ft southward. This tree is usually found on deep, rich, moist soils near lakes, rivers, and streams, as well as on protected flats, benches, and slopes of ravines and canyons.

General Description.   The Pacific yew is usually a small tree or large shrub 20 to 50 ft tall and 12 to 24 in. dbh (max. 84 by 4.7 ft), although larger trees are occasionally found on the best sites. The crown is large and conical, and even forest trees produce "limby," often fluted, and occasionally contorted boles; the root system is deep and wide-spreading. It ordinarily occurs as an occasional understory tree in old-growth, mixed coniferous forests throughout its range. It is occasionally found as a much dwarfed and often sprawling shrub at or near timberline with whitebark pine and Alaska-cedar. The foliage and seeds are poisonous, but the red, juicy aril is edible. The bark contains a compound, taxol, which is a treatment for certain cancers. The wood is used for making archery bows, canoe paddles, and cabinetwork.

Pacific yew produces some seed every year, and large crops occur at irregular intervals. The seeds are largely bird-disseminated, and those falling on wet moss, decaying wood, or leaf litter ordinarily germinate. This species is probably the most tolerant of northwestern forest trees. Growth proceeds very slowly, and maturity is attained in about 250 to 350 years. The largest trunks are usually hollow-butted and often exhibit spiral grain.

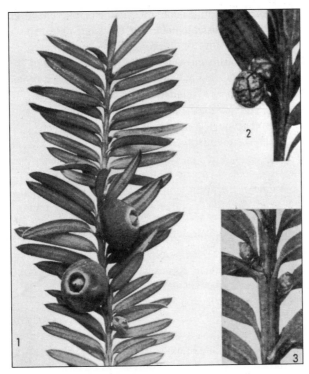

FIGURE 8-2   *Taxus brevifolia,* Pacific yew. (1) Twig with arillate seeds and 2-ranked leaves ×1. (2) Pollen cone ×3. (3) Young ovules in leaf axils ×3.

Pinales   Conifers

The conifers, with six families, about 50 genera, and over 500 species, are of primary value for forest products, ornamentals, and Christmas trees. The Pinaceae and Cupressaceae s.l. are most noteworthy in the Northern Hemisphere and the only two native in North America (Table 8-2). The Podocarpaceae and Araucariaceae are preeminent in the Southern Hemisphere. *Podocárpus* is cultivated in the southern United States. In the

TABLE 8-2   COMPARISON OF CONIFER FAMILIES IN NORTH AMERICA

Family	Leaves	Immature ovulate cones	Seed cones
Pinaceae	Alternate; acicular or linear	Bract and scale distinct and flat; ovules inverted, 2 per scale; monoecious or rarely dioecious	Woody; seeds terminally winged or wingless
Cupressaceae s.l.	Alternate, opposite, or, whorled; linear, subulate, or scale	Bract and scale fused; flat to peltate; ovules erect or inverted, one to many per scale; monoecious or dioecious	Woody, leathery, or semifleshy; seeds laterally winged or wingless

Araucariaceae, the monkey puzzle tree and Norfolk Island pine (both species of *Araucària*) are popular ornamentals, and the kauri-pine, *Ágathis austràlis* Hort. ex Lindl. of New Zealand, is one of the world's most massive trees reaching 200 ft tall and ca. 20 ft dbh. Below a bushy crown, the nearly cylindrical trunk free of branches yields a remarkable amount of valuable wood. The Cephalotaxaceae and Sciadopityaceae are natives of Asia but both *Cephalotáxus* spp., Japanese plum yew, and *Sciadópitys verticellàta* (Thunb.) Sieb. & Zucc., Japanese umbrella-pine, are cultivated in the United States. The umbrella-pine is also cultivated in southern Canada.

Besides the large amounts of timber and paper pulp produced, the conifers furnish such products as turpentine and rosin (naval stores), pine straw, bark, and Christmas trees, and serve to control erosion on the steep slopes of mountains. Also, especially in the cooler portions of the north temperate zone, vast, nearly pure forests of coniferous species are typical and greatly enhance the value of these regions for recreational purposes.

The various members of this group are known as "conifers," "evergreens," or "softwoods." In the first instance, the common name is obvious, although two families (i.e., Cephalotaxaceae and Podocarpaceae) have fleshy seeds not in cones. The name *evergreen* is less satisfactory because at least five genera (*Larix, Pseudolarix, Taxodium, Metasequoia,* and *Glyptostrobus*) are deciduous, whereas some species of broad-leaved trees (dicotyledons), are evergreen. The term *softwood* is commonly used in distinguishing between the conifers (also Taxales) and the broad-leaved trees, or "hardwoods." It often happens, however, that the wood of a "softwood" (conifer) is harder than that of a "hardwood" (broad-leaved tree); for example, wood of longleaf pine is harder than that of basswood. Even though most of these terms are none too satisfactory, they are all useful provided their limitations are understood.

Pollination is by wind, and typical of wind-pollinated plants, there is an overabundance of pollen. Pollination occurs mostly in the spring, but in some genera like *Cedrus,* pollination is in the fall. Some conifer pollen is allergenic, causing "hay fever" symptoms in sensitive individuals.

A world checklist of extant conifers (Farjon, 1998), with descriptions, line drawings, geographic ranges, ecology, synonymy, and references, is an example of an ongoing effort to document global biological diversity.

Pinaceae: The Pine Family

This is the largest and most important timber-producing family of the gymnosperms. It includes 10 genera and about 200 species, mostly distributed through the Northern Hemisphere from the forests of the tropics to the northern limits of tree growth beyond the Arctic Circle.

The genera are placed in two groups on the basis of morphology and anatomy of fossil and extant material and seed proteins (Price et al., 1987). The Pinoid group includes *Pinus, Larix, Picea, Pseudotsuga,* and *Cathàya*; the Abietoid group includes *Tsuga, Abies, Cédrus, Keteleèria,* and *Pseudolárix.* The native conifers of our area include the first four of the Pinoid group and the first two of the Abietoid group (Table 8-3). These

TABLE 8-3   COMPARISON OF NATIVE GENERA IN THE PINACEAE

Genus	Leaves	Mature cones
Pinus pine	Persistent; acicular; mostly in fascicles of 2 to 5	Pendent or erect; bracts shorter than scale; mostly maturing in 2–3 years
Larix larch	Deciduous; acicular; triangular or 4-sided; singly on terminal long shoots or on lateral, indeterminate short shoots	Erect; bracts shorter or longer than scales; maturing in 1 year
Picea spruce	Persistent; linear, usually 4-sided (flattened in some species); sessile, borne on conspicuous peglike projections from the twig	Pendent; bracts shorter than scales; maturing in 1 year
Pseudotsuga Douglas-fir	Persistent; linear, flattened; petiolate; borne on moderately raised leaf cushions	Pendent; bracts longer than scales and 3-lobed; maturing in 1 year
Tsuga hemlock	Persistent; linear, flattened; petiolate and borne on a slender peglike projection from the twig; often appearing 2-ranked	Pendent; bracts shorter than scales; maturing in 1 year
Abies fir	Persistent; linear, flattened (2 species 4-sided), sessile; may or may not appear 2-ranked	Erect; scales and bracts deciduous from central axis; bracts longer or shorter than scales; maturing in 1 year

are described subsequently in that order. Several of the others are very important as ornamentals. *Cedrus,* particularly, is commonly cultivated. It is an evergreen, having needles spirally arranged on long shoots and indeterminate short shoots, and large, erect, smooth seed cones, and essentially dioecious with pollination occurring in the fall. *Cédrus atlántica* (Endl.) G. Man. ex Carr. is the Atlas cedar, a native of the Atlas Mountains in northern Africa; *C. deodára* (D. Don) G. Don is the very popular deodar cedar, also a commercially important tree in the western Himalayas; and *C. libáni* A. Rich. is the cedar-of-Lebanon. These species and several cultivars are popular ornamentals. *Pseudolárix amábilis* (J. Nels.) Rehd., golden larch, of eastern China, is a fine ornamental.

There is a recent argument that *Pinus* should be alone in the monotypic Pinaceae and that the other genera be classified in the Abietaceae. We follow here the conservative, broader concept of the Pinaceae as treated in FNA vol. 2 (1993).

Botanical Features of the Family

Leaves deciduous or persistent; spirally arranged, in certain genera recurring in false whorls on spur shoots developed on older growth; solitary or in fascicles (*Pinus*); acicular (needlelike) or linear.

Cones (juvenile ovulate) with bract and scale distinct, flat; ovules inverted, 2 at the base of each scale; most species monoecious.

Mature seed cones woody, stalked or sessile, pendent or upright, maturing in one, two, or rarely three seasons; in some genera, disintegrating at maturity; seeds terminally winged or wingless.

PINUS L.   pine

This genus, the most important timber-producing taxon of the conifers, includes just over 100 species widely scattered through the Northern Hemisphere from near the northern limits of tree growth in North America, Europe, and Asia southward to northern Africa, Asia Minor, Malaysia, and Sumatra (here one species crosses the equator [Critchfield and Little, 1966]). In the New World, pines are found as far south as the West Indies and Nicaragua. Mexico has ca. 43 species of pines. Although of primary importance in the production of timber, the wood of most pines is also suitable for pulp and paper manufacture. Turpentine, pitch, pinewood oils, wood tars, and rosin, collectively known as *naval stores,* are obtained from the wood of several species, notably *P. palustris* and *P. elliottii* of North America, *P. pináster* Ait. of the Mediterranean basin, and *P. roxbúrghii* Sarg. (*P. longifòlia* Roxb.) of India. The leaf oils of several species are used in the manufacture of medicines, and the seeds ("pine nuts") of several others (e.g., *P. edulis* and *P. pinèa* L.) are used for food. Needles, gathered as "pine straw," are valuable as mulch.

Natural enemies of the pines are numerous. Because of the resinous nature of the foliage, bark, and wood, fire is always a serious menace, and sawflies, weevils, bark beetles, and tip moths cause considerable damage. White pine blister rust, a fungous disease from Europe, not only destroys true white pines but other five-needle soft pines as well. Root-rotting fungi are of regional importance.

About 37 species of *Pinus* are native to the United States, and nearly all the principal timber-producing regions contain one or more important pines. Pine is the state tree of Arkansas and North Carolina; several other states and provinces have named specific pine species. Fifteen species are native to the eastern United States, and the remaining ones are found in the West. Nine are native of Canada. Interspecific hybridization and introgression occur naturally and rather extensively among related, sympatric species. See Little (1979) and Burns and Honkala (1990) for the particular hybrids and references. McCune (1988) offers an interesting ecological comparison of the North American pines.

Many native, Asian, and European pine species, with many cultivars, are used as cultivated urban trees, and several have become naturalized out of their native habitats. Three of these, *P. nigra, P. sylvestris,* and *P. thunbergiana,* are mentioned later. The abundant pine pollen, so obvious in the spring, is often falsely blamed for causing allergies. It is seldom a problem, although a few cases are due to *P. contorta.*

Botanical Features of the Genus

Leaves of two sorts: (1) primary leaves, scalelike, solitary, spirally arranged, often deciduous or drying within a few weeks after their appearance, bearing in their axils (2) the secondary fascicled, acicular leaves (needles).* *Needles* triangular, semicircular, or rarely circular in transverse section, borne in fascicles of 2 to 5 (rarely as many as 8, or

---

*In seedlings, the first foliage to appear consists of a whorl of cotyledons (seed leaves); the growing shoot then develops linear, spirally arranged, serrulate juvenile leaves the first season, and in certain species for several seasons before the adult fascicled needles ultimately develop.

solitary) usually with several lines of stomata on each surface; *apex* acute; *margin* often sharply serrulate; *basal sheath* deciduous or persistent, composed of 6 to 12 bud scales; *vascular bundles* 1 or 2; *resin canals* 2 to many (rarely obscure); needles persistent for 2 to several (exceptionally 17) years, with a pungent aroma when bruised.

Cones (juvenile) unisexual; *pollen cones* axillary, red, orange, or yellow, in clusters of few to many at the base of the season's growth, consisting of a number of spirally arranged, 2-celled, sessile pollen sacs; *ovulate cones* subterminal or lateral, erect, composed of several to many spirally arranged bracts, each subtending an ovuliferous scale bearing 2 inverted, basal, adaxial ovules; all species monoecious.

Mature seed cone woody, *cone scales* armed or unarmed; *seeds* obovoid, with a terminal wing or wingless; *cotyledons* 3 to 18. The rather remarkable opening mechanism of pinecone scales was investigated by Harlow et al. (1962). Seed dispersal is by wind, birds, or mammals.

Buds scaly, the scales appressed, or free at the ends and then fringed; extremely variable in size, shape, and color, resinous or nonresinous.

Growth habit. Although the growth of a terminal leader and a "whorl" of side branches each year is a feature of several coniferous genera, it is especially typical of the pines. Certain species usually produce a single set of side branches each year, whereas others are multinodal and develop 2 or 3 such arrangements on the elongating leader during the season. Because the leaf arrangement in *Pinus* is alternate (spiral), so too is that of the branches, and the term *false whorl* should be used for these arrangements.

The pines may be separated into two main groups (subgenera), namely, (1) the "soft pines" and (2) "hard pines" (see Table 8-4 for comparison). For the complete classification of pines into sections and subsections, see Little and Critchfield (1969). The arrangement here is more geographical for convenience to the student. See Strauss and Doerksen (1990) for a discussion of pine phylogeny.

Subgenus *Strobus*   soft pines

strobus, monticola, lambertiana, flexilis, strobiformis, albicaulis, cembroides, edulis, monophylla, quadrifolia, balfouriana, longaeva, aristata

TABLE 8-4   COMPARISON OF SOFT AND HARD PINES

Feature	Soft pines	Hard pines
Needles	In fascicles of 1 to 5; cross section with 1 vascular bundle	In fascicles of 2 to 3 (rarely 5 to 8); cross section with 2 vascular bundles
Fascicle sheath	Early deciduous	Persistent and falling with fascicle, rarely deciduous
Cone scales	Usually thin at the apex; mostly unarmed except in foxtail pines	Usually thick at the apex; mostly armed with prickle
Wood	Soft; transition from early to late wood gradual	Hard or rarely soft; transition from early to late wood abrupt

Subgenus *Pinus*   hard pines (pitch pines or yellow pines)

> **Western hard pines:** *ponderosa, jeffreyi, engelmannii, sabiniana, coulteri, torreyana, radiata, attenuata, muricata, leiophylla, contorta*
>
> **Eastern hard pines:** *banksiana, virginiana, clausa, resinosa, palustris, echinata, taeda, glabra, rigida, serotina, pungens, elliottii, nigra, sylvestris, thunbergiana*

The Soft Pines

There are over a dozen soft pines—one (*P. strobus*) in the East, the others in western forests (Table 8-5).

Pìnus stròbus L.   eastern white pine, northern white pine

Distinguishing Characteristics (Fig. 8-3)

Needles $2\frac{1}{2}$ to 5 in. (6 to 13 cm) long, 5 per fascicle, dark bluish green, slender and flexible; persistent until the end of the second season or the following spring.

TABLE 8-5   COMPARISON OF THE SOFT PINES

Species	Needles	Seed cones
P. strobus eastern white pine	5 per fascicle; $2\frac{1}{2}$–5 in. (6–13 cm) long, slender, flexible	Long-stalked, cylindrical, 3–8 in. (8–20 cm) long; scales thin, umbo terminal and unarmed; seed wing long
P. monticola western white pine	5 per fascicle; 2–4 in. (5–10 cm) long, stout, rigid	Long-stalked, cylindrical, 4–10 in. (10–25 cm) long; scales thin, umbo terminal and unarmed; seed wing long
P. lambertiana sugar pine	5 per fascicle; $2\frac{3}{4}$–4 in. (7–10 cm) long, often twisted, stout, rigid	Long-stalked, cylindrical, 11–20 in. (28–50 cm) long; scales relatively thin, umbo terminal and unarmed; seed wing long
P. flexilis limber pine	5 per fascicle; 2–$3\frac{1}{2}$ in. (5–9 cm) long, clustered at ends of twigs, stout, rigid	Short-stalked, cylindrical to ovoid, 3–6 in. (7.5–15 cm) long; scales relatively thin, umbo terminal and unarmed; seed wing short or lacking
P. albicaulis whitebark pine	5 per fascicle; 1–$2\frac{3}{4}$ in. (2.5–7 cm) long, clustered at ends of twigs, stout, rigid	Short-stalked, ovoid-globose, $1\frac{1}{2}$–$3\frac{1}{4}$ in. (4–8 cm) long; serotinous; scales thick, umbo terminal and armed; seeds wingless
Pinyon or nut pines	1-5 per fascicle; $\frac{3}{4}$–$2\frac{1}{2}$ in. (2–6 cm) long, incurved, slender and flexible or stout and rigid	Short-stalked, ovoid to globose, $\frac{3}{4}$–3 in. (2–7.5 cm) long; scales thick, umbo subcentral, unarmed or with small prickle; seed barely winged or usually wingless
P. balfouriana foxtail pine	5 per fascicle; $\frac{3}{4}$–$1\frac{1}{2}$ in. (2–4 cm) long, incurved, clustered at ends of twigs, stout, rigid	Short-stalked, ovoid to cylindrical, $3\frac{1}{2}$–5 in. (9–13 cm) long; scales thick, umbo subcentral, armed with minute prickle; seed wing long
P. aristata bristlecone pine	5 per fascicle; $\frac{3}{4}$–$1\frac{1}{2}$ in. (2–4 cm) long, incurved, clustered at ends of twigs, stout, rigid	Short-stalked, subcylindrical, $2\frac{1}{2}$–$3\frac{1}{2}$ in. (6–9 cm) long; scales thick, umbo subcentral, armed with long, stiff prickle; seed wing long

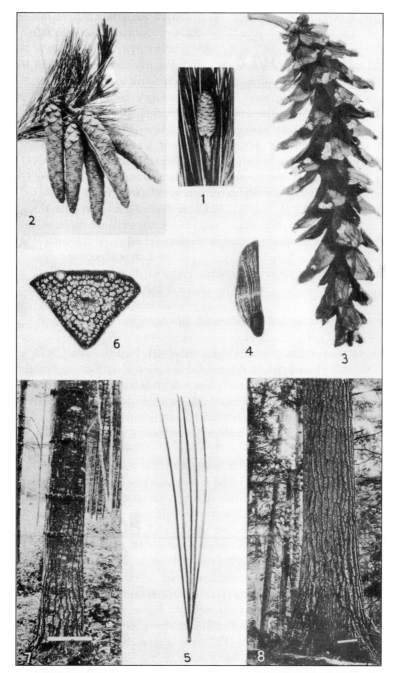

FIGURE 8-3 *Pinus strobus,* eastern white pine. (1) First year's ovulate cone $\times\frac{3}{4}$. (2) Mature seed cones and foliage $\times\frac{1}{3}$. (3) Opened seed cone $\times\frac{3}{4}$. (4) Seed $\times1$. (5) Fascicle of needles $\times\frac{3}{4}$. (6) Cross section of needle $\times35$. (7) Bark of young tree. (8) Bark of old tree.

Cones 3 to 8 in. (8 to 20 cm) long, narrowly oblong-conic, often slightly curved, stalked, with thin scales; usually falling from the tree during the winter or following spring; *apophysis* nearly smooth or slightly raised; *umbo* terminal, unarmed; *seeds* ca. $\frac{1}{4}$ in. (5 to 8 mm) long, ovoid, reddish or grayish brown, mottled with dark spots, wings ca. $\frac{3}{4}$ in. (1.5 to 2 cm) long, lined.

Twigs orange-brown, glabrous or sparingly puberulent; *buds* covered with thin reddish or orange-brown scales.

Bark on young stems thin and smooth, dark green, soon furrowed; on old trees 1 to 2 in. (2.5 to 5 cm) thick, deeply and closely fissured into narrow, roughly rectangular blocks, minutely scaly on the surface.

Range

Southeastern Canada, northeastern United States, and southern Appalachians (Map 8-1). Local disjuncts in central North Carolina, central Tennessee, eastern Kentucky, west central Indiana, and northern Illinois. It is the only soft pine in the east. Elevation: sea level northward to 4950 ft in the southern Appalachians. Eastern white pine is found on many different sites, even including such extremes as dry, rocky ridges, floodplains, and wet sphagnum bogs. Best development, however, is made on moist, well-drained, sandy loam soils of ridges.

General Description

Eastern white pine is the largest of northeastern conifers, and from the beginning of logging in this country it has been a most valuable species, first in New England, then in Pennsylvania and New York, and finally in the lake states where magnificent stands once covered large areas. It is likely that no other tree species has been more important in New England, relative to the settlement of the region, its industrial growth, and international relationships.

This species commonly varies from 80 to 100 ft in height and 2 to $3\frac{1}{2}$ ft in diameter (max. 220 by 6 ft). In the open, young trees produce broadly conical crowns with living branches persisting nearly to the ground, but under forest competition, especially that of hardwoods, a tall cylindrical bole is developed, which is often clear, on older trees, for two-thirds of its length. The crown of middle-aged and old trees consists of several nearly horizontal or ascending branches, gracefully plumelike in out-

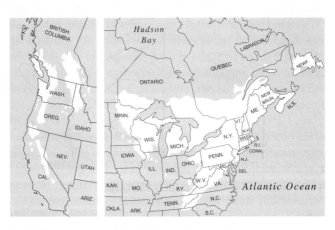

MAP 8-1   **(Left)** *Pinus monticola.* **(Right)** *Pinus strobus.*

FIGURE 8-4   Old-growth eastern white pine.

line and very distinctive in comparison with those of associated conifers. The root system is wide-spreading and moderately deep with only the vestige of a taproot; with age, root development nearer the surface may occur.

Especially in the lake states, moderately moist soils support extensive pure forests, whereas in drier areas, mixtures with red pine or even jack pine are quite common. On heavier soils, the hardwoods compete more successfully, and this pine occurs only as a scattered tree, often reaching, however, large proportions and excellent form in such mixtures.

In Canada, pure stands, or mixtures with red pine and red spruce, frequently occur, whereas in the northeastern United States, in addition to limited pure groups, eastern white pine occurs on sandy loams with the northern hardwoods, red spruce, and eastern hemlock; here the scattered pines usually tower 40 to 50 ft above the crowns of their associates (Fig. 8-4). Farther south, although reaching large size, they are not so abundant and are found with many of the central hardwoods and eastern hemlocks.

Although extensive logging has removed most of the original pine forests, this species has always been aggressive in reproducing itself. When the exodus from eastern farms began in the late 1860s, many abandoned fields quickly seeded in with this tree, and nearly pure stands have since developed. This process has continued and earned for this species, as well as some others, the local name of "old field pine."

Eastern white pine is intermediate in tolerance, more tolerant than such species as paper birch, the aspens, red pine, and jack pine, but less so than sugar maple, American beech, and hemlock.

This pine begins to produce cones at an early age, sometimes when not more than 5 to 10 years old, but good seed production can rarely be expected until the trees are 20 to 30 years old, and under forest conditions from one to two decades later. Large crops usually occur at 3- to 10-year intervals, but a few cones may be produced nearly every year.

After the first two to five seasons, the rate of growth of young trees increases, and when established, this species grows rapidly. On the best sites and during favorable seasons, young white pine may grow nearly 4 ft in height and 1 in. in diameter per year. This, however, is exceptional, and average trees usually attain not more than one-half this rate of increase. Between the ages of 50 and 100 years, height growth, especially, decreases, but maturity is not reached for another 100 years, and very old trees have at least 460 growth rings.

This pine develops one horizontal false whorl of side branches each year, a useful feature in estimating the age of the tree.* In contrast, the pitch and jack pines, as well as some of the southern species, often produce two or three such whorls in a single growing season.

---

*To the count of whorls, add 2 or 3, because as seedlings side branches are not at first developed; this method obviously cannot be used on old trees where the bark has completely obliterated the branch scars.

Because of the demand for the wood, the ease in handling of nursery stock, and the high percentage of survival when properly planted, this tree has been established by the millions throughout the Northeast; in fact, it has been the principal species used in reforestation for many years. Two serious enemies threaten it, however, especially in pure stands: (1) the white pine blister rust caused by a fungus and (2) the white pine weevil. The former is a bark disease that eventually kills the tree, and the latter deforms the trunk by killing, often repeatedly, the terminal shoots. Open-grown, scattered pasture trees are sometimes successively weeviled to such an extent that they become bushy and totally unfit for good lumber, and even trees in denser stands are badly deformed. Control measures have been applied to each of these enemies, and planting plans were modified in an endeavor to lessen the amount of damage. White pine is often damaged by high concentrations of atmospheric sulphur dioxide and ozone and by overbrowsing by whitetailed deer.

Although eastern white pine shows certain geographical variations, these are not sufficient to warrant the recognition of races, with one possible exception (Mergen, 1963).

Eastern white pine is the provincial tree of Ontario, the state tree of Maine and Michigan, and the state flower (cone and tassel) of Maine.

Pinus montícola Dougl. ex D. Don   western white pine, Idaho white pine

Distinguishing Characteristics (Fig. 8-5)

Needles 2 to 4 in. (5 to 10 cm) long, 5 per fascicle, blue-green, glaucous, slender, flexible, persisting 3 to 4 years.

Cones 4 to 10 in. (10 to 25 cm) long, narrowly cylindrical, often curved, stalked; *apophysis* yellowish brown to reddish brown, inner surface of scale deep red; *umbo* terminal, unarmed; *seeds* ca. $\frac{1}{3}$ in. (7 to 9 mm) long, reddish brown, frequently mottled with black, with a terminal wing ca. 1 in. (2 to 3 cm) long.

Twigs moderately slender, often with orange-brown pubescence during the first season but ultimately becoming dark reddish brown to purplish brown and glabrous; *buds* ca. $\frac{1}{2}$ in. (12 mm) long, cylindrical, blunt, covered by several closely appressed, imbricated scales.

Bark smooth, gray-green to light gray on young trees, thin even on old trunks where it breaks up into nearly square or rectangular dark gray or purplish gray blocks separated by deep fissures.

Range

Western (Map 8-1). Disjunct in eastern Oregon. Elevation: sea level to 3500 ft northward, 6000–9900 ft southward. Rich, porous, mountain soils of mixed conifer forests and occasionally in pure stands.

General Description

Western white pine was first observed along the banks of the Columbia and Spokane rivers in 1831 by the Scottish botanist David Douglas. Reaching a height of 150 to 180 ft and a dbh of $2\frac{1}{2}$ to $3\frac{1}{2}$ ft at maturity (max. 231 by 8.4 ft), this tree develops a long, slightly tapered shaft, often clear for 70 to 100 ft or more of its length, and a usually short, symmetrical, and somewhat open crown. Seedlings produce a prodigious taproot that is supplemented in later life by a deep, wide-spreading system of lateral roots, which makes the trees unusually windfirm.

Except in the Puget Sound basin, on the Olympic Peninsula of western Washington, and on Vancouver Island, where it occurs at or near sea level, this species is typically a mountain species.

FIGURE 8-5   *Pinus monticola,* western white pine. (1) Mature seed cone and foliage $\times\frac{3}{4}$. (2) Seeds $\times 1$. (3) Bark of old tree (*Courtesy of British Columbia Forest Service*).

Best development is attained in northern Idaho in moist valleys and on middle and upper slopes and flats of northerly exposure. Here, where nearly 80% of the commercial timber of the species is found, it is very common. Elsewhere, western white pine occurs largely on poorer and drier soils as an occasional tree or in small groves in mixture with other softwoods, and on such sites it rarely constitutes more than 5% of the stand.

This species rarely produces large seed crops, but as a rule some seed is released at intervals of 2 or 3 years after the fortieth to sixtieth year. Reproduction is best on a fresh mineral soil, typically following fires, although seeds lodging in humus often germinate, and the young trees continue to develop provided they are amply supplied with moisture during the growing season. Rated as intermediate in tolerance, young trees are shade-enduring, but full light is required in later life, and mature trees usually dominate mixed stands. Growth is moderately fast but decreases noticeably after the 125th to 150th year, and maturity is attained in about 200 to 350 years. The oldest recorded western white pine was a tree in British Columbia, with an age of 615 years.

This pine is one of the first tree species to invade the peat bogs of the Puget Sound basin. Here the young trees develop with such rapidity that they frequently outstrip those standing on firmer, drier soils nearby.

Western white pine has a number of natural enemies. The bark, even of mature trees, is extremely thin (rarely more than $1\frac{1}{2}$ in. [3 cm] thick); consequently, the trees are easily damaged by even a light ground fire. White pine blister rust and bark beetles also take a heavy annual toll.

This is an important timber tree. Its wood is used for matches, toothpicks, lumber (primarily molding and trim), and softwood furniture. It is the state tree of Idaho. Hoff et al. (1987) provide an annotated bibliography on western white pine.

Pìnus lambertiàna Dougl.   sugar pine

Distinguishing Characteristics (Fig. 8-6)

Needles $2\frac{3}{4}$ to 4 in. (7 to 10 cm) long; spirally twisted, 5 per fascicle, blue-green to gray-green, often silvery, persisting 2 to 4 years.

Cones 11 to 20 in. (28 to 50 cm) long, 4 to 5 in. (10 to 13 cm) in diameter when open, cylindrical, stalked; *apophysis* somewhat thickened, yellowish brown, inner surface of scale brown or reddish brown; *umbo* terminal, unarmed; *seeds* $\frac{1}{2}$ to $\frac{5}{8}$ in. (1.2 to 1.5 cm) long, dark brown to nearly black, lustrous, with a dark brown or black terminal wing 1 to $1\frac{1}{2}$ in. (2.5 to 4 cm) long.

Twigs rather stout, at first covered with glandular pubescence but later becoming orange-brown to purplish brown and glabrous; *buds* ca. $\frac{1}{3}$ in. (8 mm) long, ovoid, sharp-pointed, covered with several closely appressed, chestnut-brown, imbricated scales.

Bark dark green, thin, and smooth on young stems, grayish brown to purplish brown on old trunks, $1\frac{1}{2}$ to 4 in. (3.5 to 10 cm) or more in thickness, and broken into regular, superficially scaly ridges separated by deep fissures (Fig. 8-7).

Range

Western (Map 8-2). Elevation: 1100 to 5400 ft northward, 2000 to 7800 ft in the Sierra Nevada, 4000 to 10,500 southward. Found in moist, cool, mountain soils in mixed conifer forests.

General Description

Sugar pine, another of David Douglas' discoveries, derives its name from a sweet resinous substance called pinita that exudes from wounds and that was eaten by early Native Americans. It is the tallest and largest of all the pines, and 300-year-old trees on good sites commonly vary from 170 to 180 ft in height and from 30 to 42 in. or more dbh (max. 262 by 11.5 ft). The boles of forest trees are usually long, clear, and cylindrical, and support short crowns composed of several

FIGURE 8-6   *Pinus lambertiana,* sugar pine. (1) Mature seed cone $\times\frac{1}{2}$. (2) Open seed cone $\times\frac{1}{3}$. (3) Seeds $\times 1$. (4) Cross section of needle $\times 35$.

FIGURE 8-7   Bark of old-growth sugar pine.

MAP 8-2   *Pinus lambertiana.*

massive, horizontal, or occasionally contorted branches. It is an important source of highly valued lumber. The root system is similar to that of western white pine.

Sugar pine never forms pure stands, except in small scattered areas, and while it may occasionally comprise up to 75% of a mixed stand, throughout most of its range it seldom occupies more than 25% of a forested area. It attains its best development on the granitic and andesitic soils of the west slopes of the Sierra Nevada between the San Joaquin and Feather Rivers between elevations of 4500 and 5500 ft. Forest associates within these limits include ponderosa and Jeffrey pines, Douglas-fir, California red, Shasta, and white firs, giant sequoia, and incense-cedar. In southern Oregon and northern California, sugar pine is often found with Digger and ponderosa pines, Douglas-fir, tanoak, California black oak, bigleaf maple, and Pacific dogwood, usually at elevations of less than 4000 ft. Toward the southern limit of its range, this species invades subalpine forests and has been found at elevations of 10,000 ft. This pine is usually found on somewhat cooler and wetter sites than those occupied by most other pines, and to the east of the Sierran crest, where drought and extreme fluctuations in temperature frequently occur, it is very limited.

Sugar pine is not a prolific seeder. Small amounts of seed are usually produced annually, but large crops are normally released only at irregular intervals of from 2 to 7 years after the eightieth to hundredth year. A moist mineral soil covered by a thin layer of humus makes a good seedbed. Pine seeds are eaten by squirrels and other rodents, making natural reproduction in some sections very sparse. Although sunlight is beneficial to the seedlings, protection from extremes of temperature is needed for continued growth, and the most vigorous young trees are usually found under partial shade. The tolerance of this species is intermediate to intolerant with maturity.

Sugar pine is the most rapidly growing tree in the Sierra Nevada (with the exception of the giant sequoia), and on good sites trees 100 years old may be 140 ft tall and nearly 30 in. dbh. It is long-lived and occasionally attains an age of 500 to 760 years.

Sugar pine is unusually windfirm and is reasonably free from fungous diseases and insect pests, although the white pine blister rust and the mountain pine beetle sometimes cause serious losses. Dwarf mistletoe and fire cause considerable damage to young trees, and fire scars are a major source of culls in mature timber.

Pinus fléxilis James   limber pine, Rocky Mountain white pine

Distinguishing Characteristics (Fig. 8-8).   *Needles* 2 to $3\frac{1}{2}$ in. (5 to 9 cm) long, 5 per fascicle, clustered near the branch ends, dark green, stout, rigid, not toothed, persisting 5 to 6 years. *Cones* 3 to 6 in. (7.5 to 15 cm) long, ovoid, the scales thickened, and slightly reflexed at the apex; *seeds* large, with rudimentary wings or wingless. *Bark* on young stems smooth, silvery white to light gray or greenish gray; that on old trunks dark brown to nearly black, separated by deep fissures into rectangular to nearly square, superficially scaly plates or blocks.

Range.   East slopes of the Rocky Mountains in southeastern British Columbia and southwestern Alberta, south along the mountains to northern Arizona and New Mexico; west to the mountains of central and southern California, and east through Nevada, Idaho, and Montana; local in northeastern Oregon, southwestern North Dakota, Black Hills of South Dakota, and western Nebraska. Elevation: 3300 to 12,000 ft. Found on dry rocky slopes and ridges to timberline, often in pure stands.

FIGURE 8-8 Stone pines. (1) *Pinus flexilis,* open seed cone ×$\frac{3}{4}$. (2) Seeds ×1. (3) *Pinus albicaulis,* fascicle of needles ×$\frac{3}{4}$. (4) Mature seed cone ×$\frac{3}{4}$. (5) Seeds ×1.

General Description.   Limber pine was first observed near Pike's Peak by Dr. Edwin James, an army surgeon attached to Long's mountain expedition of 1820. Like other relatively inaccessible trees of high altitudes, limber pine is primarily important in the protection of valuable watersheds. Ordinarily the tree attains but small proportions, varying from 30 to 50 ft in height and from 15 to 36 in. dbh (max. 85 by 7.5 ft). The bole is stout, noticeably tapered, and supports a number of large, plumelike, often drooping branches. The result is an extensive crown which may reach to within a few feet of the ground. Young trees develop a long, sparsely branched taproot which is later supplemented by several laterals. It is intolerant, slow growing, and long lived (to 1670 yrs). There is damage from white pine blister rust, bark beetles, budworms, and several species of dwarf mistletoe.

Numerous rodents and birds (especially Clark's nutcracker) eat and disperse limber pine's large, basically wingless seeds. Clark's nutcracker can carry up to 125 seeds to at least 14 miles away from the parent tree.

It is not of prime commercial importance, although used for construction lumber, railroad ties, and poles. It also has a high ornamental value for the landscape.

This species hybridizes with *P. strobiformis* where the two overlap in distribution.

Pìnus strobifòrmis Engelm.   southwestern white pine

Distinguishing Characteristics.   *Needles* 2$\frac{1}{2}$ to 3$\frac{1}{2}$ in. (6 to 9 cm) long, 5 per fascicle, slender, serrulate at least near tip, bright green, persisting 3–5 years. *Cones* 6 to 9 in. (15 to 23 cm) long, cylindrical, yellow-brown, short-stalked, the scales slightly thickened, long, the thin apex curving back; *seeds* large, very short-winged, edible. *Bark* gray and smooth, becoming dark gray or brown and deeply furrowed into narrow, irregular ridges.

Range.   West Texas to central New Mexico, east central Arizona, and into northern Mexico. Elevation: 6270 to 9900 ft. Found in canyons and on dry, rocky slopes in high mountains.

General Description.   Straight, conical tree 50 to 80 ft tall and to 3 ft dbh. This is not abundant enough to be of commercial value, although it is used locally for fuel and posts. The large seeds were eaten by Native Americans of the Southwest.

It is considered by some as *P. flexilis* var. *reflexa* Engelm. See Steinhoff and Andresen (1971) for additional discussion of the variation. Hybrids are formed with *P. flexilis* where the two overlap.

Pìnus albicaùlis Engelm.   whitebark pine

Distinguishing Characteristics (Fig. 8-8).   *Needles* 1 to $2\frac{3}{4}$ in. (2.5 to 7 cm) long, 5 per fascicle, similar to those of limber pine, clustered toward the ends of the branchlets, persisting 5 to 8 years. *Cones* $1\frac{1}{2}$ to $3\frac{1}{4}$ in. (4 to 8 cm) long, ovoid, purplish brown, with thickened apophyses and terminally armed umbos; serotinous. *Bark* on immature trees brownish white to creamy white, smooth or superficially scaly at the base, rarely more than $\frac{1}{2}$ in. (12 mm) thick.

Range. Western British Columbia in the vicinity of Lake Whitesail south along the Olympic and Cascade Mountains of Washington, the Siskiyou, and Blue Mountains of Oregon, to the Sierra Nevada of California; south through Alberta and eastern British Columbia to northern Montana, northwestern Wyoming, and the Bitterroot Mountains of Idaho. Elevation: 4300 to 7500 ft northward, 8000 to 12,200 ft southward. It occurs in dry, rocky soils on slopes and ridges in the subalpine zone to timberline with other conifers or sometimes in pure stands.

General Description. This alpine tree, the only North American representative of the "stone pines," rarely attains a height of more than 30 to 50 ft or 12 to 24 in. dbh (max. 86 by 8.8 ft). In sheltered places, whitebark pine produces an erect, rapidly tapering bole that supports a crown of long, willowy-appearing limbs; these are extremely tough and are rarely damaged by heavy snow or even by the cyclonic winds characteristic of alpine sites. Trees on exposed areas are usually sprawling, or even prostrate and fail to develop anything remotely resembling a central stem. They are generally intolerant, slow growing, and long lived (to 880 years). Damage occurs from the mountain pine beetle, white pine blister rust, and dwarf mistletoe.

The seeds were eaten by Native Americans and are important food for black and grizzly bears. The jaylike Clark's nutcracker is the primary dispersal agent. It is of limited commercial importance, but it is valuable for watershed protection and aesthetics. Recently, there has been much concern about the health of whitebark pine ecosystems (Schmidt and McDonald, 1990).

Pinyons or Nut Pines

Dominating more than 65,000 square miles in the intermountain region and over 90,000 square miles of the arid southwestern United States, between the deserts below and the forests above, is a major vegetation type composed of small coniferous trees and shrubs (Fig. 8-9) growing in scattered or rather dense mixtures—the "pinyon-juniper woodlands." These woodlands are best developed between 4500 and 6500 ft elevation, although individual species of pinyon pines may be found between 200 and 10,000 ft. This community is expanding into adjacent shrub communities, grasslands, and aspen groves. See Miller and Wigand (1994) and Johnston (1995) for more thorough discussions.

FIGURE 8-9   Pinyon-juniper woodland. (*Courtesy of U.S. Forest Service.*)

Throughout human history, the pinyon component has made life possible, for both humans and wildlife, in an otherwise harsh environment. The large wingless seeds of these pines were the principal food for the Native Americans who gathered them each autumn in great numbers for winter storage. They are still used for winter food, and these nutritious "pine nuts," "Indian nuts," "pinyon nuts," or "pinyones" find their way to local and gourmet markets and are exported to foreign countries. A large tree in a good seed year may yield about 20 lbs of seed; a stand, 300 lbs per acre. In addition, the settlers depended upon the pinyon and juniper for fuel, fence posts, and building materials. The making of charcoal for Nevada ore smelters was a leading industry of the 1800s. Today, pinyon pine is increasingly used for Christmas trees.

Pinyons are recognized by having 1 to 4(5) needles per fascicle and stiff and incurved, open seed cones depressed-ovoid to nearly globose, and the seeds wingless.

There are four or five recognized species of pinyons in the southwestern United States and seven more in Mexico, making up a variable group with hybrids. See Lanner (1975) and Malusa (1992) for more detail. The principal species are as follows:

P. cembroìdes Zucc. Mexican pinyon (needles in 3s, rarely 2s or 4s)

P. édulis Engelm. pinyon or Colorado pinyon (needles in 2s, rarely 1 or 3). This is the state tree of New Mexico, and can reach an age of 973 years.

P. monophýlla Torr. & Frem. singleleaf pinyon (Fig. 8-10) (needles 1 per sheath, rarely 2) (incl. *P. californiàrum* D.K.Bailey). This is the state tree of Nevada.

P. quadrifòlia Parl. ex Sudw. Parry pinyon (needles in 4s, rarely 3 or 5)

FIGURE 8-10   *Pinus monophylla, singleleaf pinyon. Foliage, open seed cone, and wingless seed* ×½.

Foxtail Pines

The foxtail pines, so-called because of the bushy nature of the foliage on young branches, comprise a small group of alpine trees, three of which are found in the western United States. They retain their foliage for many years, the needles are curved, the cone scales are dorsally armed, and the seeds have long terminal wings. The trees are very small, rarely attaining a height of more than 30 to 40 ft and 12 to 24 in. dbh (max. 71 by 8 ft). They develop a short, stocky, often malformed trunk that is commonly covered for the greater part of its length in a dense, narrow, irregular crown. Owing to their inaccessibility, they contribute little or nothing to the nation's timber supply; nevertheless, they provide valuable cover on many of our high western watersheds.

Pìnus balfouriàna Grev. & Balf.   foxtail pine

Distinguishing Characteristics.   *Needles $\frac{3}{4}$ to $1\frac{1}{2}$ in. (2 to 4 cm) long, in fascicles of 5, bright blue-green, lacking a strong median groove on the abaxial side, persisting for 10 to 30 years. Cones $3\frac{1}{2}$ to 5 in. (9 to 13 cm) long,* ovoid to cylindrical, tapering to a conic base, dark reddish brown, the umbo depressed and armed with a minute (0–1 mm), incurved, weak, deciduous prickle (Fig. 8-11).

Range.   Foxtail pine is found locally in northern and central California (Sierra Nevada) between 5000 and 11,550 ft elevation. It occurs on exposed, dry, rocky slopes, ridges, and high peaks in the subalpine zone to timberline.

General Description.   This tree reaches a height of 20 to 50 ft and 1 to 2 ft dbh (max. 118 by 8.4 ft) with an irregular crown of short, spreading branches. It is shrubby at timberline. It can live for 2100 years.

FIGURE 8-11   Foxtail pines. (1) *Pinus balfouriana,* open seed cone ×$\frac{3}{4}$. (2) Fascicle of needles ×$\frac{3}{4}$. (3) *Pinus aristata,* open seed cone ×$\frac{3}{4}$. (4) Seed ×1.

Pìnus longaèva D.K. Bailey (*P. aristata var. longaeva* [D. K. Bailey] Little)
    Intermountain bristlecone pine

Distinguishing Characteristics.   *Needles* $\frac{5}{8}$ to $1\frac{3}{8}$ in. (1.5 to 3.5 cm) long, 5 per fascicle, upcurved, lacking the strong, median groove and with few resin deposits, persisting 10 to 30 years. *Cones* $2\frac{3}{8}$ to $3\frac{3}{4}$ in. (6 to 9.5 cm) long, a rich, rusty red-brown, with a rounded base, and the umbo with a raised, stiff, slender prickle 1 to 6 mm long.

Range.   These occur in Utah, Nevada, and eastern California at 5600 to 11,200 ft elevation on rocky ridges and slopes.

General Description.   It is shrubby at timberline or may attain a height of 20 to 40 ft and diameter of 1 to 2 ft (max 12.5 ft) with an irregular crown.

These are the oldest known trees, with individuals reaching 3000 to ca. 5100 years (Currey, 1965; Schulman, 1958). They have made a significant contribution to our knowledge of dendrochronology, carbon-14 dating, and archeology.

Pìnus aristàta Engelm.   Rocky Mountain or Colorado bristlecone pine

Distinguishing Characteristics.   *Needles* $\frac{3}{4}$ to $1\frac{1}{2}$ in. (2 to 4 cm) long, in fascicles of 5, bright blue-green with small drops and scales of white resin, and with a strong, abaxial, median groove, persisting 10 to 17 years. *Cones* $2\frac{1}{2}$ to $3\frac{1}{2}$ in. (6 to 9 cm) long, subcylindrical, rounded at the base, brown to gray-brown, the umbo armed with a long (6 to 10 mm), slender, stiff, incurved prickle (Fig. 8-11).

Range.   This occurs on the high ridges and slopes in Colorado, northern New Mexico, and northern Arizona. Elevation: 7500 to 11,200 ft.

General Description.   These are among the oldest living trees known, with individual trees reaching 1500 to 2400 years; however, *P. longaeva* can live twice as long. It is a small tree 20 to 40 ft tall and 1 to $2\frac{1}{2}$ ft dbh. It occurs on exposed, dry, rocky ridges in the subalpine zone to timberline, with mixed conifers or in pure stands. It is intolerant, slow growing, and long lived.

The Hard Pines

The arrangement of the hard pines is geographical (western and eastern) for convenience.

Western Hard Pines

There are at least a dozen hard or "yellow" pines scattered through the timbered areas of western United States, but only three of them, ponderosa, Jeffrey, and lodgepole pine, are of primary importance (Table 8-6).

Pìnus ponderòsa Dougl. ex Laws.   ponderosa pine

Distinguishing Characteristics (Fig. 8-12)

Needles 3 to 10(12) in. (7 to 25[30] cm) long, 3, 2 and 3, or 4 and 5 per fascicle on the same tree, dark gray-green to yellow-green, flexible, persisting 2 to 7 years; crushed needles have a turpentine odor similar to that of most other pines.

Cones 2 to 6 in. (5 to 15 cm) long, ovoid to ellipsoidal, sessile, solitary or clustered; usually leaving a few basal scales attached to the twig when shed; *apophysis* dark reddish brown to dull brownish yellow, transversely ridged and more or less diamond-

TABLE 8-6   COMPARISON OF IMPORTANT WESTERN HARD PINES

Species	Needles	Seed cones
P. ponderosa ponderosa pine	2 and 3(5) per fascicle; 4–8(11) in. (10–20[28] cm) long, stout, dark green	Ovoid, 2–6 in. (5–15 cm) long, short-stalked; umbo with short, stout prickle pointed outward
P. jeffreyi Jeffrey pine	3 per fascicle, 5–10 in. (13–25 cm) long, twisted, stout, gray-green	Ovoid, 5–10(14) in. (13–25[35] cm) long, short-stalked; umbo with long, slender, reflexed prickle
P. contorta lodgepole pine	2 per fascicle; $\frac{3}{4}$–3 in. (2–8 cm) long, stout, twisted, yellow-green to dark green	Ovoid, $\frac{3}{4}$–$2\frac{1}{4}$ in. (2–6 cm) long, sessile; umbo with small, slender prickle; usually serotinous and persistent

shaped; *umbo* dorsal, with a slender, often deciduous prickle; *seeds* ca. $\frac{1}{4}$ in. (6 to 8 mm) long, ovoid, slightly compressed toward the apex, brownish purple; wings moderately wide, ca. 1 in. (2.5 cm) long.

Twigs stout, with a turpentine odor when bruised; *buds* usually covered with droplets of resin.

Bark brown to black and deeply furrowed on vigorous or young trees ("bull pines"); yellowish brown to a cinnamon-red and broken up into large flat, superficially scaly plates separated by deep irregular fissures on slow-growing and old trunks, with an odor of vanilla.

Range

Western (Map 8-3). It is one of the most widely distributed pines in North America. Elevation: sea level to 9900 ft; best developed at 4000 to 8000 ft. Found in the mountains in open, pure stands or mixed conifer forests.

General Description

This is the most important pine in western North America (Fig. 8-13), and in the United States it is found in commercial quantities in every state west of the Great Plains. It furnishes more timber than any other single species of American pine, and in terms of total annual production of lumber by species, it is second only to Douglas-fir.

Ponderosa pine is a large tree 150 to 180 ft tall and 3 to 4 ft dbh (max. 256 by 9 ft). Even though this species commonly forms open parklike forests, mature boles are ordinarily symmetrical and clear for one-half or more of their length; short conical or flat-topped crowns are characteristic of old trees. Four-year-old trees may have taproots 4 to 5 ft long. Moderately deep wide-spreading laterals develop as the trees get older. Ponderosa pine is not exacting in its soil requirements, but trees on thin, dry soils are usually dwarfed. Its occurrence on dry sites with the nut pines and certain of the junipers is indicative of its great resistance to drought. This species attains its maximum development, however, on the relatively moist but well-drained western slopes of the Siskiyou and Sierra Nevada Mountains of southern Oregon and California. Excellent pure forests are found in the Black Hills of South Dakota, the Blue Mountains of Oregon, the Columbian Plateau northeast of the Sierra Nevada, and northern Arizona and New Mexico. It is also commonly the most abundant tree in mixed coniferous stands; east of the summit of the Cascade Range in Washington and Oregon, it occurs with western larch, Douglas-fir, and occasionally lodgepole pine; in the central

FIGURE 8-12 *Pinus ponderosa,* ponderosa pine. (1) Mature seed cone ×$\frac{3}{4}$. (2) Open seed cone ×$\frac{3}{4}$. (3) Seed ×1. (4) Pollen cones ×1. (5) Young ovulate cones ×1. (6) Cross section of needle ×35. (7) Bark (*Courtesy of U.S. Forest Service*).

FIGURE 8-13   Ponderosa pine. (*Courtesy of U.S. Forest Service.*)

Rocky Mountains with Douglas-fir; and in California with Jeffrey and sugar pines, incense-cedar, Douglas-fir, and white fir. On the Fort Lewis plains in western Washington near Puget Sound, ponderosa pine is occasionally found in association with Douglas-fir and Oregon white oak.

Small quantities of seed are produced annually, but there is no dependable periodicity for bumper crops. Over much of its range, effective regeneration depends on a bumper crop of seeds in conjunction with above-average summer rainfall. Seedlings can exist under the canopy of the parent trees, even though they grow quite slowly, and in such situations often attain a height of only 3 to 4 ft during the first 15 to 20 years. Seedling growth is best in clearings made by fire or logging. Overstocked stands and subsequent stagnation on poor sites are a problem. The seedlings will grow on sterile sites and have been planted extensively in the Nebraska sand hills and elsewhere. Ponderosa pine is classed as intolerant.

The rapidity of growth has a marked effect on the general appearance of these trees. Young, vigorous specimens commonly develop dense crowns of dark green foliage and bark that is dark brown to nearly black, more or less corky, and deeply furrowed. In contrast, the foliage of old-growth or slow-growing trees is yellow-green, and the bark is yellow-brown to cinnamon-red and plated. Those of the first type are commonly called *bull* or *blackjack* pines, and to some people ponderosa pine and bull pine are different trees. Fast-growth bull pines 150 years of age found near Cle Elum, Washington, measured 30 to 40 in. dbh, whereas more typical ponderosa pines occurring in the same vicinity were only 10 to 14 in. dbh at the same age. Thus, growth of this species varies considerably with locality. In California, trees 120 years of age averaged 23 in. dbh, whereas in Arizona trees of the same age were only 16 in. and in the Black Hills merely $10\frac{1}{2}$ in. Trees over 500 years of age are seldom encountered, but ponderosa pine can reach 1000 years of age.

Ponderosa pine is attacked by over 100 different species of insects. Of these, some 16 species of bark beetles may cause widespread damage. The mountain pine beetle (*Dendroctonus ponderosae*) is especially serious. Fires kill seedlings and cause considerable damage even to large trees. Severe fires in the past have completely destroyed hundreds of thousands of acres of ponderosa pine forest. Other destructive agents include dwarf mistletoe, numerous fungi, and air pollution, particularly ozone. Repeated light surface fires cause little if any damage. Actually, frequent low-intensity fires have been a major factor in promoting and maintaining healthy ponderosa pine forests. Due to fire-suppression policies of the twentieth century, many ponderosa pine forests are showing many signs of stress (Yazvenko and Rapport, 1997).

Previously called western yellow pine, logs of ponderosa pine were also sold under such names as Arizona white pine, California white pine, and western soft pine because the wood resembles that of the white pines rather than that of the hard, moderately heavy wood of the southern yellow pines. The name "ponderosa pine" was adopted by the U.S. Forest Service and is now accepted by the industry.

Within the broad geographical range of this species, three varieties are recognized. The typical **var. *ponderosa*** is found from sea level to 7600 ft elevation in the mountains of the Pacific Coast region from southern British Columbia to southern California and western Nevada; it has relatively long needles (12 to 25 cm) mostly in threes, twigs not glaucous, and large cones (8 to 15 cm). **Var. *arizonica*** (Engelm.) Shaw (Arizona pine) of 6900 to 8200 ft elevation in southeastern Arizona has short needles (7 to 17 cm) mostly in fives (or fours and fives) on the same tree, twigs glaucous, and small cones (5 to 8 cm). **Var. *scopulorum*** Engelm. (Rocky mountain ponderosa pine) of 3300 to 9900 ft elevation in the Rocky Mountains of southwestern North Dakota, Montana, Idaho, south to Arizona, New Mexico, and western Texas has short needles (7 to 17 cm) in twos or threes, twigs rarely glaucous, and small cones (5 to 10 cm).

Ponderosa pine is the state tree of Montana. It forms hybrids with *P. engelmannii* and *P. jeffreyi*. Besides its tremendous importance for timber, ponderosa pine has potential as an ornamental in the East.

Pìnus jéffreyi Grev. and Balf.   Jeffrey pine

Distinguishing Characteristics (Fig. 8-14)

Needles 5 to 10 in. (13 to 25 cm) long, in fascicles of 3, or 2 and 3 on the same tree, blue-green, somewhat twisted, persistent until the sixth to ninth seasons. Crushed needles have a pineapple odor.

Cones 5 to 10 (14) in. (13 to 25 [35] cm) long, ovoid; *apophysis* chestnut-brown, rhomboidal to broadly elliptical, often transversely ridged and occasionally wrinkled;

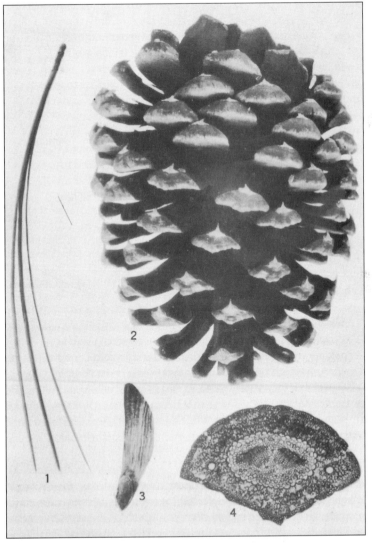

FIGURE 8-14   *Pinus jeffreyi*, Jeffrey pine. (1) Fascicle of needles ×¾. (2) Open seed cone ×¾. (3) Seed ×1. (4) Cross section of needle ×35.

FIGURE 8-15   Bark of old-growth Jeffrey pine. (*Photo by Emanuel Fritz.*)

umbo dorsal, often chalky white, terminating in a long, stout, incurved, occasionally deciduous prickle; *seeds* ca. $\frac{1}{3}$ in. (8 mm) long, yellowish brown, mottled with purple; wings $\frac{3}{4}$ to $1\frac{3}{4}$ in. (2 to 4.5 cm) long, broadest below the middle.

Twigs tinged with purple, often glaucous, with a pineapple aroma when crushed; *buds* only slightly resinous, never covered with resin droplets (a common feature of ponderosa pine).

Bark similar to that of ponderosa pine, but darker cinnamon-red and commonly tinged with lavender or purple on old trunks (Fig. 8-15).

Range

Western (Map 8-3). Elevation: generally 6000 to 9000 ft, rarely either down to 3500 or up to 10,000 ft. It occurs on dry slopes in deep, well-drained soils and either in mixed conifer forests or often in pure stands.

General Description

Throughout its range, Jeffrey pine occurs with ponderosa pine and many of its associates. It can endure greater extremes of climate (Fig. 8-16) than ponderosa pine, and east of the Sierras where conditions are drier and more rigorous, this species occurs in some abundance. The habits of Jeffrey pine are quite similar to those of ponderosa. Seedlings, relative to ponderosa, are somewhat frost-hardy and more suitable for reforesting on sites where low temperatures often prevail.

Although a stately tree, Jeffrey pine never reaches the proportions of ponderosa. Under favorable growing conditions, mature trees will attain a height of 90 to 100 ft and 36 to 60 in. dbh (max. 207 by 8 ft). It is customary to log these two species together, and because of the great similarity of their timbers, no attempt is made to separate them in the trade. Jeffrey pine hybridizes with *P. coulteri* and *P. ponderosa*.

Damage occurs from a needle blight, various rusts, and dwarf mistletoe.

Pìnus engelmánnii Carr.   Apache pine

Distinguishing Characteristics.   *Needles* 8 to 18 in. (20 to 45 cm) long, borne in fascicles of 2 to 5 (mostly 3), stout, dull green, often drooping, persisting 2 years. *Cones* 4 to $5\frac{1}{2}$ in. (10 to 14 cm) long, ovoid or conical, light brown; umbo with a short, straight or curved prickle.

Range.   Apache pine is principally a Mexican species but is found locally in southeastern Arizona and extreme southwestern New Mexico. Elevation: 5000 to 8200 ft. It occurs on rocky ridges and slopes.

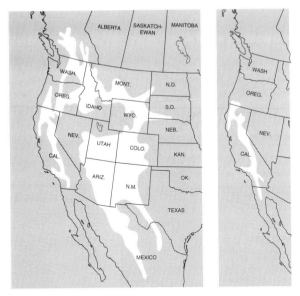

FIGURE 8-16 Windform of Jeffrey pine, growing on Sentinel Dome, Yosemite National Park, California.

MAP 8-3 **(Left)** *Pinus ponderosa.* **(Right)** *Pinus jeffreyi.*

General Description. This is a medium-sized tree, 50 to 70 ft tall and to 2 ft dbh, often called the "longleaf pine" of the Southwest. From a distance, the mature trees appear quite similar to those of ponderosa pine, a species with which it is often associated; young trees, however, are very distinctive and there is a tendency to form a grass stage. The seedling sends down a prodigious taproot, which descends to a depth of 6 ft or more during the first few years. In the meantime, the aerial portion grows very slowly and is seldom branched.

Hybrids are formed with *P. ponderosa.*

Pìnus sabinàna Dougl. ex D. Don   Digger pine, foothill pine, gray pine

Distinguishing Characteristics. *Needles* 6 to 13 in. (15 to 32 cm) long, in fascicles of 3, gray-green, flexible, persisting 3 to 4 years. *Cones* 6 to 10 in. (15 to 25 cm) long, heavy, subglobose, their reddish brown scales with irregularly rhomboidal apophyses, each terminating in a stout dorsal claw (Fig. 8-17); *seeds* oblong with very short abortive terminal wings and thick hard coat. *Bark* dark brown with flattened anastomosing ridges.

Range. The coast ranges and west slopes of the Sierra Nevada in California from the Siskiyou Mountains south to the Tehachapi and Sierra de la Liebre Mountains. Elevation: 1000 to 3000 ft, rarely down to 100 and up to 6000 ft. It occurs in high valleys, foothills, and lower mountain slopes and on dry slopes and ridges in the foothills and low mountains.

General Description. Digger pine is restricted to the low fringe forests of California. As a rule, it is a small- to medium-sized tree 40 to 50 ft tall and 12 to 24 in. dbh (max. 160 by 5 ft). The bole is usually short, the crown open and irregular, and the root system moderately deep and spreading. Scattered through the dry foothills and lower mountain slopes, it occurs either in pure, open parklike stands or in mixture with one or more of the native oaks. At higher elevations, it is occasionally associated with ponderosa pine. It is very intolerant and very susceptible to fire. It is damaged by gall rust and dwarf mistletoe.

FIGURE 8-17   (Left) *Pinus coulteri,* Coulter pine. (1) Open seed cone $\times\frac{1}{5}$. (2) Seed $\times\frac{2}{3}$. (Right) *Pinus sabiniana,* Digger pine. (1) Open seed cone $\times\frac{1}{4}$. (2) Seed $\times\frac{2}{3}$.

The common name refers to the Digger Indians who dug roots for food. Because of this derogatory term, many prefer the name foothill pine. Its large seeds have a high nutritional value and were once important in the diet of California valley Native Americans.

Pìnus coùlteri D. Don   Coulter pine

Distinguishing Characteristics.   *Needles* 6 to 12 in. (15 to 30 cm) long, 3 per fascicle, blue-green, rigid, persisting 3 to 4 years. *Cones* 8 to 14 in. (20 to 35 cm) long, the heaviest of the conifer cones, when green often weighing 4 to 6 pounds, subcylindrical to oblong-ovoid, short-stalked, persistent; *cone scales* with light brownish yellow, irregularly rhomboidal apophyses, each terminating in a long, sharp, flat claw, those near the base of the cone the longest and exhibiting the greatest curvature (Fig. 8-17); *seeds* ellipsoidal, with extremely thick hard coats, the wings longer than the seeds. *Bark* dark brown to nearly black, with broad rounded ridges separated by deep anastomosing fissures.

Range.   Central coastal California to northern Baja California, Mexico. Elevation: 3000 to 6000 ft, rarely 1000 to 7000 ft. It occurs on dry, rocky mountain slopes and ridges in foothills with other conifers.

General Description.   Coulter pine is a medium-sized tree, 40 to 50 ft tall and 15 to 30 in. dbh (max. 144 by 5 ft), largely restricted to the mountains of southern coastal California and northern Baja California. At low elevations (3000 ft), Coulter pine invades chaparral. Here the trees develop a typical open-grown form, the large crowns often extending to within a few feet of the ground. At higher elevations (3500 to 7000 ft), it occurs on dry, gravelly, or loamy soils in mixture with live oaks, incense-cedar, ponderosa and sugar pine, and bigcone Douglas-fir. It is intolerant and slow growing. Some populations have a high proportion of serotinous cones; otherwise, most cones open within a few years following maturation.

It is used for lumber and fuel. It hybridizes with *P. jeffreyi.*

Pìnus torreyàna Parry ex Carr.   Torrey pine

Distinguishing Characteristics. *Needles* 6 to 12 in. (15 to 30 cm) long, and unlike most other western yellow pines, are borne in fascicles of 5, stout, dark gray-green, persisting 3 to 4 years. *Cones* 4 to 6 in. (10 to 15 cm) long, with thick scales with a straight prickle; *seeds* large, ovoid, thick-shelled, and short-winged.

Range.   Torrey pine has the most restricted range of any American pine, occurring only in a narrow strip of coastal land in San Diego County, southern California (subsp. *torreyana*), and adjacent Santa Rosa Island (subsp. *insularis* J.R. Haller). Elevation: 0 to 500 ft. It grows on dry, sandy bluffs and slopes in pure stands.

General Description.   Torrey pine is a small- to medium-sized tree, 30 to 50 ft tall and 1 to 2 ft dbh (max. 133 by 6.5 ft), with an open, spreading crown; on exposed sites it becomes shrubby. It does well in cultivation and has been propagated with some degree of success in New Zealand for sawtimber. It is intolerant and short-lived.

Pìnus radiàta D. Don   Monterey pine

Distinguishing Characteristics. *Needles* 4 to 6 in. (10 to 15 cm) long, in fascicles of 3 or sometimes 2, rich, dark green, slender, flexible, persisting 3 or 4 years. *Cones* $2\frac{3}{4}$ to $5\frac{1}{2}$ in. (7 to 14 cm) long, asymmetrical, and serotinous (Fig. 8-18).

FIGURE 8-18   *Pinus radiata*, Monterey pine. (1) Mature seed cone $\times\frac{3}{4}$. (2) Seed $\times1$.

Range.   Monterey pine is restricted to three, small, separated coastal areas of central California in the fog belt to ca. 6 mi inland; also on two islands off the coast of Baja California, Mexico. Elevation: 100 to 1325 ft in California. It grows in dry, coarse soils and sandy loams of slopes in mixed or pure stands.

General Description.   This is a moderately tolerant, medium-sized tree, 50 to 100 ft tall and 1 to 3 ft dbh (9.2 ft max.). It is of little or no commercial value within its range, but it is a common decorative tree in many cities along the Pacific slope and appears to be quite hardy as far north as the Puget Sound basin. It is the most widely planted pine in the world and has been established in New Zealand, Australia, South Africa, Spain, and South America including Chile, where it grows very rapidly and produces sawtimber in a relatively short time. Thirty-year-old plantation trees in New Zealand may be from 100 to 140 ft high and 30 in. or more dbh; tallest individuals in New Zealand are over 200 ft high. For a discussion of this pine, see Scott (1960). It is estimated that worldwide, more than $1\frac{1}{2}$ million acres have been forested with Monterey pine.

It is damaged mainly by gall rusts and dwarf mistletoe. It hybridizes with *P. attenuata*.

Pìnus attenuàta Lemm.   knobcone pine

Distinguishing Characteristics. *Needles* 3 to 7 in. (7.5 to 18 cm) long, 3 per fascicle, slender, stiff, and yellow-green, persisting 4 to 5 years. *Cones* 3 to 6 in. (8 to 15 cm) long, asymmetric, clustered, narrowly top-shaped, knobby, and serotinous, the umbo with a short prickle; *seeds* with a narrow wing.

Range. Largely confined to the mountainous regions of southwestern Oregon and south along the coast ranges and Sierra Nevada to central (sporadically southern) California. Elevation: 1000 to 2000 ft northward, 1500 to 4000 ft southward. It frequents poor soils of dry slopes, rocky spurs, and sun-baked ridges where it may be in pure stands (especially in the north) or mixed with sugar, ponderosa, and Coulter pines and several chaparral oaks.

General Description. This is a small- to medium-sized tree, 30 to 80 ft tall and to $2\frac{1}{2}$ ft dbh. It is intolerant, fast growing, and short lived. Its seeding habits are similar to those of lodgepole pine. Following a fire, scores of viable seeds are released, and the scorched earth is soon again covered with a green mantle of seedlings. Generally, it occurs in pure stands.

The serotinous cones may remain closed for 20 or more years and sometimes become embedded in the expanding trunk. It hybridizes with *P. muricata* and *P. radiata*.

Pìnus muricàta D. Don   bishop pine

Distinguishing Characteristics. *Needles* 3 to 6 in. (8 to 15 cm) long, 2 per fascicle, thick, rigid, dark yellow-green, persisting 2 or 3 years. *Cones* small, 2 to 4 in. (5 to 10 cm) long, oblique, commonly borne in clusters of 3 to 5, and with stout, flattened, spurlike prickles, serotinous and persistent for several years.

Range. In seven widely separated coastal areas from northern California to northern Baja California. Elevation: sea level to 1000 ft. It grows on low hills and plains in the coastal fog belt. A preference is shown for swampy sites and peat bogs, although it also grows on much drier soils.

General Description. This is a medium-sized tree 40 to 80 ft tall and 2 to 3 ft dbh with a rounded or irregular crown. It is moderately tolerant, short lived, and fast growing. It often forms pure stands but never occurs in sufficient quantity to be of more than local importance. It hybridizes with *P. attenuata*.

Pìnus leiophýlla var. chihuahuàna (Engelm.) Shaw   Chihuahua pine

Distinguishing Characteristics. *Needles* $2\frac{1}{2}$ to 6 in. (6 to 15 cm) long, 3 per fascicle, stout, stiff, blue-green with white lines on all surfaces, persisting 2 years, differing from all other American yellow pines in that the fascicle sheaths are deciduous. *Cones* small, $1\frac{1}{2}$ to $2\frac{1}{2}$ in. (4 to 6 cm) long, armed with minute, often deciduous prickles, and unlike all other American pines, require three seasons to mature.

Range. This pine attains its best development in northern Mexico and is found in the United States only in scattered areas of southwestern New Mexico and east central and southeastern Arizona. Elevation: 5000 to 8200 ft. It occurs on dry, rocky ridges and slopes with other pines.

General Description. Chihuahua pine is a small- to medium-sized tree, 30 to 80 ft tall and 1 to 2 ft dbh with an open, spreading crown of upturned branches. It produces vigorous stump sprouts after logging, another feature rare among pines. The typical variety (var. *leiophýlla*), with 5 needles per fascicle, occurs in Mexico.

Pinus contórta Dougl. ex Loud.   lodgepole pine

Distinguishing Characteristics (Fig. 8-19)

Needles $\frac{3}{4}$ to 3 in. (2 to 8 cm) long, 2 per fascicle or rarely solitary, dark green to yellow-green, often twisted, persistent for 3 to 8 years.

Cones $\frac{3}{4}$ to $2\frac{1}{4}$ in. (2 to 6 cm) long, subcylindric to ovoid, often asymmetrical at the base, either opening at maturity (especially vars. *contorta* and *murrayana*) or remaining closed for many years; *apophysis* tawny to dark brown, flattened, or those toward the base knoblike; *umbo* dorsal, terminating in a long, recurved, often deciduous prickle; *seeds* ca. $\frac{1}{8}$ in. (2 to 3 mm) long, ovoid, reddish brown, often mottled with black; wings ca. $\frac{1}{2}$ in. (1 to 1.5 cm) long.

Twigs moderately stout, dark red-brown to nearly black; *buds* ovoid, slightly resinous.

Bark of coastal trees $\frac{3}{4}$ to 1 in. (2 to 2.5 cm) thick, deeply furrowed and transversely fissured, reddish brown to black, and superficially scaly; that on mountain trees ca. $\frac{1}{4}$ in. (6 mm) thick, orange-brown to gray, covered by thin, loosely appressed scales.

Range

Western (Map 8-4). Local disjuncts in the Black Hills of South Dakota and northeastern New Mexico. Elevation: sea level to 2000 ft along the coast, 1300 to 3000 ft inland northward, 7000 to 11,500 southward. It occurs in the high mountains in well-drained soils either with other conifers or often in pure stands; shore pine varieties occur in peat bogs, muskegs, and dry, sandy sites.

General Description

Lodgepole pine is a cosmopolitan tree of wide distribution through western North America. It has the broadest ecological amplitude of any conifer in North America. It hybridizes with *P. banksiana*. Three or four varieties of the species may be recognized (Critchfield, 1957; Kral, 1993).

Shore pine.   Two of the four varieties, *P. contorta* var. *contorta* and *P. contorta* var. *bolanderi* (Parl.) Vasey (shrubby), are collectively known as shore pine. Kral (1993) does not recognize var. *bolanderi*. The latter differs mainly from the former in its heavy, knobby, serotinous cones, its complete lack of needle resin canals, and its isolation. It occurs only on a few square miles of highly acidic soils in Mendocino County, California. In other respects, it is similar to the typical variety, *P. contorta* var. *contorta*. Shore pine is a small tree ordinarily 25 to 30 ft tall and 12 to 18 in. dbh. It is characterized by a short, often contorted bole and a dense, irregular crown of twisted branches, many of which extend nearly to the ground; the root system is deep and wide-spreading and includes a persistent taproot, even when growing in bogs or muskegs. The tree is one of the first to invade the peat bogs of Alaska and British Columbia, as well as those of the Puget Sound basin in western Washington where it may form pure stands. Farther south it is found most abundantly on dry sandy and gravelly sites near the Pacific Ocean to northern California. Here it sometimes mingles with Sitka spruce and occasionally with grand fir. Because of their small size and poorly formed boles, the trees of the coastal form contribute little or nothing to the nation's timber supply. Large stands occasionally retard the migration of sand dunes, but smaller ones have been completely buried by shifting sands.

FIGURE 8-19 *Pinus contorta*, lodgepole pine. (1) Fascicle of needles $\times\frac{3}{4}$. (2) Var. *latifolia*, mature seed cones $\times\frac{3}{4}$. (3) Var. *contorta*, open seed cone $\times\frac{3}{4}$. (4) Mature seed cone $\times\frac{3}{4}$. (5) Cluster of pollen cones $\times1$. (6) Young ovulate cone and young fascicles $\times1$. (7) Seeds $\times1$. (8) Bark (*Photo by H. E. Troxell*). (9) Bark of shore pine.

MAP 8-4 *Pinus contorta.*

Lodgepole pine. The two inland varieties are collectively known as "lodgepole pine."* One, Sierra lodgepole pine, *P. contorta* var. *murrayana* (Grev. & Balf.) Engelm., occurs in the Cascade, Blue, and Sierra Nevada Mountains. It is characterized by lightweight, symmetrical, nonserotinous cones that are shed soon after opening. It is the tallest and best-formed variety. Rocky Mountain lodgepole pine, *P. contorta* var. *latifolia* Engelm., on the other hand, is found in the Rocky Mountains and has hard, heavy, asymmetrical, serotinous cones. Aside from other differences in needle width, seedling color, and stomate shape, the two varieties are, in most gross characteristics, quite similar. The Rocky Mountain lodgepole pine is the principal source of lodgepole pine timber.

Lodgepole pine is a medium-sized tree 70 to 80 ft tall and 15 to 30 in. dbh (max. 150 by 7 ft), with a long, clear, slender, cylindrical bole and short, narrow, open crown. Best development is attained on a moist but well-drained sandy or gravelly loam, although trees reach commercial proportions on a variety of soil types. Unlike the shore pine, which is seldom found far from tidewater, the lodgepole pine occurs from 1300 to 11,500 ft elevation in either pure dense even-aged stands or in mixture with several other conifers. At the lower limits of its elevational range, its associates are ponderosa and other western pines, Douglas-fir, and western larch. At higher levels it is found chiefly with Engelmann spruce, subalpine fir, and limber pine in the Rockies, and with limber pine, Jeffrey pine, and California red fir in the Sierra Nevada.

Rocky Mountain lodgepole pine is the most wide-ranging and commercially utilized variety, and it is the provincial tree of Alberta. It is one of the most aggressive and hardy of western forest trees, and under favorable conditions it is capable of fully restocking cutover lands in a remarkably short time. Following fire, it quickly forms dense stands and occasionally usurps areas formerly occupied by Douglas-fir, or at higher levels by Engelmann spruce.

The gregarious habit of this species is traceable to a number of factors. The trees are prolific seeders and often produce fertile seed before they are 10 years of age. Heavy seed crops occur at intervals of 2 or 3 years, but instead of releasing all of the seeds at maturity, many of the cones, especially in the Rocky Mountains, remain closed and attached to the branches for as long as 15 to 20 or more years. When the cones remain closed, large quantities of seed are gradually accumulated. The heat of a forest fire sweeping through the stands starts the opening of the cones. After the fire has passed, the scales open fully and release their seeds upon the freshly exposed mineral soil. The subsequent reproduction is often so dense that it quickly stagnates. Under normal conditions, growth is rather slow but persistent, and maturity is attained in about 200 years with a maximum of 500 to 600 years. Trees 100 years of age in the Blue Mountains of southern Oregon average 70 to 80 ft in height and 12 in. dbh, whereas trees of the same age in the Sierras are 90 to 100 ft tall and 15 to 18 in. dbh. Lodgepole pine is rated as intolerant.

Bark beetles (especially the mountain pine beetle) inflict heavy damage in stagnated stands. Dwarf mistletoe causes severe damage, and Comandra blister rust is a serious fungal disease.

---

*So-called because of its use for poles by the Plains Indians. The lodge, or tipi, with its movable smoke flaps to control ventilation and its symbolic decorations, is perhaps the most functional and beautiful dwelling ever designed by nomadic people.

Because of the thin bark of the tree, fire is always a serious menace even though heavy new stands result. Grewal (1987) summarizes over 1400 references on lodgepole pine.

  Pinus contorta and the next three species are closely related, 2-needle pines, and constitute subsection Contortae. Their distribution forms a broad, inverted U shape—*P. contorta* in the West, *P. banksiana* in the North, *P. virginiana* in the Northeast and East, and *P. clausa* in the Southeast. Although *P. contorta* and *P. banksiana* are the only pair that are sympatric (in Alberta, Canada) and hybridize naturally, successful *P. virginiana* × *clausa* crosses can be made artificially.

Eastern Hard Pines

There are 15 hard or "yellow" pines in eastern United States, nine of which are the most important (Table 8-7). Bark beetles are the most serious pests of the southern yellow pines. Southern pine beetles (*Dendroctonus frontalis*) is the most damaging of this group of beetles.

Pìnus banksiàna Lamb.   jack pine

Distinguishing Characteristics (Fig. 8-20)

  Needles $\frac{3}{4}$ to 2 in. (2 to 5 cm) long, in fascicles of 2 (also 3 near the tip of the leader on young trees), yellow-green, flat or slightly concave on the inner surface, divergent, stout, often twisted; persistent for 2 or 3 years.

TABLE 8-7   COMPARISON OF IMPORTANT EASTERN HARD PINES

Species	Needles	Seed cones
P. banksiana jack pine	2 per fascicle; $\frac{3}{4}$–2 in. (2–5 cm) long, stout, rigid, often twisted, divergent	Oblong-conic, 1–2$\frac{1}{4}$ in. (2.5–5.5 cm) long, short-stalked; umbo mostly unarmed; serotinous and persistent
P. virginiana Virginia pine	2 per fascicle; $\frac{3}{4}$–3 in. (2–8 cm) long, stout, strongly twisted	Ovoid, 1$\frac{1}{4}$–2$\frac{1}{2}$ in. (3–7 cm) long, short-stalked; umbo with long, slender prickle; persistent, not serotinous
P. clausa sand pine	2 per fascicle; 2–3$\frac{1}{2}$ in. (5–9 cm) long, slender, slightly twisted	Narrowly ovoid, 1$\frac{1}{4}$–3$\frac{1}{2}$ in. (3–9 cm) long, short-stalked; umbo with short, stout prickle; serotinous or not
P. resinosa red pine	2 per fascicle; 4–6$\frac{1}{4}$ in. (10–16 cm) long, slender, brittle	Ovoid, 1$\frac{1}{2}$–2$\frac{3}{8}$ in. (4–6 cm) long, short-stalked; umbo unarmed
P. palustris longleaf pine	3 per fascicle; 8–18 in. (20–46 cm) long, stout, flexible; buds with silvery-white fimbriate scales	Conical-cylindrical, 6–10 in. (15–25 cm) long, sessile; umbo with small prickle
P. taeda loblolly pine	3 per fascicle; 4–9 in. (10–23 cm) long, stout, rigid	Conical-ovoid, 2$\frac{1}{2}$–6 in. (6–15 cm) long, short-stalked; umbo with stout prickle; some persisting
P. echinata shortleaf pine	2 and 3 per fascicle; 2$\frac{3}{4}$–4$\frac{3}{4}$ in. 7–12 cm) long, slender, flexible	Narrowly oblong, 1$\frac{1}{2}$–2$\frac{1}{2}$ in. (4–6 cm) long, short-stalked; umbo with small prickle; persistent
P. rigida pitch pine	3(4) per fascicle; 3–6 in. (7.5–15 cm) long, twisted, rigid	Ovoid, 1$\frac{1}{4}$–2$\frac{3}{4}$ in. (3–7 cm) long, nearly sessile; umbo with slender prickle; persistent
P. elliottii slash pine	2 and 3 per fascicle; 6–9 in. (15–23 cm) long, stout, rigid; buds with red-brown, white-ciliate scales	Ovoid, 2$\frac{1}{2}$–7 in. (6–18 cm) long, short-stalked; umbo with short, stout prickle; persistent for 1 year

FIGURE 8-20   *Pinus banksiana,* jack pine. (1) Male cones shedding pollen ×1. (2) Mature seed cones and foliage ×$\frac{1}{2}$. (3) Open seed cone and foliage ×$\frac{3}{4}$. (4) Seed ×1. (5) Fascicle of needles ×$\frac{3}{4}$. (6) Cross section of needle ×35.

Cones 1 to $2\frac{1}{4}$ in. (2.5 to 5.5 cm) long, oblong-conic, sessile, light brown, usually pointing forward, often strongly incurved, with the scales well developed only on the outer face; opening tardily and often persistent for many years; *apophysis* rounded, smooth; *umbo* dorsal, armed with a minute prickle; *seeds* ca. $\frac{1}{8}$ in. (3 to 5 mm) long, triangular, black, and roughened, with wings ca. $\frac{3}{8}$ in. (8 to 10 mm) long.

Bark thin, brown slightly tinged with red, or dark gray; irregularly divided into scaly ridges.

Range

Northern (Map 8-5). Southern disjuncts in Maine, New Hampshire, Vermont, northern New York, northern Indiana, and northern Illinois. Elevation: sea level to 2600 ft. It occurs on well-drained, sterile, acid, sandy soils of dunes and sand ridges; also rock outcrops, woodlands, and savannas, often in pure stands.

MAP 8-5 *Pinus banksiana.*

General Description

Jack pine is essentially a Canadian species that reaches its best development north and west of Lake Superior. It is the most widely distributed pine in Canada and extends farthest north of any North American pine. It extends southward into the United States in northern New England and the lake states. Jack pine is one of the most important pulpwood species within its range. Under the most favorable conditions, this pine varies from 70 to 80 ft in height and 12 to 15 in. dbh (max. 100 by 2.4 ft), but is usually much smaller. In the open, an irregular rounded crown is produced, the lower branches of which soon die but persist on the trunk for many years; this gives the tree a scraggly appearance unlike that of other conifers of the region (Fig. 8-21). When grown in dense stands, the crown is smaller, but the dead lower branches still prune very poorly. The root system, at least after the first few years, is wide-spreading and only moderately deep.

Serotinous cones are typical of this species, lodgepole pine, knobcone pine, and a few others. In jack pine, there is considerable variation from one tree to another in the degree of serotiny observed. On some trees, nearly all mature cones remain closed; on others, most of them open. At some locations, especially near its range limits, relatively few cones are serotinous. Serotiny is more common in regions subjected to naturally occurring, stand-replacing fires. When closed, cones are subjected to temperatures of 140°F; they subsequently open. Cones extracted with alcohol-benzene also open, and it is presumed that resins and perhaps other substances "glue" the scales together. Scales treated in this way, or whose bonds have been fractured mechanically, behave normally, opening when dried, closing when wet.

Jack pine is very intolerant and occurs in pure stands or open mixtures with quaking aspen and paper birch on dry, sandy, acidic, infertile soils too poor for such species as red pine or eastern white pine, with which, however, it is often found on the better sites. In fact, much of the land formerly occupied by these two species in the Great Lakes region supports pure stands of jack pine that originated after logging and fire had destroyed the layer of humus, leaving only bare sand. This species serves as a valuable pioneer tree on such areas, but, except on the very poorest, it is eventually replaced by red pine or white pine. On better sandy loam soils, these species may in

FIGURE 8-21   Jack pine. (*Courtesy of U.S. Forest Service.*)

turn yield to hardwoods such as sugar maple, northern red oak, and basswood. Further north, in the boreal forest, black spruce is a common associate of jack pine. Jack pine even occurs on rocky outcrops and in bogs.

Like the lodgepole pine of the west, which it resembles in many ways, jack pine is very prolific and begins to produce yearly crops of cones at an early age (3 to 15 years) with larger amounts at

3- or 4-year intervals. Cones that would remain closed until heat is applied are often engulfed by the trunk or branch to which they are attached. This habit of retaining so many unopened cones explains the rapidity of natural seeding after a fire has swept through a jack pine forest. Interestingly, at the peak of the conflagration, when the cones are on fire and the resin oozing from between the scales is blazing, they do not open (at least not fully). Some time later, when the crown fire has passed, the scales open, and the seeds spiral down or are windblown and land in the cooled ashes.

Perhaps because of its extensive range, jack pine can be host to a very large number of insects and diseases. Jack pine is often impacted by the spruce budworm.

Young trees make rapid growth, usually developing, like pitch pine, two or three false whorls of branches each year. Jack pine matures in about 60 years and subsequently begins to deteriorate, although trees 150 years of age or older have been found (maximum 230 years).

Jack pine is the territorial tree of the Northwest Territories. It hybridizes with *P. contorta* in areas of sympatry. Besides its importance as a pulpwood species, jack pine lumber is used for boxes, crates, and shipping containers. The federally listed bird, the Kirtland's warbler, depends on young stands of jack pine in central Michigan for their breeding habitat.

Pìnus virginiàna Mill.   Virginia pine

Distinguishing Characteristics (Fig. 8-22)

Needles $\frac{3}{4}$ to 3 in. (2 to 8 cm) long, in fascicles of 2, yellow-green to gray-green, twisted, often divergent, persistent until the third or fourth year.

Twigs at first green, becoming purplish glaucous by the end of the season and through the first winter.

Cones $1\frac{1}{4}$ to $2\frac{3}{4}$ in. (3 to 7 cm) long, usually sessile, ovoid-conic, usually not persistent for more than 3 or 4 years, the scales thin; *apophysis* reddish brown; *umbo* dorsal and armed with a slender prickle; *seeds* ca. $\frac{1}{4}$ in. (5 to 7 mm) long, oval, light brown, wings terminal, usually broadest near the middle, ca. $\frac{3}{8}$ in. (7 to 9 mm) long.

Bark thin and smooth, eventually scaly-plated, reddish brown.

Range

Eastern (Map 8-6). Local disjuncts in northeastern Mississippi, central Alabama, and eastern North Carolina. Elevation: 50 to 2800 ft. Well-drained sites in mixed forest or in pure stands; often in old abandoned fields, after fires, or in poor or severely eroded areas.

General Description

Virginia pine is usually a small, often unkempt-appearing tree about 40 ft tall and 12 in. dbh (max. 120 by 3 ft), with persisting side branches and a shallow root system. This pine is found on a wide variety of soils, but seems to do best on clay, loam, or sandy loam. It is intolerant and is commonly

MAP 8-6   *Pinus virginiana.*

FIGURE 8-22 *Pinus virginiana,* Virginia pine. (1) Open seed cone and foliage ×$\frac{3}{4}$. (2) Seed ×1. (3) Cross section of needle ×35. (4) Bark.

found as a pioneer tree and eventually replaced by hardwoods; in the meantime, it may produce more pulpwood per acre than other pines of the region on such adverse sites. Once considered a "forest weed," it is now considered important due to its rapid spread over extensive areas of neglected and abandoned agricultural land, spoil banks, and strip mines. The tree is also being grown in plantations for pulpwood, lumber, and Christmas trees. It is damaged by heart rot, pitch canker, and a variety of insects.

Pìnus claùsa (Chapm. ex Engelm.) Vasey ex Sarg.   sand pine

Distinguishing Characteristics.   *Needles* 2 to $3\frac{1}{2}$ in. (5 to 9 cm) long, 2 per fascicle, slender, flexible, persisting 2 to 3 years. *Cones* persisting, open or closed for several years, $1\frac{1}{4}$ to $3\frac{1}{2}$ in. (3 to 9 cm) long, narrowly oblong-conic, the scales armed with a short, stout prickle. *Bark* gray, relatively smooth.

Range.   Southern Georgia to central Florida, west to southern Alabama. Elevation: sea level to 200 ft. Deep, well-drained sands of ridges, coastal dunes, and flatwoods, often in pure stands.

General Description.   Sand pine is generally a small tree reaching 30 to 70 ft tall and with a trunk 1 to $1\frac{1}{2}$ ft dbh (max. 103 by 2 ft). It is cut for pulpwood and grown for Christmas trees.

Two varieties have been recognized: *P. clausa* var. *clausa* has mostly serotinous cones and occurs in dense, even-aged stands in northeastern to southern peninsular Florida and infrequently in northwestern Florida; *P. clausa* var. *immuginata* D. B. Ward is called the Choctawhatchee sand pine and differs by having the cones consistently open at maturity. It occurs in open, uneven-aged stands of northwestern Florida and southern Alabama. The two occur intermixed in some localities (Ward, 1963). The two varieties are not recognized by Kral (1993).

Pìnus resinòsa Ait.   red pine, Norway pine

Distinguishing Characteristics (Fig. 8-23)

Needles 4 to $6\frac{1}{4}$ in. (10 to 16 cm) long, in fascicles of 2, serrulate, dark yellow-green, straight, brittle and breaking cleanly when doubled between the fingers, persistent until fourth or fifth season.

Cones $1\frac{1}{2}$ to $2\frac{3}{8}$ in. (4 to 6 cm) long, ovoid-conic, subsessile; *apophysis* chestnut-brown, rounded; *umbo* dorsal and unarmed; *seeds* ca. $\frac{1}{4}$ in. (5 to 7 mm) long, somewhat mottled, with wings $\frac{1}{2}$ to $\frac{3}{4}$ in. (12 to 18 mm) long, oblique, chestnut-brown.

Twigs orange-brown, lustrous; *buds* covered with thin, ragged, orange-brown or reddish brown scales, often grayish on the margin.

Bark on young trees flaky, orange-red, eventually breaking up into large, flat, reddish brown, superficially scaly plates irregularly diamond-shaped in outline.

FIGURE 8-23 *Pinus resinosa,* red pine. (1) Fascicle of needles $\times\frac{3}{4}$. (2) Cross section of needle $\times$35. (3) Open seed cone $\times\frac{3}{4}$. (4) Mature seed cone $\times\frac{1}{2}$. (5) Seed $\times$1. (6) Bark of old tree.

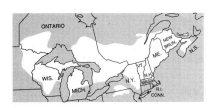

MAP 8-7 *Pinus resinosa.*

Range

Northeastern (Map 8-7). Local disjuncts in Newfoundland, northern New Jersey, eastern West Virginia, northern Illinois, and southeastern Minnesota. Elevation: 650–1400 ft northward, to 2700 ft in the Adirondacks, 3100–4200 ft in West Virginia. Well-drained soils of flatwoods, ridges, outcrops, and dunes, either in mixed or pure stands.

General Description

Red pine is one of the most distinctive of northern conifers (Fig. 8-24). Even at a considerable distance, the symmetrically oval crown with its tufted dark-green foliage appears very different from that of the ragged, unkempt jack pines or the plumelike tops of the eastern white pine. The bole of red pine is well formed, long, and cylindrical, and it is supported by a spreading root system and poorly developed taproot. Although attaining a maximum size of 116 ft tall and 5 ft dbh, most trees vary from 50 to 80 ft in height and 2 to 3 ft dbh.

FIGURE 8-24 Old-growth red pine left after logging.

Best development in the United States was made in the upper Great Lakes region where magnificent pure and mixed stands of this species occurred. On light-acid sandy soils too poor for white pine, red pine grew in abundance, and on the better sites was mixed with the former species; occasional trees, often of large size, were found in hardwood mixtures on heavier soils. Following logging and fire, much of the land that had supported red pine became too poor for anything but jack pine, which over large areas established itself as a pioneer tree. In a similar way, red pine occupied former eastern white pine land.

In its soil and moisture requirements and ability to grow under forest competition, red pine is intermediate between jack pine and eastern white pine. Arranged in order of tolerance, eastern white pine is first and jack pine third, whereas in adaptability to dry sandy soils, this order is reversed. On cutover land, red pine appears to follow jack pine and is often found in mixture with it. Both trees experience great variations in seasonal temperatures, with extremes of -40 to -60°F in winter to 90 to 105°F in summer. The growing season for red pine is from 80 to 160 days.

Seeds may be produced in small quantities each year, but good crops occur only at intervals of from 3 to 7 years. Growth of established unshaded young trees is fast, and during the first two decades they will often outstrip those of eastern white pine. Some 30-year plantations of red pine in New England contain trees with a height of 35 ft and a diameter of 6 in. The maximum age for the species is about 350 years.

Red pine is grown extensively for reforestation purposes and used for lumber, poles, cabin logs, pulp, and Christmas trees. It is also important in ornamental plantings, particularly on light sandy soils. Compared with many other pines, it is lacking in genetic diversity. It occurs mainly in highly fragmented, small populations. At one time, it was the most important timber producer in the Great Lakes area. It is the state tree of Minnesota.

This species is sometimes attacked by the Nantucket pine-tip moth, the European pine-shoot moth, sawflies, red pine scale, red pine adelgid, and Scleroderris canker. Red pine, growing on acidic soils, is remarkably free of serious pathogen damage.

Pinus palústris Mill.   longleaf pine

Distinguishing Characteristics (Fig. 8-25)

Needles 8 to 18 in. (20 to 46 cm) long, in fascicles of 3 or rarely 4 or 5, bright green, densely tufted at the ends of stout branch tips, persistent until the end of the second season; fascicle sheath to $1\frac{3}{8}$ in. (3.5 cm) long.

Cones 6 to 10 in. (15 to 25 cm) long, narrowly ovoid-cylindric, sessile or nearly so, usually leaving a few of the basal scales attached to the twig when shed; *apophysis* reddish brown, weathering to an ashy gray, wrinkled; *umbo* armed with a dorsal prickle that curves toward the base of the scale; *seeds* ca. $\frac{1}{2}$ in. (1 to 1.5 cm) long, somewhat ridged, pale with dark blotches; wings ca. $1\frac{3}{4}$ in. (4 to 5 cm) long, striped, oblique at the ends; cotyledons 5 to 10.

Twigs stout, orange-brown; *buds* large and very conspicuous, covered with silvery white, fringed scales.

Bark rough, black when young and with large reddish plates when old.

Range

Southeastern (Map 8-8). Elevation: sea level to near 2200 ft in northern Alabama. Well-drained, sandy ridges and flatwoods to wetter flatwoods and savannas, often in pure stands.

General Description

Longleaf pine is one of the most distinctive and important southern conifers. However, only fragments remain of what was one of the most extensive upland forest types in the southeast. It is estimated that only 3% of the original presettlement 90 million acres remain in the late 1990s, and nearly all this is second growth. Original longleaf pine stands were heavily impacted by naval stores, grazing, disruption of the natural fire regime, logging, and more recently by loss of available habitat due to agriculture, development, and conversion to loblolly plantations. There is now increasing interest and effort in preserving, restoring, and reintroducing longleaf pine forests on public and private lands through both natural and artificial regeneration. Natural longleaf pine stands harbor an extraordinarily high number of vascular plant species including many endemics (i.e., they are among the most species-rich communities outside the tropics). Of all the southern pines, it is most resistant to fire, insects, and disease (except brown-spot needle blight).

Although it grows best on deep well-drained sands, this species is found on a variety of sites from dry, sterile ridges to wet, low flatwoods, and in sand, loam, or clay. On the poorest sites where a hardpan is near the surface, it maintains itself, even though growth is very slow. Flat land of this character is poorly drained and in spring may be covered with shallow pools of standing water

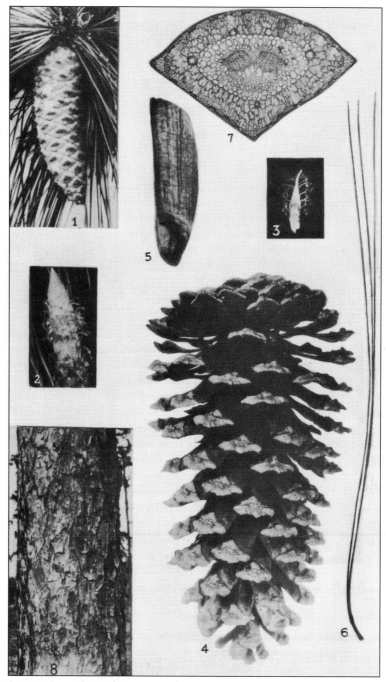

FIGURE 8-25 *Pinus palustris,* longleaf pine. (1) Mature seed cone and foliage $\times\frac{1}{2}$. (2) Terminal bud $\times\frac{3}{4}$. (3) Fringed bud scale $\times1\frac{1}{2}$. (4) Open seed cone $\times\frac{3}{4}$. (5) Seed $\times1$. (6) Fascicle of needles $\times\frac{1}{2}$. (7) Cross section of needle $\times35$. (8) Bark (*Photo by J. C. Th. Uphof*).

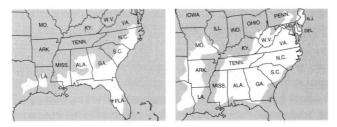

MAP 8-8   **(Left)** *Pinus palustris.* **(Right)** *Pinus echinata.*

(hence possibly the origin of the name *palustris* [of marshes] to longleaf pine). During the summer, however, such areas are exceedingly dry, and only trees like blackjack oak or turkey oak are seen scattered through the open nearly pure forests of pine. On these shallow soils underlain by hardpan, the taproot is poorly developed, and wind-thrown trees are common. As previously indicated, best growth is made on better-drained sandy soils. This growth habit is especially common in the southern part of its range where longleaf often occupies low ridges or knolls while slash or pond pines preempt the more moist depressions. It can reach 80 to 100 ft tall and with a dbh of 2 to $2\frac{1}{2}$ ft.

Longleaf is very intolerant but regenerates well under mature trees after fires because pure stands are typically open, with a small accumulation of needles or short grasses on the forest floor. Frequent summer fires, most probably caused by lightning strikes, are common in the range of the species and are thought to have been the primary driving force for development of the reproductive system of longleaf. Seed crops are irregular, with exceptionally high yields every 5 to 7 years. Seedling survival and growth are enhanced by germination on bare mineral soil. Under natural conditions, germination takes place within 2 to 5 weeks after the seeds are released. Fall germination, characteristic of longleaf, is rare in the other southern pines.

It is the nature of longleaf pine to develop very little above ground for the first 3 to 6 years (exceptionally, 20 or more years), but during this time the roots get a firm foothold. Moreover, the young tree does not exhibit annual growth rings but seems to grow intermittently when conditions permit. A dense bunch of green needles is all that appears on the soil surface, and to the inexperienced eye this is often mistaken for grass (hence the so-called grass stage of longleaf pine [Fig. 8-26]). By the end of this period, the root system is well developed, but the aerial stem shows practically no elongation.[*]

[*]This unusual growth habit is also found in South Florida slash pine and Apache pine.

FIGURE 8-26   "Grass stage" of longleaf pine.

Height and diameter growth then increase rapidly, and trees 25 years old average 45 ft in height and 6 in. dbh. Annual height growth declines markedly after 40 to 50 years. Even at maturity longleaf pine is not often a large tree. Maturity is reached in about 150 years, but old trees sometimes show more than 300 annual rings (Fig. 8-27).

Two-year-old seedlings, so well insulated by the dense cluster of needles, are highly fire-resistant. So too is the thick-barked mature tree. Only during the first 2 or 3 years of height growth is fire a threat. Some ecologists believe the species to be a fire climax throughout much of its range. Observations of hardwood encroachment and failure of longleaf regeneration where fire is excluded support this conclusion. Although longleaf pine is not considered to propagate by sprouting, one reference indicates that some 4-year-old trees cut off just above the root collar produced vigorous sprouts (Garin, 1958). Longleaf pine seedlings may be stunted or killed by brownspot disease. Lightning, followed by bark beetle attack are the most common damaging agents.

FIGURE 8-27   Longleaf pine stand with an understory of shrubs and scrub oaks as a result of excluded fire. (*Courtesy of U.S. Forest Service.*)

This pine furnishes high-quality lumber, pulp, chipped bark, and pine straw. Mature longleaf stands provide the most desirable habitat for the federally listed (endangered) red-cockaded woodpecker, and considerable effort is expended and regulations defined to manage the ecosystem for this purpose. It is the state tree of Alabama. It hybridizes with *P. taeda,* forming *P.* ×*sondereggeri* H. H. Chapm., the only named southern pine hybrid. Much additional information can be found in Farrar (1990) and Landers et al. (1995).

Pìnus echinàta Mill.   shortleaf pine

Distinguishing Characteristics (Fig. 8-28)

Needles $2\frac{3}{4}$ to $4\frac{3}{4}$ in. (7 to 12 cm) long, mostly in fascicles of 2, but also in threes on the same tree (occasionally with the latter number predominant), dark yellow-green, slender, flexible, persisting 3 to 5 years, not infrequently occurring from dormant buds along the bole.

Cones $1\frac{1}{2}$ to $2\frac{1}{2}$ in. (4 to 6 cm) long, ovoid-oblong to conical, nearly sessile, usually persistent for several years; *apophysis* reddish brown, rounded; *umbo* dorsal and armed with a small sharp, straight or curved, sometimes deciduous prickle; *seeds* ca. $\frac{3}{16}$ in. (4 to 6 mm) long, brown, with black markings; wings ca. $\frac{1}{2}$ in. (8 to 12 mm) long, broadest near the middle.

Twigs at first green and tinged with purple, eventually reddish brown and becoming flaky in 3 to 4 years; buds with red-brown scales.

Bark on small trees nearly black, roughly scaly, with small surface pockets or holes; later reddish brown and broken into irregular flat plates, scaly on the surface; phellogen layers ivory-white.

Range

Southeastern (Map 8-8). Elevation: 650 to 3000 ft. Dry, rocky, mountain ridges, old fields, and floodplains, in mixed or pure stands.

General Description

Shortleaf pine is a medium-sized to large tree 80 to 100 ft in height and 2 to 3 ft dbh (max. 146 by 4 ft). The clear well-formed bole supports a small narrowly pyramidal crown and terminates underground in a very deep taproot. Its root system is usually larger than that of the other southern pines.

Although found on many different sites, this species is most common in pure or mixed stands on dry upland soils that are neither highly acid nor strongly alkaline. Its associates include loblolly and Virginia pines, eastern redcedar, black, blackjack, post, and chestnut oaks, and mockernut hickory. In addition, especially on soils containing more moisture, bitternut hickory and sweetgum are included. West of the Mississippi River, shortleaf and longleaf pines are mixed and often attain maximum development together.

In many localities where the ranges of shortleaf and loblolly pines overlap, they are mixed together on dry soils, and because the needles and cones of the latter become somewhat dwarfed under such conditions, identity of the two species may be difficult. Generally, however, these two trees tend to be complementary, with loblolly pine favoring the heavier, wetter soils, and shortleaf pine on the lighter and drier soils. Its best growth is made on well-drained alluvial soils along streams and small rivers.

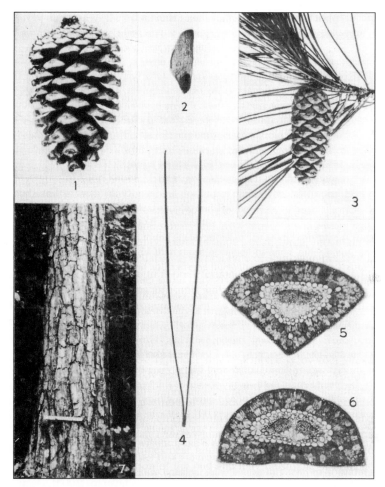

FIGURE 8-28 *Pinus echinata,* shortleaf pine. (1) Open seed cone $\times\frac{3}{4}$. (2) Seed $\times1$. (3) Mature seed cone and foliage $\times\frac{1}{2}$. (4) Fascicle of needles $\times\frac{3}{4}$. (5, 6) Cross sections of needles (3 and 2 per fascicle) $\times35$. (7) Bark of old tree.

Shortleaf pine may produce some seed nearly every year (after the fifth to twentieth), with good to excellent amounts at 3- to 10-year intervals. In terms of food sources for wildlife, shortleaf pine seed production can vary from 2 to 20 lbs of seeds per acre per year. Growth of the young trees is moderately rapid, and as in several of the other southern pines, the terminal shoots are multinodal, developing from two to four false whorls of side branches each year; false growth rings are often concomitant and may be mistaken for true seasonal rings. On good sites, trees 35 years old may be 60 ft tall and 8 in. dbh; those 60 years old can be 80 ft tall and nearly 12 in. dbh.

A remarkable feature of shortleaf pine trees up to 8 or 10 years old is their ability to sprout after their main stems have been destroyed by either fire or cutting. This ability results from the same peculiar juvenile habit mentioned for pitch pine. As reported by Stone and Stone (1943), the stem of shortleaf seedlings may become prostrate as shoot growth begins. It then turns upward from just above the cotyledons. This produces the second part of a double crook and leaves the cotyledonary part of the stem near the ground. If the main axis of the young tree is destroyed, dormant buds in the axils of primary leaves above the cotyledons start to grow and develop new sprouts.

This species may be less tolerant than loblolly pine, but it will endure suppression for many years and shows greatly accelerated growth when released; it also surpasses longleaf pine in this respect. Maturity is reached in about 170 years, whereas very old trees may reach 400.

Young trees are damaged by the Nantucket pine-tip moth, and the southern pine beetle is a serious enemy. The greatest threat is posed by the littleleaf disease. Its high resistance to the damaging fusiform rust is unique among the southern pines.

Shortleaf is used for lumber, plywood, and pulp. It hybridizes with *P. glabra, P. rigida, P. serotina,* and *P. taeda.*

Pìnus taèda L.   loblolly pine

Distinguishing Characteristics (Fig. 8-29)

Needles 4 to 9 in. (10 to 23 cm) long, in fascicles of 3 (occasionally 2), slender but somewhat stiff, yellow-green, sometimes twisted, persisting 3 years.

Cones $2\frac{1}{2}$ to 6 in. (6 to 15 cm) long, ovoid-cylindric to narrowly conical, sessile; *apophysis* rather flattened, wrinkled; *umbo* dorsal and armed with a stout sharp prickle; *seeds* ca. $\frac{1}{4}$ in. (6 to 9 mm) long, dark brown, roughened, with black markings; wings ca. $\frac{3}{4}$ in. (18 mm) long, usually broadest above the center.

Twigs yellow-brown or reddish brown; *buds* covered with reddish brown scales, free at the tips and not fringed.

Bark variable, on young trees scaly and nearly black, later $\frac{3}{4}$ to 2 in. (1.8 to 5 cm) thick, with irregular, brownish blocks, or on very old trees, with reddish brown scaly plates similar to those of shortleaf pine or even red pine; phellogen layers slate-gray.

Range

Southeastern (Map 8-9). Elevation: near sea level to 2300 ft. Deep, poorly drained flood-plains and old fields to well-drained slopes in mixed or pure stands.

General Description

Loblolly pine, the leading commercial timber species in the southern United States, is a medium-sized to large tree 90 to 110 ft in height and 2 to $2\frac{1}{2}$ ft dbh (max. 163 by 6.8 ft). The bole is long and cylindrical, and the crown, although open, is denser than that of the other southern pines. In youth, a short taproot is developed, but except on the driest of soils, it soon ceases growth in favor of an extensive lateral root system.

This species grows on a very wide variety of soils but does best on those with deep surface layers having plenty of moisture but poor surface drainage and on fine-textured subsoils. On such Coastal Plain sites, pure stands are extensive, especially on river bottoms; however, pure stands also develop even on the drier soils of the Piedmont and inland areas. Because this pine occurs on so many different sites from wet to dry, the list of associated species is large, including the other important southern pines. Loblolly-hardwood mixtures are also common and include sweetgum, oaks, hickories, and many others. This pine is less tolerant than its hardwood associates but more tolerant than longleaf and slash pines.

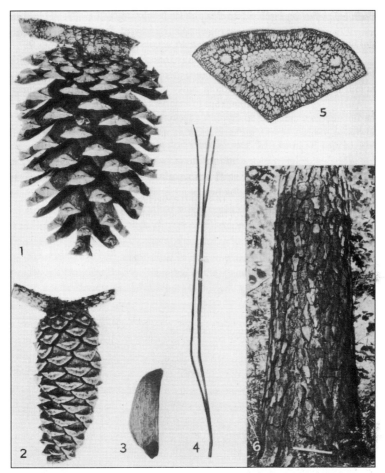

FIGURE 8-29 *Pinus taeda,* loblolly pine. (1) Open seed cone $\times\frac{3}{4}$. (2) Mature seed cone $\times\frac{1}{2}$. (3) Seed $\times 1$. (4) Fascicle of needles $\times\frac{1}{2}$. (5) Cross section of needle $\times 35$. (6) Bark of old tree.

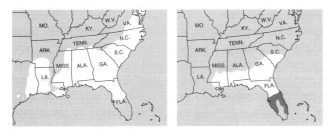

MAP 8-9   **(Left)** *Pinus taeda.* **(Right white)** *Pinus elliottii.* **(Right dark gray)** *Pinus elliottii* var. *densa.*

On cutover or abandoned farmlands in the South, this early successional species has spread to a remarkable degree and is especially aggressive in forming pure stands on old fields (hence the local name of "old field pine"). However, most commercial stands are of planted origin. In the South, nearly 2 billion seedlings are planted each year. Additional loblolly pines are planted on severely degraded sites for restoration purposes. Loblolly pine is an excellent tree for the southern landscape, and old-growth loblolly pine stands are very important to the existence of the federally listed (endangered) red-cockaded woodpecker.

Seeds may be produced on 5- to 10-year-old trees and are abundant after about the twenty-fifth year. When 1- to 3-year seedlings are cut off, they sprout, but older ones do not retain this ability. Especially on old fields, open-grown young trees make rapid growth, frequently averaging 3 ft in height and $\frac{1}{2}$ in. or more dbh per year for the first decade, and annual rings nearly 1 in. wide are sometimes produced. Trees 25 years old may be nearly 70 ft tall and 8 to 10 in. dbh; those 70 years of age can be about 90 ft tall and 24 in. dbh. Loblolly pine matures in 80 to 100 years, whereas old trees may exceed 300.

The problem in producing more loblolly pine in plantations is to extend its original range and to control competing hardwoods, which continually invade the pine stands. Besides its importance for lumber, this is a major species in paper making. More than half of U. S. wood pulp comes from the southern pines, with a large percentage from loblolly.

Widespread pests include the fusiform rust, black root rot, other fungi, and the full range of pine bark beetles, especially the southern pine beetle.

Loblolly pine hybridizes naturally with longleaf, shortleaf, pitch, slash, and pond pines. For a listing of some hybrids additionally produced by controlled pollination, see Burns and Honkala (1990). Much of the forest-tree genetic research in the South seeks to improve the growth and quality of this species. Such efforts are having a significant positive impact on southern forestry. For the definitive summary on loblolly pine biology, ecology, and management, see Schultz (1997).

Pìnus glàbra Walt.   spruce pine

Distinguishing Characteristics.   *Needles* $1\frac{1}{2}$ to 4 in. (4 to 10 cm) long, 2 per fascicle, very slender, flexible, persisting 2 to 3 years. *Cones* $1\frac{3}{8}$ to $2\frac{3}{4}$ in. (3.5 to 7 cm) long, similar to those of *P. echinata,* but more globose when open, and the scales nearly unarmed (small weak prickles). *Bark* not plated like other southern pines, finely furrowed, and with cross fissures, similar to that of southern red oak and spruce.

Range.   Eastern South Carolina to northern Florida, west to southeastern Louisiana. Elevation: sea level to 500 ft. Moist, lowland swamp forests in mixed stands.

General Description.   Spruce pine is a medium-sized tree 80 ft tall and 2 ft dbh (max. 125 by 4.2 ft). Nowhere abundant, it is more shade tolerant than most other yellow pines and rarely forms pure stands. It is usually seen as a scattered tree among hardwoods and loblolly pine, along stream banks or rich moist hammocks of the Coastal Plain. It is of very limited commercial value, although it is used sparingly for lumber, pulp, and Christmas trees. It forms hybrids with *P. echinata.*

Pìnus rígida Mill.   pitch pine

Distinguishing Characteristics (Fig. 8-30)

Needles 3 to 6 in. (7.5 to 15 cm) long, in fascicles of 3, yellow-green, stiff, usually somewhat twisted, standing out at nearly right angles to the twig, often produced on short "water sprouts" or in tufts along the trunk; mostly deciduous during the second season.

FIGURE 8-30 *Pinus rigida,* pitch pine. (1) Fascicle of needles ×$\frac{3}{4}$. (2) Cross section of needle ×35. (3) Open and mature seed cones ×$\frac{3}{4}$. (4) Seed ×1. (5) Bark of old tree.

Cones $1\frac{1}{4}$ to $2\frac{3}{4}$ in. (3 to 7 cm) long, nearly sessile, ovoid, usually persistent for many years and serotinous or not; *apophysis* light brown, smooth; *umbo* dorsal and armed with a rigid prickle; *seeds* ca. $\frac{1}{4}$ in. (4 to 6 mm) long, triangular to oval, dull black, sometimes mottled with gray, with brownish terminal wings.

Bark at first dark and very scaly; eventually 1 to 2 in. (2.5 to 5 cm) thick at the base of old trees and smoother with brownish yellow, flat plates separated by narrow, irregular fissures.

Range

Northeastern (Map 8-10). Local disjuncts in western Kentucky, northcentral North Carolina, and northcentral Alabama. Elevation: sea level to 2000 ft northward, 1400 to 4600 ft in the southern Appalachians. Slopes and ridges of the mountains, river valleys, and swamps elsewhere; found in mixed or pure stands.

MAP 8-10 *Pinus rigida.*

General Description

Naval stores, which the French harvested first in the Northeast, came originally from *P. rigida,* hence the common name. The industry then moved south to *P. palustris,* which was much more extensive and more resinous, then later further south to *P. elliottii* when *P. palustris* was all but depleted.

Pitch pine is a tree of great diversity in form, habit, and development. In the northern part of its range, through New England, it is commonly a small tree growing on the poorest of acidic, sandy, sterile soils with gray birch and scrub oak (*Q. ilicifolia* Wangenh.). Farther south, especially in Pennsylvania, it reaches its best development and varies from 50 to 60 ft in height and 1 to 2 ft dbh (max. 135 by 4 ft). Populations of dwarf trees (8 to 12 ft tall) occur in the New Jersey pine plains, where cone serotiny is common, and in West Virginia.

The form of the tree is exceedingly variable, and on exposed sites it is often very grotesque, whereas in better situations a tall columnar bole and small open crown are produced; the root system is deep, and the tree is relatively windfirm until overmature, when shallower roots predominate.

Although pitch pine is extremely hardy in maintaining itself on the driest, most unproductive sites, best growth is made on sandy loam soils with moderate amounts of moisture. Particularly near the coast (New Jersey), this species may be found on wet peat soils of Atlantic white-cedar swamps. Very little pitch pine is found on the more favorable areas because of its pronounced intolerance to forest competition, especially that of the hardwoods, which usually preempt the better sites. This pine is found naturally in fairly open, pure stands, or in mixture with such hardwoods as scarlet, black, and chestnut oak, various hickories, black tupelo, and red maple; best growth is attained in the hardwood mixture.

The production of cones begins very early, at 3 to 4 years, and 12-year-old trees often bear quantities of viable seeds. Good crops may occur at 4- to 9-year intervals; many of the cones remain unopened until midwinter and then gradually release their seeds upon the snow, in this way providing food for birds and small rodents. However, the time of seed release varies. On some trees, it occurs soon after the cones mature, but on others, the cones may be serotinous. A serotinous-cone ecotype occurs in New Jersey as a result of frequent fires. Serotiny in these cones affords the seeds protection from damaging high temperatures during the fires (Fraver, 1992).

The stems of seedlings often lie nearly flat on the ground. Subsequent growth to an upright position leaves the tree with a double curvature at the base. Evidence of this juvenile habit disappears as the tree becomes older.

On good sites, the growth of pitch pine can average 1 ft per year for the first 50 to 60 years. By age 90, height growth essentially ceases (max. age 450 yrs). Diameter growth is often made in installments, with false growth rings resulting. Estimates of age made by counting whorls or rings are, therefore, liable for error unless this habit is understood.

Fire is the most serious enemy of this species, even though it is remarkably resistant. Young trees commonly produce sprouts from dormant buds of the stump or main stem after burning or other injury, and these persist for many years, a feature rare among the conifers. Pitch pine looper can cause significant losses of pitch pine.

Pitch pine seedlings have been produced in large quantities for forest planting, especially in Pennsylvania, and this species may be of considerable importance, especially on soils too poor for other trees. It hybridizes with *P. echinata*, *P. serotina*, and *P. taeda*.

This species is used for construction lumber and pulp and was used for ship building, railroad ties, and naval stores, as mentioned earlier.

Pìnus serótina Michx.   pond pine

Distinguishing Characteristics.   *Needles* 5 to 8 in. (13 to 20 cm) long, 3 (rarely 4) per fascicle, and flexible, often in short tufts along the trunk, particularly after fire, persisting 2 to 3 years. *Cones* more nearly globose than those of pitch pine, 2 to $2\frac{1}{2}$ in. (5 to 6 cm) long, and the scales are armed with weak, sometimes deciduous prickles.

Range.   Southern New Jersey and Delaware, south to central and northwestern Florida, and southeastern Alabama; local disjuncts in central Alabama. Elevation: sea level to ca. 660 ft. Swamps, pocosins, bays, pond margins, poorly drained flatwoods, often in pure stands.

General Description.   This tree is considered by some writers to be a variety or subspecies of *P. rigida* (Smouse and Saylor, 1973a, b). The two become sympatric and hybridize in the Delmarva peninsula. The features of serotinous cones and tufted needles along the trunk are shared by both pitch and pond pines, and the latter distinguishes these pines from most other southern pines. It produces prolific sprouts even at advanced ages and can recover from complete defoliation by fire.

Pond pine is an intolerant, small- to medium-sized tree, 40 to 70 ft tall and 1 to 2 ft dbh (max. 94 by 3 ft), found in pure stands on wet sites including poorly drained boggy areas on low hills between streams. Such areas of acid organic soils were called "pocosins" (swamp-on-a-hill) by the Native Americans, and the name persists. This species is increasing in importance as a pulpwood and saw timber producer. The most serious disease of pond pine is red heart.

It forms hybrids with *P. echinata, P. rigida,* and *P. taeda.*

Pìnus púngens Lamb.   Table Mountain pine

Distinguishing Characteristics.   *Needles* $1\frac{1}{4}$ to $2\frac{1}{4}$ in. (3 to 6 cm) long, 2 per fascicle, dark yellow-green, rigid, often twisted, persisting 3 years. *Cones* persisting for many years, 2 to $3\frac{1}{2}$ in. (5 to 9 cm) long, ovoid, heavy, the scales armed with large conspicuous sharp spurs, serotinous. *Bark* reddish brown, plated.

Range.   Pennsylvania south to northeastern Georgia and east Tennessee; local disjuncts in New Jersey, Delaware, and the Piedmont of Virginia, North Carolina, and South Carolina. Elevation: 1650 to 4500 ft. Dry, rocky, southwest-facing slopes, knobs, and ridges, in mixed or pure stands.

General Description.   This intolerant pine is a small- to medium-sized tree, 20 to 40 ft tall (max. 98 ft) and 1 to 2 ft dbh, and occurs sparingly on dry, often rocky slopes, and granite outcrops of the central and southern Appalachians. Especially on the southern tablelands, the tree attains sufficient size and quantity (pure stands) to furnish pulp and timber, although the quality of the lumber is low. Stands of Table Mountain pine, often mixed with pitch pine, depend on periodic medium- to high-intensity fire for regeneration and maintenance. It is most important for watershed protection. Seed germination occurs with sunlight and a thin layer of humus over mineral soil.

Pìnus ellióttii Engelm.   slash pine

Distinguishing Characteristics (Fig. 8-31)

Needles 6 to 9 in. (15 to 23 cm) long, in fascicles of 2 and 3, dark glossy green, tufted at the ends of tapering branches but extending back some distance along the branch (compare with longleaf pine); persistent until the end of the second season; *fascicle sheath* ca. $\frac{1}{2}$ in. (9 to 12 mm) long.

Cones $2\frac{1}{2}$ to 7 in. (6 to 18 cm), ovoid-conic, stalked, base of open cone truncate; usually persistent until the following summer; *apophysis* reddish brown, lustrous, commonly full and rounded; *umbo* dorsal and armed with a sharp prickle; *seeds* ca. $\frac{1}{4}$ in. (6 to 8 mm) long, ovoid, black, ridged; wings ca. 1 in. (2 to 2.5 cm) long, thin and transparent.

Twigs orange-brown; *buds* covered with reddish brown white-ciliate scales, free at the tips and not fringed.

Bark on young trees deeply furrowed, later becoming plated and ca. 1 in. (2 to 3.2 cm) thick, with thin, papery, purplish layers; phellogen layers ivory-white.

FIGURE 8-31   *Pinus elliottii,* slash pine. (1) Fascicle of needles ×$\frac{1}{2}$. (2) Cross section of needle ×35. (3) Mature seed cone ×$\frac{1}{2}$. (4) Open seed cone ×$\frac{3}{4}$. (5) Bark of old tree (*Photo by W.R. Mattoon, U.S. Forest Service*). (6) Seed ×1.

Range

Southeastern (Map 8-9). Elevation: sea level to 500 ft. Low flatwoods, pond margins, swamps, also uplands and old fields.

General Description

Slash pine varies from 60 to 100 ft in height and averages about 2 ft dbh (max. 150 by 4 ft). The trunk is straight and erect with a narrow ovoid crown. The root system may reach a depth of 9 to

15 ft or more, unless hampered by the presence of a hardpan near the surface. It is the fastest grow-ing of the southern yellow pines.

In virgin forests, this tree was found on sandy soils, in depressions, around ponds, or on other low sites with an abundance of moisture in the surface layers. Longleaf pine occupied the drier knolls, and over large areas this alternation of species was frequently seen. Such distribution may have been due in part to the relative susceptibility of slash pine to fire damage. With more recent fire protection, this pine invades the drier soils in mixture with the more fire-resistant longleaf pine (Squillace, 1966). On cutover areas, slash pine is very aggressive and, like loblolly pine, quickly preempts abandoned land, especially where the soil is moist.

Slash pine is less tolerant than loblolly, but more so than longleaf. Trees first produce seed abundantly when about 20 years old, with heavy yields at about 3-year intervals. By the end of the first year, young trees are from 8 to 16 in. tall, and subsequent height growth on good sites can average 3 to 4 ft per year for the first 8 to 10 years. Trees of slash pine 25 years of age average 65 ft in height and about 10 in. dbh. Like loblolly pine, young fast-growing trees of this species some-times show annual rings from $\frac{1}{2}$ to 1 in. wide. Such rapid growth coupled with high-quality fiber has resulted in the establishment of many thousands of acres of plantations, some of which are well outside the natural range of this species in warm-temperate and subtropical climates, particu-larly in southern Brazil (Zobel et al., 1987).

This species, because of its aggressiveness and value for pulp, timber, and naval stores, will undoubtedly remain important in the future, despite the increasing importance of loblolly and longleaf pine. Enemies include red-brown butt rot, pitch canker, and bark beetles, but fusiform rust (*Cronartium quercuum*) has the most serious impact.

Pinus elliottii var. dénsa Little and Dorman   South Florida slash pine

Distinguishing Characteristics. This variety differs from the typical slash pine in several ways: The needles are in twos (rarely threes); base of the open cone is rounded rather than trun-cate; the seedlings spend 2 to 6 years in a dwarf "grass stage" similar to that of longleaf pine; the bark is somewhat scaly, and the trunk divides into large spreading branches forming a flat-topped or rounded crown; and the wood is very hard and heavy, with very wide summerwood, and is usu-ally denser than that of the other southern yellow pines.

Range. The range of South Florida slash pine is south Florida and northward in a narrow strip along both coasts. Over much of its range (Map 8-9), this is the only native pine, occurring both in pure stands and scattered through the grasslands at an elevation of sea level to 35 ft. Typically, this variety is found on dry sites, either sandy flat lands or limestone outcrops (Little and Dorman, 1954).

General Description. This variety is not used for naval stores, and it is not commercially planted. Langdon (1963) showed that there is an area in the central part of the state where the ranges of typical slash pine and South Florida slash pine overlap. Squillace (1966) indicated that the variation is essentially continuous or clinal, both within and between varieties. Also, in the northern part of its range, South Florida slash pine may be found on wet sites. This variety is more tolerant of fire than typical slash pine. The South Florida slash pine rocklands outside of Miami and into the Everglades harbor many endemic plant species.

Naturalized Hard Pines

Pìnus nìgra Arnold   Austrian pine, European black pine

This species, introduced from southern Europe in 1759, has been planted extensively in the United States and Canada as an ornamental, and it has escaped cultivation locally. Superficially, it is sim-ilar to the native red pine, with needles 2 per fascicle, but the needles are stouter and do not snap when bent as do those of red pine; the bark is dark brown to black; the buds silvery; and the cones larger, 2 to 3 in. (5 to 8 cm) long, sessile, round-based, yellowish brown with a short deciduous prickle. Dothistroma needle blight has caused much damage to and mortality of this species. The trees can attain a height of 150 ft and a dbh of 3 ft.

Pìnus sylvéstris L.   Scotch pine, Scots pine

Distinguishing Characteristics. *Needles* $1\frac{1}{2}$ to $2\frac{3}{4}$ in. (4 to 7 cm) long, in fascicles of 2 (rarely 3), mostly blue-green, slightly twisted, rigid, and sharp-pointed, persisting 2 to 4 years. *Cones* oblong-conical and $1\frac{1}{4}$ to $2\frac{1}{2}$ in. (3 to 6 cm) long, stalked, reflexed; the apophysis flat, or raised and pyramidal, with a small scarcely armed umbo (Fig. 8-32). *Twigs* slender, dull green to orange-brown, aging gray-brown and rough, buds red-brown and resinous. *Bark* soon becoming orange in color, and this feature is persistent until the tree reaches nearly full size, at which time it becomes dark and furrowed on the trunk but remains orange on the branches and the upper bole.

Range and General Description. This is the most widely distributed pine in the world, spanning much of Europe and Asia and from the Arctic Circle to the Mediterranean Sea (Critchfield and Little, 1966; Mirov, 1967). Because of its centuries of management in European forests, it has been widely used in reforestation for pulpwood, for erosion control, and more recently in Christmas-tree plantations in the United States. In fact it is the most harvested Christmas tree in the United States. Long cultivated for landscape purposes, it has become naturalized locally in southeastern Canada and in the United States from Maine to Iowa, especially on very well-drained soils in large openings. It reaches 115 ft tall and a dbh of 2 ft with an irregular open crown.

P. thunbergiàna Franco (*P. thunbergii* Parl.)   Japanese black pine

This species has become naturalized in Massachusetts and along the Atlantic coast. It is similar to *P. resinosa* with 2 needles per fascicle but differs in the needles bending rather than breaking when

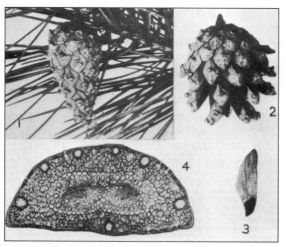

FIGURE 8-32   *Pinus sylvestris,* scotch pine. (1) Mature seed cone and foliage $\times\frac{3}{4}$. (2) Open seed cone $\times\frac{3}{4}$. (3) Seed $\times1$. (4) Cross section of needle $\times35$.

bent, in silvery rather than reddish brown buds, and in the minutely armed cone scales. It differs from *P. nigra* in having stalked, truncate-based cones and nonresinous terminal buds.

Introduced in 1855, this has become a frequent urban tree along the Atlantic coastal areas because it is more resistant of salt spray and is more cold hardy than loblolly pine. Usually less than 25 ft tall, it is attractive with an open crown with a few long, spreading, or pendulous branches with the tips upturned showing the conspicuous silvery, candlelike buds. There are several cultivars.

LARIX Mill.   larch

Although much larger during previous geological periods, this genus now includes only about 10 species of deciduous trees widely scattered through the forests of eastern and western North America, Europe, and Asia. *Lárix decídua* Mill., European larch, is an important continental tree used for reforestation in the eastern United States and Canada, where it has now become naturalized. The foliage is yellow-green, the twigs yellowish or straw-colored, and the upright cones are ca. $1\frac{1}{4}$ in. (2 to 3.8 cm) long (Fig. 8-33). This species, with several cultivars, also finds use as an ornamental in the United States. The Japanese larch, *L. kaémpferi* (Lamb.) Carr. (*L. leptolèpis* [Sieb. and Zucc.] Gord.), also with several cultivars, is another exotic similarly cultivated. An interesting garden hybrid of these two species, the Dunkeld larch, *L. ×marschlínsii* Coaz., promises to be of some importance; this form grows very rapidly and appears to be more resistant to insect or fungal attack than either of the parental species.

There are three species of *Larix* native to the United States, one in the Northeast and two in the West. Two are important (Table 8-8). The ecological aspects of larches are examined in Gower and Richards (1990) and Schmidt and McDonald (1995).

Botanical Features of the Genus

Leaves deciduous, solitary, and spirally arranged on new growth (long shoots) and on older growth in dense spirals only on lateral spurs (indeterminate short shoots); *shape* acicular, more or less flattened, triangular, or less frequently 4-angled in cross section,

FIGURE 8-33   *Larix decidua,* European larch. (1) Pollen cone ×1. (2) Ovulate cone ×1. (3) Open seed cones ×$\frac{3}{4}$. (4) Bark of old tree.

TABLE 8-8   COMPARISON OF IMPORTANT LARCHES

Species	Leaves	Mature cones	Young twigs
L. laricina tamarack	$\frac{3}{4}$–1 in. (2–2.5 cm) long, blue-green	$\frac{1}{2}$–$\frac{3}{4}$ in. (1.2–1.9 cm) long, oblong-ovoid bracts shorter than scales	Glabrous or glaucous
L. occidentalis western larch	1–1$\frac{1}{2}$ in. (2.5–4 cm) long, lustrous green	1–1$\frac{1}{2}$ in. (2.5–4 cm) long, oblong; bracts exserted	Pale pubescent

keeled below or occasionally above and below, with numerous lines of stomata on all surfaces but most abundant below; *apex* pointed or rounded; base decurrent; fall color yellow.

Cones (juvenile) solitary, terminal, appearing with the leaves; *pollen* cones globose, ovoid, or oblong, sessile or pedicelled, yellow to yellowish green; *ovulate* cones subglobose, erect, consisting of few to many generally scarlet, sharp-pointed bracts, each subtending a short suborbicular to rectangular scale with two inverted ovules; all species monoecious.

Mature seed cones short-stalked, erect, subglobose to oblong, maturing in one season; *cone scales* thin, persistent, longer or shorter than the bracts; *bracts* long-acuminate; *seeds* triangular, terminally winged; cotyledons 6.

Twigs with buds small, subglobose, covered by several imbricated scales, non-resinous; leaf scars raised on decurrent base, most occurring on spur shoots, bundle scar 1.

Lárix laricina (Du Roi) K. Koch   tamarack, eastern larch

Distinguishing Characteristics (Fig. 8-34)

Leaves deciduous, linear, $\frac{3}{4}$ to 1 in. (2 to 2.5 cm) long, flexible, 3-angled, bright blue-green; fall color golden yellow.

Cones $\frac{1}{2}$ to $\frac{3}{4}$ in. (1.2 to 1.9 cm) long, erect, oblong-ovoid, short-stalked; *scales* slightly longer than broad, sparingly erose on the margin; *bracts* not visible, except near the base of the cone; *seeds* ca. $\frac{1}{8}$ in. (3 mm) long, the wings $\frac{1}{4}$ in. (6 mm) long, light chestnut-brown.

Twigs orange-brown, glabrous, marked by many small leaf scars, on older growth showing conspicuous short spurs; *buds* globose, dark red, glabrous.

Bark thin and smooth on young stems, later becoming $\frac{1}{2}$ to $\frac{3}{4}$ in. (1.2 to 2 cm) thick, gray to reddish brown, scaly.

Range

Northern (Map 8-11). Local disjuncts in western Maryland, southwestern Pennsylvania, northern West Virginia, westcentral Ohio, and southeastern Minnesota. Elevation: sea level to 4000 ft. Bogs and swamps, lake shores, beach thickets, and well-drained or drier uplands northward; in mixed or pure stands.

FIGURE 8-34   *Larix laricina,* tamarack. (1) Mature seed cones ×$\frac{3}{4}$. (2) Seed ×1. (3) Foliage showing long shoot above and spur shoots (indeterminate short shoots) below, both with needles ×$\frac{1}{2}$. (4) Bark of old tree. (5) Spur shoot in winter ×$\frac{1}{2}$. (6) First-year twig (long shoot) in winter ×1$\frac{1}{2}$.

General Description

Tamarack is a small- to medium-sized tree 40 to 80 ft tall and 1 to 2 ft dbh (max. 100 by 4 ft) with a long clear, cylindrical bole, open pyramidal crown, and shallow wide-spreading root system (Fig. 8-35). In the southern part of its range, it is usually restricted to cool swamps or sphagnum bogs, but farther north it grows best on moist benches and better-drained uplands. Its chief tree associate in bogs is black spruce, whereas on drier sites it is often found in open mixtures with this species and also balsam fir, quaking aspen, paper birch, and jack pine. Associates on better organic soils include northern white-cedar, black ash, red maple, and balsam fir. Tamarack is exceedingly intolerant and must be a dominant tree to survive under forest conditions.

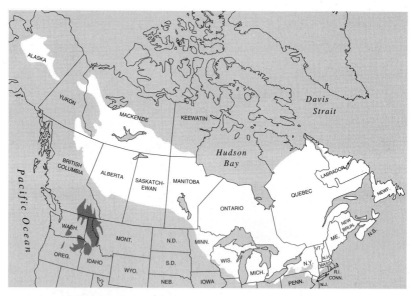

MAP 8-11   **(Dark gray)** *Larix occidentalis.* **(White)** *Larix laricina.*

Seeds are produced annually, with larger crops at 2- to 6-year intervals. Growth of young trees on favorable sites is rapid but decreases after 40 to 50 years; maturity is reached in 100 to 200 years. On Isle Royale in Lake Superior, very long-lived tamaracks have been found, the oldest with an age of 335 years.

Tamarack is not often damaged by fire because of its wet surroundings. Trees of this species past the juvenile stage, in spite of their thin bark and shallow roots, may be fairly fire-resistant even on upland sites. The rare bog fire usually kills or heavily damages trees growing there. Two serious insect enemies are the larch sawfly and the introduced (Europe) casebearer. Where defoliation is severe and repeated from year to year, tree death results.

Tamarack is used as an ornamental and a source of pulp, lumber, posts, railroad ties, fuel, and dogsled runners.

Lárix occidentàlis Nutt.   western larch

Distinguishing Characteristics (Fig. 8-36)

Leaves 1 to 1½ in. (2.5 to 4 cm) long, pale green, lustrous, 3-angled; fall color yellow.

Cones 1 to 1½ in. (2.5 to 4 cm) long, purplish red to reddish brown, oblong, short-stalked; *cone scales* broader than long, occasionally finely toothed at the tip of the reflexed apex; *bracts* exserted, shouldered, terminating in a long awn; *seeds* ¼ in. (6 mm) long; the wings ½ in. (13 mm) long, thin, and fragile.

Twigs stout, at first pale pubescent, but soon becoming orange-brown and glabrous; *buds* chestnut-brown.

Bark reddish brown to cinnamon-red, scaly on young stems; 4 to 6 in. (10 to 15 cm) thick on old trunks and then with flat-plated ridges separated by deep irregular fissures.

FIGURE 8-35 Tamarack. (*Courtesy of U.S. Forest Service.*)

Range

Northwestern (Map 8-11). Elevation: 1650 to 5300 ft northward, to 7000 ft southward. Mountain slopes and valleys, usually mixed with other conifers.

General Description

This tree was discovered in 1806 by the Lewis and Clark expedition on the upper watershed of the Clearwater River in western Montana. David Douglas also found this larch in northeastern Washington in 1827. He erroneously concluded, however, that he had encountered a natural extension of the European larch into the New World. Thus it was not until 1849, when Thomas Nuttall recognized that the larches in the Blue Mountains of northeastern Oregon did indeed comprise a new species, that the identity of this tree was correctly established.

Western larch is the largest of the American larches and one of the most important trees of the Inland Empire. The boles of mature trees are clear for a considerable length, characteristically

FIGURE 8-36 *Larix occidentalis,* western larch. (1) Foliage on long shoot and spur shoots $\times\frac{1}{2}$. (2) Ovulate cones $\times 1$. (3) Pollen cones $\times 1$. (4) Mature seed cone $\times\frac{3}{4}$. (5) Open seed cone $\times\frac{3}{4}$. (6) Seeds $\times 1$. (7) Bark (*Courtesy of British Columbia Forest Service*).

tapered, and vary from 140 to 180 ft in height and 3 to 4 ft dbh (max. 233 by 8 ft). The crowns are usually short, open, and essentially pyramidal; in the summer, they are distinguishable at a considerable distance by their light, lustrous green foliage, and in winter by their bareness. Excellent anchorage is afforded by a deep wide-spreading root system, and the trees are seldom wind-thrown. Western larch attains maximum size on deep, moist, porous soils in high valleys and on mountain slopes of northern and western exposure, although thrifty trees often occur on drier gravelly soils. Nearly pure open forests of western larch are found in northwestern Montana, northern Idaho, and northeastern Washington. It is often the most abundant species in the western larch–Douglas-fir forests of the northern Rockies, where it is also associated with western white pine, and at higher elevations with lodgepole pine, Engelmann spruce, and subalpine fir. Its other forest associates include western hemlock, grand fir, ponderosa pine, and occasionally western redcedar.

Small amounts of seeds are released annually, and heavy crops are produced at irregular intervals. Seed germination and initial seedling growth are favored by mineral soils exposed by fire or mechanical scarification. Seedling growth is extremely vigorous, and plants only 4 years of age are often 4 ft or more in height. Considering the relatively short growing season, this seems all the more remarkable because Douglas-fir and ponderosa pine on the same site rarely exceed a height of 20 in., and Engelmann spruce seedlings of the same age are usually less than 1 ft high. This height growth advantage of the larch can continue for 10 to 20 years. Western larch is very intolerant throughout life and is dominant in old-growth mixed stands. Maturity is ordinarily reached in 300 to 400 years, but trees from 700 to 915 years of age have been reported. Commercially, this species attains its best development in the Priest River country of northern Idaho.

The bark of old trees is thick and moderately fire-resistant, but young trees are severely damaged by only a light ground fire. Wood-destroying fungi commonly attack overmature timber, and dwarf mistletoe reaches epidemic proportions in some localities. This tree, the most important of the larches, is used for fine veneer, lumber, and pulp. It was also split (rived) for shakes and shingles in earlier times if western redcedar was not available.

Lárix lyállii Parl.   subalpine larch

Distinguishing Characteristics.   *Needles* $\frac{3}{4}$ to $1\frac{1}{4}$ in. (2 to 3.5 cm) long, 4-angled, pale blue-green. *Cones* erect, elliptical, 1 to 2 in. (2.5 to 5 cm) long, with exserted, 3-toothed bracts. *Twigs* densely white woolly.

Range.   Southeastern British Columbia and southwestern Alberta south to western Montana and west to northeastern Washington. Elevation: 5000 to 8000 ft. Rocky soils near timberline; locally in pure stands and also mixed with other conifers.

General Description.   Subalpine larch is a small timberline tree, 30 to 50 ft tall and 1 to 2 ft dbh (max. 101 by 7 ft), with long, willowy, often pendulous branches, found on several high watersheds in the Cascade, Bitterroot, and northern Rocky Mountains. It can reach nearly 800 years in age. It occurs in pure open groves or intermingled with whitebark pine, mountain hemlock, subalpine fir, and Engelmann spruce. Subalpine larch is intolerant and slow growing. Because of its small size, poor form, and inaccessibility, it is of no commercial importance. However, it is aesthetically attractive in the remote, high mountain terrain and important for watershed protection. It hybridizes with western larch.

PICEA A. Dietr.   spruce

The genus *Picea* includes 35 to 40 species of trees found in the cooler regions of North America, Mexico, and Eurasia. The wood of the spruces is strong for its

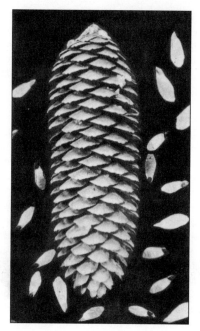

FIGURE 8-37 *Picea abies,* Norway spruce. Open cone and seeds ×$\frac{3}{4}$.

weight, moderately long-fibered, odorless, but slightly resinous and is of primary importance in the manufacture of pulp and paper. The most important European species is *Pìcea àbies* (L.) Karst., Norway spruce. This is a tall tree, readily identified by its dark yellow-green, usually drooping branchlets, somewhat laterally compressed leaves, and large, pendent cones 4 to 7 in. (10 to 18 cm) long (Fig. 8-37). This species, with many cultivars, is a popular ornamental in many sections of the United States and southeastern Canada. Norway spruce has also been widely used in forest plantations, especially in the east where it grows more rapidly than the native spruces. It has escaped from cultivation in various localities and is naturalized locally. The resinous bark exudations of this tree furnish the so-called Burgundy pitch, which is the basic material for a number of varnishes and medicinal compounds, and the new leafy shoots, combined with flavorings and sugar, are often used in brewing spruce beer. Spruce chewing gum has been obtained from red and black spruce.

In Asia the Himalayan spruce, *Picea smithiàna* (Wall.) Boiss., is one of the most important of indigenous softwood species. *Picea jezoénsis* (Sieb. and Zucc.) Carr., yezo or jeddo spruce, *P. omorìka,* Serbian spruce, *P. orientális,* Oriental spruce, and *P. toràna* (K. Koch.) Koehne (*P. polita* Carr.), the tigertail spruce, are four oriental species suitable for ornamental planting in the United States. Serbian and Oriental spruce are especially highly regarded for the landscape.

The natural enemies of the spruces are rather numerous, and some of them cause considerable damage to immature timber. Because of the thin bark, poor pruning, and flammable foliage, fire is always a serious menace, and damage by windthrow is common. The trees are also subject to periodic attacks of the eastern spruce gall aphid, which lays its eggs at the base of partially developed leaves near the tips of twigs. A large conelike gall soon develops, beyond which all growth ceases. The spruce budworm often destroys whole stands of young timber, and the white pine weevil may damage certain species. Trees growing in large industrial cities are frequently affected by smoke and flue gases.

Seven spruces are native to North America north of Mexico. The Brewer spruce, *Picea breweriàna* S. Wats., a narrowly endemic tree found only on the high ridges of the Siskiyou Mountains of southern Oregon and northern California, is the only one of little or no commercial value. The other six will be described.

Botanical Features of the Genus

Leaves spirally arranged, extending from all sides of the twigs or massed toward the upper surface; *shape* linear or nearly acicular; flattened, laterally compressed, or more commonly 4- (rarely 3-) angled, with numerous stomatiferous lines on the lower surfaces, or in many forms on all surfaces; *apex* usually sharp-pointed; persistent for 7 to 10 years, sessile upon conspicuous peglike projections of the *twig*; when bruised, often emitting an aromatic, or fetid odor.

Cones (juvenile): *pollen cones* erect or pendent, stalked, consisting of numerous spirally arranged pollen sacs, appearing from axillary buds on the shoots of the previous season or terminal; *ovulate cones* terminal, erect or nearly so, composed of numerous purple or green scales, each of which bears two inverted ovules and is subtended by a bract; all species monoecious.

Mature seed cones pendent (rarely nearly erect), woody, comprising a number of persistent, unarmed scales, often erose along the margins and much longer than the bracts; *seeds* two to each cone scale, ovoid to cylindrical, often compressed, terminally winged; cotyledons 4 to 15.

Buds conic to ovoid, with numerous imbricated scales, nonresinous or sometimes resinous.

The Eastern and Transcontinental Spruces (Table 8-9)

Pìcea rùbens Sarg.   red spruce

Distinguishing Characteristics (Fig. 8-38)

Leaves $\frac{1}{2}$ to $\frac{5}{8}$ in. (12 to 15 mm) long, linear, 4-sided, shiny yellow-green, apex blunt or acute.

Cones $1\frac{1}{4}$ to $1\frac{3}{4}$ in. (3 to 4.5 cm) long, ovoid-oblong, chestnut-brown at maturity, falling during the first winter or following spring; *scales* rigid, rounded, entire on the margin or very lightly erose; *seeds* $\frac{1}{8}$ in. (3 mm) long, dark brown; wings $\frac{1}{4}$ in. (6 mm) long.

TABLE 8-9   COMPARISON OF EASTERN AND TRANSCONTINENTAL SPRUCES

Species	Leaves	Mature cones	Twigs
P. rubens red spruce	Shiny, dark yellow-green, sharp-pointed	$1\frac{1}{4}$–$1\frac{3}{4}$ in. (3–4.5 cm) long, ovoid-oblong; scales rounded, margin entire or sparingly erose	More or less pubescent; acute hairs only
P. mariana black spruce	Dull blue-green, more or less glaucous, blunt-tipped	$\frac{5}{8}$–$1\frac{1}{4}$ in. (1.5–3 cm) long, ovoid; scales rounded, margin erose	Pubescent; acute and capitate hairs
P. glauca white spruce	Blue-green, glaucous, sharp-pointed, more or less fetid when crushed	$1\frac{1}{4}$–$2\frac{1}{2}$ in. (3–6 cm) long, oblong-cylindrical; scales truncate, margin entire	Glaucous and glabrous

FIGURE 8-38 *Picea rubens,* red spruce. (1) Foliage $\times\frac{1}{2}$. (2) Dead twig showing pegs $\times 2$. (3) Mature seed cone $\times\frac{3}{4}$. (4) Seed $\times 1$. (5) Open seed cone $\times\frac{3}{4}$.

Twigs orange-brown, more or less pubescent with acute hairs, the decurrences rounded; *buds* ovoid, reddish brown.

Bark $\frac{1}{4}$ to $\frac{1}{2}$ in. (6 to 12 mm) thick, separating into close, irregular, grayish brown to reddish brown scales; inner layers reddish brown (Fig. 8-39).

Range

Northeastern (Map 8-12). Disjuncts in northwestern Connecticut, eastern Long Island, and northern New Jersey. Elevation: sea level in the North to 6600 ft in the southern Appalachians. Swamps and bogs northward to mountain tops, ridges, and bogs southward; often in pure stands or with fir.

General Description

Red spruce is one of the most important of northeastern conifers and is especially characteristic of the mountainous regions in northern New York and New England. In old-growth stands, trees 60 to 70 ft tall and 12 to 24 in. dbh are common (max. 152 by 4.8 ft). Maximum development of the species occurs in the southern Appalachians where humidity and rainfall are especially high during the growing season.

Open-grown trees develop a broadly conical crown that extends nearly to the ground; under forest conditions, the crown is restricted to the upper portion of the tree and is somewhat pagoda-shaped. The boles of forest-grown trees are long and cylindrical; the root system is shallow and wide-spreading.

FIGURE 8-39   Bark of old-
growth red
spruce.

MAP 8-12   *Picea rubens.*

Red spruce occurs in pure stands and in mixture with many other species. It is found in swamps or bogs with black spruce, tamarack, balsam fir, and red maple but grows poorly on such sites. Faster growth is made on adjacent better-drained flats in company with such species as balsam fir, eastern hemlock, eastern white pine, and yellow birch; scattered trees occur throughout the neighboring hardwood mixture (sugar maple, yellow birch, and beech), on higher ground, and here development is probably best. Red spruce was likely often the dominant tree in many of these stands, prior to heavy logging in the late 1800s and early 1900s. This species is also found in pure stands on upper slopes where the soil is very thin and rocky. In general, acid sandy loam soils with considerable moisture support the best spruce. In the southern Appalachians, it is often associated with Fraser fir or occurs in pure stands at slightly lower elevations than the fir.

Abundant moisture is also essential for good reproduction, and the most favorable conditions are found under mixed stands of hardwoods and conifers. Open-grown spruce may begin to yield good seed crops when 30 to 40 years old, but in dense stands it is usually much later. Good seed years occur at 3- to 8-year intervals. The young trees are very tolerant, more so than those of the associated species, with the possible exception of eastern hemlock, sugar maple, and American beech, but they grow slowly under forest cover (1 ft in height may represent 15 years' growth), and even in the open they do not make rapid growth.

Red spruce is long-lived and scarcely reaches maturity in less than 200 years, and trees with 350 to 430 annual rings have been confirmed. Because of the shallow root system, red spruce is susceptible to windthrow; it is also severely damaged by fire and defoliated by the spruce budworm. Mature trees are especially susceptible to eastern spruce beetle (*Dendroctonus rufipennis*). Its decline and mortality since the 1960s is probably due to an interaction of insects, fungi, atmospheric deposition, and climatic aberrations.

This is an important tree for paper pulp, lumber, musical string instruments, plywood, and composite board. It is the provincial tree of Nova Scotia. It hybridizes with *P. mariana* in eastern Canada.

Pìcea mariàna (Mill.) B.S.P.   black spruce

Distinguishing Characteristics (Fig. 8-40)

Leaves $\frac{1}{4}$ to $\frac{5}{8}$ in. (6 to 15 mm) long, 4-sided, dull blue-green, blunt-pointed; more or less glaucous.

Cones $\frac{5}{8}$ to $1\frac{1}{4}$ in. (1.5 to 3 cm) long, ovoid, purplish, turning brown at maturity; *scales* brittle, rigid, rounded, erose on the margin; *seeds* $\frac{1}{8}$ in. (3 mm) long, dark brown;

FIGURE 8-40 *Picea mariana*, black spruce. (1) Mature seed cone ×$\frac{3}{4}$. (2) Open seed cone ×$\frac{3}{4}$. (3) Bark of old tree (*Courtesy of Canadian Forest Service*).

wings $\frac{1}{4}$ to $\frac{3}{8}$ in. (6 to 9 mm) long. The cones persist for many years (20 to 30) and often form large clusters easily seen at a distance.

Twigs brownish, pubescent with acute and capitate hairs, the decurrences flat; *buds* ovoid, somewhat puberulous.

Bark $\frac{1}{4}$ to $\frac{1}{2}$ in. (6 to 12 mm) thick, broken into thin, flaky, grayish brown to reddish brown scales; freshly exposed inner layers somewhat olive-green.

Range

Northern (Map 8-13). Several disjunct populations occur south of the general range. Elevation: sea level to 5000 ft. Acid bogs and swamps, lake shores, and dune ridges; less common in neutral or alkaline swamps.

General Description

Black spruce is a small- to medium-sized tree 30 to 40 ft tall and 6 to 12 in. dbh (max. 100 by 3 ft), with a long, straight, tapering bole, irregularly cylindrical crown, and a shallow, spreading root system. In the southern part of its range, this species is commonly restricted to cool sphagnum bogs; in the far north, it is also found on dry slopes but makes its best growth on moist, well-drained alluvial bottoms. Among the conifers, this species is of interest because with white spruce and tamarack

MAP 8-13 *Picea mariana*.

it marks "the northern limit of tree growth" and assumes there a prostrate or shrubby habit. Black spruce is especially characteristic as a pioneer tree on the floating mats extending outward from the shores of small ponds that are slowly becoming bogs ("muskegs") through the deposition of plant litter. In such situations, growth is exceedingly slow, and "trees" 2 in. dbh may show from 80 to 90 or more growth rings in cross section.

Black spruce occurs in dense pure stands or in mixture with species adapted to similar sites. Like most of the other spruces, it is tolerant, but less so than balsam fir and northern white-cedar. Natural pruning is poor, and unless the tree is grown in very dense stands, the side branches persist for many years.

Although most of the good seed is released during the first 4 years following cone maturation, viable seed may come from 15-year-old cones. Fire may start the opening of these semiserotinous cones, and large amounts of seed are then released. Some seed may be produced each year (after the tenth) with larger crops at irregular intervals. Propagation on wet sites also commonly takes place by "layering"; the lower branches become embedded in the moist sphagnum and sprout roots, which eventually support the erect branch without aid from the parent tree. Black spruce may attain an age of 280 years.

This is the most important pulpwood species in Canada but is also used for lumber and Christmas trees. It is damaged by several fungous diseases, spruce budworm, sawflies, and dwarf mistletoe. It is the provincial tree of Newfoundland. It hybridizes with *P. rubens* in eastern Canada and reportedly with *P. glauca.*

Picea glaùca (Moench) Voss   white spruce

Distinguishing Characteristics (Fig. 8-41)

Leaves $\frac{1}{2}$ to $\frac{3}{4}$ in. (12 to 19 mm) long, linear, blue-green, and glaucous; apex acute but not sharp to the touch; tending to be crowded on the upper side of the branch by a twist at the base of those below; crushed foliage often with a pungent odor, hence the common name "cat spruce."

Cones $1\frac{1}{4}$ to $2\frac{1}{2}$ in. (3 to 6 cm) long, narrowly oblong, light brown; *scales* thin and flexible, usually truncate, rounded, or slightly emarginate at the apex; *seeds* $\frac{1}{8}$ in. (3 mm) long, pale brown; wings $\frac{1}{4}$ to $\frac{3}{8}$ in. (6 to 9 mm) long.

FIGURE 8-41   *Picea glauca,* white spruce. (1) Mature seed cone and foliage $\times\frac{3}{4}$. (2) Open seed cone $\times\frac{3}{4}$. (3) Bark of old tree (*Courtesy of Canadian Forest Service*).

Twigs glabrous, somewhat glaucous, orange-brown to gray; *buds* ovoid, with sometimes reflexed scales, ragged, or occasionally ciliate on the margin.

Bark thin, flaky or scaly, ashy brown; freshly exposed inner layers somewhat silvery.

Range

Northern (Map 8-14). Local disjuncts in central Montana, western South Dakota, northeastern Wyoming, and southeastern Michigan. Elevation: sea level to 4000 ft or timberline. Mesic to wet sites in conifer forests of muskegs, bogs, mountain slopes, and edges of streams or lakes.

General Description

White spruce is one of the most important and widely distributed conifers of Canada and extends southward into the northern United States. When grown in the open, it develops a handsome conical crown, which extends nearly to the ground. Forest trees are from 60 to 70 ft in height and 18 to 24 in. dbh (max. 184 by 4 ft). The maximum age is about 600 years.

This species forms extensive pure stands but also occurs in other local forest types. Best growth is made on moist loam or alluvial soils; although found on many different sites, white spruce is especially typical of stream banks, lake shores, and adjacent slopes. It requires soils with higher nutrients than does black spruce.

In comparison with the preceding species, white spruce grows faster and is much more tolerant. It is important for reforestation both in the United States and Canada and is used for lumber and pulp, house logs, and musical string instruments. It is also a widely planted ornamental. The pliable roots of white spruce were used by Native Americans for lacing birchbark canoes and making woven baskets.

MAP 8-14   *Picea glauca.*

Along the Canadian Rockies where the ranges of white spruce and Engelmann spruce overlap, certain varieties of white spruce have been recognized, including western white spruce *P. glauca* var. *albertiàna* (S. Brown) Sarg. In this region, so much natural hybridization is taking place between white and Engelmann spruces that perhaps varieties should not be recognized (Taylor, 1993; Wright, 1955). There is considerable variation in cone size and scale structure on these spruces. White spruce also hybridizes with *P. mariana* and *P. sitchensis.*

Although white spruce is host to a very wide variety of damaging insects and diseases, wildfire is the primary cause of loss. Mature trees are especially susceptible to spruce beetles.

White spruce is the provincial tree of Manitoba and the state tree (as Black Hills spruce) of South Dakota.

The Western Spruces (Table 8-10)

Pìcea sitchénsis (Bong.) Carr.   Sitka spruce

Distinguishing Characteristics (Fig. 8-42)

Leaves $\frac{5}{8}$ to 1 in. (15 to 25 mm) long, flattened, bright yellow-green above, bluish white, glaucous below, very sharp-pointed, tending to extend outward at right angles from all sides of the twig.

Cones 2 to $3\frac{1}{2}$ in. (5 to 9 cm) long, ovoid-oblong, falling in the late autumn and early winter; *scales* thin, papery but stiff, oblong or elliptical, erose at the apex; *seeds* $\frac{1}{8}$ in. (3 mm) long, reddish brown, the wings $\frac{1}{3}$ to $\frac{1}{2}$ in. (8 to 13 mm) long.

Twigs glabrous, orange-brown; *buds* ovoid, the scales obtuse.

Bark thin, rarely as much as 1 in. (2.5 cm) thick even on the largest trunks, divided at the surface into thin, loosely appressed, concave, elliptical, silvery gray to purplish gray scales.

TABLE 8-10   COMPARISON OF IMPORTANT WESTERN SPRUCES

Species	Leaves	Seed cones	Twigs
P. sitchensis Sitka spruce	Flattened, bright yellow-green above, blue-green below, sharp-pointed but not prickly; spreading	2–$3\frac{1}{2}$ in. (5–9 cm) long; scales oblong or elliptical, apex truncate, erose, stiff; persisting 1 year	Glabrous; bud scales reflexed
P. engelmannii Engelmann spruce	4-angled, blue-green, acute but not sharp-pointed; usually pointing forward; not pungent to taste	$1\frac{1}{4}$–$2\frac{3}{4}$ in. (3–7 cm) long; scales rhombic-oblong, apex appressed, truncate to acute, erose, flexible; persisting 1 year	Minutely pubescent; bud scales appressed or slightly reflexed
P. pungens blue spruce	4-angled, blue-green, sharp-pointed and prickly, stiff; spreading; pungent to taste	$2\frac{1}{2}$–4 in. (6–10 cm) long; scales rhombic-oblong, apex spreading, truncate or emarginate, erose, flexible; persisting 2 years	Glabrous; bud scales reflexed

FIGURE 8-42   *Picea sitchensis,* Sitka spruce. (1) Foliage $\times\frac{1}{2}$. (2) Pollen cone $\times$1. (3) Ovulate cone $\times$1. (4) Open seed cone $\times\frac{3}{4}$. (5) Seed $\times$1. (6) Bark of old-growth tree (*Photo by J.D. Cress, Seattle*).

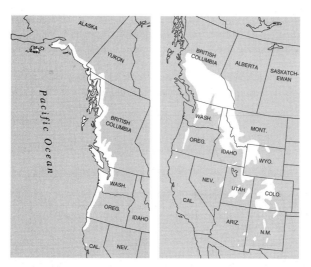

MAP 8-15   **(Left)** *Picea sitchensis.* **(Right)** *Picea engelmannii.*

Range

Northwestern (Map 8-15). Elevation: sea level to 2800 ft or timberline in Alaska, below 1200 ft in California. Coastal fog belt; peat bogs; in mixed or pure stands.

General Description

Sitka spruce, which is also known in the lumber trade as tidewater spruce, is one of the major timber-producing species of the Pacific Northwest. It is the largest spruce in the world. Mature trees on the Olympic peninsula of Washington vary from 180 to 200 ft in height and from 4 to 5 ft dbh (max. 315 by nearly 19 ft). Many individuals yield from 6000 to 8000 board feet of lumber, and an occasional tree contains as much as 40,000 ft B.M.

Forest-grown trees feature long, clear, cylindrical boles that terminate in short, open crowns with horizontal and ascending branches terminating in pendulous branchlets. The typically shallow wide-spreading root system provides only moderate resistance to windthrow.

Sitka spruce forms extensive pure forests in many parts of its range and also occurs in mixture with several other softwoods and hardwoods. In Alaska and elsewhere, its chief associate is western hemlock. In British Columbia, Washington, and northcentral Oregon, it mingles with Douglas-fir, western redcedar, Pacific silver and grand firs, red alder, bigleaf maple, and black cottonwood. Occasional associates in southern Oregon include redwood and Port-Orford-cedar. Sitka spruce is moderately tolerant. The largest trees throughout its range occur on deep loams of high-moisture retention. This species also invades peat bogs of the Puget Sound region. Trees on such sites, however, are of poor form and rarely attain commercial size. Sitka spruce is a coastal species usually found no more than 30 miles from tidewater and often much closer because the coastal plain is narrow and the mountains sometimes rise almost directly from the sea. An interesting growth form of this species is found on the dry sandy soils and dunes along the coasts of Oregon and northern California. Here, shaped by the ceaseless westerly winds, individuals may be dwarfed, contorted, sprawling, or prostrate.

Sitka spruce ordinarily produces some seed each fall (after the twentieth year) and usually releases copious crops every third or fourth year. Seedlings become readily established on a variety of soil types. Dependable moisture is required for seedling survival, and those growing on moist organic layers often die during droughts. Growth is vigorous in youth, and the trees soon overtop their forest associates. Its production of epicormic branches is rare among conifers. This spruce is rated as tolerant, but not as much so as western hemlock. Maturity is reached in about 500 years, but occasional forest monarchs 700 to 882 years old have been reported.

Sitka spruce is a source of lumber and pulp. During World War II, the famous British "Mosquito" bomber's airframe was crafted of strong, lightweight Sitka spruce. It is being widely planted on the acid, organic soils of Ireland where its growth is outstanding. Windthrow and the white pine weevil each cause serious losses.

Sitka spruce is the state tree of Alaska. It hybridizes with *P. glauca* in British Columbia and Alaska.

Pìcea engelmánnii Parry ex Engelm.   Engelmann spruce

Distinguishing Characteristics (Fig. 8-43)

Leaves $\frac{5}{8}$ to $1\frac{1}{4}$ in. (15 to 30 mm) long, 4-sided, blue-green, flexible, the apex often blunt; with a rank odor when crushed; often somewhat appressed and tending to point toward the tip of the twig.

Cones $1\frac{1}{4}$ to $2\frac{3}{4}$ in. (3 to 7 cm) long, ovoid-oblong; *scales* thin and somewhat papery, rhombic-oblong, and commonly erose at the apex, appressed; *seeds* $\frac{1}{8}$ in. (3 mm) long, nearly black; wings about $\frac{1}{2}$ in. (13 mm) long, oblique.

Twigs more or less pubescent, light brown to gray; *bud scales* more often appressed than in blue spruce.

Bark very thin, broken into large purplish brown to russet-red, thin, loosely attached scales.

Range

Western (Map 8-15). Local disjuncts from eastern Washington to northern California and from northcentral Montana to southeastern Arizona. Elevation: 2000 to 3500 ft northward, 4000 to 12,000 ft southward. Subalpine zone to timberline (Fig. 8-44); in mixed or pure stands.

General Description

The name of this spruce commemorates Dr. George Engelmann, noted German-American physician and botanist of the middle nineteenth century.

Engelmann spruce is typically a mountain species, and under favorable conditions for growth, it attains a height of from 100 to 120 ft and a dbh of 18 to 30 in., although somewhat larger trees (max. 223 by 8 ft) occur on the best sites. Its general habit is quite similar to that of Sitka spruce, and, like that species, it reaches its maximum size on deep, rich, loamy soils of high-moisture content.

Besides occurring in extensive pure stands, Engelmann spruce is found with other species comprising some 14 recognized forest types. The most common associate is subalpine fir. Through the central Rocky Mountains, lodgepole, limber, and whitebark pines, Douglas-fir, and quaking aspen may also be included (Burns and Honkala, 1990).

Engelmann spruce produces large crops of seed every 2 or 3 years (after the twentieth). Seedling development is good in moist duff soils covering the floor of virgin forests. However, seedling development is poorest in both deep shade and full sun. Reproduction by layering occurs

FIGURE 8-43 *Picea engelmannii,* Engelmann spruce. (1) Foliage $\times\frac{1}{2}$. (2) Mature seed cone $\times\frac{3}{4}$. (3) Open seed cone $\times\frac{3}{4}$. (4) Seed $\times 1$. (5) Bud $\times 1$. (6) Bark of old tree (*Courtesy of Western Pine Association*).

FIGURE 8-44 Windform of Engelmann spruce near timberline.

especially near timberline, but individuals produced in this way never attain commercial proportions. In fact, at timberline it is reduced to a mound less than head high.

This spruce is tolerant, but among its common associates is less so than subalpine fir and the hemlocks. Trees of all ages are often found under the canopy of old trees, and individuals often suppressed for 50 to 100 years quickly recover upon being released. Growth is restricted by a short summer season, and trees 16 to 22 in. dbh are often 350 to 450 years of age. The average maximum age for Engelmann spruce appears to be in the neighborhood of 400 years. Occasional trees over 600 years of age have been reported (max. 852 years).

Periodic outbreaks of the Engelmann spruce bark beetle have been extremely damaging to mature stands in the central Rocky Mountain region. The western spruce budworm, several wood-rotting fungi, and dwarf mistletoe also cause damage. The bark is thin even on old trunks, and fires cause extensive damage. Windthrow, particularly in partially cut old-growth stands, results in heavy loss.

This spruce is a source of construction lumber, plywood, pulp, and wood for musical string instruments.

Where the ranges of Engelmann and white spruce overlap, a confusing array of natural hybrids is found. It also hybridizes with *P. pungens*.

Pìcea púngens Engelm.   blue spruce, Colorado blue spruce

Distinguishing Characteristics.   *Leaves* 4-sided, $\frac{3}{4}$ to $1\frac{1}{4}$ in. (20 to 31 mm) long, sharp-pointed, stiff, spreading, dull blue-green or bluish with whitish lines and glaucous when young, with a pungent resinous odor when crushed. *Cones* $2\frac{1}{2}$ to 4 in. (6 to 10 cm) long, cylindrical, sessile or short-stalked, light brown, the scales long and narrow, rhombic-oblong, erose, more or less flexible and spreading. *Twigs* glabrous, yellow-brown, stout, bud scales usually reflexed. *Bark* gray to reddish brown, furrowed with scaly round ridges.

Range.   Rocky Mountains of southern and western Wyoming and eastern Idaho, south to northern and eastern Arizona and southern New Mexico. Elevation: 6000 to 9000 ft in the North; 7000 to 10,000 ft in the South.

General Description.   Those who are familiar only with the shapely, dense, and beautiful silvery-blue ornamental cultivars of this species would scarcely recognize the blue spruce in its central Rocky Mountain domain. Sculptured by gale-force winds and heavy snows, the trees often bear little resemblance to the carefully nurtured specimens of parks and yards. Even the foliage, except for a brief time in early spring when the needles are coated with a powdery waxy bloom, is dull dark green to blue-green with only an occasional tree silvery blue. Blue spruce is often 70 to 90 ft tall and 2 ft dbh (max. 148 by 5 ft). It resembles Engelmann spruce, with which it is sometimes associated at lower elevations. However, much blue spruce occurs on relatively rich and moist soils along streams and lower slopes. It is moderately tolerant. Limited hybridization occurs between these two species.

Although not an important timber tree, this species and many of its cultivars, because of their habit, beautiful foliage, and ability to withstand drought and extremes of temperatures, are highly prized as ornamentals.

Blue spruce is the state tree of Utah and (as Colorado blue spruce) of Colorado and the provincial tree of Newfoundland and Labrador.

PSEUDOTSUGA Carr.   Douglas-fir

This genus includes five species of trees widely scattered through the forests of western North America, southwestern China, Japan, and Taiwan. Although these were originally included under *Abies,* they are readily distinguished by their pointed buds, petiolate leaves, pendent seed cones, persistent cone scales, and exserted 3-lobed bracts. Douglas-fir, *P. menziesii* (Mirb.) Franco, is the only important species in the genus and is one of two found in the western United States.

Botanical Features of the Genus

Leaves spirally arranged; *shape* linear, flattened, grooved on the upper surface and with a broad band of stomata on each side of the midrib below; *apex* blunt to rather sharply pointed; *base* constricted into a very short petiole; persistent for 5 to 8 years and leaving a suborbicular, slightly raised leaf scar upon removal or when shed; with a characteristic odor when crushed.

Cones (juvenile) solitary; *pollen cones* axillary, consisting of a number of spirally arranged, short-stalked, subglobose pollen sacs; *ovulate cones* conical, terminal or in the axils of the upper leaves, composed of several spirally arranged, imbricated, 3-lobed bracts each subtending a small ovate scale with two basal ovules; all species monoecious.

Mature seed cones ovoid-cylindrical, pendant, maturing at the end of the first season; the conspicuous 3-lobed *bracts* longer than the rounded *cone scales*; *seeds* more or less triangular, with a large, rounded terminal wing that partially envelops the seed; cotyledons 6 to 12.

Buds fusiform, sharp-pointed, covered by several imbricated, shiny brown scales.

Pseudotsùga menzièsii (Mirb.) Franco (*P. taxifolia* [Lamb.] Britt.)   Douglas-fir

Distinguishing Characteristics (Fig. 8-45)

Leaves $\frac{5}{8}$ to $1\frac{1}{4}$ in. (1.5 to 3 cm) long, yellow-green or blue-green, standing out from all sides of the twig or with a tendency to be somewhat 2-ranked; *apex* rounded-obtuse or rarely acute, stomatiferous below, persistent for 8 or more years.

FIGURE 8-45 *Pseudotsuga menziesii*, Douglas-fir. (1) Foliage $\times\frac{3}{4}$. (2) Buds $\times$1. (3) Mature seed cone (var. *menziesii*) $\times\frac{3}{4}$. (4) Ovulate and pollen cones $\times$1. (5) Open seed cone (var. *glauca*) $\times\frac{3}{4}$. (6) Seed $\times$1. (7) Bark of old tree.

Cones $1\frac{5}{8}$ to 4 in. (4 to 10 cm) long, pendent, ovoid-cylindric, with exserted, 3-lobed, forklike, appressed or strongly reflexed bracts; *seeds* triangular, terminally winged.

Buds fusiform, sharp-pointed, lustrous brown.

Bark on young stems smooth except for resin blisters; becoming 6 to 12 in. (15 to 31 cm) thick on old trees, and then divided into thick, reddish brown ridges separated by deep irregular fissures. In a few instances, the bark is "tight" (fine-textured) on old trees and corky on others, particularly those of the mountain form.

Range

Western (Map 8-16). Elevation: sea level to 2000 ft northward and to 6000 ft southward; inland, above 2000 ft northward, 8000 to 10,000 ft southward. Moist soils along the coast; rocky soils on mountain slopes.

General Description

Douglas-fir, monarch of Pacific Northwest forests, was first observed by Menzies on Vancouver Island when he accompanied the British naval captain Vancouver on an expedition to the Pacific coast in the early 1790s. For more than a quarter of a century, this tree was variously classified as a spruce, hemlock, true fir, and even as a pine; in fact, logs exported by the Hudson's Bay outpost near the mouth of the Columbia River were listed in European ports as "Oregon-pine," a name that has persisted in the trade to this day, especially in Australia. It remained for David Douglas, a Scottish botanist sent out by the Royal Horticultural Society in 1825, to study this tree and show that it was sufficiently different to be considered as distinct from other previously described conifers; later, Carriére coined the new generic name *Pseudotsuga*. This name is a rather unfortunate choice because it literally means "false hemlock." The common name, Douglas-fir, commemorates Douglas and serves to distinguish this species from the true firs (*Abies*).

MAP 8-16   *Pseudotsuga menziesii.*

Douglas-fir comprises about 50% of the standing timber of our western forests, producing more timber than any other American species. It is also important as a Christmas tree. It is the state tree of Oregon.

Douglas-fir is a dimorphic species with two more or less distinct varieties. One of these is restricted to the forests of the Pacific slope, and the other to the Rocky Mountain region. However, the two types intergrade where they are sympatric in northeastern Washington and southern British Columbia. Usually the foliage of the Rocky Mountain tree is typically blue-green to dark green, and that of the coast form is yellow-green. However, atypical colors are found in both areas. The principal botanical difference between these two varieties lies in the structure of their cones. Rocky Mountain trees have small cones rarely $1\frac{1}{2}$ to $2\frac{3}{4}$ in. (4 to 7 cm) long, with spreading, often reflexed bracts. By contrast, the cones of the coastal form are $2\frac{1}{2}$ to 4 in. (6 to 10 cm) long with straight, appressed bracts. To complicate matters further, there is a "Fraser River" or intermediate Douglas-fir found in interior British Columbia, which is a form of the Rocky Mountain variety.

P. menziesii var. menziesii   coast Douglas-fir

This is the largest tree in the Northwest, and in the United States it is second only to the giant sequoias of California. Trees in old-growth forests average 180 to 250 ft in height and 4 to 6 ft dbh, although heights of 325 ft and diameters of 8 to 10 ft may be attained. The tallest reported specimen in the United States, near Mineral, Washington, was supposedly 393 ft in height and 15 ft dbh. Some believe that the height is a great exaggeration. In Olympic National Park, Washington, one tree is reported to be 298 ft tall, and another has a confirmed dbh of 15.9 ft.

Old trees are characterized by an exceptionally clear, long, cylindrical bole (Fig. 8-46) and either a rounded or an irregularly flat-topped crown. However, in closed stands of young trees, early natural pruning is poor, and a clear butt log may not begin to develop before the eightieth year. Anchorage is provided by a strong, well-developed, wide-spreading lateral root system. Trees are found on a variety of soils but make their best growth on deep, rich, well-drained, porous loams in regions where there is an abundance of both soil and atmospheric moisture. In youth, Douglas-fir forms extensive, pure, even-aged stands, but these are later commonly invaded by other species. Although rated as intermediate in tolerance, Douglas-fir is less so than its usual associates in the Pacific northwest, with the exception of noble fir. In the principal timber belt of the Sierra Nevada, it occurs chiefly with white fir and ponderosa pine, in the northern coast ranges of California with ponderosa pine, tanoak, and Oregon white oak, and in the redwood belt with redwood.

Seed is produced in quantity after about the twenty-fifth year, with heavy crops at irregular intervals. Dense thickets spring up rapidly after logging, provided fires are excluded. Under favorable conditions, trees 10 years of age may be 12 to 15 ft tall and 1 to 2 in. dbh. Pole size is often attained in 30 to 40 years, and saw logs may be harvested in less than 80 years. The trees attain great age and are commonly found with over 700 (max. 1375) growth rings. Douglas-fir is extensively planted in the Pacific Northwest.

Principal enemies are fire, the Douglas-fir bark beetle, certain heart rots and other fungi, and dwarf mistletoe. Large tracts of old-growth stands dominated by Douglas-fir provide critical habitat for the federally listed northern spotted owl and other organisms.

P. menziesii var. *glaùca* (Beissn.) Franco   Rocky Mountain Douglas-fir

This variety rarely exceeds a height of more than 130 ft (max. 200 ft) or a diameter of 3 ft. It occurs in both pure and mixed stands. Douglas-fir is more tolerant than its many coniferous Rocky Mountain associates except western hemlock and Engelmann spruce.

Although most abundant on moist sites, it is quite drought-resistant and is often found in arid areas with ponderosa pine. Its maximum age is 1275 years. It is frost-resistant and hardy in the East and is a common ornamental of that region. The trees are grown for timber in Europe and have been planted successfully in many parts of the world.

Pseudotsùga macrocárpa (Vasey) Mayr   bigcone Douglas-fir

Distinguishing Characteristics.   *Leaves* pointed. *Cones* large, 4 to 8 in. (10 to 20 cm), with rigid, woody scales subtended by longer trident bracts.

FIGURE 8-46   Old-growth Douglas-fir. (*Photo by J.D. Cress, Seattle.*)

Range. Southern California from eastern Santa Barbara County south along the mountains to Baja California. Elevation: 900 to 7875 ft. Dry slopes and canyons.

General Description. This tree never attains the grandeur or stature of Douglas-fir. Its rapidly tapering bole (max. 173 by 7.6 ft), usually clothed for the greater part—if not all—of its length in a massive pyramidal crown, is seldom used for the production of lumber in commercial quantities. This tree contributes little or nothing to the nation's timber supply, but it is a valuable cover tree and aids materially in erosion control. This species can develop a new crown following complete defoliation by fire. Bigcone Douglas-fir can live for 600 years.

TSUGA (Endl.) Carr.　　hemlock

This genus includes about 10 species of evergreen, usually pyramidal-shaped trees confined to the forests of eastern and western North America, Japan, Taiwan, China, and the Himalayas. Hemlock pollen has been identified in several French and Polish peat bogs, and fossil wood has been unearthed in other sections of Europe, but hemlocks are not found in the modern forests of that continent.

The Japanese hemlocks, *T. sieboldii* Carr. and *T. diversifòlia* (Maxim.) Mast., and the Chinese species, *T. chinénsis* (Franch.) Pritz., are Oriental timber-producing species. They are frequently used for ornamental purposes in other parts of the world, including North America.

Four species of *Tsuga* are indigenous to North America (two in the East and two in the West) (Table 8-11). Of these, the western hemlock is at present the most important timber-producing species in this genus. Hemlock bark contains from 7 to 12% tannin, and in the United States, bark of the eastern hemlock was one of the principal commercial sources of tannin for many years. Hemlocks are an important source of pulp.

Botanical Features of the Genus

Leaves spirally arranged, but often apparently 2-ranked by a twist of the petioles of the lowest leaves, and those on the upper side of the branchlet short and appressed; *shape* linear, flattened or less frequently angular or semicircular in cross section, generally grooved above, with two broad bands of stomata below; *margin* either entire or serrulate above the middle; *apex* blunt; conspicuously petioled; persistent for 3 to 6 years; each borne on a short peglike projection from the twig.

TABLE 8-11　COMPARISON OF IMPORTANT HEMLOCKS

Species	Leaves	Mature cones	Twigs
T. canadensis eastern hemlock	Tapering from base to apex, bands of stomata below well defined	Stalked, $\frac{5}{8}-\frac{3}{4}$ in. (1.5–2 cm) long, oblong-ovoid; scale margins not undulate	Densely pubescent; buds $\frac{1}{16}-\frac{1}{8}$ in. (1.5–2.5 mm) long
T. heterophylla western hemlock	Uniform width from base to apex, bands of stomata below poorly defined, upper surface grooved	Sessile, $\frac{3}{4}$–1 in. (2–2.5 cm) long, ovoid; scale margins undulate	Finely pubescent; buds $\frac{1}{8}-\frac{3}{16}$ in. (2.5–3.5 mm) long

Cones (juvenile) solitary; *pollen cones* axillary, globose, consisting of a number of spiral pollen sacs; *ovulate cones* erect, terminal on the lateral branchlets, consisting of a number of membranous bracts, each subtending a nearly orbicular scale with two basal ovules; *bracts* and *scales* of nearly the same length; all species monoecious.

Mature seed cones pendent, globose to ovoid or oblong, subsessile, maturing in one season; *scales* orbicular to ovate-polygonal, mostly with entire margins, several times longer than the bracts; *bracts* inconspicuous (rarely exserted); *seeds* small, ovoid-oblong, dotted with minute resin vesicles, with obovate, terminal wings; cotyledons 3 to 6.

Buds small, ovoid to globose, nonresinous.

Tsùga canadénsis (L.) Carr.   eastern hemlock, Canada hemlock

Distinguishing Characteristics (Fig. 8-47)

Leaves $\frac{3}{8}$ to $\frac{5}{8}$ in. (8 to 15 mm) long, tapering from base to apex, dark yellow-green, rounded or slightly emarginate at apex, margin serrulate along upper half, marked below with 2 white lines of stomata; persistent until the third season.

FIGURE 8-47 *Tsuga canadensis,* eastern hemlock. (1) Ovulate cone ×2. (2) Pollen cones ×2. (3) Mature seed cone ×$\frac{3}{4}$. (4) Open seed cone ×$\frac{3}{4}$. (5) Seed ×1. (6) Foliage ×$\frac{1}{2}$. (7) Leaf showing two white bands below ×2. (8) Bark of old tree.

Cones $\frac{5}{8}$ to $\frac{3}{4}$ in. (1.5 to 2 cm) long, short stalked, oblong-ovoid; *scales* suborbicular, smooth-margined; shedding their seeds during the winter and often persistent during the next season; *seeds* ca. $\frac{1}{16}$ in. (1 to 2 mm) long, with several resin vesicles; wings terminal, broad.

Bark on young trees flaky or scaly; some with wide, flat ridges; on old trees heavily and deeply furrowed; freshly cut surfaces showing purplish streaks.

Range
Northeastern (Map 8-17). Local disjuncts in central North Carolina, northwestern Alabama, western Ohio, and central Indiana. Elevation: sea level to 2400 ft northward, mostly 2000 to 6000 ft southward. Cool, moist valleys, coves, and slopes, low ridges, bluffs, lake shores; in mixed or pure stands.

General Description
Eastern hemlock is usually a medium-sized tree 60 to 70 ft tall and 2 to 3 ft dbh, although in the mountains of West Virginia, North Carolina, and Tennessee, it has attained a maximum height of 170 ft and a dbh of 6 to 7 ft. This species is exceedingly graceful in youth, and open-grown trees develop a dense, pyramidal crown, with the lower branches often touching the ground. The slender terminal leader in hemlocks is rare among conifers in that the new growth droops or curves away from the vertical as much as 90 degrees or more. During the first season, compression wood develops on the undersurface of the curve, and straightening begins. Several years may pass before this is completed, and the curved tip is at any season an excellent mark of identification. The crowns of old trees are inclined to be ragged in outline; even in dense stands, the lower branches remain alive for many years and, when dead, prune poorly. This causes knots, which in this species are often remarkably hard and flintlike. The bole, except in old trees, shows considerable taper and

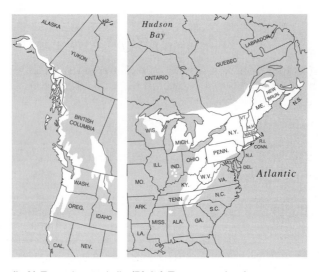

MAP 8-17   **(Left)** *Tsuga heterophylla.* **(Right)** *Tsuga canadensis.*

is supported below by a very superficial, wide-spreading root system. The shallow root system makes it very susceptible to drought damage.

This tree is found on many types of soil, and it reaches its best development in cool, moist situations. Eastern hemlock is remarkably tolerant, perhaps more so than any of its associates. Growth rings of trees under heavy shade may be so narrow that a hand lens is needed to see them. When such a suppressed tree is liberated by the removal of others around it, the sudden increase in growth is often amazing. Rings 10 times the width of the previous ones may be produced.

Eastern hemlock is associated with most of the species common to mesic and submesic sites of the northeastern forests and the Appalachian Mountains. Over long periods of time, it tends to replace its associates; pure stands, however, are of limited extent.

Abundant seed is produced at 2- to 3-year intervals, and seedling reproduction is plentiful on moist, shaded humus; in fact, the seeds even germinate on old, partially rotted stumps or logs, and occasionally in the axils of tree branches. In this respect, yellow birch is the chief competitor, both beginning as epiphytes. Although some reproduction may occur on bare, exposed soil, this hemlock usually starts under the shade of other species and, growing very slowly, forces its way, during a period of a century or more, into the crown cover overhead. Many hemlocks 2 to 3 in. dbh may be 100 to 200 years old. A maximum age of 539 years has been confirmed. There is one report of an eastern hemlock 988 years old.

Hemlocks are highly valued as ornamentals, and young trees may show great variation in form and foliage. About 70 cultivars of the eastern hemlock have been described. This tree is also used for lumber and newspaper pulp. Several fungi, such as root rots and rusts, fire, and many insects may cause damage. The hemlock looper has impacted eastern hemlock recently, especially in Maine. The most recent and extreme damage is from the hemlock woolly adelgid (*Adelges tsugeae*), which started in the Northeast and has spread southward into the Virginia and North Carolina mountains. It may have the potential for extreme destruction to the natural stands. Even where hemlock woolly adelgid is not a threat, there has been great concern about the substantial decline in this forest type, due to its regeneration requirements and heavy browsing by white-tailed deer.

Eastern hemlock is the state tree of Pennsylvania.

Tsùga caroliniàna Engelm.   Carolina hemlock

This is endemic to the upper, dry, rocky slopes, ridges, and ledges of the southern Appalachian Mountains from southwestern Virginia to northeastern Georgia; elevation 2300 to 4000 ft. The cones (Fig. 8-48) and leaves are longer (cone $\frac{3}{4}$ to $1\frac{3}{8}$ in. [2 to 3.5 cm]; leaves $\frac{3}{8}$ to $\frac{3}{4}$ in. [10–20 mm]) than those of eastern hemlock, the leaves spread in all directions, and the margins are entire. Carolina hemlock is well adapted to relatively drier sites in comparison to eastern hemlock. It is sometimes used as an ornamental. It reaches 40 to 60 ft tall and up to 2 ft dbh.

FIGURE 8-48   *Tsuga caroliniana,* Carolina hemlock. Open seed cone $\times\frac{3}{4}$.

Tsùga heterophýlla (Raf.) Sarg.   western hemlock, Pacific hemlock

Distinguishing Characteristics (Fig. 8-49)

Leaves $\frac{1}{4}$ to $\frac{3}{4}$ in. (6 to 20 mm) long, flattened, of uniform width from base to apex, dark shiny green, grooved above, with 2 poorly defined bands of stomata below; mostly rounded or blunt at the apex, margin serrulate, persistent for 4 to 7 years.

FIGURE 8-49   *Tsuga heterophylla,* western hemlock. (1) Foliage and seed cones ×$\frac{3}{4}$. (2) Seed ×1. (3) Bark of old tree.

Cones $\frac{3}{4}$ to 1 in. (1.8 to 2.5 cm) long, sessile, ovoid, light brown; *scales* suborbicular, often more or less wavy along the margin; *seeds* ca. $\frac{1}{16}$ in. (2 mm) long, ovoid, nearly surrounded by a large straw-colored wing.

Bark thin even on the largest trees, separated by deep, narrow fissures into broad, flat, russet-brown ridges; inner bark dark red streaked with purple.

Range

Northwestern (Map 8-17). Elevation: sea level to 2000 ft along the coast; to 6000 ft inland. Deep, moist, acid soils on flats and also drier lower slopes.

General Description

Western hemlock is one of the major timber-contributing species in the Pacific Northwest and is the most important timber hemlock in North America. Largest of the four North American hemlocks, this tree varies from 125 to 175 ft in height and 2 to 4 ft dbh (max. 258 by 9.4 ft). Forest trees produce a long, clear, symmetrical bole and a shallow, wide-spreading root system. The crown is short, open, and pyramidal, and, like the eastern hemlock, features a drooping terminal leader. An abundance of both soil and atmospheric moisture is requisite for rapid growth, and the largest trees are invariably found on moist porous soils in regions where the annual precipitation is at least 70 in. per year. In drier situations, the trees are somewhat smaller and slower-growing, but they do eventually reach commercial proportions. Western hemlock occurs in pure, dense, even-age forests and as an occasional tree in mixed hardwood and coniferous stands, then commonly as an understory species. Nearly pure, extensive forests of western hemlock occur in southeastern Alaska, coastal British Columbia, and western Washington. Those restricted to the middle-upper western slopes of the Cascade and Olympic Mountains usually occur above the Douglas-fir belt. At low elevations, hemlock commonly invades forests chiefly composed of Douglas-fir, Pacific silver and grand firs, western redcedar, black cottonwood, bigleaf maple, and red alder. At higher levels, its associates include noble fir, Alaska-cedar, mountain hemlock, and western white pine. Western hemlock is a major component of five forest types and is a minor species in eleven others (Burns and Honkala, 1990).

Western hemlock begins to produce seed after the twenty-fifth to thirtieth year, and large crops are borne every 2 to 8 years thereafter. Nearly any sort of soil provides a satisfactory seedbed. Partially rotted stumps and logs are often literally covered with hemlock seedlings, and fresh mineral soil, moist duff, or even sphagnum peat are suitable for seedling development. This species often usurps cutover and burned-over areas formerly occupied by other species when moisture is not a limiting factor. The trees are very tolerant throughout life. Rate of growth compares favorably with that of Douglas-fir, and trees 90 to 100 years of age are commonly 110 to 130 ft high and 15 to 18 in. dbh. The largest trees rarely attain an age of more than 500 years, but they can reach 1238 years old. Western hemlock is susceptible to fire injury, butt rot is prevalent in overmature trees, and dwarf mistletoe is a common parasite.

Western hemlock timber was virtually unknown until the close of World War I, except locally. However, since then, it has found a ready market in many of the world's leading lumber centers. It is also a major pulping species in the Northwest. The bark of this tree contains tannin (9 to 16%), but it is rarely used commercially.

Western hemlock is the state tree of Washington.

Tsùga mertensiàna (Bong.) Carr.   mountain hemlock

Distinguishing Characteristics (Fig. 8-50).   *Leaves* $\frac{1}{4}$ to 1 in. (6 to 25 mm) long, blue-green, semicircular in cross section, margin entire, the upper surface often keeled or grooved; extending from all sides of the twig or crowded toward the upper side, stomatiferous on all surfaces. *Cones* 1 to 3 in. (2.5 to 7.5 cm) long (the largest cones are found on trees toward the southern limits of the range), oblong-cylindrical, yellow-green to purple, the scales often strongly reflexed after shedding their seeds; *seeds* $\frac{1}{8}$ in. (3 mm) long, with one or two resin vesicles; the wings terminal, about $\frac{1}{2}$ in. (13 mm) long. *Bark* dull purplish brown to reddish brown, divided into narrow flattened ridges by deep, narrow fissures.

Range. Southern Alaska south along the coast ranges of British Columbia, Washington, and Oregon, and inland on the upper slopes of the Sierra Nevada to central California; also in southeastern British Columbia, and south in the Rocky Mountains to northeastern Oregon, northern Idaho, and northwestern Montana. Elevation: sea level to 3500 ft northward, 5500 to 11,000 ft southward. Sheltered valleys, alpine meadows to exposed slopes and ridges, in mixed or pure stands.

General Description. Mountain hemlock, although one of the largest of alpine trees, is often a low, sprawling, or even prostrate shrub on windswept ridges at timberline. Forest trees, however, average 75 to 100 ft in height and $2\frac{1}{2}$ to $3\frac{1}{2}$ ft dbh (max. 194 by 7 ft) and are characterized by a shallow root system, which supports a long, clear, or limby bole with a narrow pyramidal crown of drooping or even pendulous branches. Trees on the sides of canyons and on very steep slopes are

FIGURE 8-50   *Tsuga mertensiana,* mountain hemlock. Foliage and mature seed cones $\times\frac{1}{2}$.

commonly pistol-butted (Fig. 8-51). In open situations, particularly in high-alpine meadows, the boles are excessively tapered. Here mountain hemlock forms pure, parklike stands or occurs with other regional subalpine species. The best stands of this species, however, are found on moist slopes, flats, and heads of ravines of northerly exposure. Although pure stands are seldom extensive, mountain hemlock commonly constitutes 85% or more of certain mixed coniferous forests. This species reaches its maximum development in southern Oregon, and well-stocked stands are found in the vicinity of Crater Lake National Park.

Seeds are borne at an early age, and regular crops are produced annually after the twenty-fifth to thirtieth year. Seedlings are very tolerant, comparing favorably in this respect with those of western hemlock. Dense shade causes suppression, but the trees recover rapidly upon being released. Growth is at no time rapid, and trees 200 to 300 years of age are seldom ever more than 18 to 24 in. dbh on the best sites. Trees 500 years of age or more are rarely found. The laminated root rot that kills in expanding circles is a virulent pathogen of mature stands.

This species is used as an ornamental in a number of European countries, and several cultivars are known. The bark of mature trees contains large quantities of tannin, but owing to the relative inaccessibility of the trees, it is not used at the present time. Like the subalpine fir, it could perhaps be a more important timber and pulpwood tree in the future.

ABIES Mill.   fir

The genus *Abies* includes 42 species of trees, widely scattered through the forests of North and Central America, Europe, Asia, and northern Africa. Trees growing in southern latitudes are usually restricted to the upper slopes of mountains and are commonly found on sites at or near timberline; those in boreal forests, however, are largely confined to regions of relatively low elevations, with only a few forms ascending to the upper limits of tree growth.

FIGURE 8-51   Mountain hemlock. (*Courtesy of U.S. Forest Service.*)

This genus comprises a number of important species, although the timber produced is often somewhat inferior to that of the spruces. Several species, including balsam fir, furnish pulpwood. Other forest products traceable to certain of the firs include the oleoresins known to the trade as Canada balsam (Strasburg turpentine) and leaf oils. The former is obtained from superficial pitch blisters on the bark and used chiefly in the manufacture of medicinal compounds and varnishes, as a mounting medium in the preparation of microscope slides, and for cementing lenses in optical instruments. Leaf oils are largely used in the manufacture of pharmaceutical compounds. Several species are very important for Christmas greenery.

Abies pìndrow Royle, Himalayan silver fir, is a valuable species indigenous to India. *Abies álba* Mill., silver fir, or "Swiss pine," is an important timber tree of southern and central Europe. The Japanese momi and nikko firs, *A. fírma* Sieb. and Zucc. and *A. homolèpis* Sieb. and Zucc., are two noteworthy trees of Japan, the former being one of its principal pulpwood species. In Central America, the wood of the sacred fir, *A. religiòsa* (HBK) Schlecht. and Cham., is used, particularly in Mexico and Guatemala, for structural purposes. The Siberian fir, *A. sibírica* Ledeb., forms extensive forests through northern and eastern Russia, Siberia, and Turkestan. Some of the timber produced by this species has been offered to the world's trade, and it may become a competitor of the American pulpwood species.

All native species and a number of introduced firs are highly prized for decorative purposes. The Spanish fir, *A. pinsápo* Boiss., a tree with sprucelike needles, is a common ornamental in the Pacific Northwest and in Canada. The stately Greek fir, *A. cephalónica* Loud., the Nordmann fir, *A. nordmanniàna* (Steven) Spach., and the nikko fir previously mentioned are other trees favored for ornamental plantings.

Eleven species of *Abies* are included in the coniferous flora of North America (FNA vol 2, 1993). Nine of these are scattered through the forests of the West (Table 8-12), and two are found in the East. See Robson et al. (1993) for a discussion of species relationships.

Botanical Features of the Genus

Leaves spirally arranged, often twisted and appearing 1-ranked (all pointing upward) or 2-ranked (pointing to the sides), sessile; *shape* linear, flattened in cross section, or

TABLE 8-12   COMPARISON OF IMPORTANT WESTERN FIRS

Species	Leaves	Mature cones	Bark
A. amabilis Pacific silver fir	$\frac{3}{4}$–$1\frac{1}{2}$ in. (2–4 cm) long, flattened, bands of stomata only below	3–6 in. (7.5–15 cm) long; bracts shorter than scales	Silvery white to ashy gray
A. magnifica California red fir	$\frac{3}{4}$–$1\frac{3}{8}$ in. (2–3.5 cm) long, 4-angled, bands of stomata on all sides	6–9 in. (15–23 cm) long; bracts shorter (or longer) than scales	Red-brown
A. procera noble fir	1–$1\frac{5}{8}$ in. (2.5–4 cm) long, 4-angled, grooved above, bands of stomata on all sides	4–7 in. (10–18 cm) long; bracts longer than scales	Dark gray, tinged with purple
A. grandis grand fir	$\frac{3}{4}$–2 in. (2–5 cm) long, flattened, bands of stomata only below; 2-ranked	2–$4\frac{1}{4}$ in. (5–11 cm) long; bracts shorter than scales	Gray-brown to reddish brown
A. concolor white fir	$1\frac{1}{2}$–3 in. (4–7.5 cm) long, flattened, bands of stomata on both sides; obscurely 2-ranked	3–5 in. (7.5–13 cm) long, bracts shorter than scales	Dark ashy gray to nearly black, corky ridges

rarely 4-angled and then grooved above, with numerous lines of stomata on the lower or occasionally all surfaces; *apex* acute, rounded, or shallowly to deeply notched or bifid; persistent for 7 to 10 years; usually with a balsamic, or turpentinelike odor when bruised.

Cones (juvenile) solitary; *pollen cones* appearing from buds of the previous season, and borne on the underside of the lower crown branches in the axils of leaves, pollen sacs spirally arranged; *ovulate cones* erect, composed of many bracts each subtending a large scale with two inverted basal ovules; strobili axillary on the upper crown branches of the previous season's growth; all species monoecious.

Mature seed cones erect, maturing at the end of the first season; *scales* thin, broadly fan-shaped, particularly near the center of the cone, longer or shorter than the bracts; *bracts* apiculate, often shouldered, their margins erose; both scales and bracts together deciduous at maturity, leaving a mostly bare spikelike axis that frequently persists through the winter; *seeds* ovoid to oblong, with conspicuous resin vesicles; the broad terminal wings often delicately tinted with shades of pink, rose, lavender, or brown; cotyledons 5 to 7.

Twigs with ovoid to oblong buds, blunt, less commonly pointed, resinous or rarely nonresinous; leaf scars smooth, circular, and light colored.

Eastern Firs

Àbies balsàmea (L.) Mill.   balsam fir

Distinguishing Characteristics (Fig. 8-52)
Leaves $\frac{1}{2}$ to 1 in. (12 to 25 mm) long, flattened, dark shiny green above, silvery banded below, blunt, or slightly notched at the apex, 2-ranked, tending to crowd toward the upper surface of the twig.

Cones $1\frac{1}{2}$ to $3\frac{1}{4}$ in. (4 to 8 cm) long, oblong-cylindrical, green tinged with purple; *bracts* usually shorter than the scales, with rounded erose shoulders and short, pointed tips; *seeds* ca. $\frac{1}{8}$ in. (3 mm) long, with broad purplish brown wings.

Buds $\frac{1}{8}$ to $\frac{1}{4}$ in. (3 to 6 mm) long, subglobose, resinous, with orange-green scales.

Bark $\frac{1}{2}$ in. (13 mm) thick, dull green, later with grayish patches, smooth except for numerous raised resin blisters; eventually breaking up into small reddish brown irregular scaly plates.

Range
Northern (Map 8-18). Local disjuncts in southwestern Wisconsin, northeastern Iowa, West Virginia, and western Virginia. Elevation: sea level to timberline northward, to 5700 ft southward. Swamps to well-drained uplands.

General Description
Balsam fir is a small- to medium-sized tree 40 to 60 ft tall and 12 to 18 in. dbh (max. 125 by 3 ft). It is perhaps the most symmetrical of northeastern trees with its dense, dark green, narrowly pyramidal crown, terminating in a stiff, spikelike tip (Fig. 8-53). The moderately tapering bole usually retains small dead, persistent branches and is supported by a shallow, wide-spreading root system.

Balsam fir is a cold-climate tree, requiring abundant moisture for best development. In swamps, it often forms pure stands but does best in association with white spruce (boreal region)

FIGURE 8-52 *Abies balsamea,* balsam fir. (1) Foliage $\times\frac{1}{2}$. (2) Mature seed cone $\times\frac{1}{2}$. (3) Dead twig showing smooth, round leaf scars $\times 1\frac{1}{2}$. (4) Seed cone with scales partly dehisced $\times\frac{3}{4}$. (5) Seed cone axis remaining after nearly all scales have dehisced $\times\frac{3}{4}$. (6) Abaxial view of scale with subtending bract $\times 1\frac{1}{4}$. (7) Adaxial view of scale with 2 winged seeds $\times 1\frac{1}{4}$. (8) Bark of old tree.

MAP 8-18   *Abies balsamea.* The disjunct area in Virginia and West Virginia is *A. balsamea* var. *phanerolepis.*

and red spruce (northeastern United States) on the adjacent flats that are better drained; even here, pure stands or groups are commonly found. On higher ground, it occurs as a scattered tree in mixture with other northern species. It appears again in dwarfed, matted, pure stands, or entangled with black spruce near the windswept summits of northern mountains where great extremes in temperature occur. In pure stands at high elevations in northeastern mountains, mortality occurs in crescent-shaped bands, known as "fir waves."

Good seed years occur at about 2-year intervals after the fifteenth, and large quantities of seeds are then produced. A single balsam fir tree can produce nearly 20,000 seeds. However, within a given year, a significant proportion of seeds may be empty due to inadequate pollination. It is aggressive in restocking cutover lands, and it makes rapid growth if not shaded. Propagation of new fir trees by layering occurs from lower branches in contact with moist soil. Balsam fir is very shade-tolerant. For a complete list of associates and tolerances, see Bakuzis and Hansen (1965).

This fir is short-lived, and although it may reach an age of nearly 200 years, trees over 90 years old show a high percentage of heart rot caused by wood-destroying fungi. The chief insect enemies of balsam fir are the spruce budworm and the balsam woolly adelgid; fire also causes severe loss.

Balsam fir is an important pulpwood species and is also used for lumber. Plantation-grown balsam fir, Fraser fir, and Douglas-fir are in high demand for Christmas trees. Balsam fir is the provincial tree of New Brunswick.

Abies balsamea var. *phanerolepis* Fern., bracted balsam fir, intermediate between *A. balsamea* and *A. fraseri,* in both cone morphology and geography, has been variously interpreted as either a relict hybrid from an earlier time (Pleistocene) when the species were sympatric in the Virginia/West Virginia area, an ecotype or variety of balsam fir, or a relict of an ancient north-south cline (Clarkson and Fairbrothers, 1970; Jacobs et al., 1984; Thor and Barnett, 1974). More recent evidence (Clark, 1998) indicates two distinct populations, one in West Virginia more like balsam fir, and one in Virginia more like Fraser fir. This may indicate relicts of broader distributions of both species. Trees of the coastal regions in eastern Canada, identified as this variety, may indicate a more northern distribution of Fraser fir in earlier times. The West Virginia population is locally known as Canaan fir, and it is grown commercially for Christmas trees in that area.

FIGURE 8-53   Balsam fir. (*Courtesy of U.S. Forest Service.*)

Àbies fràseri (Pursh) Poir.   Fraser fir, Fraser balsam fir

This species is very similar to *A. balsamea* except for the cone bracts, which are longer than the scales and strongly reflexed. It is considered as *A. balsamea* var. *fraseri* Spach by some (Thor and Barnett, 1974). Typical trees have a characteristic appearance and reach a height of 30 to 50 ft. It is a tree of the high mountains in southwestern Virginia, western North Carolina, and eastern Tennessee at 4500 to 6684 ft elevation. It is very shade tolerant.

This species is of particular importance in the Christmas tree trade and is cultivated extensively for this purpose. Nearly all of the natural stands have been severely damaged, with mortality of mature trees at 44–91%, by the balsam woolly adelgid (*Adelges piceae*) first found on Mount Mitchell, North Carolina, in 1957, and in combination with acid precipitation. The once magnificent stands in such locations as Mount Mitchell and elsewhere in the high mountains of North Carolina and Tennessee are now only dead snags. Reproduction is extensive, however, and the future status of Fraser fir in natural stands is extremely uncertain.

Western Firs

Àbies amábilis Dougl. ex Forbes   Pacific silver fir, Cascades fir

Distinguishing Characteristics (Fig. 8-54)

Leaves $\frac{3}{4}$ to $1\frac{1}{2}$ in. (2 to 4 cm) long, lustrous dark green above, silvery white (stomatiferous) below; flattened in cross section, deeply grooved above; crowded toward the upper side of the twig, those above often twisted at the base in such a way that they appear to be appressed along the twig; notched or pointed at the apex; those from cone-bearing branches occasionally somewhat thicker and often also stomatiferous on the upper surface at the apex.

Cones 3 to 6 in. (7.5 to 15 cm) long, cylindrical to barrel-shaped, deep purple; *bracts* shorter than the scales, with rounded, poorly defined shoulders, and a broad, wedgelike spinose tip; *seeds* ca. $\frac{1}{2}$ in. (13 mm) long, yellowish brown, the wings very broad, straw-colored.

Twigs stout, yellowish brown, puberulous; buds $\frac{1}{4}$ in. (5 to 7 mm) long, subglobose, dark purple, resinous.

Bark ashy gray, with large, irregular chalky-colored blotches and resin blisters on stems up to 3 ft dbh; superficially scaly on the largest trunks.

Range

Northwestern (Map 8-19). Local disjunct in northwestern California. Elevation: sea level to 1000 ft northward, to 6600 ft southward. Coastal fog belt and interior mountain valleys and slopes.

General Description

Pacific silver fir is the most abundant fir in the Pacific Northwest and forms extensive pure stands in many localities. It attains its largest proportions in the Olympic Mountains in western Washington, where mature trees vary from 140 to 160 ft in height and 2 to 4 ft dbh (max. 236 by 8.4 ft). The bole and root systems are similar to those of other firs of low altitudes, but the crown, particularly of old trees, is usually pyramidal or even spikelike; hence, it is readily distinguished from that of either the noble or grand fir, which are usually domelike. The trees are most abundant on deep moist soil covering slopes of southern and western exposure and grown on a variety of soil types. In mixed stands at

MAP 8-19   **(Left)** *Abies amabilis.* **(Right)** *Abies magnifica.*

FIGURE 8-54 *Abies amabilis,* Pacific silver fir. (1) Foliage $\times\frac{1}{2}$. (2) Seed $\times1$. (3) Mature seed cone $\times\frac{3}{4}$. (4) Abaxial view of cone scale and subtending bract $\times1$. (5) Bark of old tree.

low levels, this species is commonly associated with Sitka spruce, Douglas-fir, grand fir, western hemlock, and western redcedar, and at higher altitudes it occurs with Engelmann spruce, subalpine and noble firs, Alaska-cedar, western larch, western white pine, and mountain hemlock.

Large seed crops are borne every 3 to 6 years (after the thirtieth). Seedlings become readily established in either a duff or mineral soil and are very tolerant. In fact, this species is considered as tolerant as western hemlock and western redcedar, and it may be that the true climax forest west of the Cascades and in the Olympics is one of Pacific silver fir. Although excessive shade results in extremely slow growth, excellent recovery is made upon release. Even so, growth is not rapid, and forest trees 150 to 250 years of age are often only 15 to 24 in. dbh. Maturity is attained in about 250 years; trees of a much greater age are seldom found (max. 725 years).

Because of its beautifully shaped crown and dense lustrous foliage, the silver fir is commonly used ornamentally, particularly in Europe. It is also used for construction lumber, plywood, and pulp. Pacific silver fir and noble fir have long been known as "larch" because that name was once commonly applied to lumber from these two species and shipped to the Orient.

Outbreaks of balsam woolly adelgid recently have reached epidemic proportions and have taken a heavy toll among all age and vigor classes. Wood-destroying and root-rotting fungi are common in overmature timber. Damage also occurs from other fungi, several insects, and dwarf mistletoe. Fire and windthrow also cause much damage.

Àbies magnífica A. Murr.   California red fir, red fir

Distinguishing Characteristics (Fig. 8-55)

Leaves $\frac{3}{4}$ to $1\frac{3}{8}$ in. (2 to 3.5 cm) long, silvery blue to dark blue-green, on new shoots often silvery white; 4-angled, stomatiferous on all sides, usually pointed at the apex, crowded toward the upper side of the branchlets.

Cones 6 to 9 in. (15 to 23 cm) long, 2 to $3\frac{1}{4}$ in. (5 to 8 cm) in diameter, cylindrical to barrel-shaped, purplish brown; *bracts* shorter than the scales, with more or less parallel edges and terminating rather abruptly in a spikelike tip; *seeds* $\frac{1}{2}$ to $\frac{3}{4}$ in. (13 to 19 mm) long, dark brown, with large, broad, reddish purple terminal wings. *Abies magnifica* var. *shasténsis* Lemm., Shasta red fir, has bracts longer than the cone scales, but there are intermediate forms.

Twigs light brown; *buds* $\frac{1}{4}$ in. (6 mm) long, ovoid, dark brown, nonresinous, or slightly resinous.

Bark smooth and chalky on young stems; 4 to 6 in. (10 to 15 cm) thick on old trunks and divided by deep furrows into rounded or plated reddish-colored ridges.

Range

Western (Map 8-19). Elevation: sea level to 4500 ft northward, 5300 to 9000 ft southward. Mountain slopes, ravines, and ridges.

General Description

This is one of the two largest North American firs and at maturity varies from 150 to 180 ft in height and 4 or 5 ft dbh (max. 252 by 10 ft). Under forest conditions, it is similar in form to noble fir, but it is at once distinguished from that species by its reddish-colored bark. The largest trees are found on cool, moist, gravelly, or sandy soils in sheltered ravines and gulches or in protected mountain slopes, although this species also reaches commercial proportions on much poorer sites.

FIGURE 8-55 *Abies magnifica,* California red fir. (1) Foliage ×$\frac{1}{2}$. (2) Mature seed cone ×$\frac{3}{4}$. (3) Seed ×1. (4) Abaxial view of scale and subtending bract ×1. (5) Bark of old tree.

California red fir occurs in pure forests above the white fir belt and in some sections is a common tree at timberline. In mixed stands, it is found with many species; in southern Oregon and northern California, these include Douglas-fir, sugar pine, and ponderosa pine at lower altitudes, and mountain hemlock and lodgepole pine at higher levels. In the Sierra Nevada, it is found with white and Shasta firs, lodgepole, western white, ponderosa, Jeffrey, and sugar pines, incense-cedar, giant sequoia, and mountain hemlock. Trees at high elevations standing on exposed ridges frequently lose their upper crowns from the action of gale-force winds. Such destroyed crowns may be replaced by upsweeping lateral branches from just below the broken top. This replacement growth is often distinctive when silhouetted against the sky.

Red fir produces large seed crops every 2 or 3 years (after the fortieth). Seedlings are hardy and survive well. Older trees cannot meet intensive forest competition and are intermediate in tolerance. Growth is never rapid, and trees 250 to 350 years of age are commonly only 20 to 35 in. dbh. The longevity of this species is not too well known, but trees in excess of 500 years of age have been reported.

This species is damaged by several insects and fungi, but dwarf mistletoe is the most serious pest of California red fir. It is more sensitive to drought, but less sensitive to frost damage, than white fir.

California red fir is used for construction lumber. It is also an excellent Christmas tree.

Where California red fir and noble fir are sympatric, in southern Oregon and northern California, many populations are intermediate and have been interpreted as either hybrids or ancestral to the two species (Hunt, 1993).

Àbies prócera Rehd. noble fir

Distinguishing Characteristics (Fig. 8-56)

Leaves 1 to $1\frac{5}{8}$ in. (2.5 to 4 cm) long, blue-green, stomatiferous on all surfaces, grooved above, pointed at the apex, and crowded toward the upper side of the twig; those on sterile branches flexible, somewhat flattened; those on cone-bearing shoots stout, stiff, and nearly equally 4-sided.

Cones 4 to 7 in. (10 to 18 cm) long, $1\frac{5}{8}$ to $2\frac{3}{8}$ in. (4 to 6 cm) in diameter, cylindrical, olive-brown to purple; *bracts* exserted, strongly reflexed, and imbricate, thus completely ensheathing the cone; *seeds* $\frac{1}{2}$ in. (12 mm) long, dull brown, often tinged with red; wings lustrous light brown to straw-colored.

Twigs slender, reddish brown, finely pubescent; *buds* $\frac{1}{8}$ in. (3 mm) long, blunt, oblong-conical, resinous.

Bark 1 to 2 in. (2.5 to 5 cm) thick, gray and smooth for many years, with prominent resin blisters; eventually dark gray, often tinged with purple and broken up into thin, nearly rectangular plates separated by deep fissures on old trunks.

Range

Northwestern (Map 8-20). Elevation: mainly 3500 to 7000 ft, infrequently 200 to 8900 ft. Mountain ridges, slopes, and high valleys.

General Description

Noble fir, the largest (max. 295 by 9 ft) of the Cascade firs and the one producing the best timber, has the smallest range. It was first discovered in 1825 by David Douglas in the Oregon Cascades near the Columbia River. The finest stands have long since been logged, and those that remain are

FIGURE 8-56 *Abies procera*, noble fir. (1) Foliage $\times\frac{1}{2}$. (2) Mature seed cone $\times\frac{3}{4}$. (3) Seed $\times 1$. (4) Abaxial view of scale and subtending bract $\times 1$. (5) Bark of old tree (*Photo by C.F. Brockman*).

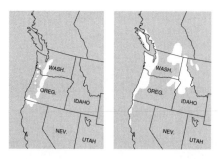

MAP 8-20   **(Left)** *Abies procera.* **(Right)** *Abies grandis.*

for the most part relatively inaccessible at the present time. Mature trees produce a long, clear, columnar bole with an essentially domelike crown and are anchored by a moderately deep and spreading root system. A deep, moist, cool soil is preferred, although good growth is also made on thin, rocky soils if provided with an abundance of moisture. Noble fir usually occurs singly or in small groups in association with other conifers. Douglas-fir, western hemlock, Pacific silver and subalpine firs, western white pine, and western redcedar are its principal forest associates in northerly portions of its range, whereas toward the southern limit it also occurs with white fir and sugar pine.

Noble fir is not a prolific seeder. Small quantities of seed may be borne annually, but heavy crops are released only at 3- to 6-year intervals (after the fifteenth). Of all the firs in the Pacific Northwest, noble fir is least tolerant. Growth in both height and diameter, although very slow in the seedling stage, is moderately rapid during the first century, as is shown by the fact that trees 100 to 120 years of age are commonly 90 to 120 ft in height and 20 to 24 in. dbh. Maturity is reached in about 350 years, although trees with 600+ annual rings have been reported.

The bark of noble fir is very thin, and trees of all age classes are easily killed by fire. Overmature trees are frequently damaged by one or more species of decay-producing fungi, including the Indian paint fungus. Damage by insects is rarely severe, but damage occurs from dwarf mistletoe in the southern part of the range. This fir is used for lumber, as Christmas trees, and as an ornamental. Because of its high value, it is being extensively planted within its range.

Hybrids are formed with *A. lasiocarpa* and *A. magnifica.*

Àbies grándis (Dougl. ex D. Don) Lindl.   grand fir

Distinguishing Characteristics (Fig. 8-57)

Leaves $\frac{3}{4}$ to 2 in. (2 to 5 cm) long, flattened, lustrous, dark yellow-green above, silvery (stomatiferous) below, those on sterile side branches 2-ranked, the ones from the upper side shorter than the laterals or those from underneath; often crowded on the upper side of cone-bearing branches; blunt or emarginate at the apex.

Cones 2 to $4\frac{1}{4}$ in. (5 to 11 cm) long, cylindrical, yellowish green to greenish purple; *bracts* shorter than the scales, the shoulders erose, rounded-truncate or cordate, terminating in a short spikelike tip; *seeds* $\frac{3}{8}$ in. (9 mm) long, pale yellow-brown; wings about $\frac{3}{4}$ in. (19 mm) long, straw-colored.

Twigs slender, brown, puberulous; *buds* $\frac{1}{4}$ in. (6 mm) long, subglobose, resinous.

Bark smooth, gray-brown with resin blisters and chalky white blotches on young stems; reddish brown, plated, or more commonly deeply furrowed or divided into flat ridges on old trunks.

Range

Northwestern (Map 8-20). Elevation: sea level to 1500 ft along the coast, to 6000 ft inland. Moist coastal forests, valleys, and mountain slopes.

FIGURE 8-57 *Abies grandis,* grand fir. (1) Foliage $\times\frac{1}{2}$. (2) Mature seed cone $\times\frac{3}{4}$. (3) Adaxial side of scale with seeds (above) and abaxial side with subtending bract (below) $\times$1. (4) Seeds $\times$1. (5) Bark of old tree.

General Description

Grand fir is one of the two firs found in the northern Rocky Mountain region and one of four common to the forests of the Pacific Northwest. Trees in coastal forests reach 140 to 160 ft in height and 2 to 4 ft dbh (max. 267 by 7 ft), but in the Rocky Mountains they are rarely more than 120 ft high or 3 ft dbh. Mature trees develop long, clear, columnar boles and domelike crowns (Fig. 8-58); the roots are deep and spreading. Grand fir occurs most frequently on deep, moist alluvial soils in gulches, along streams, and on gentle mountain slopes. Although it occasionally forms limited pure

FIGURE 8-58   Grand fir. (*Courtesy of U.S. Forest Service.*)

stands (northcentral Idaho), it is much more abundant in mixed hardwood and coniferous forests. Douglas-fir, western larch, and ponderosa and lodgepole pines are its common associates in the Rocky Mountains; in the Bitterroot Mountains to the west it occurs with western larch, western white pine, Engelmann spruce, and subalpine fir. At sea level and low elevations in the Cascade Mountains of eastern Washington and Oregon, grand fir occurs with Sitka spruce, Pacific silver fir, western redcedar, western hemlock, Douglas-fir, Oregon ash, red alder, bigleaf maple, and black cottonwood. Redwood is an associated species at the southern limit of its range, and at higher elevations to the south, grand fir occurs with Shasta red, white, noble, and subalpine firs and western white pine.

The seedling habits of grand fir are similar to those of white fir. Grand fir is a tolerant tree, but it is less so than western hemlock or western redcedar. Early height growth in Idaho often compares favorably with that of western white pine and on the Pacific coast with that of Douglas-fir. Maturity is reached in about 200 years.

Overmature trees are frequent hosts of the brown stringyrot and Indian paint fungi. Young trees are commonly attacked by the eastern spruce budworm and several other insects. It is used for lumber and pulp and is a preferred Christmas tree in the Pacific Northwest.

Grand fir hybridizes with *A. concolor.* See Hunt (1993) for a discussion.

Àbies cóncolor (Gord. and Glend.) Hildebr.   white fir

Distinguishing Characteristics (Fig. 8-59)

Leaves $1\frac{1}{2}$ to 3 in. (4 to 7.5 cm) long, silvery blue to silvery green, extending at nearly right angles from all sides of the twig or more or less obscurely 2-ranked; flattened, stomatiferous above and below, rounded or acute at the apex.

Cones 3 to 5 in. (7.5 to 13 cm) long, oblong, olive-green to purple; *bracts* shorter than the scales, with short, broad erose shoulders, and a spikelike tip; *seeds* $\frac{1}{2}$ in. (13 mm) long, yellow-brown, the wings rose-tinted.

Twigs moderately stout, light brown to gray; *buds* ca. $\frac{1}{4}$ in. (6 mm) long, subglobose, yellowish brown, resinous.

Bark 4 to 7 in. (10 to 18 cm) thick on old trunks, ashy gray and divided by deep irregular furrows into thick, corky, flattened ridges; young stems with conspicuous resin blisters.

Range

Western (Map 8-21). Elevation: sea level to 2000 ft northward, 5500 to 11,200 ft southward. Mountain slopes and ridges.

General Description

White fir has the largest, though fragmented, range of any of the commercial western firs. It attains its maximum development in the Sierra Nevada, where it becomes 130 to 150 ft in height and 3 to 4 ft dbh (max. 246 by 9 ft). The Rocky Mountain trees are usually much smaller and rarely attain a height of 100 ft or a diameter of 15 to 30 in. In youth, the crown is often elongated and covers a slightly tapered bole from one-half to two-thirds of its length. Older trees usually have a domelike crown unless malformed by the fir dwarf mistletoe. The root system is shallow and wide-spreading. White fir occurs most abundantly on deep, rich, moist but well-drained, gravelly, or sandy loam-covered slopes and benches with a northerly exposure. However, this species requires less moisture than other western firs and possesses a remarkable ability to exist

FIGURE 8-59 *Abies concolor,* white fir. (1) Foliage $\times\frac{1}{2}$. (2) Mature seed cone $\times\frac{3}{4}$. (3) Adaxial view of scale with two winged seeds $\times 1$. (4) Abaxial view of scale with subtending bract $\times 1$. (5) Bark of old tree (*Photo by Emanuel Fritz*).

MAP 8-21   Abies concolor.

on dry, thin layers of partially decomposed granite or nearly barren rocks. Trees on such sites, however, are small and often malformed.

Pure stands of white fir are seldom encountered, except as second growth, but forests comprising up to 80% of this species are fairly numerous in some parts of the Sierra Nevada. Sugar, ponderosa, and Jeffrey pines, incense-cedar, California red fir, and Douglas-fir are common associates. Forests composed of sugar pine and white fir are fairly common, and a red fir–white fir combination occurs in some areas. White fir–ponderosa pine stands may be found east of the Sierra Nevada. White fir is also found with grand fir in the Cascade Mountains of southern Oregon, where intermediate forms exist between the two species and where identification of either is sometimes difficult.

White fir is extremely variable as discussed by Hunt (1993). Little (1979, 1980b) recognized two varieties: the Rocky Mountain white fir (var. *concolor*) and Sierra white fir or California white fir (var. *lowiana* [Gord.] Lemm.), the Pacific coast variety, with shorter notched leaves. The latter is treated as a distinct species (*A. lowiàna* [Gord.] A. Murr.) by Hunt (1993).

Good seed crops occur at about 2- to 4-year intervals. The seeds show transient viability, and germination averages about 37%. Nearly any soil except heavy clay makes a satisfactory seedbed.

Seedlings develop readily under the canopy of old trees and are even tolerant in later life provided they receive ample moisture. Late spring freezes often decimate seedlings, and reproduction is usually sparse on burns and logged-off lands, due to high insolation, until a cover of shrubs and numerous herbaceous species has become well established. Growth in both height and diameter is usually slow for the first 25 to 30 years, but after this period it is comparable to or even faster than that of its associated species. Growth rate gradually diminishes after the first century, and maturity is attained in about 300 years. Trees more than 350 years old are seldom encountered.

White fir is severely damaged by a true and a dwarf mistletoe, and the fir engraver beetle causes extremely heavy damage in some areas; heart rots are quite prevalent in others.

This fir is commonly used as an ornamental in many parts of the United States and Canada, particularly in the East where it is generally the best of all firs for the landscape due to its relative drought tolerance. Some of the more glaucous cultivars produce a beautiful effect when planted formally with certain of the more somber evergreens. It is also used for construction lumber, plywood, and Christmas trees.

Àbies lasiocárpa (Hook.) Nutt.   subalpine fir

Distinguishing Characteristics (Fig. 8-60).   *Leaves* $\frac{3}{4}$ to $1\frac{1}{2}$ in. (2 to 4 cm) long, flattened, pale blue-green, stomatiferous on both surfaces, spreading nearly in 2 horizontal ranks. *Cones* $2\frac{1}{4}$ to 4 in. (6 to 10 cm) long, cylindrical, purplish gray to nearly black; *bracts* shorter than the pubescent scales, with erose rounded shoulders and long spinelike tips. *Bark* smooth and chalky on young stems; furrowed and scaly on old trunks.

Range.   Central Yukon and southeastern Alaska, southeast to central Oregon, south to southern New Mexico and southern Arizona; local disjuncts in northwestern California. Elevation: sea level to 3000 ft in Alaska; 2000 to 7000 ft in British Columbia; 2000 to 7900 ft in Washington and Oregon; 8000 to 12,000 ft in the southern Rocky Mountains. Mountain slopes and ridges to timberline.

FIGURE 8-60   *Abies lasiocarpa*, subalpine fir. (1) Foliage $\times\frac{1}{2}$. (2) Mature seed cone $\times\frac{3}{4}$. (3) Abaxial view of scale with subtending bract $\times1$. (4) Seeds $\times1$. (5) Bark of old tree (*Courtesy of British Columbia Forest Service*).

FIGURE 8-61   Narrow, pyramidal crowns of subalpine firs.

General Description.   Subalpine fir was first observed by Lewis and Clark when they crossed the Bitterroot Mountains in Montana and Idaho in September 1805. It is largely restricted to high western forests, and on exposed sites at timberline it is often reduced to a grotesque shrub of sprawling or prostrate habit. On protected sites of good quality, however, trees attain a height of 60 to 100 ft and 18 to 24 in. dbh (max. 173 by 7 ft); under average conditions, they develop dense, narrowly pyramidal crowns (Fig. 8-61) and symmetrical, moderately tapering boles that are often clear for 30 to 40 ft. Open-grown trees, particularly in alpine meadows, feature excessively tapered trunks and dense spire-like crowns of great beauty that commonly extend to the ground. In such open sites, subalpine fir frequently reproduces by layering.

Subalpine fir forms limited pure stands but also occurs in mixed forests. It is an almost constant associate of Engelmann spruce in the Rocky Mountains. It is also found with other high-elevation species throughout its range. It is very shade tolerant.

Because of inaccessibility, subalpine fir, along with Engelmann spruce, was until recently of little commercial importance, except locally, but serves principally as a cover tree on high watersheds. It can live for almost 500 years. Subalpine fir is damaged by several insects and fungi and is susceptible to windthrow.

The more interior populations of the Rocky Mountains have been recognized by Hunt (1993) as *A. bifòlia* A. Murray and called Rocky Mountain subalpine fir. Introgression occurs between the two, and *A. bifolia* hybridizes with *A. balsamea* in north-central Alberta. *Abies lasiocarpa* hybridizes with *A. procera*.

Abies lasiocarpa var. *arizónica* (Merriam) Lemm. (*Abies arizonica* Merriam)   corkbark fir

Corkbark fir is a small tree of the southern Rocky Mountains and occurs on thin gravelly or rocky soils at elevations of from 8000 to 10,000 ft. It is readily separated from other firs by its thick, white, corky bark, which is free of resin blisters. The foliage is similar to that of subalpine fir, but in subalpine fir the cone scales are mostly wedge-shaped, whereas in corkbark fir they are often halberdlike at the base. Also, the bud scales of subalpine fir have crenate or dentate margins, whereas those of corkbark fir are entire. There appear to be a number of intergrading forms wherein these differences are not consistent and hence cannot always be relied upon for separation.

In southern Colorado, corkbark fir replaces subalpine fir as an associate of Engelmann spruce and as such is logged and distributed as white fir. However, the relative inaccessibility of this species, together with its smaller size and limited distribution, has prevented it from becoming commercially important except locally.

The relationships and taxonomy of the corkbark fir are questionable. According to Hunt (1993), it should probably be treated as a southern segregate of *A. bifolia* rather than as a variety of *A. lasiocarpa*. However, the latter is retained here until this question is resolved.

Àbies bracteàta D. Don   bristlecone fir

Distinguishing Characteristics.   The needles are flat, $1\frac{1}{2}$ to $2\frac{1}{4}$ in. (4 to 6 cm) long, stiff, sharp-pointed, and are arranged in several ranks. The most distinctive features of this species, however, are the cones, $2\frac{3}{4}$ to 4 in. (7 to 10 cm) long with spiny, long-exserted and strongly reflexed bracts with a very long bristle, and the ovoid, nonresinous, tan buds $\frac{3}{4}$ to 1 in. (18 to 24 mm) long.

Range.   This tree is the rarest of the firs. It is scattered through the dry forests of Santa Lucia Mountains of Monterey County, California, between elevations of 2000 and 5000 ft.

General Description.   Bristlecone fir is a small- to medium-sized tree, 40 to 100 ft tall (max. 182 ft) and to 3 ft dbh. The very aromatic resin from the trunk has been used as incense in the Spanish missions.

Cupressaceae s. lat.: The Redwood or Cypress Family

The Taxodiaceae (redwood family) and Cupressaceae (cypress family) have long been recognized as distinct (Little, 1979). More recent studies, however, have shown that a more realistic classification combines them into one (Brunsfeld et al., 1994; Eckenwalder, 1976; Judd et al., 1999; Price and Lowenstein, 1989). The monophyletic Cupressaceae s. str. are derived from within the old Taxodiaceae. Cupressaceae s. lat. include 25 to 30 genera (*Sciadopitys* is in a separate family, Sciadopityaceae) and 110 to 130 species widely scattered throughout the world and in both Northern and Southern Hemispheres. Eight genera are native to North America (Table 8-13).

As treated here, *Sequoia, Sequoiadendron,* and *Taxodium* would have been included in the earlier recognized Taxodiaceae; the others would be in the Cupressaceae s. str.

Several genera, not native to the United States or Canada, have become popular ornamentals. Examples are the China-fir (*Cunninghàmia lanceolàta* [Lamb.] Hook.), Japanese-cedar (*Cryptoméria japónica* [L.f.] D. Don) with numerous cultivars, Oriental arborvitae (*Platyclàdus orientális* [L.] Franco) with numerous cultivars, and the dawn-redwood (*Metasequòia glyptostroböides* H.H.Hu & Cheng). The dawn-redwood is the most recently discovered. Although known as a widely distributed fossil, it was found growing in central China in the 1940s (Hu, 1948; Li, 1964), introduced into cultivation around the world, and is now widely grown throughout the temperate Northern Hemisphere. It has the appearance of baldcypress, but the leaves appear opposite rather than alternate on the determinate short shoots, and the seed cone is cylindrical with four vertical rows of persistent scale/bracts.

The common name of "cedar" applied to many species in this family is confusing. True cedars belong to the genus *Cedrus* of the Pinaceae.

Pollen from these trees can cause allergic reactions in susceptible individuals.

TABLE 8-13   COMPARISON OF NATIVE GENERA IN THE CUPRESSACEAE S. LAT.

Genus	Leaves	Branchlets	Seed cones
Sequoia sequoia	Alternate, linear, $\frac{3}{8}$–1 in. (10–25 mm) long, 2-ranked; juvenile scalelike, many-ranked; persistent	Persistent, spreading	Elliptical to subglobose, $\frac{1}{2}$–$1\frac{3}{8}$ in. (1.2–3.5 cm) long, remaining intact at maturity; seeds with 2 lateral wings
Sequoiadendron giant sequoia	Alternate, scalelike, $\frac{1}{8}$–$\frac{1}{4}$ in. (3–6 mm) long, many-ranked; persistent	Persistent, appressed to spreading	Ovoid or oblong, $1\frac{3}{4}$–$3\frac{1}{2}$ in. (4.5–9 cm) long, remaining intact at maturity; seeds with 2 lateral wings
Taxodium baldcypress	Alternate, linear and 2-ranked, $\frac{1}{4}$–$\frac{5}{8}$ in. (5–17 mm) long; or, subulate and many-ranked; on deciduous short shoots	Deciduous; horizontal with linear leaves or erect with subulate leaves	Globose, $\frac{3}{4}$–1 in. (2–2.5 cm) diameter, shattering at maturity; seeds triangular, ± wingless at the angles
Calocedrus incense-cedar	Scales opposite and decussate, appearing 4-whorled, facial and lateral leaves different; persistent	Persistent, flattened and frondlike	Oblong, $\frac{3}{4}$–1 in. (2–2.5 cm) long, with 6 thin, flattened, woody scale/bracts; seed with 2 unequal lateral wings
Thuja thuja	Scales opposite and decussate, facial and lateral leaves different; persistent	Persistent, flattened and frondlike	Elliptical, ca. $\frac{1}{2}$ in. (12 mm) long, with 8–12 thin, woody scale/bracts; seeds with 2 equal, lateral wings
Cupressus cypress	Scales opposite and decussate; persistent	Persistent, spreading	Globose, $\frac{3}{4}$–$1\frac{3}{8}$ in. (2–3.5 cm) diameter, with 6–12 leathery, peltate scale/bracts; seeds laterally winged; maturing in 2 years
Chamaecyparis white-cedar	Scales opposite and decussate, facial and lateral leaves ± similar; persistent	Persistent, flattened and frondlike	Globose, $\frac{1}{4}$–$\frac{1}{2}$ in. (6–12 mm) diameter, with 4–8 leathery or dry, peltate scale/bracts; seeds laterally winged; maturing in 1 year
Juniperus juniper	Dimorphic; scales opposite and decussate; or, subulate and mostly ternate; persistent	Persistent, spreading	Globose, $\frac{1}{4}$–$\frac{5}{8}$ in. (6–15 mm) diameter, with 3–8 fleshy, peltate scale/bracts, fused into a berrylike cone; seeds wingless

Botanical Features of the Family

Trees and shrubs evergreen or deciduous, aromatic, resinous.

Leaves alternate, opposite and decussate, or whorled; linear, subulate, or scale.

Cones (juvenile) with bracts and scales partially to completely fused into a scale/bract complex (see Fig. 7-10); ovules erect, 2 to 12 on each scale; plants monoecious or dioecious; pollination by wind from late summer to spring.

Mature cones *pollen cones* very small; *seed cones* elongated or globose, woody, leathery, or semifleshy; composed of flat, peltate, or wedge-shaped scale/bracts; maturing in 1–2 years; *seeds* with 2 or 3 lateral wings, or wingless.

In most members of the Cupressaceae s. str., scale leaves are the typical mature leaf type, and subulate leaves are the juvenile type. Cultivars of several genera, particularly *Thuja* and *Chamaecyparis,* maintain the juvenile forms. At one time, the genus

Retinispora or *Retinospora* was used for these scale-leaved conifers in the juvenile subulate-leaved stage. In a few species, such as *Juniperus communis,* the subulate leaf is the mature form.

***SEQUOIA* Endl.**   sequoia

This group is of ancient lineage and flourished during the Cretaceous and Tertiary periods; about 40 fossil forms have been described. Although many forms, now extinct, were widely scattered through the forests of the Northern Hemisphere, the modern genus consists of but one species, the coast redwood of California and southwestern Oregon. The name *Sequoia* commemorates Sequoyah, a famous Cherokee Indian who developed an alphabet for his tribe. This was a brilliant achievement and enabled the Cherokee to publish a newspaper and book—something no other Native American tribe could do in 1821.

Sequòia sempervìrens (D. Don) Endl.   redwood, California redwood, coast redwood

Distinguishing Characteristics (Fig. 8-62)
    Leaves dimorphic, yellow-green; (1) those on lateral branches stiff, $\frac{3}{8}$ to 1 in. (10 to 25 mm) long, mostly linear, 2-ranked by a twist of the narrowed leaf base, decurrent,

FIGURE 8-62   *Sequoia sempervirens,* redwood. (1) Foliage $\times\frac{3}{4}$. (2) Pollen cones $\times 1\frac{1}{2}$. (3) Terminal ovulate cone $\times 1\frac{1}{4}$. (4) Open seed cone $\times\frac{3}{4}$. (5) Seeds $\times 1\frac{1}{4}$.

with two lighter-colored bands of stomata below; *apex* acute, usually sharp-pointed; (2) leaves on tips of leaders and fertile branchlets ca. $\frac{1}{4}$ in. (6 mm), long, keeled, oblong to somewhat scalelike, several-ranked.

Cones (juvenile) solitary; *pollen cones* ovoid to oblong, composed of a number of spirally arranged pollen sacs; *ovulate cones* consisting of 15 to 20 peltate scale/bracts, each with 2 to 5 erect ovules in a single row; plants monoecious.

Mature seed cones $\frac{1}{2}$ to $1\frac{3}{8}$ in. (1.2 to 3.5 cm) in diameter, ovoid to subglobose, reddish brown, with 15 to 20 peltate, wrinkled scale/bracts, maturing at the end of one season; *seeds* $\frac{1}{8}$ in. (3 mm) long, light brown, with narrow lateral wings.

Buds globose, covered by several acute, imbricate scales.

Bark reddish brown to cinnamon red, deeply furrowed, fibrous, 3 to 12 in. (8 to 30 cm) thick.

Range

Western (Map 8-22). Elevation: sea level to 3000 ft. Alluvial soils on flats, benches, or terraces in the coastal fog belt.

General Description

Redwood, probably the tallest tree in the world and certainly the tallest conifer, is also a massive tree (Fig. 8-63) commonly 200 to 275 ft in height and 8 to 12 ft dbh (max. 369 by 25 ft). From a relatively shallow wide-spreading root system rises a clear, impressively tall, buttressed, and somewhat tapering trunk that supports a short, narrow, irregularly conical crown.

Redwood is restricted to a number of separated areas in a narrow strip of coast about 450 miles in length and from 20 to 40 miles in width. Its eastern limits appear to be governed by atmospheric moisture; inland, some distance beyond points where the ocean fogs are cut off by the high coastal mountain ridges, redwood occurs along riparian corridors. Because of redwood's architecture, including a long, dense canopy, it intercepts tremendous amounts of fog, which then drips into the soil below.

Redwood is the dominant species throughout this belt; forests containing 80% or more of redwood are extensive, and on moist benches and alluvial bottoms 90% or even more of the standing timber is redwood. Douglas-fir, Sitka spruce, grand fir, western hemlock, tanoak, Pacific madrone, red alder, and California-laurel are common associates.

Redwood is a prolific annual seeder (after the fifth to fifteenth year). Seeds are very small (120,000 per lb). Although seeds will germinate on moist duff, the young trees can scarcely penetrate it and usually die from drying out. Fresh mineral soil is best. This may result when a large tree falls and the upturned roots strip away the overlying litter and humus. On alluvial flats where the best stands of redwood are found, periodic floods may deposit from several inches to 2 ft of fresh silt, which is soon covered with a thick growth of seedlings. Mature trees may develop new and higher root systems with better aeration in this new, fertile soil. Logging and fire also uncover extensive areas available for new reproduction. Unlike nearly all other conifers, redwood stumps produce vigorous sprouts that reach tree size in a remarkably

MAP 8-22 **(Dark Gray)** *Sequoia sempervirens.* **(White)** *Sequoiadendron giganteum.*

FIGURE 8-63   Redwoods in the Jedediah Smith Redwoods State Park, Del Norte County, California. (*Photo by Richard St. Barbe Baker.*)

short time. On the best sites, 20-year-old sprouts attain an average height of 50 ft and a dbh of 8 in., and in 50 to 80 years they commonly reach merchantable proportions. Where coppice growth is removed, a second generation of equally vigorous sprouts soon makes its appearance around the collar of the new stumps, and ultimately they too reach commercial size.

The original redwood forests covered 1,950,000 acres. About 12% remains as old growth, much in state and national parks. Redwood lumber is very valuable, especially where decay resistance is important.

The heaviest stands of timber in the world are found in the redwood belt. On alluvial bottom-lands bordering the Smith and Eel Rivers through Humboldt and Del Norte counties in north coastal California, where redwood attains its maximum development, mature stands average 125,000 to 150,000 ft B.M. to the acre, and some river flats have yielded as high as 2,000,000 ft B.M. per acre. Jepson (1910) reported a single tree that, upon being sawn into lumber, yielded 480,000 ft B.M. excluding waste.

The rate of growth of redwood both in height and in diameter measurably exceeds that of most American timber-producing species, and research in the second-growth forests indicates that no other species, with the possible exception of eastern cottonwood, is capable of producing as large a tree or as great a volume of wood in the same period of time. Jepson (1910) stated that the lead-ers of open-grown trees 4 to 10 years of age commonly elongate 2 to 6 ft a year and that the annual terminal shoots for these same trees are rarely less than 2 ft in length during the next 20 to 30 years. Trees can reach over 200 ft in height within the first 100 years. Redwood is remarkably tol-erant, and trees that have been suppressed for protracted lengths of time quickly recover once they are released. This even applies to old trees. It is difficult to ascertain at what age redwood matures because old trees exhibit from 400 to 2200 growth rings.

Most of the damage in redwood forests is traceable either directly or indirectly to fire. The bark on old trees is very thick (to 12 in.) and offers considerable resistance, but young reproduc-tion is readily killed even by a moderately light ground fire. Most, if not all, of the dry rot found in the basal portions of old trees had its inception near fire scars or seams. Recurring fires are responsible for hollow butts or "goose pens" so common in old growth, and although neither heart rot nor the formation of goose pens actually kill trees, both serve to weaken them materi-ally, and much high-grade saw timber is also destroyed. Significant damage also comes from wind and black bears.

Burls, those small to very large excrescences that appear along the bole of many trees, are also a feature of this species. The large ones, if solid, are removed from logs and cut into beautifully figured veneers. Smaller ones are frequently turned on a lathe into trays, bowls, and similar objects or, because of their ability to sprout when placed in water, are offered for sale along with other redwood novelties. A perfume extracted from the leaves is another product of this species offered to the tourist trade. The bark is a popular garden mulch.

California redwood is the state tree of California. It is planted throughout the United States and Canada as an ornamental, yet it has not escaped cultivation.

SEQUOIADENDRON Buchholz   giant sequoia

This genus along with *Sequoia* should be considered as "descendants of widely sepa-rated genera belonging to a larger group or family of worldwide distribution" (Buchholz, 1939) during the Cretaceous and Tertiary periods.

Sequoiadéndron gigantèum (Lindl.) Buchholz   giant sequoia, bigtree, Sierra-redwood

Distinguishing Characteristics (Fig. 8-64)

Leaves blue-green; on lateral and lower branches, $\frac{1}{8}$ to $\frac{1}{4}$ in. (3 to 6 mm) long, ovate, scalelike, appressed; on leaders and young trees, lanceolate, up to $\frac{1}{2}$ in. (12 mm) long, spreading.

Cones (juvenile) similar to those of redwood; *ovulate cones* with peltate to wedge-shaped scale/bracts each with 3 to 9 ovules in one or two rows.

FIGURE 8-64   *Sequoiadendron giganteum,* giant sequoia. (1) Mature seed cones and foliage of scale leaves $\times\frac{3}{4}$. (2) Terminal twig with lanceolate leaves $\times 1\frac{1}{2}$. (3) Seeds $\times 1\frac{1}{4}$. (4) Bark of old tree (*Courtesy of U.S. Forest Service*).

Mature seed cones $1\frac{3}{4}$ to $3\frac{1}{2}$ in. (4.5 to 9 cm) long, ovoid-oblong, reaching full size the first season after emerging from the bud but requiring an additional year for the seed to mature, and persistent, mostly unopened, for a number of years; *cone scale/bracts* woody, 25 to 40, peltate or more nearly wedge-shaped; *apophyses* rugose, depressed; *seeds* $\frac{1}{8}$ to $\frac{1}{4}$ in. (3 to 6 mm) long, straw-colored, laterally broad winged; usually not shed until the next October after maturity, and then irregularly for many years.

Buds naked.

Bark 12 to 31 in. (30 to 60 cm) thick on old trunks, cinnamon-red, fibrous, divided into broad rounded ridges by deep fissures; very fire resistant.

Range

Western (Map 8-22). Elevation: 5000 to 8000 ft; infrequently 3000 to 8900 ft. Slopes, canyons, and valleys.

General Description

The giant sequoias, or bigtrees, are numbered among the largest and oldest of living organisms, and some of them may be more than 2500 to 3200 years old. In their presence, all sense of proportion is lost, and trees 4 to 10 ft in diameter appear small by comparison. It is difficult to really appreciate the size of the giant sequoias because the associated trees are also so large. It is small wonder, therefore, that a feeling of reverence comes over one upon entering a grove of these patriarchs whose gigantic red trunks are like the supports of some vast outdoor cathedral. The emotions aroused by the silent ageless majesty of these great trees are akin to those of primitive people for whom they would have been objects of worship, and it is unlikely that centuries of scientific training will ever completely efface this elemental feeling.

Giant sequoia does not attain the height of the redwood but greatly surpasses it in diameter. The General Sherman tree in Sequoia National Park, for example, averages 32 ft in diameter at its base, 27.4 ft at a point 8 ft from the ground, and 17.5 ft at a height of 60 ft. The total height, however, is "only" 275 ft. In general, the larger sequoias commonly attain a diameter of 10 to 15 ft and a height of 250 to just over 300 ft (max. dbh 35.7 ft). Perhaps a better appreciation of their size may be gained when it is discovered that a person of normal height could lie crosswide on the largest branch (7 ft in diameter) of the General Sherman tree and not be seen from the ground 130 ft below. The General Grant tree is over 40 ft in diameter.

Giant sequoias are restricted to 75 groves that vary from 1 to 4000 acres. About 35,000 acres of giant sequoia forests remain. Most of this land is protected from logging. How much longer the giant sequoia might maintain itself is problematical, as is also the effect of humans in clearing the lesser vegetation around certain individual trees. Disturbance of the forest litter under the bigtrees may result in the damage of many small fibrous roots; their death, if in large numbers, is sooner or later reflected in the effectiveness of the whole anchorage system. Even in the primeval forest, it is not certain but that the final overthrow of the largest trees is caused by the gradual shifting of the center of gravity until the balance of the massive tree is upset, because trees have fallen on days when there was little or no wind.

Like redwood, insects and fungi cause but minor damage. The bark is extremely thick (to 31 in.), which is an effective protection from fire. Lightning, however, causes spectacular damage to certain individuals (Fry and White, 1930). Fire exclusion during the twentieth century seems to have had the most adverse impact on giant sequoia regeneration and maintenance.

Rarely does giant sequoia occur alone. Its principal associates include sugar, ponderosa, and Jeffrey pines, California red and white firs, and incense-cedar. Giant sequoia is regarded as intolerant.

Buchholz (1938), in working out the life history of the giant sequoia seed cone, noted that pollination, fertilization, and growth to full size all take place the first season after emergence of the ovulate cone from the bud (the former is present in embryonic form the previous season). While full size is thus attained in one season, a second year elapses before the embryos are developed and the seed matures. Perhaps the most remarkable feature, however, is that the cone remains green and usually closed for as many as 20 years after maturity. During this time, the peduncle develops growth rings (with occasional lapses). Few seeds are shed until the cone dies or is detached from the tree.

For so large a tree, the seeds might seem to be very small, furnishing a good illustration of the fact that there is no correlation between the sizes of trees and their seeds. There are about 91,000 seeds per lb. Another interesting feature is the presence in the cones of a powdery dark red pigment, which is released together with the seeds when the cones are dried artificially. When dissolved in water, this substance furnishes a nonfading writing fluid or ink. The natural function of the pigment, which has a high tannin content, is not well known, but seeds shed normally are always more or less coated with a thin water-deposited layer of this substance. Experiments seem to indicate a slightly higher percentage of germination of seeds treated naturally or artificially with solutions of the pigment, whereas the vigor of seedlings developed from such seeds is somewhat greater than from untreated samples.

Although the giant sequoia produces large quantities of seeds almost every year (after the twentieth), the percentage of germination is low, and bare mineral soil is necessary for seedling development. In the 1930s, the trail- and road-building activities of the Civilian Conservation Corps in the giant sequoia belt did much to stimulate natural reproduction, and where mineral soil was laid bare, many young trees appeared.

The seedlings develop a short taproot, but after a few years this ceases to develop; therefore, the tree is characterized by a relatively shallow, wide-spreading system of laterals. The crown of young trees is dense and conical, with the branches reaching to the ground. Later the much-tapered bole becomes clear of lower limbs, and the crown still maintains its juvenile shape; eventually, old trees become conspicuously buttressed with a rounded or irregular flat-topped crown.

This species is commonly used ornamentally along the Pacific coast and in some of the eastern states and foreign countries. It is frequently planted in Canada for landscape purposes. It has not escaped cultivation.

The wood of old-growth trees is exceedingly brittle but, like that of the redwood, durable almost beyond belief, and trees that fell more than a thousand years ago are still perfectly sound, with the exception of the sapwood, which soon decays. The oldest records of forest fires and early climatic conditions determined by dendrochronological studies have been obtained from trees of this sort.

Although the giant sequoia contributes little lumber, there is increasing interest in managing giant sequoia for uses similar to the coast redwood. Weatherspoon et al. (1986) provide many additional, interesting details about giant sequoia.

TAXODIUM Rich.   baldcypress

This genus was once widely distributed in the prehistoric forests of Europe and North America. Now, however, there are but one or two species: *Taxodium distichum* (L.) Rich., baldcypress (deciduous), native to the southeastern United States, and *T. mucronàtum*

Ten. (*T. distichum* var. *mexicànum* Gordon), Montezuma baldcypress (leaves persistent), found from Guatemala through Mexico to extreme southeastern Texas. A tree, known locally as *El Gigante,* of the latter species standing near Oaxaca, Mexico, has been estimated to be 2000 years old and to have a dbh of ca. 37 ft, although it is thought to be a natural graft of three trees. A number of trees of *T. distichum* in North Carolina have been determined to be 1000 to 1700 years old (Stahle et al., 1988).

Botanical Features of the Genus

Leaves alternate, deciduous or persistent, either (1) linear and 2-ranked, sometimes conspicuously stomatiferous below, or (2) lanceolate (subulate) appressed and overlapping; individual leaves dropping from the terminal long shoots or persisting on and dropping with the short shoots.

Cones (juvenile) unisexual; *pollen cones* in drooping racemes or panicles, each small and globose and composed of several pollen sacs opposite in 2 ranks; *ovulate cones* subglobose, comprising several spirally arranged peltate scale/bracts, each bearing 2 erect, basal ovules; plants monoecious.

Mature seed cones nearly globose, $\frac{3}{4}$ to 1 in. (2 to 2.5 cm) in diameter, woody, maturing and shattering in one season, comprising 5 to 10 nearly peltate scale/bracts, each 4-angled in cross section, resinous glandular on the inner surface, and terminating in a short, transverse mucro; *seeds* irregularly 3-angled and wingless or barely winged at the angles.

Buds subglobose, covered with several sharp-pointed, imbricate scales.

Lateral twigs (determinate short shoots) seasonally or annually dehiscent with the leaves.

Taxòdium dístichum (L.) Rich.

This species is treated here as comprising two varieties following Watson (1983, 1993). However, Godfrey (1988) has a description of the variation and argument for recognizing them as separate species. Duncan and Duncan (1988) also treat them as species. Although the varieties are morphologically and ecologically distinct as extremes, numerous individuals and populations exist that exhibit morphological and/or ecological intermediacy or that show curious admixtures of characters representing the extremes (Watson, 1983). Specimens from saplings, stump sprouts, fertile branchlets, terminal vegetative branchlets, or late-season growth may be very similar and not identifiable to variety. See Liu et al. (1996) for an annotated bibliography on baldcypress and pondcypress.

Taxodium distichum var. distichum   baldcypress

Distinguishing Characteristics (Figs. 8-65, 8-66)

Twigs (determinate short shoots) mostly 2-ranked, pendent or horizontally spreading from the branch.

Leaves decurrent, narrowly linear, $\frac{1}{4}$ to $\frac{5}{8}$ in. (5 to 17 mm) long, laterally divergent, the free base contracted and twisted, appearing 2-ranked on the twigs.

FIGURE 8-65   *Taxodium distichum* var. *distichum,* baldcypress. (1) Drooping racemes of pollen cones $\times\frac{3}{4}$. (2) Young clusters of pollen cones and ovulate cones $\times1$. (3) Mature seed cone and foliage of distichous, linear leaves on spreading twigs $\times\frac{3}{4}$. (4) Seeds $\times1$.

FIGURE 8-66   Old-growth baldcypress with buttress and knee. (*Courtesy of U.S. Forest Service.*)

MAP 8-23 *Taxodium distichum.*

Bark fibrous, $1\frac{1}{4}$ to $1\frac{1}{2}$ in. (3 to 4 cm) in thickness, or scaly and thinner; dark reddish brown to gray with shallow furrows.

Range

Southeastern (Map 8-23). Elevation: sea level to 530 ft, locally in Texas to 1750 ft. Brownwater rivers, lake margins, swamps, coastal marshes, and river bottoms.

Taxodium distichum **var. *imbricárium*** (Nutt.) Croom (*T. ascéndens* Brong.)   pondcypress

Distinguishing Characteristics (Fig. 8-67)

Twigs (determinate short shoots) mostly ascending vertically from the branch.

Leaves spirally arranged in 5–8 or more ranks, narrowly lanceolate (subulate), $\frac{1}{8}$ to $\frac{3}{8}$ in. (3 to 10 mm) long, appressed and overlapping, not contracted or twisted at base.

Bark brown to light gray, thicker and more deeply furrowed than in var. *distichum.*

Range

Southeastern (Map 8-23). Elevation: sea level to 330 ft. Blackwater rivers, lake and pond margins, swamps, pocosins, shallow depressions in pine flatwoods.

FIGURE 8-67   *Taxodium distichum* var. *imbricarium,* pondcypress. Nearly scalelike, appressed, imbricated leaves on more or less erect twigs. (*Photo by J. W. Hardin.*)

General Description of the Species

This species is the most distinctive of southern conifers. In the dim light of an old-growth cypress or cypress–gum forest, the massive, often ashy gray columnar trunks with their peculiarly swollen, fluted bases, knees, and branches bearded with Spanish moss produce an unforgettable impression.

Baldcypress is a large tree, 100 to 140 ft tall and 3 to 5 (to 17) ft dbh. The trunk of young trees shows considerable taper and supports an open, narrowly pyramidal crown; with age, the bole becomes more cylindrical and the crown irregularly flattened. The root system is a very distinctive feature of the tree. It consists of several descending roots providing anchorage and many shallow, wide-spreading roots from which rise, especially where the water level fluctuates, the peculiar conical structures called "knees." What function these outgrowths may have (if any) is not fully understood. Like the buttresses, they may be a passive development due to the stimulation of a fluctuating water level. The height of both is a good indicator of the periodic, local, high water level.

Best growth of this species is made on deep, fine, sandy loams with plenty of moisture in the surface layers and moderately good drainage. Cypress is rarely found, however, on such areas, presumably because of hardwood competition, and as a result is typical of permanently flooded areas, where it forms extensive pure stands. In such situations, it has few associates, but probably the most common is water tupelo; on slightly higher ground the bottomland hardwood species such as sweetgum, green ash, red maple, American elm, and certain of the oaks are more aggressive, and the cypress eventually disappears. This species, although preeminently adapted to freshwater swamps, extends into the coastal region of brackish tidewater, where it makes poor growth.

Reproduction of seeds is persistent but not widespread because their weight precludes distribution by wind.

Trees up to 40 to 60 years old sprout vigorously from the stump, and even much older trees often produce a few sprouts. Cypress is considered to be intermediate in tolerance. In deep shade, seeds will germinate but survival is poor. Growth in partial shade is much slower than in full sun.

Cypress has been considered a slow-growing tree, but studies of old stands do not indicate what may be expected of young trees, which often reach a foot in height the first season, nearly 2 ft the second, and as much as 18 ft in 10 years. Second-growth trees 100 years old are about 100 ft tall and 15 to 20 in. dbh. Because of its great durability (resistance to decay), the wood of cypress is often known in the trade as the "wood everlasting." Baldcypress is the longest lived tree in eastern North America, with maximum ages exceeding 1700 years.

Baldcypress and pondcypress have been planted as ornamentals far north of their natural range, in several of the northern states, southern Canada, and Europe. It is the state tree of Louisiana.

Small, white, mushroom-shaped growths sometimes seen on leaves are caused by the gall midge, *Taxodiomyia cupressi.* A brown pocket rot is found in very mature specimens and causes an interesting pattern known as "pecky cypress." The exotic nutria causes extensive losses of planted baldcypress in the southern United States.

CALOCEDRUS Kurz   incense-cedar

For many years, *Libocèdrus* was the generic name given to a group of some 10 species scattered widely around the perimeter of the Pacific Ocean. Only one species, the incense-cedar, *L. decúrrens* Torr. of California and Oregon, was found in North America. The others were in Chile and Argentina, New Zealand, New Caledonia, New Guinea, southern China, Burma, and Taiwan. Further studies through the years produced evidence for splitting this group into five genera. *Libocedrus* Endl. is now applied to five species of New Zealand and New Caledonia; *Calocedrus* Kurz is now the generic name for our native incense-cedar. For the names, distribution, and importance of the other genera, consult Dallimore and Jackson (1967). Little (1979, 1980b) still recognized *Libocedrus* s. lat.

Calocedrus, in addition to the North American incense-cedar, has two other species in south China and along the Burmese border and in Taiwan. *Calocèdrus macrolèpis* Kurz, the south China species, now rare, was more common in past geological ages, and perfectly preserved buried tree trunks are "mined" and sawn into boards for making coffins.

Calocèdrus decúrrens (Torr.) Florin   incense-cedar

Distinguishing Characteristics (Fig. 8-68)

Leaves $\frac{1}{8}$ to $\frac{1}{2}$ in. (3 to 12 mm) long scales, lustrous, dark yellow-green, oblong-ovate, 4-ranked, decurrent and adnate to the twigs except at the tips; glandular, having an aromatic odor when crushed; laterals keeled, almost wholly ensheathing the facial pairs; branchlets flattened into sprays.

Cones (juvenile) terminal; *pollen cones* with 12 to 16 decussate, filamentous, 4-celled pollen sacs; *ovulate cones* composed of 6 scale/bracts, 2 bearing 2 erect basal ovules; trees monoecious.

FIGURE 8-68   *Calocedrus decurrens*, incense-cedar. (1) Foliage showing scale leaves ×2. (2) Mature seed cones ×$\frac{3}{4}$. (3) Seeds ×1. (4) Open seed cone and flattened spray ×$\frac{3}{4}$. (5) Pollen cones ×$\frac{3}{4}$. (6) Bark of old tree (*Photo by Paul Graves*).

MAP 8-24   *Calocedrus decurrens.*

Mature seed cones $\frac{3}{4}$ to 1 in. (2 to 2.5 cm) long, pendent, leathery, the inner 2 of the 6 scale/bracts becoming greatly enlarged and giving to the partially open cone the appearance of a duck's bill; *seeds* $\frac{1}{8}$ to $\frac{1}{2}$ in. (3 to 12 mm) long, unequally laterally winged, straw-colored; cotyledons 2, rarely 3.

Bark thin, smooth, and gray-green or scaly and tinged with red on young stems; on old trunks 3 to 8 in. (7.5 to 20 cm) thick, yellowish brown to cinnamon-red, fibrous, deeply and irregularly furrowed.

Range

Western (Map 8-24). Elevation: 1000 to 9200 ft. Cool, moist, mountain soils.

General Description

Fremont first discovered incense-cedar on the south fork of the American River in California in 1844. Specimens that he collected 2 years later at the headwaters of the Sacramento River were used by Torrey in compiling the original botanical description of this species.

At maturity, this tree attains a height of from 75 to 110 ft and a diameter of 3 to 4 ft (max. 229 by 12.9 ft). Old trees are usually broadly buttressed at the base and have rapidly tapering trunks, which are often fluted and clothed for nearly one-half of their length in crowns of lustrous foliage. In youth, the crown is typically conical, but that of old trees is more irregular and often deformed by dwarf mistletoe and witches' brooms. Seedlings develop profusely branched root systems, and old trees possess well-developed, moderately deep lateral roots.

Incense-cedar grows best on cool, moist sites in the transition zone. Pure stands are virtually unknown, although it may form up to 50% of forests chiefly composed of sugar pine, ponderosa pine, and white fir. It is also a nearly constant associate of the giant sequoia. Ponderosa and Jeffrey pines, together with small amounts of bigcone Douglas-fir and white fir, are its principal associates in southern California; it also occurs in the coast ranges east of the redwood belt with both hardwoods and softwoods. In the Cascades of southern Oregon, incense-cedar invades the mixed coniferous forests composed largely of sugar, ponderosa, and western white pines, white fir, and Douglas-fir.

Good seed years occur about every 3 to 6 years. Reproduction is abundant in moist, organic litter, although good survival may also be found on moist mineral soils. Seedling root development is very slow; consequently, this tree is rare where long droughts are common. Established individuals are quite drought tolerant. Incense-cedar exhibits greater shade tolerance than most of the conifers with which it is associated and grows quite persistently under moderate cover with plenty of soil moisture. Sudworth (1967) observed that reproduction was usually good under old trees and in open situations, but that it was especially abundant in dense thickets under thinned stands where the more tolerant species could not compete. Growth is at no time rapid; trees 200 years of age vary from 90 to 95 ft in height and 25 to 30 in. dbh, and trees twice as old are rarely more than 100 ft tall and 40 in. dbh. Maturity appears to be attained in about 300 years, and trees more than 500 years old are rarely encountered; maximum ring count is 933.

Pocket dry rot is responsible for considerable damage in old trees. Three species of root-rotting fungi account for more mortality than all other pests. A rust fungus and witches' brooms are also prevalent in some sections, but they are relatively unimportant. Incense-cedar mistletoe causes considerable damage. Young trees are quite susceptible to fire.

These trees are used for lumber (especially for exterior siding), as pencils, and as ornamentals grown throughout the United States.

THUJA L.   thuja, arborvitae

This is a small genus consisting of five species of trees or large shrubs found in the forests of China, Taiwan, Korea, Japan, and North America. The Oriental arborvitae and the Japanese "hiba," *Platyclàdus orientális* (L.) Franco (*Thuja orientalis* L.)* and *Thujópsis dolabràta* (L.f.) Sieb. and Zucc., respectively, were formerly included in this genus; however, they are each now considered distinct and monotypic.

Two native species, one a small- to medium-sized northeastern tree and the other a large northwestern tree, are the only representatives of *Thuja* in North America (Table 8-14).

TABLE 8-14   COMPARISON OF NORTH AMERICAN THUJAS

Species	Leaves	Mature cones
T. occidentalis northern white-cedar	Dull green, glandular-pitted	With 4 fertile scale/bracts, the tips smooth or with minute, inconspicuous prickle
T. plicata western redcedar	Dark glossy green, essentially eglandular	With 6 fertile scale/bracts, the tips usually prickly

Botanical Features of the Genus

Leaves persistent, small scales, decussate; the facial leaves flattened, grooved, and often glandular, the lateral leaves rounded or keeled; on leading shoots ovate to lanceolate and somewhat larger than those on the lateral branchlets; branchlets flattened into sprays and held more or less horizontally or drooping.

Cones (juvenile) terminal; *pollen cones* arising from branchlets near the base of the shoot, individual strobili ovoid to globose, consisting of several pairs of decussate pollen sacs; *ovulate cones* appearing at the tips of short terminal branchlets, and comprising 8 to 12 scale/bracts, each bearing 2 or 3 erect ovules; plants monoecious.

Mature seed cones erect, ovoid-cylindrical, leathery to semiwoody, composed of several thin scale/bracts only a few of which are fertile; cones solitary for the most part and maturing in one season; *seeds* small, laterally 2-winged; cotyledons 2.

Thùja occidentàlis L.   northern white-cedar, eastern arborvitae

Distinguishing Characteristics (Fig. 8-69)

Leaves yellow-green, not lustrous; on leading shoots nearly $\frac{1}{16}$ to $\frac{1}{8}$ in. (1.5 to 3 mm) long, glandular, and long-pointed; on lateral branchlets scalelike and flattened, $\frac{1}{8}$ in. (3 mm) long, obscurely glandular-pitted, the foliage sprays often fanlike.

Cones $\frac{3}{8}$ in. (10 mm) long, erect, oblong; releasing their seeds in the fall, but persisting during the winter, tips of the scale/bracts rounded or with a very minute,

---

*The Oriental arborvitae has numerous cultivars and is much used as an ornamental in the United States and Canada. It may be distinguished by somewhat "lacy" foliage with often vertical sprays, globose cones $\frac{5}{8}$ in. (15 mm) in diameter, leathery becoming woody, the scale/bracts thick with a distinct curved hook, and seeds wingless.

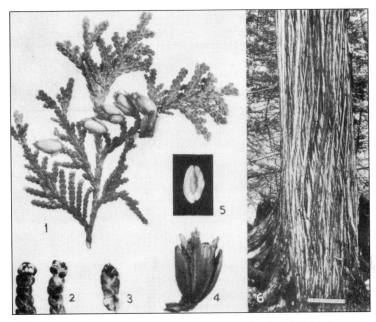

FIGURE 8-69 *Thuja occidentalis,* northern white-cedar. (1) Foliage in flattened spray and mature seed cones ×$\frac{3}{4}$. (2) Pollen cones before and after shedding pollen ×2. (3) Ovulate cone ×2. (4) Open seed cone ×2. (5) Seed ×2. (6) Bark of old tree.

inconspicuous prickle; usually 4 scales fertile; *seeds* $\frac{1}{8}$ in. (3 mm) long, wings as wide as the body.

Bark $\frac{1}{4}$ to $\frac{3}{8}$ in. (6 to 8 mm) thick, reddish to grayish brown, fibrous, forming a more or less close network of connecting ridges and shallow furrows, grayish on the surface.

Range

Northeastern and southern Appalachians (Map 8-25). Disjuncts in northcentral Illinois, western Pennsylvania, Rhode Island, and eastern Massachusetts. Elevation: sea level to 3000 ft. Swamps, neutral or alkaline soils on dry, limestone uplands, low stream borders, in mixed or pure stands.

General Description

Northern white-cedar commonly reaches 40 to 50 ft tall and 2 to 3 ft dbh (max. 125 by 6 ft). The crown, narrowly pyramidal in youth, later becomes irregularly oblong, and the tapering bole is supported by a shallow, wide-spreading root system. This moderately tolerant species occurs on a wide variety of soils but reaches its best development in outlying stations both in the North and in the South, on dry limestone outcrops or wet, organic soils overlying limestone bedrock. Even though it is in this respect a calcicole, northern white-cedar is also characteristic of shallow sphagnum-covered basins such as those in the Adirondack Mountains, where, however, the growth is very slow, about 1 ft dbh in 200 years. Growth may be from two to three times faster on the best sites.

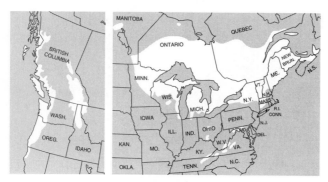

MAP 8-25   **(Left)** *Thuja plicata.* **(Right)** *Thuja occidentalis.*

Some seeds are produced annually (after the twentieth to thirtieth year) with large crops at 3- to 5-year intervals. In suitable locations, especially in swampy or dry limestone areas, extensive pure stands develop. On wet sites, this tree forms new individuals by layering. In mixed stands, balsam fir, eastern hemlock, eastern white pine, spruces, tamarack, black ash, yellow birch, and the northern maples are common associates. An age of 400 years or more is probably attained; however, ring counts are not satisfactory because about 80% of old trees are rotten at the heart.

In much of its range, it is an important source of winter browse, in addition to being a source of lumber, poles, fencing, and shingles. More than 100 cultivars are known (Den Ouden and Boom, 1965), and these may be easily propagated by cuttings. There are no serious insect or fungous pests, but it is susceptible to fire.

This species was probably the first North American tree introduced into Europe—about 1566. Northern white-cedar and its cultivars are important landscape plants in eastern North America.

Thùja plicàta Donn ex D. Don   western redcedar, giant arborvitae

Distinguishing Characteristics (Fig. 8-70)

Leaves lustrous dark yellow-green, usually without glandular pits, similar to those of *T. occidentalis* but somewhat coarser; the foliage sprays long, drooping, and often fern-like, or stringy.

Cones also similar to those of the preceding species, but more often with prickly scale/bracts, 6 of which are fertile; *seeds* $\frac{1}{8}$ in. (3 mm) long, the lateral wings about as wide as the seed.

Bark $\frac{1}{2}$ to 1 in. (12 to 25 mm) thick, fibrous, cinnamon-red on young stems, gray on old trunks, and forming a closely interlacing network.

Range

Northwestern (Map 8-25). Elevation: sea level to 3000 ft near the coast and to 6600 ft in the Rockies. Slightly acid soil of moist flats and slopes, banks of rivers, swamps, and bogs.

General Description

Western redcedar, the larger of our two North American species of arborvitae, was first observed by the Malaspina expedition on the west side of Vancouver Island, British Columbia, near Nootka

FIGURE 8-70 *Thuja plicata,* western redcedar. (1) Foliage in flattened sprays and mature seed cones $\times\frac{3}{4}$. (2) Seed $\times 2$. (3) Open seed cones $\times\frac{3}{4}$. (4) Ovulate cones $\times 2$. (5) Pollen cones $\times 2$. (6) Bark of old tree.

Sound in 1791. Today, it is one of the four most important timber species of the Pacific Northwest and is the principal timber tree used in shingle manufacture both in the United States and Canada.

Under favorable conditions for growth, forest trees attain gigantic proportions and vary from 150 to 200 ft in height and 4 to 8 ft dbh (max. 277 by 21 ft). A shallow, wide-spreading root system supports a broadly buttressed, often fluted base and rapidly tapering bole. The crown is typically irregular and is usually composed of numerous more or less horizontal or drooping branches that bend upward near their tips to form a distinct appearance.

It is seldom found in dry soils, although stunted growth is occasionally found on such sites. Western redcedar seldom occurs in pure stands but often constitutes up to 50% of mixed forests throughout its range. Where moisture is plentiful, western redcedar is a tree of great tolerance.

Western redcedar releases large amounts of seed at about 3-year intervals (after the fifteenth to twenty-fifth year), with smaller amounts almost every year. Seedling mortality from drought, fungi, birds, and insects is often excessive, and only a few trees may result from the tremendous number of seeds released. Tree growth is relatively slow, although individuals 80 years old may average 20 in. dbh and 100 ft in height. Maturity is reached in about 350 years, but trees with more than 1200 growth rings have been confirmed.

Because of its shallow root system and thin bark, western redcedar is readily killed by fire. Pecky heart rot is commonly found in overmature trees, and almost all very old trees are hollow-butted. There is little insect damage.

Western redcedar wood is extremely durable. One old wind-thrown "nurse tree" on the Olympic peninsula was found supporting 14 hemlocks that were over 100 years of age, and although the sapwood of the redcedar had rotted away, the heartwood was sound in every respect.

Several of its cultivars are grown as ornamentals, and this species is managed for timber in Europe and New Zealand. It is also used for siding, poles, lumber, boats, fences, pulp, and the spectacular totem poles and lodges so characteristic of the Northwest. It is the provincial tree of British Columbia. Additional information is given by Minore (1983).

CUPRESSUS L.   cypress

Cupressus consists of 10 to 26 species of trees and shrubs found in western North America, Mexico, Central America, the Mediterranean basin, the Himalaya Mountains, and western China. Disagreements on the number and rank of taxa reflect the patterns of variation and are the basis for the wide range of the number of species noted earlier.

The Italian or Mediterranean cypress, *C. sempervìrens* L., is a large and important tree of slender, columnar habit widely scattered through forests of the Mediterranean basin. There, it is also planted extensively as an ornamental, especially along avenues for formal effect. Several species of *Cupressus,* including the Italian cypress, are widely used ornamentals in the warmer portions of North America.

Seven species are native to North America. Some produce timber and Christmas trees used locally, and some are used as ornamentals, particularly cultivars of Monterey cypress, *C. macrocárpa* Hartweg. Only one is of significant economic importance.

Botanical Features of the Genus

Leaves persistent, scales (often subulate on juvenile or vigorous growth), decussate in 4 uniform ranks, or the lateral pairs boat-shaped and the facial pairs flattened; finely serrate on the margins and commonly glandular on the back; with irregularly disposed 4-angled branchlets.

Cones (juvenile) terminal; *pollen cones* cylindrical, composed of numerous decussate pollen sacs; *ovulate cones* consisting of 6 to 12 peltate scale/bracts, each with a terminal mucro and 6 to many basal ovules arranged in 2 or more rows; plants monoecious.

Mature seed cones globose, woody, or leathery, maturing at the end of the second season; scale/bracts peltate, each with a central boss or mucro; *seeds* 6 to many on each scale; compressed, with narrow lateral wings; cotyledons 2 to 5.

Cupréssus arizónica Greene　　Arizona cypress

Distinguishing Characteristics (Fig. 8-71). *Leaves* gray-green, having a fetid odor when bruised. *Cones* $\frac{3}{4}$ to $1\frac{1}{4}$ in. (2 to 3 cm) in diameter, subglobose, dark reddish brown, with 6 to 8 bossed scale/bracts. *Bark* breaking into thin irregular scales on young stems, furrowed and fibrous on old trunks.

Range. Local and rare, western Texas, west to southwestern New Mexico, Arizona, and southern California. Elevation: 2475 to 6600 ft. Coarse, gravelly, rocky soils on mountain slopes and coves of northerly exposure.

General Description. Under favorable growing conditions, this tree is 50 to 65 ft tall and 15 to 30 in. dbh (max. 102 by 5 ft). The bole of mature trees exhibits considerable taper and is covered for half of its length in a dense conical crown composed of short, stout, horizontal branches (Fig. 8-72). Trees on dry, sterile, rocky mountain slopes and canyon walls are small or even dwarfed but may persist on such sites for many years. This species occurs most abundantly in open pure forests, but it is occasionally observed in mixture with Arizona pine and live oaks.

Seeds are borne in abundance every year, but owing to the fact that much of it never finds conditions suitable for germination, reproduction is usually scanty. The trees are tolerant throughout life. Growth is seldom rapid, and trees 6 to 12 in. dbh frequently show 70 to 100 growth rings; the largest trees are rarely more than 400 years old.

The Arizona cypress is used as an attractive ornamental eastward to the Atlantic coast.

FIGURE 8-71　*Cupressus arizonica,* Arizona cypress. (1) Foliage and mature seed cones nearly open $\times\frac{3}{4}$. (2) Seeds ×2.

FIGURE 8-72   Arizona cypress. (*Courtesy of U.S. Forest Service.*)

CHAMAECYPARIS Spach   white-cedar, false-cypress

This genus includes six or seven species. Three are indigenous to North America (Table 8-15); one of these, *C. thyoides,* is a coastal species of the East; *C. nootkatensis* and *C. lawsoniana* are important trees of the Pacific coast. The Japanese Sawara-tree and Hinoki-cypress, *C. pisífera* (Sieb. and Zucc.) Endl. and *C. obtùsa* (Sieb. and Zucc.) Endl., respectively, together with many of their culivars, are favored ornamental trees in many sections of the United States and Canada. The remaining species, *C. formosénsis* Matsum., is a massive tree native to Taiwan. This tree is said to attain a maximum height of 195 ft, a diameter of 23 ft, and an estimated age of 1500 years (Dallimore and Jackson, 1967).

An artificially made, fertile, intergeneric hybrid between *Chamaecyparis nootkatensis* and *Cupressus macrocarpa* is known as ×*Cupressocypàris leylándii* (A. B. Jackson & Dallim.) Dallim. This Leyland cypress, with several cultivars, has become very popular as a fast-growing, attractive ornamental and Christmas tree.

TABLE 8-15   COMPARISON OF THE NATIVE *CHAMAECYPARIS*

Species	Leaves	Mature cones
C. thyoides Atlantic white-cedar	$\frac{1}{16}$–$\frac{1}{8}$ in. (1.5–3 mm) long, with circular glands	$\frac{1}{4}$–$\frac{3}{8}$ in. (4–9 mm) diameter; somewhat fleshy, maturing in 1 year; scale/bracts terminating in a short point; seeds 1 or 2 per scale
C. lawsoniana Port-Orford-cedar	$\frac{1}{16}$–$\frac{1}{8}$ in. (1.5–3 mm) long, with linear glands	$\frac{3}{8}$–$\frac{1}{2}$ in. (8–12 mm) diameter; woody, maturing in 1 year; scale/bracts blunt tipped; seeds 2–4 per scale
C. nootkatensis Alaska-cedar	$\frac{1}{16}$–$\frac{1}{8}$ in. (1.5–3 mm) long, usually eglandular	$\frac{3}{8}$–$\frac{1}{2}$ in. (8–12 mm) diameter; woody, maturing in 1–2 years; scale/bracts ending in a long point; seeds 2–4 per scale

Botanical Features of the Genus

Leaves persistent, scales decussate, ovate, acuminate, entire on the margin, the lateral pairs boat-shaped, the facial pairs flattened; on terminal shoots linear-lanceolate or needlelike, spreading; usually turning brown after the second or third season, but persistent for many years; branchlets flattened into sprays.

Cones (juvenile) of both sexes terminal but on separate branches; *pollen cones* cylindrical, composed of numerous decussate pollen sacs; *ovulate cones* subglobose, usually consisting of 6 to 12 decussate, peltate, ovuliferous scale/bracts, each bearing 2 to 5 erect ovules; plants monoecious.

Mature seed cones erect, globose, leathery or semifleshy becoming woody when open, maturing in 1 to 2 years; *cone scale/bracts* peltate, each with a central boss; *seeds* 2 or rarely up to 5 per scale, slightly compressed, laterally winged; cotyledons 2.

Chamaecýparis thyoìdes (L.) B.S.P.   Atlantic white-cedar, white-cedar

Distinguishing Characteristics (Fig. 8-73)

Leaves $\frac{1}{16}$ to $\frac{1}{8}$ in. (2 to 3 mm) long, keeled, with circular glands on the back, dark blue-green, turning brown the second year, but persistent for several years.

Cones $\frac{1}{4}$ to $\frac{3}{8}$ in. (4 to 9 mm) in diameter, somewhat fleshy; at maturity bluish purple, glaucous, later turning brown; scale/bracts short pointed; *seeds* 1 or 2 for each fertile scale, $\frac{1}{8}$ in. (3 mm) long or smaller.

Bark thin, but on old trunks $\frac{3}{4}$ to 1 in. (19 to 25 mm) thick, ashy gray to reddish brown, somewhat similar in appearance to that of northern white-cedar.

Range

Eastern (Map 8-26). Disjuncts in westcentral Georgia; extending further inland in the North Carolina sandhills. Elevation: sea level to 1650 ft. Peaty, acidic soils of low woodlands and swamp forests.

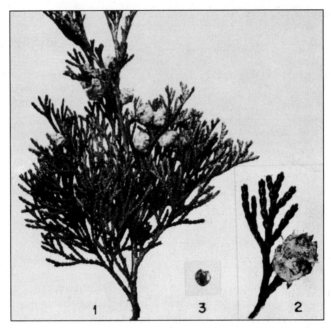

FIGURE 8-73 *Chamaecyparis thyoides,* Atlantic white-cedar. (1) Foliage in flattened sprays and mature seed cones $\times\frac{3}{4}$. (2) Mature seed cone $\times2$. (3) Seed $\times2$.

General Description

Atlantic white-cedar on good sites averages 80 to 85 ft in height and 10 to 14 in. dbh (max. 120 by $5\frac{1}{2}$ ft). The bole of mature trees grown in moderately dense stands is long, cylindrical, and clear of branches for about three-quarters of its length, making it an ideal pole or post timber. The crown of forest trees is small, narrowly conical, and composed of slender limbs with somewhat drooping branchlets; the root system is superficial and wide-spreading, and mature trees are very susceptible to windthrow.

MAP 8-26 *Chamaecyparis thyoides.*

This species is characteristic of freshwater swamps and bogs, wet depressions, or stream banks, and is rarely found except on such sites; extensive pure stands (often extremely dense) are the rule, occurring on shallow peat-covered soils underlain by sand. Because of the extensive north-south distribution of Atlantic white-cedar, the list of associated trees is large. Some of the more common are hemlock, eastern white pine, gray birch, black tupelo, and red maple in the North, and slash and pond pines, baldcypress, sweetbay, and loblolly bay in the South. Young trees are nearly as tolerant as balsam fir or hemlock, and much more so than the pines or baldcypress; however, they do not survive under the dense cover of older growth, especially hardwoods.

The production of seed in dense stands begins between the ages of 10 and 20 years (earlier in open-grown trees), and good crops are borne nearly every year. The small winged seeds are widely disseminated by wind. Following fires, pure, even-aged stands occur, from which merchantable timber can be harvested in 75 to 100 years or less. Much of the crop is used for posts and prefabricated log cabins. The wood is resistant to decay.

For a thorough discussion of the distribution, ecology, and variability of Atlantic white-cedar and recognition of *C. thyoides* var. *henryae* (Li) Little, of the southeastern Gulf coast, see Clewell and Ward (1987), Laderman (1989), and Ward and Clewell (1989).

Chamaecýparis lawsoniàna (A. Murr.) Parl.   Port-Orford-cedar, Lawson false-cypress, Oregon white-cedar

Distinguishing Characteristics (Fig. 8-74)

Leaves $\frac{1}{16}$ to $\frac{1}{8}$ in. (2 to 3 mm) long, yellow-green to blue-green, blunt, with linear glands, stomatiferous on the lower facial leaves; forming fine, flattened, feathery or lacy sprays; persistent until the third or fourth season.

Cones $\frac{3}{8}$ to $\frac{1}{2}$ in. (8 to 12 mm) in diameter, reddish brown, often glaucous, globose, composed of 3 pairs of peltate scale/bracts each bearing 2 to 4 seeds and blunt-tipped, maturing in one season; *seeds* $\frac{1}{8}$ in. (3 mm) long, reddish brown, with 2 broad lateral wings.

Bark 6 to 10 in. (15 to 25 cm) thick on old trees, silvery brown, fibrous, divided into thick, rounded ridges separated by deep irregular furrows.

Range

Northwestern (Map 8-27). Elevation: sea level to 5000 ft. Rocky ridges in the coastal fog belt; local disjuncts in the Californian Siskiyou Mountains and on Mt. Shasta.

General Description

Port-Orford-cedar is a large tree, 140 to 180 ft in height and 4 to 6 ft dbh (max. 229 by 16 ft). The boles of large trees are sometimes buttressed and commonly clear for 150 ft or more of their length. The crown is characteristically short, conical, and composed of numerous more or less horizontal or somewhat pendulous branches; the roots are shallow to moderately deep and spreading.

This species does best in regions where there is an abundance of both soil and atmospheric moisture. It is less exacting in this respect, however, than redwood, and it frequently occurs on drier ridges 30 to 40 miles inland. Except in the vicinity of Coos Bay, Oregon, Port-Orford-cedar is rarely found in pure stands. Although found in a relatively small area, this species spans the floristic transition zone between the trees of California and those of the Pacific Northwest. Common associates in coastal Oregon include Douglas-fir, Sitka spruce, western hemlock, and western red-cedar. Farther inland and southward, it is found with such species as western white and sugar pines, incense-cedar, red fir, and red alder.

Port-Orford-cedar reproduces readily on burned or unburned sites with about equal aggressiveness. Large seed crops occur every 3 to 5 years (after the fifth to twentieth year) with smaller amounts annually. Seedling trees grow persistently in either shade or full sunlight, although under very dense cover they are usually suppressed and eventually die. This species is moderately tolerant. Growth continues at a moderate rate, and the trees mature in about 300 to 350 years. The oldest trees are often 500 or more years of age.

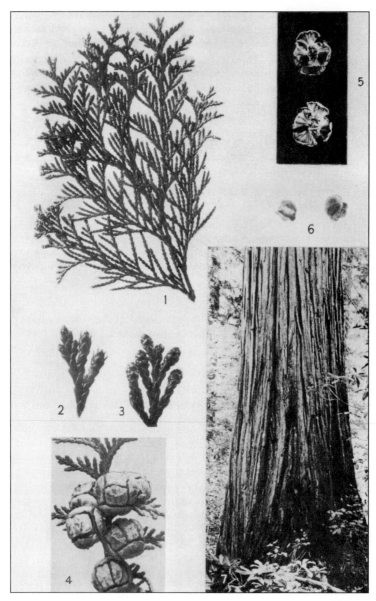

FIGURE 8-74   *Chamaecyparis lawsoniana,* Port-Orford-cedar. (1) Foliage in flattened sprays $\times\frac{3}{4}$. (2) Ovulate cones $\times$2. (3) Pollen cones $\times$2. (4) Mature seeds cones nearly open $\times$2. (5) Open seed cones $\times$2. (6) Seeds $\times$2. (7) Bark of old tree (*Courtesy of U.S. Forest Service*).

Port-Orford-cedar is a commonly used tree for ornamental planting, both in North America and Europe. Great variation in color and form have been found, and some 198 cultivars are listed by Den Ouden and Boom (1965). As a timber species, it is one of the most valuable of North American conifers. The aromatic wood is used for woodenware, and much is shipped to Japan for use in homes and temples, as a replacement for the Japanese Hinoki-cypress (*C. obtusa*).

This species is relatively free from insect and fungal diseases, except a root rot (caused by *Phytophthora lateralis*) is a great threat. Fire causes considerable damage to the thin-barked young trees, and an introduced root rot causes serious damage. Windthrow is also a hazard. Zobel et al. (1985) summarize the ecology, pathology, and management of Port-Orford-cedar.

Chamaecýparis nootkaténsis (D. Don) Spach   Alaska-cedar, Nootka false-cypress, yellow-cedar

Distinguishing Characteristics (Fig. 8-75)

Leaves about $\frac{1}{8}$ in. (3 mm) long, acute, blue-green to gray-green, usually eglandular or occasionally glandular on the keel, appressed, but the apices of the lateral pairs often free, turning brown during the second season but persistent until the third.

Cones $\frac{1}{3}$ to $\frac{1}{2}$ in. (8 to 12 mm) in diameter, purplish brown to reddish brown, with 4 to 6 peltate scale/bracts, long-pointed, each bearing 2 to 4 seeds; maturing at the end of the second season; *seeds* about $\frac{1}{4}$ in. (6 mm) long, the lateral wings about twice as wide as the seed.

Bark thin, grayish brown, scaly on young stems; forming thin, narrow, flattened, interlacing ridges on old trunks.

MAP 8-27   (White) *Chamaecyparis nootkatensis.*
(Dark gray) *Chamaecyparis lawsoniana.*

Range

Northwestern (Map 8-27). Local disjuncts farther inland in British Columbia and Oregon. Elevation: sea level northward, 2000 to 7000 ft southward and inland. Wet mountain soils in mixed or pure stands.

General Description

This species, known in the trade as Alaska-cypress, yellow-cypress, or yellow-cedar, is typical of the Pacific Northwest region. It is a medium-sized tree, 60 to 90 ft high and 2 to 3 ft dbh (max. 200 by 13.7 ft). In the forest, it usually develops a broadly buttressed, often fluted base and rapidly tapering bole, which is often clear for about one-half of its length. The conical crown is composed of numerous drooping branches with long, pendulous, flattened sprays of foliage. The root system varies with the site; in moist soils it is shallow, but in drier situations it penetrates to much greater depths.

Alaska-cedar reaches its maximum development on the islands of southeastern Alaska and British Columbia near tidewater, where both soil and atmospheric moisture are abundant. Much of the timber in this region has been logged, however, and the best stands are now located in the humid forests of the

FIGURE 8-75 *Chamaecyparis nootkatensis,* Alaska-cedar. (1) Foliage in flattened sprays and mature seed cones ×$\frac{3}{4}$. (2) Seed ×2. (3) Bark (*Photo by Kenneth Walin*).

Olympic and Cascade Mountains in northern Washington. Here it forms limited pure stands but usually occurs in mixture with other conifers. At low elevations, these include western hemlock, Sitka spruce, Pacific yew, Pacific silver and grand firs, and western redcedar; at high altitudes are found subalpine fir, mountain hemlock, western white, and whitebark pines, Engelmann spruce, and western larch. Timberline trees are often reduced to a sprawling shrublike habit, or they may become even prostrate on wholly unprotected sites. This species is considered tolerant, more so than western white pine or western larch, but less than Engelmann spruce, subalpine fir, the hemlocks, Sitka spruce, or western redcedar.

Alaska-cedar may produce a small amount of seed annually, with large crops at irregular intervals of 4 or more years. The seeds have transient viability, low germination (about 12%), and the percentage of seedling survival is never very great. The seeds are able to germinate in rocky, gravelly, or clay soil, leaf litter, or even moss, provided there is an abundance of soil moisture. The seedling, once it has become established, continues to grow very slowly but persistently both in height and in diameter. Pole-sized trees may be 100 to 200 years old. This species is one of the longest-lived of western conifers. An age of 1000 years is not uncommon, and one old tree was reported to show 3500 growth rings (questionable; oldest accurate age is 1834 yrs). Hundreds of thousands of acres of Alaska-cedar in southeastern Alaska have been killed by an unknown cause, although abiotic factors appear to be responsible (Hennon and Shaw, 1997).

The lumber (with bright yellow heartwood that is extremely durable) is used for furniture, boats, flooring, marine pilings, and paneling. Most is now exported to Japan. Hennon and Harris (1997) have compiled an annotated bibliography on Alaska-cedar.

JUNIPERUS L.   juniper

The junipers constitute about 60 species of trees and shrubs widely scattered through North and Central America, Japan, Taiwan, China, the Himalayas, the Mediterranean basin, northern Africa, Abyssinia, the Canary Islands, the Azores, and the West Indies.

Juniperus commùnis L., common juniper, is a shrub or small tree that is circumpolar in the Northern Hemisphere and is the most widely distributed native conifer in both North America and the world. The var. *depréssa* Pursh, oldfield common juniper, is usually a sprawling shrub common to many sections of eastern and western United States and Canada. It is readily recognized by its peculiar bushy habit of growth, its long-subulate, ternate leaves and no scale leaves, and its glaucous blue berrylike cones. Several other varieties of *J. communis* as well as cultivars of many other species are used as ornamentals.

Oil obtained from both the wood and leaves of certain species is used in the manufacture of perfumes and medicines. The leaf oils possess diuretic properties, and cattle raisers exercise particular care in making sure that their grazing livestock do not forage on junipers. Gin derives its characteristic taste from juniper "berries," and they are also used as a flavoring in some cooking and are an important food for wildlife.

Thirteen species of *Juniperus* are native to the United States; the most important is *J. virginiana.*

Interspecific hybridization, introgression, clines, and ecotypes have been documented in the junipers. See Flake et al. (1978) and Adams (1986, 1993) for a discussion.

Botanical Features of the Genus

Leaves persistent, opposite or ternate; always subulate on juvenile growth; on older plants (1) wholly subulate and ternate, (2) wholly scalelike and decussate, (3) both 1 and 2 on the same plant; leaves with or without glands; the subulate ones frequently conspicuously stomatiferous and with a single medial resin canal.

Cones (juvenile) terminal or axillary; *pollen cones* composed of several decussate pollen sacs; *ovulate cones* composed of 3 to 8 decussate or ternate, pointed scale/bracts, tightly coalesced, some or all bearing 1 or 2 basal ovules; plants dioecious or rarely monoecious.

Mature seed cones red-brown, blue, or blue-black, often glaucous, berrylike and remaining closed, maturing in one, two, or three seasons; *seeds* 1 to 12 ovoid, terete, or angular, unwinged, usually requiring two or more seasons to germinate; cotyledons 2, or 4 to 6.

Juníperus virginiàna L.   eastern redcedar

Distinguishing Characteristics (Fig. 8-76)

Leaves about $\frac{1}{16}$ in. (1.5 mm) long, opposite and decussate (rarely ternate), dark green scales, arranged to form a 4-sided branchlet; the long sharp-pointed (subulate) juvenile leaves predominate for a number of years and are usually present in small numbers even on old trees.

FIGURE 8-76 *Juniperus virginiana,* eastern redcedar. (1) Juvenile foliage $\times\frac{3}{4}$. (2) Mature foliage and mature seed cones $\times\frac{3}{4}$. (3) Bark of old tree (*Courtesy of* U.S. *Forest Service*).

Cones $\frac{1}{8}$ to $\frac{5}{16}$ in. (3 to 7 mm) in diameter, subglobose, ripening the first season, pale green turning to dark blue at maturity, glaucous; *seeds* 1 or 2, rarely 3 or 4, requiring 2 or 3 years to germinate; cotyledons 2; dioecious.

Bark $\frac{1}{8}$ to $\frac{1}{4}$ in. (3 to 6 mm) thick, light reddish brown, fibrous, separating into long, narrow, fringed scales; often ashy gray on exposed surfaces.

MAP 8-28 *Juniperus virginiana,* not including var. *silicicola.*

Range

Eastern (Map 8-28). Local disjuncts in northern Texas, southwestern Kansas, and southwestern North Dakota. Elevation: sea level to 4600 ft in the southern Appalachians. This species has an extremely broad ecological amplitude from sand dunes, dry uplands, old fields, pastures, and fence rows to lake borders, floodplains, hammocks, and swamps; and from acidic soils to limestone outcrops.

General Description

Eastern redcedar is a small- to medium-sized tree 40 to 50 ft tall and 12 to 24 in. dbh (max. 120 by 5 ft). The crown is dense and pyramidal, columnar (Fig. 8-77), or rather broadly bushy.

FIGURE 8-77   Eastern redcedar, pyramidal (left) and columnar (right) growth forms at approximately the same age. (*Photo by J. W. Hardin.*)

Although best growth is made on deep alluvial soils, hardwood competition eliminates most redcedars from such sites. The tree is most common on the poorest of dry soils in pure stands or open mixtures with shortleaf and Virginia pines, or the dry soil oaks, hickories, and other hardwoods. The so-called cedar barrens of Tennessee and northern Alabama have supplied much of the commercial timber produced.

As treated here, *J. virginiana* includes two varieties, the typical var. *virginiana* and the southern redcedar, var. *silicicola* (Small) E. Murray (*J. silicicola* [Small] L. H. Bailey, as recognized by many earlier authors). According to the study by Adams (1986, 1993), the latter is a coastal foredune and sand ridge ecotype and is mainly coastal from North Carolina to central and northwestern Florida. It differs from the typical by larger pollen cones, smaller seed cones, a more flattened crown, and the restricted habitat.

Eastern redcedar is one of the first plants to invade pastures and abandoned fields. Dispersal of the seeds is primarily by birds, which eat the berrylike cones in large numbers. Seed-bearing may begin about the tenth year. Delayed germination is the rule, and the seedlings do not appear until the second or third season after the seeds are sown. Eastern redcedar is a slow-growing, intolerant tree reaching a maximum age of about 850 years.

Seedling variation is pronounced, and many ornamental forms have been propagated vegetatively; in fact, redcedar and its cultivars are among the best of native ornamental evergreens. "Cedar apples," orange-red galls caused by the cedar-apple rust, are often conspicuous features, particularly after a rain.

The reddish, aromatic wood is used for furniture and paneling and the resistant trunks for fenceposts. It is an excellent windbreak species. Redcedar is also browsed by deer and livestock, and browse lines about 5 ft above the ground lend a unique aspect to the trees in a field or pasture. Fire is the most serious damaging agent.

Eastern redcedar hybridizes with *J. horizontalis* and *J. scopulorum.* Earlier belief of hybridization with *J. ashei* has been refuted by more recent studies (Adams, 1993; Adams and Turner, 1970). Much additional information about eastern redcedar can be found in Schmidt and Piva (1995).

Juníperus scopulòrum Sarg.   Rocky Mountain juniper

Distinguishing Characteristics (Fig. 8-78).   *Leaves* often glaucous, blue or blue-gray, scale leaves $\frac{1}{16}$ to $\frac{1}{8}$ in. (1–3 mm) long, margin entire, the gland elliptic, scarcely overlapping if at all. *Seed cones* maturing in 2 years, of 2 sizes, globose to 2-lobed, light blue and glaucous but dark blue-black below glaucous coating, resinous; seeds 1–2(3). *Bark* brown, in thin strips or in plates on larger branchlets.

FIGURE 8-78   *Juniperus scopulorum,* Rocky Mountain juniper. (1) Foliage and young seed cones $\times\frac{3}{4}$. (2) Foliage and mature seed cones $\times\frac{3}{4}$. (3) Bark of old tree.

Range. British Columbia, south to Arizona and New Mexico, east to the western Dakotas. Rocky slopes and hillsides; sea level in British Columbia to 9000 ft southward.

General Description. This has the widest distribution of any of the western junipers. It is related to *J. virginiana* and hybridizes with it in areas of overlap.

Trees are conic to rounded and single-stemmed, the branches drooping at the ends. It sometimes reaches 50 ft tall and $1\frac{1}{2}$ ft in diameter. It may occur in pure stands, but is usually found scattered with ponderosa pine, pinyon pines, and Douglas-fir.

The cones are an important food for birds and mammals. The wood is used primarily for posts, fuel, and cedar chests.

Additional Juniper Trees

Most of these species are important components of the pinyon-juniper woodlands of the Southwest (see under "pinyon pines"; also see Miller and Wigand, 1994). They are mostly large bushy shrubs or small stunted trees (sometimes much larger) on poor, dry soils in arid regions. Those that do attain tree size usually have a stout, rapidly tapering bole partially covered by a dense pyramidal to subglobose crown of more or less irregular outline. These trees are of little or no value for sawtimber, but they may be used for poles, posts, cross ties, mine props, and fuel.

These western junipers may be divided into two groups, those with blue cones and those with reddish or red-brown cones. Depending upon the species, these cones may ripen in one or two growing seasons. The states mentioned in the following summary represent the areas of greatest concentration for each species. Consult Adams (1993), Burns and Honkala (1990), and Little (1979) or local guides for greater detail.

Blue Cone

J. monospèrma (Engelm.) Sarg.   oneseed juniper

Dioecious shrub or tree to 25 ft tall with ascending lower branches; scale leaves with denticulate margins and elongated gland; cones dark reddish blue with a single seed. It is found on semiarid slopes and plateaus of the Rockies at 3000 to 7000 ft elevation, mostly in Arizona, New Mexico, west Texas, and Colorado at 3000 to 7000 ft elevation.

J. áshei Buchholz   Ashe juniper

This is a dioecious tree to 40 ft tall; leaf margins denticulate, the gland raised-hemispheric; cones dark blue and glaucous. This occurs on limestone glades and bluffs mostly in central Texas, south into Mexico, and northeastward into Oklahoma and along the Missouri-Arkansas line at 800 to 2000 ft elevation.

J. occidentális Hook.   western juniper

This is a monoecious or dioecious tree to 30 ft tall. Scale leaves denticulate with a conspicuous ovate to elliptic gland; cones of 2 sizes, maturing in 2 years, blue to blue-black and glaucous with 2(–3) seeds. It is common at high elevations from southern Washington south to southern California to 10,000 ft elevation.

Reddish, Red-Brown, or Brown Cones

J. deppeàna Steud.   alligator juniper

This gets its name from the bark that is divided into squares, or blocks, so that it resembles alligator hide. Dioecious trees to 50 ft with trunk 2–4 ft in diameter. Scale leaves with denticulate margins and conspicuous ovate to elliptic gland; cones of 2 sizes and maturing in 2 years, reddish brown, with 3–6 seeds. It is found mostly in Arizona, New Mexico, western Texas, and south into Mexico at 4500 to 8000 ft elevation.

J. osteospérma (Torr.) Little   Utah juniper

Monoecious tree to 40 ft tall with trunk 1–3 ft in diameter. Scale leaves with denticulate margins and inconspicuous glands; cones of 1 or 2 sizes, bluish brown with 1–2 seeds. This is typical of the Great Basin, mostly in Utah, Nevada, and Arizona from 3000 to 8000 ft elevation.

J. califórnica Carr.   California juniper

Monoecious tree to 40 ft tall and 1–2 ft in diameter. Scale leaves with denticulate margins and conspicuous ovate to elliptic gland; cones of 1 size, bluish brown, with 1–2 seeds. As its name suggests, it is largely restricted to the mountains of California at 1000 to 5000 ft elevation.

9

MAGNOLIOPHYTA (ANGIOSPERMS)

Angiosperms include the flowering plants, which are the most common, most complex and highly evolved, and most widely distributed vascular plants now inhabiting the earth's surface. They supply food, clothing, and shelter for both people and wildlife, and innumerable drugs, oils, dyes, and other products. Numbering nearly 250,000 species in 405 families, they are found from the luxuriant, rain-drenched forests of the tropics to the nearly barren tundra in polar regions. Many of them form the only cover on hot, dry soils in arid regions; others occur on mountains to the highest altitudes of plant growth, and still others in aquatic habitats of fresh, brackish, or marine waters. Showing great variation in size and structure, angiosperms range from small, floating disklike bodies with minute flowers to enormous trees 300 ft or more in height. Some complete their life cycle within a few weeks, while others may span many centuries.

The origin of the angiosperms was, according to the fossil record, in Lower Cretaceous, from 140 to 125 million years ago. They had a progressive and remarkable evolutionary explosion and migration and rapidly became the dominant vegetation over most of the earth. By the Eocene of the Tertiary period (50 million years ago), many modern genera had evolved.

The angiosperms are characterized as seed plants with ovules and seeds enclosed in a modified leaf (carpel). A postzygotic endosperm is the primary food-storage tissue of the developing seed. Frequently, however, cotyledons replace the endosperm as the storage tissue in the mature seed.

Angiosperms are divided into two classes, *Magnoliopsida* commonly called the "dicots," and *Liliopsida* or "monocots." Although common names for these classes indicate the number of cotyledons, there are other differences as indicated in Table 9-1.

TABLE 9-1 COMPARISON OF THE TWO CLASSES OF ANGIOSPERMS

Class	Cotyledons	Leaf Venation	Flower Parts	Vascular Bundles in Stem
Dicots	2 (rarely 1 to 4)	Pinnate or palmate, rarely parallel	Various, typically in fours or fives, seldom in threes	Separate or fused, forming a ring around a central pith
Monocots	1	Mostly parallel	Typically in threes, seldom in fours, rarely fives	Scattered throughout the stem

However, each of these characters is variable, so it is safer to base the distinction on two or more of the differences. The two classes are further divided into a total of 11 subclasses following the system proposed by Cronquist (1988). In the phylogenetic arrangement shown in Figure 9-1, the monocots arose from primitive dicots, and the relative size of the subclass balloons is roughly proportional to the number of species in each. The characteristics of the subclasses are not given here because of their complexity and because they are based mainly on reproductive morphology. See Cronquist (1988) for these details. This modern classification is being used in many recent books (Elias, 1989; FNA, 1997; Gleason and Cronquist, 1991).

Trees, defined by their habit and having secondary xylem (wood), are only found among the dicots. However, there are several monocots of tree form and size—the palms, bamboos, century plants, and yuccas—but they do not have wood. Trees, in the broad sense then, occur in all the dicot subclasses and three of the monocot subclasses. They represent both primitive and advanced evolutionary types. A list of the 47 families included here, by class and subclass, follows.

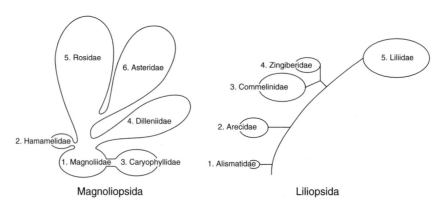

Magnoliopsida   Liliopsida

FIGURE 9-1 Two classes and 11 subclasses of angiosperms. The phylogenetic arrangement is by Cronquist. (*Reprinted from The Evolution and Classification of the Flowering Plants, 2d ed., 1988, The New York Botanical Garden, Bronx.*)

Magnoliopsida—The Dicots

Magnoliidae

Magnoliaceae	magnolia family	Lauraceae	laurel family
Annonaceae	custard-apple family		

Hamamelidae

Platanaceae	sycamore family	Juglandaceae	walnut family
Hamamelidaceae	witch-hazel family	Myricaceae	waxmyrtle family
Ulmaceae	elm family	Fagaceae	beech family
Moraceae	mulberry family	Betulaceae	birch family
Leitneriaceae	corkwood family	Casuarinaceae	beefwood family

Caryophyllidae

Cactaceae	cactus family

Dilleniidae

Theaceae	tea family	Sapotaceae	sapodilla family
Tiliaceae	linden family	Ebenaceae	ebony family
Tamaricaceae	tamarisk family	Styracaceae	storax family
Salicaceae	willow family	Symplocaceae	sweetleaf family
Ericaceae	heath family		

Rosidae

Rosaceae	rose family	Rhamnaceae	buckthorn family
Mimosaceae	mimosa family	Sapindaceae	soapberry family
Caesalpiniaceae	caesalpinia family	Hippocastanaceae	buckeye family
Fabaceae	bean family	Aceraceae	maple family
Elaeagnaceae	oleaster family	Anacardiaceae	cashew family
Myrtaceae	myrtle family	Simaroubaceae	quassia family
Cornaceae	dogwood family	Meliaceae	mahogany family
Aquifoliaceae	holly family	Rutaceae	rue family
Euphorbiaceae	spurge family	Araliaceae	ginseng family

Asteridae

Oleaceae	olive family	Bignoniaceae	trumpet-creeper family
Scrophulariaceae	figwort family		

Liliopsida—The Monocots

Arecidae

Arecaceae	palm family

Commelinidae

Poaceae	grass family

Liliidae

Agavaceae	century-plant family

MAGNOLIOPSIDA—THE DICOTS

MAGNOLIIDAE

Magnoliaceae: The Magnolia Family

This family not only includes some of the most interesting of our modern trees but also some of the most primitive. Fossils of numerous, now extinct, species have been found, some of which extended northward as far as Alaska and Greenland during the latter part of the Cretaceous period. The family comprises ca. 7 genera and 220 species of trees or shrubs found in southeast Asia, in the eastern United States southward through Central America, and from the West Indies to eastern Brazil. Two genera and nine species are native in North America (FNA, 1997) (Table 9-2).

TABLE 9-2	COMPARISON OF NATIVE GENERA IN THE MAGNOLIACEAE		
Genus	**Leaves**	**Flowers**	**Fruit**
Magnolia magnolia	Unlobed; apex acute to obtuse; base cuneate to auriculate	Stamens introrse	Aggregate of follicles; seed with brightly colored, oily aril
Liriodendron yellow-poplar	4–6-lobed; apex truncate to broadly notched; base rounded to truncate or slightly cordate	Stamens extrorse	Aggregate of samaras; seed without aril

Botanical Features of the Family

Leaves deciduous or persistent, alternate, simple, stipulate; the stipules enclose the bud, and their conspicuous scars encircle the twig.

Flowers large, actinomorphic, perfect (rarely unisexual), terminal or axillary, solitary; *sepals* 3, *petals* 6 or more; *stamens* and *pistils* many, arranged spirally on an elongated receptacle; *ovary* superior; entomophilous.

Fruit a conelike aggregate of follicles or samaras.

MAGNOLIA L.   magnolia

The trees and shrubs included in this genus are particularly well known on account of their large, showy, beetle-pollinated flowers that, in certain species, are nearly a foot or more in diameter. Many ornamental forms, both native and exotic, natural and hybridized, are widely used in the landscape. The genus was named in honor of Pierre Magnol, a celebrated French botanist of the seventeenth century. *Magnolia* numbers some 120 or more species, scattered through southern and eastern Asia, Mexico to Venezuela, and the eastern United States. Eight species are native to this country, but only two are of any importance as timber producers (Table 9-3).

Botanical Features of the Genus

Leaves deciduous or persistent, unlobed, entire margined; stipules large, early deciduous, free or fused to the petiole and leaving a scar on the petiole.

TABLE 9-3   COMPARISON OF IMPORTANT MAGNOLIAS

Species	Leaves	Flowers	Fruit	Buds
M. acuminata cucumbertree	Deciduous; stipules fused to petiole	Greenish yellow, $1\frac{1}{2}$–$3\frac{1}{2}$ in. (4–9 cm) wide	Glabrous	Silvery white, silky
M. grandiflora southern magnolia	Persistent; stipules free	Snow-white, 6–8 in. (15–20 cm) wide	Rusty tomentose	Rusty brown, pubescent

Flowers in the native, deciduous species appearing just after the leaves; *sepals* 3; *petals* 6 to 15, in series of 3; *stamens* and *pistils* numerous, spirally arranged; *pollination* by beetles, flies, and bees.

Fruit a conelike aggregate of spirally placed, 1- to 2-seeded follicles; *seeds* with a scarlet, pulpy outer layer (aril), when mature, suspended by long slender threads outside the open follicles.

Twigs somewhat bitter tasting, aromatic, moderately stout, straight or slightly zigzag; *pith* homogeneous or inconspicuously diaphragmed, terete; *terminal buds* usually large and conspicuous, with a single outer scale; *lateral buds* smaller; *leaf scars* crescent-shaped to oval; *bundle scars* conspicuous, scattered or in a double row; *stipular scars* conspicuous, encircling the twig.

Magnòlia acuminàta L.   cucumbertree, cucumber magnolia

Distinguishing Characteristics (Fig. 9-2)

Leaves deciduous, 4 to 10 in. (10 to 25 cm) long, 3 to 6 in. (7.5 to 15 cm) wide; *shape* broadly elliptical to ovate; *margin* entire or repand; *apex* acute to acuminate; *base* broadly cuneate, rounded, or cordate; *surfaces* glabrous and yellow-green above, glabrous or pubescent below; *fall color* yellow-brown or pale yellow.

Flowers $1\frac{1}{2}$ to $3\frac{1}{2}$ in. (4 to 9 cm) wide, greenish yellow to golden yellow.

Fruit aggregate $1\frac{1}{2}$ to 3 in. (4 to 9.5 cm) long, cylindrical to ovoid, glabrous, rarely over 1 in. (2.5 cm) in diameter; *seeds* $\frac{1}{2}$ in. (12 mm) long, red, after emergence suspended on slender threads.

Twigs moderately stout, reddish brown, becoming grayish brown, glabrous to pubescent; *terminal buds* $\frac{1}{2}$ to $\frac{3}{4}$ in. (12 to 18 mm) long, silvery silky; *lateral buds* smaller; *leaf scars* horseshoe-shaped; *bundle scars* 5 to 9.

Bark brownish gray, fissured into narrow flaky ridges.

Range

Eastern (Map 9-1). Elevation: 100 to 4000 ft. Rich, mesic slopes, ravines, valleys, and along streams.

General Description

Cucumbertree attains heights of 80 to 90 ft and 3 to 4 ft dbh (max. 125 by 6 ft). It has a straight, clear, slightly buttressed bole, pyramidal crown, a deep but wide-spreading root system, and is

FIGURE 9-2 *Magnolia acuminata,* cucumbertree. (1) Twig $\times 1\frac{1}{4}$. (2) Flower $\times\frac{1}{2}$. (3) Aggregate of follicles with seeds $\times\frac{3}{4}$. (4) Leaf $\times\frac{1}{2}$. (5) Bark.

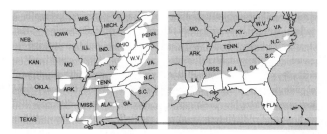

MAP 9-1 **(Left)** *Magnolia acuminata.* **(Right)** *Magnolia grandiflora;* naturalized farther northeast than shown on the map.

usually found on moist, fertile soils of loose texture, with the other hardwoods of the region, such as white and red oaks, hickories, black tupelo, American beech, ashes, and yellow-poplar. Best development is reached at the base of the Appalachians in the Carolinas, Tennessee, and Kentucky, but the tree is nowhere abundant. It is the most cold-hardy of the native magnolias but is relatively short-lived (to 150 years). Cucumbertree is moderately shade tolerant.

The wood is commonly mixed and utilized with that of yellow-poplar. The lumber is used for furniture, interior trim, doors, and boxes. It is a handsome shade tree with beautiful flowers (although often out of sight until fallen on the ground).

This is a variable species and some authors recognize two varieties.

Magnòlia grandiflòra L.   southern magnolia, evergreen magnolia

Distinguishing Characteristics (Fig. 9-3)

Leaves leathery, persistent; 4 to 8 in. (10 to 20 cm) long, $2\frac{1}{2}$ to 4 in. (6 to 10 cm) wide; *shape* narrowly oval to ovate; *margin* entire; *apex* bluntly acute or acuminate; *base* cuneate; *surfaces* bright green and very lustrous above, green and puberulent or rusty tomentose below.

Flowers 6 to 8 in. (15 to 20 cm) wide, fragrant; with 6 or 9 to 12, large creamy white petals; appearing in early summer.

Fruit aggregate $2\frac{3}{4}$ to 4 in. (7 to 10 cm) long, ovoid or short cylindrical, $1\frac{1}{2}$ to $2\frac{1}{2}$ in. (4 to 6 cm) in diameter; tan tomentose.

Twigs rusty tomentose; terminal buds 1 to $1\frac{1}{2}$ in. (2.5 to 3.5 cm) long, pale or rusty tomentose; lateral buds smaller.

Bark dark gray, smooth becoming furrowed and scaly (Fig. 9-4).

Range

Southeastern (Map 9-1). Elevation: sea level to 400 ft. Upland well-drained soils, mesic slopes, ravines, maritime forests near coasts. Its range now extends northward to Maryland and westward to higher elevations by escaping cultivation.

General Description

Southern magnolia is a medium-sized tree, commonly 60 to 100 ft tall and 2 to 3 ft dbh (max. 135 by 7 ft), with a tall straight bole and somewhat pyramidal crown. It is easily identified in the forest by its persistent (2 years) shiny, dark green leaves, which are often bronze or dark brown

FIGURE 9-3 *Magnolia grandiflora,* southern magnolia. (1) Aggregate of follicles $\times\frac{3}{4}$. (2) Flower opened before anthesis to show tight spiral of stamens and pistils $\times\frac{1}{4}$.

FIGURE 9-4 Bark of southern magnolia.

below. This magnolia is a southern bottomland species but will not survive long inundations; it is found on moist but well-drained soils in association with American beech, sweetgum, yellow-poplar, white ash, oaks, and hickories. It is becoming more abundant onto mesic upland sites with fire suppression. Winter droughts can cause dieback and mortality.

Besides the timber produced, the seeds are eaten by various birds and small mammals. Southern magnolia is a very valuable ornamental on account of its large, showy flowers, evergreen leaves, and numerous cultivars. It is planted for this purpose in North America as far north as New Jersey on the Atlantic coast and along the Pacific coast northward to British Columbia. It is one of America's great gifts to the gardens of the world.

Southern magnolia is the state tree of Mississippi and the state flower of Louisiana and Mississippi. It hybridizes with sweetbay.

Other Native Magnolias

Several other native species, although of little or no importance as timber producers, are worthy of mention on account of their conspicuousness in the forest and because some are commonly used as ornamentals throughout the mid-Atlantic to southern states.

Magnòlia virginiàna L.   sweetbay, sweetbay magnolia

Distinguishing Characteristics.   *Leaves* $2\frac{3}{4}$ to 6 in. (7 to 15 cm) long, $1\frac{3}{8}$ to 2 in. (3.5 to 5 cm) wide, oblong to elliptical, glaucous and white pubescent below, deciduous in the late fall, or persistent until spring. *Flowers* creamy white, cup-shaped, 2 to $3\frac{1}{4}$ in. (5 to 8 cm) in diameter, very attractive and aromatic. *Aggregate* of follicles ellipsoid to globose, $1\frac{1}{4}$ to 2 in. (3 to 5 cm) long.

Range. Massachusetts to Florida and west to Texas. Elevation: sea level to 500 ft. Swamps, pocosins, bays, and low woods.

General Description. Trees to a height of 80 ft and dbh of 3 ft (max. 91 ft by 50 in.). Flowering begins very early, and it is not unusual to see a slender stem only about 2 ft tall with a big flower at the top. This moderately tolerant species is variable with two growth forms: deciduous and multibranched northward; evergreen and single-trunked southward. Sweetbay is moderately flood tolerant. These trees are an important deer browse throughout the year. The wood is used for furniture, boxes, and tool handles. It is also cultivated as an ornamental lawn, park, or street tree. See the discussion of "bay trees" under *Persea.*

Magnòlia macrophýlla Michx.   bigleaf magnolia

Distinguishing Characteristics. *Leaves* deciduous, 10 to 32 in. (25 to 80 cm) long, 6 to 12 in. (15 to 30 cm) wide, broadly elliptical to oblanceolate, white below, cordate or auriculate at the base; fall color yellow. *Flowers* creamy white, often with a purple blotch inside and at the base of the inner petals, 10 to 12 in. (25 to 30 cm) or more in diameter, fragrant. *Aggregate* of follicles subglobose to globose, over 2 in. (5 cm) in diameter, with 50 or more carpels.

Range. North Carolina to southern Alabama and Louisiana, north to Kentucky and West Virginia; disjunct in southeastern South Carolina. Elevation: sea level to 1200 ft. This is found in scattered populations in moist woods of ravines, valleys, and ridges.

General Description. This is a small tree with the largest simple leaves and flowers of any North American tree or shrub. It, along with Fraser and umbrella magnolias, are occasionally planted as ornamentals for their large leaves, large flowers, and tropical appearance.

Magnòlia áshei Weatherby   Ashe magnolia

Distinguishing Characteristics. *Leaves* deciduous, 8 to 24 in. (20 to 60 cm) long, 4 to 12 in. (10 to 30 cm) wide, elliptical to oblanceolate, white below, deeply cordate to auriculate at base; fall color yellow-brown. *Flowers* $6\frac{1}{4}$ to 12 in. (16 to 30 cm) in diameter, whitish with a purple blotch at the inside base of each petal, aromatic. *Aggregate* of follicles cylindrical to ovoid, less than 2 in. (5 cm) in diameter, with less than 40 carpels.

Range. Local and rare in northwestern Florida. Elevation: sea level to 200 ft. This occurs in upland woods, in ravines, and on bluffs.

General Description. This is a small tree or shrub. It is treated by some authors as a subspecies or variety of the bigleaf magnolia.

Magnòlia fràseri Walt.   Fraser magnolia

Distinguishing Characteristics. *Leaves* deciduous, 10 to 20 in. (25 to 50 cm) long, 3 to 5 in. (8 to 15 cm) wide, oblanceolate, pale green below, auriculate at base; fall color yellow. *Flowers* creamy white, $6\frac{1}{4}$ to $8\frac{1}{2}$ in. (16 to 22 cm) in diameter, fragrant. *Aggregate* of follicles ellipsoidal, 2 to 4 in. (5 to 10 cm) long.

Range. West Virginia and western Virginia south to northern Georgia. Elevation: 800 to 5000 ft. This is a moderately tolerant, understory tree of rich woods on slopes and in mountain coves of the southern Appalachians. It can reach 110 ft tall and 32 in. dbh.

Magnòlia pyramidàta Bartr.   pyramid magnolia

> **Distinguishing Characteristics.**   *Leaves* deciduous, 7 to 10 in. (18 to 25 cm) long, 2 to $4\frac{3}{4}$ in. (5 to 12 cm) wide, panduriform to rhombic-spatulate, pale green below, deeply cordate to auriculate at base; fall color yellow. *Flowers* creamy white, $2\frac{1}{2}$ to $3\frac{1}{4}$ in. (6 to 8 cm) in diameter, fragrant. *Aggregate* of follicles ellipsoidal, $1\frac{1}{2}$ to $2\frac{1}{2}$ in. (4 to 6 cm) long.
>
> **Range.**   South Carolina to northwestern Florida, and west to northeastern Texas. Elevation: sea level to 400 ft. This is a small tree of rich woods, slopes, and river bluffs of the Piedmont and Coastal Plain.
>
> **General Description.**   This is treated by some authors as a subspecies or variety of Fraser magnolia.

Magnòlia tripétala L.   umbrella magnolia

> **Distinguishing Characteristics.**   *Leaves* deciduous, 8 to 16 in. (20 to 40 cm) long, 4 to 8 in. (10 to 20 cm) wide, oblanceolate, often clustered at the ends of branches, pale green below, acuminate at base; fall color yellow. *Flowers* creamy white, malodorous, 10 to 12 in. (25 to 30 cm) in diameter. *Aggregate* of follicles cylindric, $3\frac{1}{2}$ to 4 in. (6 to 10 cm) long.
>
> **Range.**   Southern Pennsylvania to southern Alabama and Mississippi, and to eastern Oklahoma and the Ohio River valley. Elevation: 100 to 1400 ft. This is an understory tree in moist soils of slopes and ravines and along streams.

LIRIODENDRON L.   yellow-poplar

Fossil remains indicate that this genus, with several forms, was once widely distributed over North America and the Old World. At present, however, only two species remain: *L. chinénse* (Hemsley) Sarg. of central China and *L. tulipifera* of the eastern United States. The common name "yellow-poplar" for the American species is misleading because the true poplars are in the Salicaceae or willow family and are unrelated to the Magnoliaceae.

Liriodéndron tulipífera L.   yellow-poplar, tuliptree, tulip-poplar

Distinguishing Characteristics (Fig. 9-5)

Leaves deciduous; 4 to 6 in. (8 to 15 cm) long and wide; *shape* nearly orbicular, mostly 4-lobed; *lobes* rounded to acute and sinuses shallow to deep; *margin* of the lobes entire; *base* and *apex* nearly truncate (or apex often broadly notched); *surfaces* puberulent and pale below; *petioles* slender, 5 to 6 in. (12 to 15 cm) long; *stipules* large and conspicuous, together encircling the twig; *fall color* butter-yellow.

Flowers appearing in spring just after the leaves unfold; $1\frac{1}{2}$ to 2 in. (4 to 5 cm) wide, cup-shaped or tuliplike; *sepals* 3; *petals* 6, in two whorls, light greenish yellow with a conspicuous orange spot at the inside base of each; *stamens* and *pistils* many, spirally arranged.

Fruit an erect conelike aggregate of samaras, $2\frac{1}{2}$ to 3 in. (6 to 7 cm) long, at maturity the upper samaras deciduous from the central, more or less persistent axis (these "cones" are of value in identifying the larger trees at some distance after the leaves have fallen); *samaras* 4-angled, terminally winged, 1 to $1\frac{1}{2}$ in. (2.5 to 4 cm) long, 2-seeded, but commonly 1-seeded by abortion.

FIGURE 9-5   *Liriodendron tulipifera,* yellow-poplar. (1) Twig ×1$\frac{1}{4}$. (2) Fruit (samara) ×1. (3) Aggregate of samaras ×$\frac{3}{4}$. (4) Flower ×$\frac{3}{4}$. (5) Leaf ×$\frac{1}{2}$. (6) Bark of young tree. (7) Bark of old tree.

Twigs moderately stout, reddish brown, bitter to the taste; *pith* diaphragmed, terete; *terminal buds* ca. $\frac{1}{2}$ in. (12 mm) long, flattened, "duck-billed" in appearance, the scales valvate in pairs, with only the two outer ones visible; *lateral buds* much smaller; leaf scars circular to oval; *bundle scars* small, scattered; *stipular scars* conspicuous, encircling the twig.

Bark on young trees dark greenish or orange-brown and smooth, with small white spots; soon ashy gray, breaking up into long, rough, interlacing rounded ridges separated by fissures; inner bark bitter and aromatic.

Range

Eastern (Map 9-2). Elevation: sea level to 1000 ft northward, to 4500 ft in the southern Appalachians. Bottomlands in wet to mesic woods, mountain coves, and lower slopes.

General Description

Yellow-poplar is not only one of the most distinctive trees of eastern North America, it is also one of the most valuable timber producers. In the forest, this tree presents an unmistakable outline. The tall, clear, almost arrow-straight trunk terminates in a small, rather open, oblong crown. It can reach 175 ft in height, 11 ft dbh, and 400 years of age. The root system is deep and wide-spreading, and the stumps sprout readily.

Yellow-poplar is intolerant and site-sensitive. On good sites, its juvenile height growth, following the first year, especially from stump sprouts, is so rapid that it can outstrip its competition. On poor sites, it is dominated by many other species.

Fast-growing trees on good sites may be 120 ft tall and 18 to 24 in. dbh at age 50. Maturity is reached in about 200 years, with very old trees attaining the three-century mark. Yellow-poplar is found in some 16 forest types, associated mostly with other hardwoods throughout its range. Coniferous associates are white pine and hemlock, and, in the South, loblolly pine.

An effective pioneer on disturbed sites, yellow-poplar often persists in climax stands. Seed production begins at about 15 years and continues beyond two centuries. Seed is produced annually, and bumper crops occur frequently. Pollination is by insects (tuliptree honey is a commercial product) but is often incomplete. This reproductive weakness is overcome by the ability of filled viable seeds to accumulate in the litter for nearly 10 years and then to produce multitudes of seedlings when conditions for germination are suitable.

Relatively few insects or pathogens cause widespread damage. Yellow-poplar is, however, very susceptible to impact basal wounding and subsequent decay. It forms an important deer browse with sprouts, buds, flowers, and seedlings being eaten. The fruits are an important food for birds and small mammals.

Yellow-poplar is a beautiful, but somewhat brittle, ornamental tree with its large conspicuous flowers and notched leaves on long petioles that quiver in the slightest breeze. Because of this last feature or perhaps certain wood characters, the name *poplar* became attached to this member of the Magnolia family.

Yellow-poplar is a polymorphic species with chemical, morphological, and ecological differences showing a regional and clinal pattern (Parks et al., 1994). This is the state tree of Indiana, Kentucky, and Tennessee.

MAP 9-2   *Liriodendron tulipifera.*

Annonaceae: The Custard-Apple Family

This family includes 130 genera and over 2300 species of trees, shrubs, and lianas mainly in the tropics except *Asimina,* which extends north to Michigan and southern Ontario. The family produces several well-known tropical fruits such as custard apple (*Annòna reticulàta* L.), cherimoya (*Annona cherimòya* Miller), and soursop (*Annona muricàta* L.). These fleshy aggregates form by the fusion of carpels and receptacle and have the general appearance of a fleshy cone. There are two genera and four species of trees native in the United States. Only one, pawpaw, is relatively widespread.

Botanical Features of the Family

Leaves persistent or (in ours) deciduous, alternate, simple, estipulate, distichous, entire, aromatic.

Flowers solitary, entomophilous, perfect, regular; *tepals* 6 to 9 in 2 or 3 whorls; *stamens* many; *ovary* 3-carpellate, superior; appearing before the leaves.

Fruit a berry or aggregate of berries; *seeds* large.

Asímina trilòba (L.) Dunal   pawpaw

Distinguishing Characteristics.   *Leaves* entire, 7 to 12 in. (18 to 30 cm) long, $2\frac{3}{4}$ to 5 in. (7 to 13 cm) wide, elliptical to usually oblanceolate, acuminate at apex and base, pale below; fall color yellow. *Flowers* to $1\frac{1}{2}$ in. (4 cm) across, maroon, nodding, in early spring before the leaves; pollination by carrion flies. *Fruit* a berry, 3 to 6 in. (7.5 to 15 cm) long, 1 to $1\frac{1}{2}$ in. (2.5 to 4 cm) in diameter, smooth-skinned, yellowish flecked with brown streaks, creamy soft around several large, shiny brown, oblong seeds. *Twigs* with naked buds covered with rust-colored hairs; pith diaphragmed the second year. *Bark* light gray-brown, smooth.

Range.   Southern Ontario, southern Michigan, and western New York, south to northwestern Florida, west to eastern Texas, and north to southeastern Nebraska. Elevation: sea level to 4900 ft in the southern Appalachians. Moist soils of floodplains, of stream banks, and in the understory of hardwood forests.

General Description.   Pawpaw is a shade intolerant, small tree to 30 ft tall and 10 in. or more dbh. The large leaves may appear like magnolia, but the twig lacks the stipular scars, and the hairy, dark, rust-colored, naked buds are conspicuous. The fruit has an unusual but pleasant taste, but until recently it was eaten mainly by wildlife and local people who happen to know where the trees occur. Root sprouting is common, and many trees may form small but rather dense clones. There has been a surge of new interest and attempt to develop commercially marketable fruit and ornamental cultivars. It is a beautiful small shade tree with a conical crown, attractive spring flowers, delicious fruit, and bright yellow fall color.

Lauraceae: The Laurel Family

The laurel family includes ca. 50 genera and some 2850 species of trees and shrubs, which are largely evergreen. Most of them are tropical, and although a few extend into the temperate zones, the forests of the American tropics and those of southeastern Asia support a particularly luxurious growth of lauraceous plants.

Many aromatic substances present in the leaves, stems, bark, roots, and fruits of certain species have been investigated and commercially exploited. From "bois de rose,"

Aníba rosaeodòra Ducke, is obtained linaloa oil, a compound used in the manufacture of expensive perfumes. Camphor tree, *Cinnamòmum camphòra* (L.) T. Nees & Eberm., is the source of true camphor of commerce. It is a native of Asia and is commonly planted in the southern United States as an ornamental evergreen of lawns, of parks, and along streets. Cinnamon and cassia bark are products of two other Asiatic members of this genus, namely, *C. vérum* J. Presl and *C. aromáticum* Nees, respectively. The avocado or alligator pear, now so widely cultivated throughout the tropics and available in most markets, is the fruit of *Persèa americàna* Mill., a small tree originally native to the American tropics but now rapidly becoming naturalized in many other tropical countries. *Laúrus nòbilis* L. of the Mediterranean area and cultivated in North America provides the bay leaves used in cooking.

A few commercial timbers of the Lauraceae find their way into the lumber markets of the world, but most of them are used only locally. *Endiàndra palmerstòni* C. T. White, the Oriental walnut of eastern Australia, resembles black walnut, *Juglans nigra* L. sufficiently (superficially at least) to permit substitution for this species. *Ocotèa rodiaèi* (R. Schomb.) Mez., the British Guianan greenheart, has long since proved its worth in marine construction; this is exported in quantity, particularly to European ports, and small amounts are utilized along the American seacoasts; the gates of the Panama Canal locks were originally fabricated with this wood. The South African stinkwood, *Ocotèa bullàta* E. Mey., because of its strength, durability, color, and pleasing figure, is a cabinet wood and building timber of recognized merit.

Four or five genera have arborescent forms in the United States, but these are of only minor importance. *Nectándra coriàcea* (Sw.) Griseb., Florida nectandra, with entire, persistent leaves, is native to southern Florida. *Líndera bénzoin* (L.) Blume, spicebush, is a common large shrub of the eastern states; it has deciduous, entire, aromatic leaves and twigs, yellow flowers, and small clustered buds. The remaining genera, *Umbellularia* (monotypic), *Persea,* and *Sassafras,* are worthy of a more detailed description.

Botanical Features of the Family

Leaves persistent or deciduous, alternate (rarely opposite), simple, often glandular-punctate, usually aromatic, estipulate.

Flowers perfect and imperfect (some species polygamous), regular, entomophilous; *tepals* 6-parted; *stamens* in 3 or 4 whorls of 3 each; *pistils* 1, ovary superior, 1-celled and with a solitary ovule.

Fruit a 1-seeded berry or drupe.

Umbellulària califórnica (Hook. and Arn.) Nutt.   California-laurel, Oregon-myrtle, California bay

Distinguishing Characteristics (Fig. 9-6).   *Leaves* persistent, coriaceous, 2 to 5 in. (5 to 13 cm) long, $\frac{1}{2}$ to $1\frac{1}{2}$ in. (1.2 to 4 cm) wide, lanceolate to elliptical, entire, pungent-spicy when crushed. *Flowers* perfect, yellowish green, in axillary umbels, appearing from January to March before the new leaves. *Fruit* an acrid, yellow-green to purple drupe. *Twigs* yellow-green, terminal bud present, pith homogenous. *Bark* brown, with thin appressed scales.

FIGURE 9-6   *Umbellularia californica,* California-laurel. (1) Leaf ×$\frac{3}{4}$. (2) Flowers ×$\frac{3}{4}$. (3) Fruit (drupe) ×1. (4) Bark. *(Photos 2 and 4 by W.I. Stein, U.S. Forest Service.)*

Range.   Coos Bay region of southwestern Oregon southward through the Coast Ranges and lower Sierra Nevada to southern California. Elevation: sea level to 1500 ft in the north; 2000 to 6000 ft in the south. Dry, rocky soils on slopes and bluffs; also in moist bottomlands.

General Description.   California-laurel is the only lauraceous species in western North America. In moist sites in southwestern Oregon, it is a large tree 100 to 175 ft in height and 3 to 10 ft dbh, although elsewhere in its range it is usually a small- or medium-sized tree 40 to 80 ft tall and 18 to 30 in. dbh. In drier sites, it is commonly shrubby, and along the windswept bluffs of the Pacific Ocean it often becomes prostrate and forms a spreading network of thickly matted branches. The bole, even under forest conditions, occasionally divides near the ground into several ascending limbs resulting in a broad, dense, round-topped crown; the root system is usually deep but wide-spreading.

Large burls or smaller excrescences are often formed on the trunks of older trees, and those devoid of defects command a high price in the trade. The smaller ones are often turned into many sorts of fancy articles of woodenware, and the large ones are shipped to veneer plants where they are carefully cut into thin sheets of intrinsic value and beauty. It is the most valued hardwood of the West.

The aromatic properties of all parts of the tree have been used in primitive medicine. The evergreen leaves have a very strong, camphorlike odor when crushed or bruised. In spite of this, goats are known to forage on them during the winter months when more suitable browse is extremely sparse or lacking. The fruits and seeds are eaten by various wildlife.

Seeds are produced in abundance after 30 to 40 years; they mature in one season and germinate within a few weeks after dispersal. Seedlings often form dense clumps under partial shade and develop quite rapidly; this species also sprouts from the root collar. California-laurel is shade tolerant.

SASSAFRAS Nees and Eberm.   sassafras

> There are three species of *Sassafras*; one is found in central China, another in Taiwan, and the third in the eastern United States.

Sássafras álbidum (Nutt.) Nees   sassafras

Distinguishing Characteristics (Fig. 9-7).   *Leaves* deciduous, 3 to 5 in. (7.5 to 13 cm) long, 1 $\frac{1}{2}$ to 4 in. (4 to 10 cm) wide, variable in shape. On the same tree three different forms of leaves may be found: entire and (1) somewhat elliptical and unlobed (typical of older trees), (2) mitten-shaped (either right- or left-handed), and (3) three-lobed, rarely 4–5 lobed; fall color yellow to orange or red. *Flowers* imperfect (plants dioecious), fragrant, yellowish, appearing before the leaves. *Fruit* a blue drupe on a thickened, reddish pedicel. *Twigs* green, aromatic, buds scaly, bundle scar 1, stipular scars lacking, pith homogeneous. *Bark* deeply furrowed, reddish brown, spicy, and aromatic.

FIGURE 9-7   *Sassafras albidum,* sassafras. (1) Fruits (drupes on thickened pedicels) $\times\frac{3}{4}$. (2) Three-lobed leaf $\times\frac{1}{2}$. (3) Unlobed leaf $\times\frac{1}{2}$. (4) Two-lobed or mitten-shaped leaf $\times\frac{1}{2}$. (5) Twig $\times 1\frac{1}{4}$.

Range.   Southern Ontario, east to southwestern Maine, south to central Florida, west to eastern Texas and eastern Oklahoma, north to central Michigan. Elevation: sea level to 5000 ft in the southern Appalachians. Disturbed sites and in dry to mesic forests, clearings, fence rows, and old fields.

General Description.   This is usually a small- to medium-sized tree but may reach a maximum of 100 ft tall and 6 ft dbh. It is intolerant and is common as a pioneer tree. Its best growth is on moist, well-drained, sandy soils.

The very durable wood, often mixed with that of black ash, is used for furniture. Sassafras may be of greater importance in the future, because in certain parts of its range it is, by means of prolific root sprouts as well as seeds, rapidly restocking abandoned farm lands. Oil of sassafras is used in the preparation of certain soaps and flavorings for medicines, candy, and sassafras tea. The young leaves, dried and powdered, are quite mucilaginous and are used to both thicken and flavor creole dishes.

Damage occurs from several insects and fungi. It forms an important deer browse during spring, summer, and winter, and the seeds are important food for various wildlife.

PERSEA Mill.   persea

This is a mostly tropical genus with ca. 150 species including the avocado mentioned earlier. There are two species of native trees, to 25 ft tall, in the southeastern Coastal Plains. Both are recognized by their *leaves*—alternate, simple, elliptical, 3 to 6 in. (7.5 to 15 cm) long, $\frac{3}{4}$ to $1\frac{1}{2}$ in. (2 to 4 cm) wide, persistent, entire, estipulate, and aromatic as in the "bay leaf" of commerce; *flowers* small and rather inconspicuous; *fruits* dark blue to black drupes, $\frac{1}{4}$ to $\frac{1}{2}$ in. (7 to 12 mm) in diameter. They are of some importance for browse, and the fruits are eaten by birds and small mammals. Both of the following species are shade tolerant, are significant food (fruit and foliage) to wildlife, and have wood that is used locally for cabinetwork and boat building.

Persèa borbònia (L.) Spreng.   redbay

Distinguished by its scattered straight hairs, barely visible and appressed to the lower leaf surface; the twig and petioles are sparsely pubescent or glabrous. It is a common tree in mesic to xeric woodlands and coastal dunes from North Carolina to Florida and west to southeastern Texas.

Persèa palústris (Raf.) Sarg.   swampbay

The underside of the leaves have curved hairs not appressed to the surface, and twigs and petioles are usually pubescent. It is common in wet woodlands, hammocks, bay forests, and edges of marshes from southern Delaware south to Florida and west to southeastern Texas. This species is sometimes considered a variety of *P. borbonia* (Little, 1979). The two sometimes grow together along the coast and are difficult to distinguish.

"Bay Trees"

There are three genera of bay trees that are evergreen, often grow together in "bay" forests of the southeastern Coastal Plain, and are often confused when flowers or fruits are lacking. The red and swamp bays (*Persea,* Lauraceae) have aromatic, entire leaves, green below and estipulate; sweetbay (*Magnolia virginiana,* Magnoliaceae) has aromatic, entire leaves but with stipular scars around the twig, and the leaves white below; loblolly bay (*Gordonia lasianthus,* Theaceae) has leaves that are not aromatic when crushed, pale green below, estipulate, and bluntly serrulate on the margin.

HAMAMELIDAE

Platanaceae: The Sycamore or Planetree Family

This family has a single genus, *Platanus,* with six or seven species—three are native to North America, one of the eastern states, the other two of the West. The foreign species are found in Mexico and from southeastern Europe to India.

PLATANUS L.   sycamore, planetree

Botanical Features of the Genus

Leaves deciduous, alternate, simple, palmately 3- to 7-lobed; *petioles* enlarged at the base and enclosing the lateral bud; *stipules* leaflike and very conspicuous, encircling the twig and often persisting.

Flowers imperfect (plants monoecious), anemophilous, appearing with the leaves, in separate globose heads at the ends of a long peduncle; individual flowers minute; *staminate* with 3 to 8 sepals, petals, and stamens; *pistillate* with 3 to 8 sepals, petals, and carpels. Sycamore pollen can cause allergic reactions in susceptible individuals.

Fruit a globose head (multiple) of narrowly elongated achenes, each surrounded at the base by a circle of erect brown hairs and usually bearing at the apex a minute curved spur; long persistent on a tough, fibrous peduncle until the multiple shatters.

Twigs conspicuously zigzag; *pith* homogenous, terete; *terminal buds* lacking; *lateral buds* divergent, resinous, with a caplike outer scale; *leaf scars* encircling the buds; *bundle scars* several; *stipular scars* encircling the twig.

Plátanus occidentàlis L.   sycamore, American planetree

Distinguishing Characteristics (Fig. 9-8)

Leaves 4 to 8 in. (10 to 20 cm) in diameter; *shape* broadly ovate to orbicular, 3- to 5-lobed, sinuses broad and usually shallow; *margin* of lobes sinuately toothed; *apex* acuminate; *base* cordate to truncate; *surfaces* glabrous above, at first hairy throughout the lower surface but later only along the veins; *petioles* stout, 2 to 3 in. (5 to 7 cm) long, hollow at the base; *fall color* yellow to brown.

Fruit heads $\frac{3}{4}$ to $1\frac{1}{4}$ in. (2 to 3 cm) in diameter, borne singly on slender peduncles 3 to 6 in. (7.5 to 15 cm) long; usually persistent on the tree during part of the winter.

Twigs moderately slender, dark orange-brown, conspicuously zigzag; *pith* homogeneous, terete; *leaf scars* nearly surrounding the bud; *stipular scars* distinct, encircling the twigs; *terminal buds* lacking; *lateral buds* divergent, resinous, and with a single visible caplike scale.

Bark on young branches brownish; soon characteristically mottled (brown, green, and white) by the exfoliation of the outer bark which exposes the lighter creamy white inner layers, bark on the lower trunk of older trees often entirely brown and scaly.

FIGURE 9-8 *Platanus occidentalis,* sycamore. (1) Leaf and stipules $\times\frac{1}{2}$. (2) Head of pistillate flowers $\times\frac{1}{2}$. (3) Head of staminate flowers $\times\frac{1}{2}$. (4) Head (multiple) of achenes $\times\frac{3}{4}$. (5) Achene surrounded by hairs $\times1$. (6) Twig $\times1\frac{1}{4}$. (7) Bark of young tree. (8) Bark of old tree.

MAP 9-3 *Platanus occidentalis.*

Range

Eastern (Map 9-3). Elevation: sea level to 3200 ft. Alluvial soils of river bottoms and lake shores.

General Description

Sycamore is one of the largest (perhaps the greatest dbh) of eastern hardwood trees and commonly attains a height of over 100 ft and a dbh of 3 to 8 ft (max. 153 by 11 to 14 ft). Even at a considerable distance, this tree can hardly be mistaken for any of its associated species. The mottled bark is very striking in appearance, and the open, spreading, white-barked crown with its somewhat crooked smaller branches is also characteristic. The root system is superficial. Its associates include American elm, soft maples, water and Nuttall oaks, river birch, cottonwood, sweetgum, and willows. Sycamore is usually rated from intermediate to intolerant and occurs as a pioneer tree in small, nearly pure groves that are transitional. As competition from more tolerant species increases, sycamore remains only as a scattered tree in a number of forest climax types.

Seeds are usually produced each year and are shed irregularly throughout the fall, winter, and early spring. Growth is very fast, and first-year seedlings may reach 8 ft in height. Trees less than 20 years old commonly reach 70 ft in height and 9 in. dbh. Sycamore may attain an age of 500 to 600 years.

Except for frequent defoliation by anthracnose, sycamore in natural stands is not seriously impacted by insects or diseases. However, damage by a number of diseases can be serious in intensively managed plantations.

Long used for lumber and veneer, sycamore is now much used for wood pulp. Established either as seedlings or cuttings in intensively cultivated and fertilized plantations ("short-rotation forestry"), its volume growth is outstanding. Vigorous sprouting permits two or three short coppice rotations. It is of little value to wildlife.

Although this species is often planted as an ornamental yard, park, and street tree, the London planetree, *P.* ×*acerifòlia* (Ait.) Willd., is often more commonly cultivated due to its anthracnose resistance. The London planetree is presumably a hybrid between *P. occidentalis* L., sycamore, and *P. orientális* L., the Oriental plane. It is similar in appearance to the native sycamore except that the bark is more olive-green in color, the leaf lobes longer and narrower, and the fruit heads may occur in twos and threes rather than singly.

Western Species

Two western sycamores, occurring in widely separated regions, commonly form dense cover along streams or on precipitous slopes and aid materially in controlling erosion.

Plátanus wrìghtii Wats.   Arizona sycamore

Probably the most abundant broad-leaved tree in the Southwest, it occurs generally along stream banks and canyon walls in southern Arizona and southwestern New Mexico. The tree is readily distinguished by its abaxially glabrescent, deeply 5- to 7-lobed leaves (the terminal lobe $\frac{2}{3}$ or more the length of the blade), with cordate bases, and heads of achenes borne in racemose clusters of two to four.

Plátanus racemòsa Nutt.   California sycamore

Medium-sized, broadly buttressed tree of wide distribution along stream banks in the Coast Ranges and in the foothills along the west slopes of the Sierra Nevada Mountains in California. It is identified by its abaxially tomentose, narrowly 3- to 5-lobed leaves (the lobes about $\frac{1}{3}$ to $\frac{2}{3}$ the length of the blade), with cuneate (rarely cordate) bases, and heads of achenes in racemose clusters of two to seven.

Hamamelidaceae: The Witch-Hazel Family

This family includes ca. 31 genera and ca. 100 species of trees and shrubs scattered through the forested regions of eastern North America and Mexico to Central America, also South Africa, Madagascar, Australia, Asia, and the Malayan archipelago.

Commercial storax, a balsam used in the manufacture of soaps, perfumes, and pharmaceutical preparations, is obtained from the resinous exudations of the bark of *Liquidámbar òrientalis* Mill. of Asia Minor.* Burmese storax, a similar but somewhat inferior compound, comes from the resinous constituents of *Altíngia excélsa* Noronha and is used locally in the preparation of incense and medicinal compounds. *Altíngia, Bucklándia,* and *Liquidambar* are the principal timber-contributing genera of this family.

Three genera are represented in the eastern United States: *Fothergílla,* with two shrubby species, is confined mostly to the Southeast; *Hamamelis,* with two species, is found in the East; and *Liquidambar* is an important hardwood of the South (Table 9-4).

Botanical Features of the Family

Leaves deciduous or persistent, alternate, simple, stipulate, often with stellate hairs.

Flowers perfect or imperfect (plants monoecious), anemophilous or entomophilous; *calyx* 4- or 5-parted or 0; *petals* 4, 5, or 0; *stamens* 4, 5, or more; *pistils* 1 with a 2-celled ovary enclosing 1 or more ovules in each cell.

Fruit a 2-celled capsule, borne singly, or in multiple heads; seeds winged or wingless.

LIQUIDAMBAR L.   sweetgum

The genus *Liquidambar* has three or four species; only one, *L. styraciflua,* is native to the new world; the other species are found in Asia. *Liquidambar formosàna* Hance, Formosan sweetgum, is cultivated as an ornamental in North America.

---

*Storax may also be prepared in commercial quantities by wounding trees of the native sweetgum, but for economic reasons it will probably not compete with that from abroad, except when the foreign supply is interrupted.

TABLE 9-4   COMPARISON OF IMPORTANT GENERA IN THE HAMAMELIDACEAE

Genus	Leaves	Flowers	Capsules	Buds
Liquidambar sweetgum	Palmately lobed and veined	Imperfect, trees monoecious	Globose multiple	Scaly
Hamamelis witch-hazel	Unlobed, pinnately veined	Perfect	Borne singly	Naked

Liquidámbar styracíflua L.   sweetgum, American sweetgum

Distinguishing Characteristics (Fig. 9-9)

Leaves deciduous, 3 to $7\frac{1}{2}$ in. (7.5 to 19 cm) in diameter; *shape* orbicular, star-shaped, deeply and palmately 5- to 7-lobed; *margin* finely serrate; *apex* of the lobes acuminate; *base* truncate or slightly cordate; *surfaces* bright green and lustrous above, pubescent in the axils of the veins below; somewhat fragrant when crushed; *petioles* long and slender; stipulate; *fall color* yellow or deep red to purple.

Flowers imperfect (plant monoecious), both types in heads, anemophilous; *staminate* heads in racemes, *pistillate* terminal and solitary; *staminate* flowers lacking calyx and corolla, stamens indefinite in number; *pistillate* flowers with a minute calyx, no corolla, 4 nonfunctional stamens, and a 2-celled ovary; appearing with the leaves.

Fruit a woody, globose head (multiple) of 2-celled, beaked capsules, 1 to $1\frac{1}{4}$ in. (2.5 to 3 cm) in diameter, persistent during the winter; *seeds* 2 in each capsule, ca. $\frac{3}{8}$ in. (8 to 12 mm) long, terminally winged, also with many, small, abortive ovules.

Twig shiny, green to yellowish brown, slender to stout, somewhat angled or terete, aromatic, with or without corky wings (Fig. 9-10), which sometimes appear the first year; *pith* homogeneous, stellate; *terminal buds* ovate to conical, $\frac{1}{4}$ to $\frac{1}{2}$ in. (6 to 12 mm) long, with several, shiny, orange-brown scales; *lateral buds* similar but smaller; *leaf scars* crescent-shaped to triangular; *bundle scars* 3, conspicuously white with a dark center; stipular scars slitlike and inconspicuous even with magnification.

Bark grayish brown; deeply furrowed into narrow, somewhat rounded, flaky ridges.

Range

Southeastern (Map 9-4). Elevation: sea level to 3000 ft in the southern Appalachians. Mesic, upland woods, slopes, ravines, floodplains, and stream banks.

General Description

Sweetgum at its best is a large tree 80 to 120 ft tall and 3 to 4 ft dbh (max. 155 by 6 ft) with a long, straight, often buttressed bole, small oblong or pyramidal crown, and shallow wide-spreading root system.

Sweetgum is one of the most widespread trees in the southern forest. It is an important though intolerant member of four forest-cover types and is a component of 24 others; these are mostly hardwoods, but southern pines and baldcypress are also included. It is found on a wide variety of sites, but it always develops best on rich, moist alluvial soils. Its growth is retarded on dry sites, and prolonged drought may cause extensive dieback. Maturity is reached in 200 to 300 years.

Viable seed production is normally abundant. As the buoyant seed is not exacting in seedbed requirements, sweetgum is an aggressive pioneer of disturbed sites, particularly old fields. The

FIGURE 9-9 *Liquidambar styraciflua,* sweetgum. (1) Twig ×1$\frac{1}{4}$. (2) Staminate and pistillate flowers ×$\frac{1}{2}$. (3) Multiple of capsules ×1. (4) Winged seed ×1. (5) Leaf ×$\frac{1}{2}$. (6) Bark (*Courtesy of U.S. Forest Service*).

FIGURE 9-10   Sweetgum branch with corky wings $\times \frac{1}{2}$.

MAP 9-4   *Liquidambar styraciflua.*

pure even-aged stands that often develop on such sites can be quite dense. Only on very fertile sites is individual stem growth exceptional, and here the total volume growth is very high.

Sweetgum stumps sprout prolifically. In addition, an individual stem, particularly a young one, when weakened, cut, or top-killed, can produce an abundance of fast-growing root suckers. This capacity is in part responsible for much of the regeneration of the species, and many stands are of sprout origin.

Sweetgum has few insect and disease problems. Sweetgum blight, associated with protracted drought, has caused decline and mortality throughout most of its range. Many species of rodents girdle young stems, and lush vine growth may override slow-growing saplings. It is very fire-sensitive, and top-kill or decay often follows even light fires.

This species is a desirable ornamental in the milder portions of the North because of the attractive shape and brilliant autumnal coloration of the leaves. It is hardy at least as far north as central New York and will exist even in dry, heavy clay soils. Of several cultivars, one, 'Rotundiloba', has rounded lobes and does not set fruit. It was discovered in 1930 as a wild mutant in North Carolina.

Sweetgum is one of the more important commercial hardwoods of the United States, especially in the manufacture of utility and decorative plywood panels. Its use by the southern paper industry, as an intensively managed plantation species, increases yearly. The seeds are eaten by birds and small mammals.

HAMAMELIS L.   witch-hazel

Witch-hazel, with four species, two in North America, although of little or no importance commercially for its wood, is so common and widely distributed that it deserves at least brief mention. No more distinctive shrub or small tree can be found in the eastern United States (Fig. 9-11). *Leaves* are simple and alternate, $1\frac{1}{2}$ to $6\frac{3}{4}$ in. (4 to 17 cm) long, broadly elliptical, with margin irregularly and coarsely repand or sinuate; fall color yellow. *Flowers* conspicuous, with their 4 long, yellow to reddish, straplike petals borne in late fall (often after the leaves have dropped) or winter before the leaves, fragrant, entomophilous.

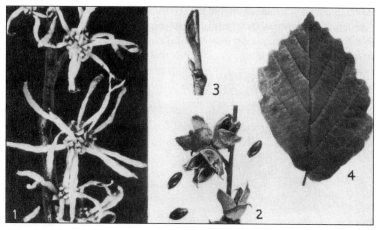

FIGURE 9-11   *Hamamelis virginiana,* witch-hazel. (1) Flowers ×1. (2) Fruit (capsule) and seeds ×$\frac{3}{4}$. (3) Bud ×2. (4) Leaf ×$\frac{1}{2}$.

Fruit maturing the following season into a 2-valved woody capsule, from which the shining, black, wingless seeds are ejected forcibly to a distance of several feet or more from the tree. *Twig* with tawny, hairy buds, naked except for two small scales; pith homogeneous; leaf scars half-round to 3-lobed with 3 bundle scars; stipular scars slitlike and inconspicuous.

Witch-hazel is said to be especially prized by "water diviners," some of whom prefer twigs cut from this tree. An alcoholic extract prepared from the bark or small branches is a well-known rubbing lotion. The seeds are eaten by various wildlife.

Hamamèlis virginiàna L. is found from New Brunswick to Iowa, south to Georgia and southern Arkansas. It flowers late in the fall. A second species, **H. vernális** Sarg., is a shrub of the Ozark region flowering in winter. Two other species are native to Japan and China. There are now numerous cultivars that are becoming popular as ornamentals.

Ulmaceae: The Elm Family

The elm family includes 18 genera and ca. 150 species of trees and shrubs, widely distributed throughout the temperate regions of both hemispheres, with a few species in the tropics. Four genera are represented in North America, but only two include common trees (Table 9-5).

TABLE 9-5   COMPARISON OF IMPORTANT GENERA IN THE ULMACEAE

Genus	Leaves	Flowers	Fruit
Ulmus elm	Doubly serrate (native species)	Perfect	Samara
Celtis hackberry	Entire or singly serrate	Perfect or imperfect	Drupe

Zelkòva serràta (Thunb.) Mak, Japanese zelkova, with several cultivars, is a frequent ornamental in the East. It was introduced about 1860 and differs from the elms by its simple serrate leaf margins and drupes.

Botanical Features of the Family

Leaves deciduous (rarely persistent), alternate, simple, stipulate, pinnately veined, usually serrate, often inequilateral at the base.

Flowers perfect or imperfect (plants polygamo-monoecious), anemophilous; *calyx* 4- to 9-lobed or parted; *corolla* none; *stamens* 4 to 6; *ovary* superior, usually 1-celled with a single ovule; *styles* 2.

Fruit a samara, drupe, or nut.

ULMUS L.   elm

This genus includes some of the most useful and well-known forest and ornamental trees of the Northern Hemisphere. The 18 to 20 species are scattered throughout eastern North America, Europe, and Asia, and in many regions are important constituents of hardwood forests. The wood is important commercially, and the tough bark fibers were sometimes used by primitive peoples for making rope and coarse cloth; the inner mucilaginous bark of certain species is used in medicine or occasionally as emergency food. The pollen is allergenic and can be a problem in the spring and fall. Many elms are highly regarded as shade and ornamental trees, and some are perhaps unsurpassed in form and general usefulness for this purpose. However, their future use must be evaluated in terms of the rapid spread of the lethal Dutch elm disease and several other diseases. Several foreign species, with many cultivars, have been introduced for ornamental planting. One is Scotch or Wych elm, *U. glàbra* Huds.; the bark of this tree remains fairly smooth even on large trunks. A common grafted form, Camperdown elm, 'Camperdownii', is the familiar "umbrella elm" often seen in parks and yards. In certain areas, *U. pùmila* L., Siberian elm, and *U. parvifòlia* Jacq., Chinese elm, are common small-leafed ornamentals, although Siberian elm is inferior to Chinese elm as an ornamental. Both have escaped from cultivation into waste places and along fence rows and roadsides across the United States. Many foresters do not distinguish between the two. Both are distinct from native elms by their singly serrate leaf margins. Of the six native species, three are of considerable importance and will be described in detail (Table 9-6).

Botanical Features of the Genus

Leaves alternate, simple, usually doubly serrate, short-petioled, inequilateral at base, distichous.

Flowers perfect, appearing before the leaves unfold, or in some species autumnal; borne on slender pedicels in fascicles or cymes (sometimes racemose); *ovary* flattened, surmounted by a deeply 2-lobed style.

Fruit maturing in the late spring, or in some species autumnal; a flattened, oblong to suborbicular samara; seed cavity encircled by a thin, membranous or papery wing,

TABLE 9-6   COMPARISON OF IMPORTANT ELMS

Species	Flowers	Samaras	Twigs
U. americana American elm	In long-pedicelled fascicles	$\frac{3}{8}-\frac{1}{2}$ in. (10–12 mm) long, deeply notched at apex, ciliate on margin	Brown, glabrous or sparingly pubescent; buds brown, acute but not sharp
U. rubra slippery elm	In subsessile fascicles	$\frac{1}{2}-\frac{3}{4}$ in. (12–18 mm) long, emarginate at apex, pubescent only on outside of seed cavity	Ashy gray, scabrous; buds nearly black, pubescent
U. thomasii rock elm	In racemose cymes on slender, drooping pedicels	$\frac{3}{8}-\frac{3}{4}$ in. (9–18 mm) long, rounded at apex, pubescent and margin ciliate	Reddish brown, glabrous or puberulent, later with corky ridges; buds brown, sharp-pointed

which is often notched above, and subtended below by the persistent remnants of the calyx; *seed* lacking endosperm.

Twigs slender to stout, in some species corky, slightly zigzag; *pith* terete, homogeneous; *terminal buds* lacking; *lateral buds* medium-sized, with the scales imbricate in 2 ranks; *leaf scars* 2-ranked, small, semicircular; *bundle scars* conspicuous, depressed, 3 or more; *stipular scars* small, at times indistinct.

Úlmus americàna L.   American elm, white elm

Distinguishing Characteristics (Fig. 9-12)

Leaves $3\frac{1}{4}$ to 6 in. (8 to 15 cm) long, 1 to 3 in. (2.5 to 7.5 cm) wide; *shape* oblong-obovate to elliptical; *margin* coarsely doubly serrate; *apex* acuminate; *base* conspicuously inequilateral; *surfaces* glabrous or slightly scabrous above; usually pubescent below; *fall color* bright yellow.

Flowers appearing before the leaf buds open, long-pedicelled, in fascicles of 3 or 4.

Fruit maturing in the spring, as the leaves unfold; $\frac{3}{8}$ to $\frac{1}{2}$ in. (10 to 12 mm) long, oval to oblong-obovate, deeply notched at apex, margin ciliate.

Twigs slender, zigzag, brown, glabrous or slightly pubescent; *lateral buds* ca. $\frac{1}{4}$ in. (6 mm) long, ovoid, acute but not sharp-pointed, smooth or sparingly downy, chestnut-brown.

Bark on older trees typically divided into grayish, flat-topped ridges, separated by roughly diamond-shaped fissures, but sometimes rough and without a definite pattern; the outer bark when sectioned shows irregular, corky, buff-colored patches interspersed with the reddish brown fibrous tissue.

Range

Eastern (Map 9-5). Elevation: sea level to 2500 ft. River bottoms, swamps, disturbed fields, road sides, cutover forests. Widely cultivated and naturalized in Idaho and Arizona.

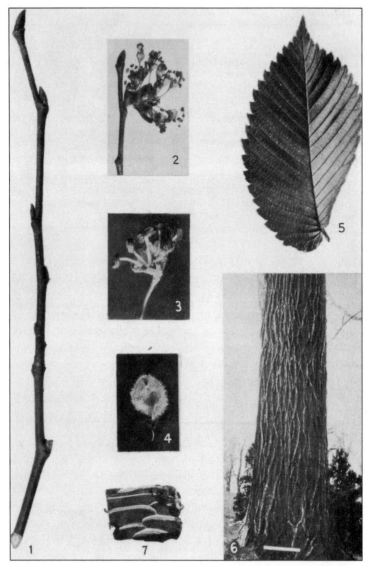

FIGURE 9-12 *Ulmus americana,* American elm. (1) Twig ×1 $\frac{1}{4}$. (2) Inflorescence ×$\frac{3}{4}$. (3) Flower ×4. (4) Fruit (samara) ×1. (5) Leaf ×$\frac{1}{2}$. (6) Bark. (7) Section through bark showing alternating brown and white layers (compare with that of slippery elm).

General Description

Probably no other North American tree is more easily recognized from a distance. When grown in the open, the trunk usually divides near the ground into several erect limbs strongly arched above and terminating in numerous slender, often drooping branchlets, the whole forming a vase-shaped crown of great beauty and symmetry (Fig. 9-13). At times the branches are more wide-spreading and give rise to the so-called oak form of this tree. In the forest, the buttressed, columnar trunk, 2

FIGURE 9-13   Open-grown form of American elm. (*Courtesy of U.S. Forest Service.*)

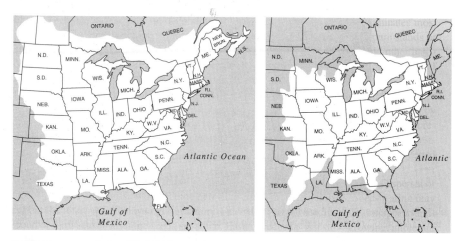

MAP 9-5   **(Left)** *Ulmus americana.* **(Right)** *Ulmus rubra.*

to 4 ft dbh, may rise to a height of 50 to 60 ft before branching and is topped by a small crown of arching branches (max. 160 by 11 ft).

The root system of American elm is extensive but shallow, and the tree is a common inhabitant of wet flats where standing water may accumulate in the spring and fall; in such places, its associates include the soft maples, some of the swamp oaks, willows, poplars, and black ash. The fastest growth and maximum size, however, are attained on better-drained, rich bottomland soils, in association with sweetgum, basswood, green ash, and other hardwoods. In the western portion of its range, American elm is a common tree along water courses in company with boxelder, green ash, and cottonwoods. This elm, when planted in the deep dry soils of the Great Plains region (shelterbelts), may send down quite deep taproots to the water table. Because it is a component of no less than 26 forest-cover types, the complete list of associates is large.

On a tolerance scale, American elm may be rated as intermediate. Seeds are produced annually (May to June) in extremely large numbers, and they germinate on moist soil within a few days of being released. They may be wind-carried for a quarter of a mile or more and may be waterborne over much greater distances. Maturity is reached in about 150 years, but some trees can reach over 400 years of age.

Much of the American elms, once the favorite shade tree of innumerable yards, parks, streets, and university campuses, has been destroyed by Dutch elm disease since 1930. This is caused by a vascular fungus vectored by bark beetles introduced to the United States presumably in elm veneer logs arriving from Europe. Other native elms are also susceptible to Dutch elm disease but to a lesser extent. Substantial losses are also caused by elm yellow (phloem necrosis). Although some cultivars are resistant to Dutch elm disease, they are not resistant to this other fatal disease. The David elm, *U. dividiàna* Schneid., has been introduced into the United States as a resistant species from northern China. Also, after 60 years of research, resistant cultivars of American elm, 'Valley Forge', 'American Liberty', and 'New Harmony' elms, are now available.

The fruits are important as wildlife food, and the saplings are frequently browsed by deer and rabbits.

American elm is the state tree of Massachusetts and North Dakota.

***Úlmus rùbra* Mühl.**   slippery elm, red elm

Distinguishing Characteristics (Fig. 9-14)

Leaves 4 to 7 in. (10 to 18 cm) long, 2 to 3 in. (5 to 7.5 cm) wide; *shape* elliptical to obovate, or oval; leaf often creased along the midrib; *margin, apex,* and *base* as for the preceding species; *surfaces* very scabrous above, soft pubescent below; *fall color* yellow.

Flowers appearing before the leaves, in short-pedicelled fascicles.

Fruit maturing when the leaves are about one-half expanded, $\frac{1}{2}$ to $\frac{3}{4}$ in. (12 to 18 mm) long, suborbicular in shape, emarginate at apex, margin and surface of wing smooth, outer surface of seed cavity pubescent.

Twigs stouter than in American elm, ashy gray to brownish gray, scabrous; *lateral buds* dark chestnut-brown to almost black, pubescent; *flower buds* larger, often with an orange tip; inner bark mucilaginous and "slippery."

Bark dark reddish brown, and, in contrast to that of both American and rock elms, it does not show buff-colored corky patches or streaks when sectioned with a knife; bark fissures not diamond-shaped in outline, ridges more nearly parallel than in American elm, often coarsely scaly or with vertical plates; with declining vigor, bark composed of flat, thin, gray plates; inner bark mucilaginous, with a somewhat aromatic flavor.

FIGURE 9-14   *Ulmus rubra,* slippery elm. (1) Twig ×1$\frac{1}{4}$. (2) Leaf ×$\frac{1}{2}$. (3) Samara ×1. (4) Bark. (5) Section through bark showing only brownish layers (compare with that of American elm).

Range

Eastern (Map 9-5). Disjuncts in southern Maine, southern Louisiana, Kansas, southwestern Nebraska, and eastern North Dakota. Elevation: sea level to 2000 ft. Mesic to moist soils of slopes and bottomlands, or on drier calcareous soils.

General Description

Slippery elm is a medium-sized tree 60 to 70 ft tall and 18 to 30 in. dbh (max. 135 by nearly 7 ft); it resembles American elm in general appearance except for a greater clear length of bole and a tendency of the twigs to be ascending.

This elm makes best growth on moist, rich bottomland soils, along streams, and on lower slopes. Although somewhat less susceptible to Dutch elm disease than American elm, it is also killed by elm yellows and has many other diseases and insect pests. It stump-sprouts readily.

Slippery elm was well known to the early pioneers on account of its mucilaginous inner bark, which was chewed to quench the thirst, and when steeped in water produced a common remedy for throat inflammations and fever; the powdered bark was recommended for poultices. Native

Americans used the wood for tomahawk handles. It is used now for furniture, paneling, and boxes. The saplings are browsed by deer, and the fruits are eaten by various wildlife.

Úlmus thómasii Sarg.   rock elm, cork elm

Distinguishing Characteristics (Fig. 9-15)

Leaves 2 to 4 in. (5 to 10 cm) long, $\frac{3}{4}$ to 2 in. (2 to 5 cm) wide; *shape* obovate to elliptical; *margin* coarsely doubly serrate; *apex* acuminate; *base* often nearly equilateral; *surfaces* usually glabrous, often somewhat lustrous above, slightly pubescent below; *fall color* yellow.

Flowers appearing before the leaves, on slender pedicels, in racemose cymes.

Fruit maturing when the leaves are about half grown, $\frac{3}{8}$ to $\frac{3}{4}$ in. (9 to 18 mm) long, oval to obovate, pubescent, more or less rounded at apex, ciliate on the margin, seed cavity indistinct.

FIGURE 9-15   *Ulmus thomasii,* rock elm. (1) Samara ×1. (2) Corky outgrowths on twig ×1$\frac{1}{4}$. (3) Leaf ×$\frac{1}{2}$. (4) Inflorescence ×$\frac{3}{4}$. (5) Twig ×1$\frac{1}{4}$. (6) Bark.

Twigs light reddish brown, glabrous or slightly puberulent, after a year or two often developing conspicuous corky ridges; *lateral buds* similar to those of American elm but longer and more sharply pointed, downy-ciliate.

Bark somewhat similar to that of American elm but often darker and more deeply and irregularly furrowed, especially on younger trees.

Range

Northcentral (Map 9-6). Disjuncts in several states around the normal range. Elevation: 200 to 2500 ft. Mesic forests, moist, well-drained uplands, rocky ridges, and flatwoods, also limestone bluffs.

General Description

Rock elm, so-called because of its extremely hard, tough wood, is a medium-sized to large tree that, in virgin stands, attains a height of some 80 ft and dbh of 3 to 4 ft (max. 103 by 6 ft). In contrast to the two preceding species, the tall straight trunk extends unbranched for some distance into the narrow oblong crown. It stump-sprouts easily. Although best development is made on moist, loamy soils of lower Michigan and Wisconsin, it is common on dry, rocky ridges and limestone bluffs. It is intermediate in tolerance but shows a dramatic ability to respond to release. Rock elm occurs with many other hardwoods. Drastic cutting and Dutch elm disease have greatly reduced the original supply of rock elm timber, which was never very large, but was used for implements and tool handles. Rock elm can regenerate from seeds as well as root and stump sprouts. It is an important wildlife food.

Southern Elms

Úlmus alàta Michx.   winged elm

Leaves $1\frac{1}{4}$ to $2\frac{1}{2}$ in. (3 to 6 cm) long; *twigs* very slender, with or without 2 light corky wings nearly $\frac{1}{2}$ in. (12 mm) wide. This species is usually a small- to medium-sized tree found equally on dry gravelly uplands and on fertile bottomlands from southeastern Virginia to southern Missouri, south to central Florida and eastern Texas. Perhaps it is the least tolerant of the native elms. Flowering in spring. Winged elm is often planted as a shade tree in southern United States and outside North America, although it is susceptible to Dutch elm disease and elm yellows.

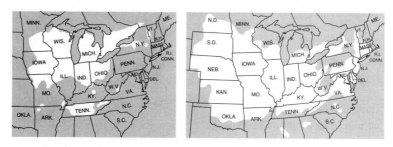

MAP 9-6   **(Left)** *Ulmus thomasii.* **(Right)** *Celtis occidentalis.*

Úlmus crassifòlia Nutt.   cedar elm

This is another small-leafed species found in Mississippi, southern Arkansas, and Texas on moist or dry limestone derived soils. The *twigs* often bear 2 corky wings that are only about half as wide as those of winged elm. Flowering is in late summer. It hybridizes with *U. serotina.*

Úlmus serótina Sarg.   September elm, red elm

Leaves 2 to $3\frac{1}{2}$ in. (5 to 9 cm) long (larger than those of the last two species); *twigs* often with 2 or 3 corky wings. September elm occurs only sporadically on alluvial soils and limestone hills from eastern Kentucky to southern Illinois, south to northeastern Georgia and eastern Oklahoma. Flowering occurs in the fall. It hybridizes with *U. crassifolia.*

CELTIS L.   hackberry

Celtis, with ca. 60 species, is found in many of the temperate and tropical regions of the world. Some of these species, because of their variability, are not well defined, and the genus is in much need of additional study. There are possibly six species native in North America. Throughout the world, several species supply valuable lumber, the drupes are consumed by wildlife and to a lesser extent by humans, and some species are cultivated as ornamental and shade trees. A number of Asian species have been introduced into North America as ornamentals. The pollen is allergenic.

Botanical Features of the Genus

Leaves deciduous (persistent in a few tropical forms), alternate, distichous, simple, singly serrate or sometimes entire, with more or less inequilateral base.

Flowers perfect and imperfect (plants polygamo-monoecious) appearing just before or with the leaves; the *staminate* borne below, in fascicles, the *perfect* and *pistillate* flowers above, solitary in the leaf axils; *calyx* 4- to 6-lobed; *stamens* 4 to 6.

Fruit a thick-skinned, thin-fleshed, ovoid or globose drupe; pit surface with a reticulate (netlike) pattern.

Twigs slender, zigzag; *pith* terete, very finely chambered at the nodes, homogeneous elsewhere; *terminal buds* lacking; *lateral buds* small, closely appressed; *leaf scars* 2-ranked, oval to crescent-shaped; *bundle scars* 3; *stipular scars* minute.

Céltis occidentàlis L.   hackberry

Distinguishing Characteristics (Fig. 9-16)

Leaves 2 to 5 in. (5 to 13 cm) long, $1\frac{1}{2}$ to $2\frac{1}{2}$ in. (4 to 6 cm) wide; *shape* ovate to ovate-lanceolate; *margin* sharply serrate (to well below the middle, then entire below); *apex* acuminate, falcate; *base* inequilaterally cordate; *surfaces* glabrous, or slightly scabrous above; smooth or sparingly hairy below; *fall color* pale yellow.

Flowers appearing in spring with or soon after the leaves.

Fruit borne on peduncles $\frac{1}{2}$ to $\frac{3}{4}$ in. (12 to 18 mm) long, drupe $\frac{1}{4}$ to $\frac{3}{8}$ in. (6 to 10 mm) in diameter, subglobose or ovoid, dark red or purple; flesh thin, edible; ripening in September and October, often persistent for several weeks; pit conspicuously reticulate.

FIGURE 9-16 *Celtis occidentalis,* hackberry. (1) Twig ×1$\frac{1}{4}$. (2) Flower ×3. (3) Fruit (drupe) ×$\frac{3}{4}$. (4) Pit showing reticulations ×1. (5) Leaf ×$\frac{1}{2}$. (6) Bark.

Twigs slender, zigzag, reddish brown, somewhat lustrous; *lateral buds* small, acute, closely appressed.

Bark grayish brown, smooth but with characteristic corky warts or ridges, later somewhat scaly.

Range

Northeastern (Map 9-6). Local disjuncts in several provinces and states around the continuous range. Elevation: sea level to 5000 ft. Stream banks, bottomlands, rocky slopes, and bluffs, also calcareous soils.

General Description

Hackberry is usually a small tree 30 to 40 ft tall and about 16 in. dbh (max. 130 by 6 ft). Best growth is made on rich, moist, alluvial soils, but especially in the Northeast the tree is commonly found on limestone outcrops and other droughty sites. Hackberry occurs as a scattered tree in mixture with other hardwoods. Seed dispersal is mainly by birds, and stump sprouts from young trees often persist. It exhibits a wide range of growth rates and tolerance ratings from intermediate to tolerant, both of which are related to site conditions.

Hackberry and its cultivars are used to some extent as ornamentals and urban trees, and succeed under adverse conditions of soil and moisture. Long overlooked by the wood-using industry, both hackberry and sugarberry are of increasing importance in the manufacture of paneling, furniture, boxes, and plywood. The fruits are eaten by birds and small mammals.

Céltis laevigàta Willd.   sugarberry, sugar hackberry

Sugarberry is a shade-tolerant, medium-sized tree 60 to 80 ft tall (max. 110 ft) and 2 to 4 ft dbh. It ranges along the Coastal Plain from Virginia to southern Florida, west to the valley of the Rio Grande, and north in the Mississippi Valley through eastern Oklahoma, Missouri, western Tennessee, and Kentucky to southern Illinois and Indiana. The leaves are narrower than those of hackberry, only slightly serrate to entire and apex sharply acute to acuminate, and the fruit is borne on shorter peduncles; the otherwise smooth bark is marked by conspicuous warty excrescences. It is a tree of bottomlands where the fruits are important to wildlife, especially birds. Sugarberry is planted extensively as a street tree in the South.

Moraceae: The Mulberry Family

This is a family of ca. 40 genera and 1100 species of trees, shrubs, vines, and herbs, distributed for the most part in the warmer regions of the world, but with a few species in the temperate zone. The sap is milky, and certain genera, notably *Fìcus* and *Castílla,* are a source of rubber. Eight genera are represented in North America including *Broussoné-tia,* paper-mulberry, and four others that have been introduced and have become naturalized. They all are small- or medium-sized trees of little value as timber producers but are often used ornamentally. Other genera are a source of valuable timbers, edible fruit, paper fibers, and dyes. The pollen is allergenic.

Botanical Features of the Family

Leaves deciduous or persistent, alternate, simple, stipulate.

Flowers imperfect, the plants monoecious or dioecious; *sepals* 4; *petals* 0; *stamens* 4; *ovary* superior or inferior, 2-carpellate; anemophilous except for the entomophilous *Ficus.*

Fruit a small drupe or achene, usually in multiples.

MORUS L.   mulberry

Of the 10 species of mulberry, two are native to the United States, and two others have become naturalized. The fruit, a multiple of drupelets, resembles a blackberry and is a very important food for birds and small mammals.

Black mulberry, *M. nìgra* L., probably a native of Persia and commonly cultivated in Europe, is occasionally cultivated in North America but is not known to have become naturalized. It has been mistaken for dark-fruited *M. alba* by several authors.

Botanical Features of the Genus

Leaves deciduous, alternate, simple, to 7 in. (18 cm) long, elliptical to suborbicular, venation pinnipalmate or palmate, unlobed or 2 or 3-lobed, the margin crenate-serrate, the base cordate or truncate; milky juice evident in broken petiole of very young leaves; stipulate; *fall color* yellow.

Flowers unisexual, the tree monoecious or dioecious, each flower small and in compact, cylindrical, or globose racemes or spikes, appearing with the leaves.

Fruit a cylindrical or nearly globose multiple of drupelets, $\frac{3}{4}$ to $1\frac{1}{4}$ in. (2 to 3 cm) long, each drupelet globose and white to purple or black in color.

Twigs with some milky juice, slender, brown, smooth; *leaf scars* semicircular or nearly circular; *bundle scars* many; pseudoterminal and lateral *buds* ovoid, acute, light brown, ca. $\frac{3}{16}$ in. (5 mm) long, appressed to the twig, and with 3 to 6 outer scales; *stipular scars* narrow and unequal; *pith* solid and homogeneous, round.

Bark thin, dark brown, scaly, and furrowed.

Mòrus rùbra L.   red mulberry

This is an eastern tree of floodplains, river valleys, and moist hillsides from southeastern Minnesota to southern Ontario and Massachusetts, south to southern Florida and Texas, and locally in southern New Mexico. It reaches to 40 ft tall and 18 in. dbh. Leaves scabrous-pubescent above and soft pubescent below, $2\frac{1}{2}$ to 5 in. (6 to 13 cm) long; fruit dark purple and edible (Fig. 9-17). It is shade tolerant, forms root sprouts, and it seldom occurs as a weed in cities and other disturbed areas, as does white mulberry.

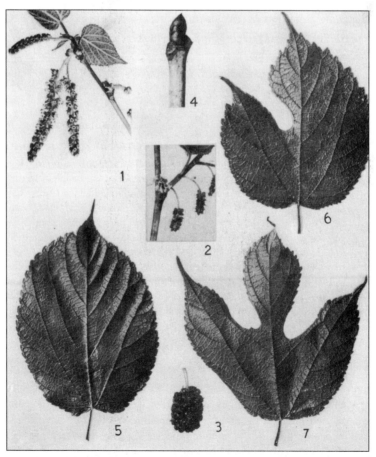

FIGURE 9-17 *Morus rubra*, red mulberry. (1) Staminate flowers $\times\frac{1}{2}$. (2) Pistillate flowers $\times\frac{1}{2}$. (3) Multiple of drupelets $\times\frac{3}{4}$. (4) Twig $\times1\frac{1}{4}$. (5–7) Leaves showing variation in lobing $\times\frac{1}{2}$.

Vegetatively, red mulberry is sometimes mistaken for basswood (*Tilia*). However, *Tilia* has 2 (rather than 3 to 6) bud scales; the twig is more zigzag and lacks milky juice; the leaf margin is more sharply serrate with a thickened tip on each tooth; the leaf surface is not scabrous and often has stellate or fasciculate hairs.

Red and white mulberries are often confused. Both are variable in leaf pubescence, and intermediates occur possibly due to hybridization.

Mòrus álba L.   white mulberry

A native of China, this tree was introduced in colonial times (Jamestown, VA in the early 1600s) in attempts to establish the silkworm industry in the United States. Although the attempt failed, the tree is now naturalized locally throughout the United States and southern Ontario and has become weedy in disturbed areas, particularly in cities and towns. The leaves are lustrous and smooth above and generally glabrous below; the fruit is white, pink, violet, or black. The Russian mulberry, *M. alba* 'Tatarica', was introduced about 1875 and has become widely distributed in the western United States. It is the hardiest of all mulberries, grows rapidly in sandy or clay soils, and is used in shelterbelts or for the production of farm timbers.

Mòrus microphýlla Buckl.   Texas mulberry, mountain mulberry

Texas mulberry grows in the southwest (Oklahoma west to eastern Arizona and south into Mexico) along streams, in mountain canyons, and on dry hills, at 2100 to 6600 ft elevation. It is similar to red mulberry, but the leaves are smooth above and smaller, 1 to $2\frac{1}{2}$ in. (2.5 to 6 cm) long, and the fruit is subglobose and nearly black.

MACLURA Nutt.   Osage-orange

This genus is monotypic, and the single species, ***Maclùra pomífera*** (Raf.) C. K. Schneid., originally restricted to southern Arkansas, southern Oklahoma, and northeastern Texas, is now planted for hedges and windbreaks or as an ornamental throughout the United States and southern Canada. *Leaves* deciduous, $2\frac{1}{2}$ to 5 in. (6 to 13 cm) long, entire, oblong-lanceolate to ovate, with a narrow, pointed apex, and dark green and shiny above. *Flowers* unisexual and trees dioecious. *Fruit* a peculiar and very distinctive, spherical, semifleshy, yellow-green multiple of drupes, $3\frac{1}{2}$ to 5 in. (9 to 13 cm) in diameter; when crushed, it exudes a bitter milky juice (Fig. 9-18). *Twigs* armed with sharp thorns. The *wood* is of a characteristic bright yellow-orange color and is used for turning and for bows (hence, the origin of the name "bois d'arc," or bowwood, which is often applied to this species); the wood, the most decay-resistant of North American timbers, yields a yellow dye when extracted with hot water. Only male trees should generally be selected for urban planting because of the large, heavy, abundant fruit produced on female trees.

FICUS L.   fig

This is one of the largest genera of flowering plants (ca. 750 species), chiefly tropical, and in the United States is restricted to two arborescent species, *Fìcus aúrea* Nutt., Florida strangler fig, and *F. citrifòlia* Mill., shortleaf fig or wild banyan tree, both natives of southern Florida. The latter species is the banyan tree often planted for shade. The commonly cultivated, edible fig, *F. cárica* L., is grown for its fruit in the southern states and sometimes escapes cultivation. Rubber may be produced from the latex of the India

FIGURE 9-18 *Maclura pomifera,* Osage-orange. (1) Leaf ×$\frac{1}{2}$. (2) Twig with thorn ×1$\frac{1}{4}$. (3) Multiple of drupes ×$\frac{1}{2}$.

rubber tree, *F. elástica* Roxb. ex Hornem., but most of it comes now from *Hèvea brasiliénsis* (Willd. ex A. Juss) Muell. Arg. (Euphorbiaceae), or by synthetic processes.

Leitneriaceae: The Corkwood Family

Although of no commercial importance, this monotypic family is of interest botanically, and because it produces the lightest wood grown in the United States (specific gravity of about 0.21; compared with cork, 0.24 and balsa wood, 0.12). The relationships of this unique family are debatable. See Bogle (1997) for additional discussion of the possible affinities.

The single species, corkwood, **_Leitnèria floridàna_** Chapm., is a rare shrub or small deciduous tree with few branches. It is native to marshes, swamps, and stream or riverbanks in scattered areas of the Gulf Coastal Plain and lower Mississippi Valley. *Leaves* alternate, 2$\frac{1}{2}$ to 6$\frac{3}{4}$ in. (6 to 17 cm) long, oblong to elliptic-lanceolate, thick, and pubescent below; *flowers* unisexual on separate trees in catkins before the leaves emerge, anemophilous; *fruit* a dry drupe; *twig* with homogeneous pith, leaf scars crescent-shaped or 3-lobed with 3 bundle scars, stipular scars lacking, bud scales imbricate and hairy; *bark* reddish brown, smooth, with conspicuous buff-colored lenticels.

Juglandaceae: The Walnut Family

The Juglandaceae include nine genera and ca. 60 species of trees and large shrubs that are widely distributed through the forests of the North Temperate Zone, and to a lesser extent in the tropical forests of both the Northern and Southern Hemispheres. The importance of this family lies chiefly in the many valuable timber trees and nut trees that it includes. The genera *Juglans, Engelhárdtia,* and *Pterocárya* produce cabinet woods, and the wood of *Carya* is extremely tough and suitable for many purposes where strength is required. The nuts of several species of *Carya* and *Juglans* are used as wildlife and human food in many localities; the bark and fruit husks of a few species are sources of yellow dyes and tannic compounds. Several species of *Carya, Juglans,* and *Pterocarya* are cultivated as ornamentals. The pollen of many species is allergenic. Two genera and 17 species of this family are found in the United States (Table 9-7).

TABLE 9-7   COMPARISON OF NATIVE GENERA IN THE JUGLANDACEAE

Genus	Flowers	Nuts	Pith
Juglans walnut	Staminate aments unbranched; stamens 7–50 per flower	Husk indehiscent; shell of nut corrugated or rugose (rarely smooth)	Chambered after the first year
Carya hickory	Staminate aments 3-branched; stamens 3–15 per flower	Husk usually dehiscent along 4 sutures; shell of nut ± smooth and often ribbed	Solid and homogeneous

Botanical Features of the Family

Leaves deciduous, mostly alternate, pinnately compound, estipulate, more or less aromatic.

Flowers imperfect (plants mostly monoecious), anemophilous, appearing with or after the leaves; *staminate* in drooping axillary aments, the individual flowers consisting of 3 to many often nearly sessile stamens surrounded by a 3- to 6-lobed calyx and subtended by a bract; *pistillate* solitary or in few-flowered spikes terminating the new growth, the individual flowers with a 1- to 4-celled pistil, short style, and 2 plumose stigmas, the ovary inferior, covered by a 3- to 5-lobed or parted calyx and subtended by an adnate involucre consisting of a bract and 2 bracteoles.

Fruit a bony nut encased in a semifleshy or woody, dehiscent or indehiscent "husk" formed from the calyx and involucre; *seed* lacking endosperm but with large, convoluted cotyledons.

JUGLANS L.   walnut

The walnuts, although numerous and widely distributed during past geological periods (Tertiary), now have only about 20 species. These are found in the forests of North, Central, and South America, the West Indies, southern Europe, and southern and eastern Asia. The French, Turkish, Italian, and Circassian walnut timbers are produced by *Juglans règia* L., which is also an important timber-contributing species of India. The so-called English walnuts are the fruits of this tree, and in recent years, large plantations have been developed in both Oregon and California.

Six species of *Juglans* are native to the United States, but only two of them (*J. nigra* and *J. cinerea*) are important for lumber (Table 9-8). *Juglans híndsii* Jeps. ex R. E. Smith and *J. califórnica* Wats. are California species. The former is used as rootstock for grafting English walnuts along the Pacific slope, and occasionally as a shade tree. Little walnut, *J. microcárpa* Berlandier, the fifth species, is a small tree native to the Southwest, and the remaining southwestern species is Arizona walnut, *J. màjor* (Torr.) Heller.

TABLE 9-8 COMPARISON OF IMPORTANT WALNUTS

Species	Leaves	Nuts	Pith	Bark
J. nigra black walnut	With 9 to 23 ovate-lanceolate leaflets	Globose; nut corrugations rounded	Buff-colored	Dark brown to black, with thin anastomosing ridges
J. cinerea butternut	With 11 to 17 oblong-lanceolate leaflets	Oblong-ovoid; nut corrugations sharp	Chocolate-colored	Ashy gray, with broad, anastomosing ridges

The walnuts and butternut represent different evolutionary lines within the genus. Therefore, the eastern walnut is more closely related to the western walnuts than to the eastern butternut.

Botanical Features of the Genus

Leaves odd- or even-pinnately compound, consisting of 9- to 23-sessile or nearly sessile leaflets, the central ones longest; *leaflets* more or less oblong-lanceolate, each with an acute to acuminate apex, inequilateral base, and finely serrate margin; *petiole* and *rachis* stout, usually pubescent.

Flowers (Fig. 9-19) imperfect (plant monoecious); *staminate* aments preformed, appearing as small, scaly, conelike buds, unbranched; stamens 7 to many; *pistillate* in 2-

FIGURE 9-19 *Juglans,* walnut. (1) Pistillate and staminate flowers $\times\frac{1}{2}$. (2) Staminate flower, face view $\times3$. (3) Pistillate flower $\times3$. (4) Staminate flower, rear view $\times3$.

to 8-flowered spikes, the individual flowers consisting of a 2- to 4-celled pistil sur-mounted by a short style and 2 divergent, plumose stigmas; appearing with or just after the developing leaves.

Fruit a nut with a semifleshy, indehiscent husk; nut thick-walled, rugose or deeply corrugated; seed sweet, often oily.

Twigs stout, with an acrid taste, pubescent or glabrous; *pith* chambered after the first season except between the seasons' growth, stellate in cross section; *terminal buds* few-scaled, often appearing naked; *lateral buds* commonly superposed; *leaf scars* obcor-date to obdeltoid, slightly 3-lobed, with three U-shaped bundle scars; *stipular scars* lacking.

Júglans nìgra L.   **black walnut, eastern black walnut**

Distinguishing Characteristics (Fig. 9-20)

Leaves 12 to 24 in. (30 to 61 cm) long, with 9 to 23 nearly sessile leaflets attached to a stout puberulent rachis, the terminal leaflet often suppressed; *leaflets* $2\frac{1}{4}$ to 6 in. (6 to 15 cm) long, 1 to $1\frac{1}{4}$ in. (2.5 to 3 cm) wide, ovate-lanceolate; *margin* finely serrate; *apex* acute to acuminate; *base* inequilateral; *surfaces* lustrous, dark yellow-green and glabrous above, pale green and pubescent below; *petiolule* very short, puberulent; *fall color* yellow.

Flowers appearing in late May and early June; *staminate* with 17 to 50 nearly sessile stamens.

Fruit $1\frac{1}{2}$ to 3 in. (4 to 8 cm) in diameter, globose or nearly so, solitary or in clusters of 2 or 3; with a thick, semifleshy, yellowish green, pubescent husk; nut corrugated, with rounded ridges; *seed* sweet, oily.

Twigs stout, light brown to orange-brown, lenticellate; *pith* chambered, with thin diaphragms, buff-colored; *terminal buds* short, blunt, covered by a few pubescent scales; *lateral buds* much smaller, often superposed; *leaf scars* elevated, obcordate, without a hairy fringe on the upper margin.

Bark dark brown to grayish black, divided by deep, narrow furrows into thin ridges, the whole forming a roughly diamond-shaped pattern.

Range

Eastern (Map 9-7). Disjuncts in western and central New York, and western Vermont. Elevation: sea level to 4000 ft. Mesic woods, slopes, and along streams.

General Description

Black walnut is not only first in importance among the native species of *Juglans,* but it is also one of the most highly valued of North American hardwoods. Although usually a medium-sized tree 70 to 90 ft tall and 2 to 3 ft dbh, it may reach a maximum height of 130 ft and a dbh of 8 ft.

This species is very intolerant, and under forest competition develops a tall, well-formed clear bole, with a small, open crown that is always dominant in the stand; the root system is deep and wide-spreading, with a definite taproot, at least in early life.

Small amounts of seed may be produced every year, with large crops at 2- or 3-year intervals; rodents are the chief means of dispersal, although they also use many of the nuts for food.

FIGURE 9-20   *Juglans nigra,* black walnut. (1) Leaf $\times\frac{1}{5}$. (2) Bark. (3) Nut with husk $\times\frac{3}{4}$. (4) Nut with husk removed $\times\frac{1}{3}$. (5) Cross section of nut with husk $\times\frac{3}{4}$. (6) Twig $\times 1\frac{1}{4}$ showing dark chambered pith.

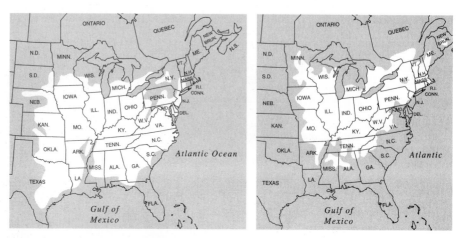

MAP 9-7   **(Left)** *Juglans nigra.* **(Right)** *Juglans cinerea.*

On deep, rich, moist soils, of alluvial origin, young seedlings may grow 3 ft in height the first season and double this figure the second year; logs 10 in. dbh may be produced in 35 years. This species, however, is very sensitive to soil conditions and grows much more slowly on poorer sites; large pure stands are rare, but walnut groves are frequently found, apparently from seeds buried by squirrels. The chief associates of black walnut include yellow-poplar, white ash, black cherry, basswood, American beech, numerous oaks, and hickories. Black walnut matures in about 150 years, but its life span may be as much as 250 years.

The supply of black walnut timber has been greatly depleted, but efforts have been made to popularize the planting of this valuable species. Besides the timber that it produces (highly prized for furniture, gunstocks, etc.), the fruit is sold locally for baked goods and ice cream, and new varieties are being developed with thinner shells.

Black walnut is one of the most notorious of allelopathic trees, inhibiting growth of some other plants under and around the trees. For additional information on this subject, see the term *allelopathy* in the glossary and also Waller (1987). A number of insects and a few diseases damage black walnut.

Black walnut hybridizes with *J. hindsii.*

Júglans cinèrea L.   butternut, white walnut

Distinguishing Characteristics (Fig. 9-21)

Leaves 15 to 24 in. (38 to 61 cm) long, with 11 to 17 nearly sessile leaflets, attached to a stout, pubescent rachis; *leaflets* 2 to $4\frac{1}{2}$ in. (5 to 11 cm) long, $1\frac{1}{4}$ to $2\frac{3}{8}$ in. (3 to 6 cm) wide, oblong-lanceolate; *margin* serrate; *apex* acute to acuminate; *base* inequilateral and rounded; *surfaces* yellowish green, rugose above, paler and soft pubescent below; *petiolules* extremely short, pubescent; *fall color* yellow or brown.

Flowers appearing in late May or early June; *staminate* with 7 to 15 nearly sessile stamens.

Fruit $1\frac{1}{2}$ to 3 in. (4 to 8 cm) long, oblong-ovoid, in clusters of 2 to 5, or solitary; with a semifleshy, greenish bronze, indehiscent husk clothed with glandular pubescence, sticky to the touch; nut deeply corrugated, with sharp ridges; *seed* sweet and very oily.

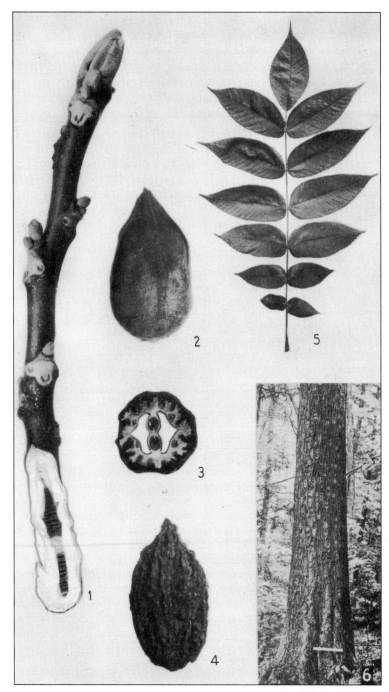

FIGURE 9-21   *Juglans cinerea*, butternut. (1) Twig $\times 1\frac{1}{4}$ showing chambered pith and velvety ridge above leaf scar. (2) Nut with husk $\times \frac{3}{4}$. (3) Cross section of nut with husk $\times \frac{3}{4}$. (4) Nut with husk removed $\times \frac{3}{4}$. (5) Leaf $\times \frac{1}{5}$. (6) Bark.

Twigs stout, greenish gray to reddish brown, lenticellate; *pith* chambered, with thick diaphragms, dark chocolate-brown to nearly black; *terminal buds* somewhat elongated, blunt, covered by a few pubescent scales, the outer ones lobed; *lateral buds* much smaller, often superposed, covered by rusty-brown tomentum; *leaf scars* elevated, obcordate with a tan velvety ridge along the upper margin.

Bark light gray, divided by shallow to moderately deep fissures into broad, flat ridges; later more closely furrowed with a roughly diamond-shaped pattern.

Range

Eastern (Map 9-7). Disjuncts in central North Carolina, northern Alabama, northern Mississippi, and Arkansas. Elevation: sea level to 4800 ft. Mesic slopes, ravines, coves, river banks, floodplains, also dry, rocky soils over limestone.

General Description

Butternut is a small- to medium-sized tree, 40 to 60 ft tall and 12 to 24 in. dbh (max. 110 by 5 ft). The bole is often short and divides into a few stout, ascending limbs that form a broad, open, somewhat irregular, flat or round-topped crown. The root system comprises a taproot and a number of deep-seated but wide-spreading laterals; however, the roots of trees growing on shallow soils are superficial and devoid of anything resembling a taproot.

Butternut never occurs in pure stands, but rather is found occasionally in mixed hardwood forests with black cherry, basswood, hickories, American beech, oaks, yellow-poplar, and elms. At the northern limits of its range, it is frequently mixed with sugar maple and yellow birch, and much more rarely with eastern white pine.

The silvical features of butternut are similar to those of black walnut, but it is shorter-lived, usually not reaching a greater age than 75 years. It is also less site demanding. Since the mid-1960s, there has been a decline in butternut throughout its range to the point where it is now so rare that it is a candidate for federal listing as threatened. A canker-producing fungus is a major cause. There is no resprouting when the crowns and trunks are killed by the canker.

In addition to the timber used for cabinetwork and novelties, the nuts are used for food, a sweet sugary syrup may be obtained by boiling down the sap, the green husks liberate a dye that colors cloth orange or yellow, and the dried inner bark of the roots contains natural substances used in medicine. Like black walnut, butternut is allelopathic to certain plant species.

CARYA Nutt.   hickory

It is known that numerous species of *Carya* were included in the ancient floras of Europe, northern Africa, Asia, and North America prior to the Glacial epoch. Since then, many of them have become extinct; today there remain only 18 species. A few are found in southeastern Asia. Thirteen range through eastern and southern United States, four of these are also found in adjacent Canada, and one is confined to Mexico.

Hickories are a somewhat difficult group due to intrinsic variability and interspecific hybridization between sympatric species, even between sections of the genus. Trichome morphology and vesture are important characters for identification of hickories (Hardin and Stone, 1984) and should be combined with bark, leaflet, twig, and fruit characters for positive identification. Several species are considered as varieties by some authors (Stone, 1997).

Hickories produce heavy, strong, especially shock-resistant wood with high fuel value; pecan hickory, besides furnishing timber, is valuable for its edible fruit. The pecan is the most important nut tree in North America. Hickory nuts of several species were used by Native Americans as an important food, and they continue to be important for various wildlife. Although hickories are messy due to their fruits, they are exceptional, long-lived shade trees with outstanding golden-yellow fall color.

Botanical Features of the Genus

Leaves odd-pinnately compound, consisting of 3 to 17 sessile or nearly sessile leaflets; in the true hickories, the terminal one is often the largest, and the lowest pair is often the smallest; in the pecan hickories, all the leaflets are more similar in size; *leaflets* ovate to obovate, with acute to acuminate apices, inequilateral bases, and finely serrate margins; variously pubescent; *petiole* and *rachis* usually stout, glabrous or pubescent.

Flowers (Fig. 9-22) imperfect (plants monoecious); *staminate* in 3-branched aments, in clusters on the new growth or near the summit of the previous season's; individual flowers with 3 to 7 stamens, and a 2- or 3-lobed calyx; *pistillate* flowers in 2- to 10-flowered terminal spikes; the individual flowers with a bract and three bracteoles and a 1-celled ovary surmounted by 2 sessile stigmas; appearing just before or just after the leaves.

Fruit an ovoid, pyriform, or globose nut, sometimes flattened, encased in a semi-woody "husk" that completely or partially splits along 4 sutures; *nut* ribbed, smooth, or slightly rugose, with a thick or thin shell; *seed* sweet or bitter.

Twigs stout to moderately slender, dark brown, gray, or orange-brown; *terminal buds* much larger than the laterals, covered by either 6 to 9 thin, imbricated scales or 4 to 6

FIGURE 9-22   *Carya,* hickory. (1) Pistillate and staminate flowers $\times\frac{1}{2}$. (2) Pistillate flowers $\times 3$. (3) Staminate flower $\times 3$.

TABLE 9-9   COMPARISON OF SECTIONS OF *CARYA*

Section	Leaves	Fruit husks	Terminal bud scales
Carya true hickories	With (3)5 to 7(9) leaflets; terminal leaflet the largest	Usually unwinged, occasionally ribbed at the sutures	Imbricate with 6 to 9 thin scales
Apocarya pecan hickories	With (5)7 to 13(17) leaflets; all leaflets ± similar	Usually broadly winged at the sutures	± valvate with 4 to 6 fleshy scales

fleshy, more or less valvate scales and appearing naked; *leaf scars* 3-lobed or obdeltoid, the *bundle scars* many, mostly in 3 U-shaped clusters; *pith* homogeneous, somewhat stellate.

The species of native hickories are grouped into two sections: "true hickories" (Sect. *Carya*) and "pecan hickories" (Sect. *Apocarya*) (Table 9-9).

True Hickories (Table 9-10)

Cárya ovàta (Mill.) K. Koch   shagbark hickory

Distinguishing Characteristics (Fig. 9-23)

Leaves 8 to 14 in. (20 to 36 cm) long, with 5 (rarely 7) ovate-lanceolate to obovate, sessile or nearly sessile leaflets; *terminal leaflets* 5 to 8 in. (12.5 to 20 cm) long, 2 to 3 in. (5 to 7.5 cm) wide, larger than the lateral leaflets; *margin* finely serrate, ciliate with a subterminal tuft of white hairs on the teeth; *apex* acute to acuminate; *base* cuneate; *surfaces*

TABLE 9-10   COMPARISON OF IMPORTANT TRUE HICKORIES

Species	Leaves	Nuts	Bark
C. ovata shagbark hickory	Usually 5 leaflets; essentially glabrous; subterminal tuft of hairs on teeth	$1\frac{1}{4}$–$2\frac{1}{2}$ in. (3–6 cm) diameter; husk $\frac{1}{4}$–$\frac{1}{2}$ in. (6–12 mm) thick; nut 4-ribbed, thick-shelled, not strongly compressed	Broken into long, thin plates, the split ends curling away from the trunk
C. laciniosa shellbark hickory	Usually 7 leaflets; velvety beneath; without subterminal tuft of hairs on teeth	$1\frac{3}{4}$–$2\frac{1}{2}$ in. (4.5–6 cm) diameter; husk $\frac{1}{4}$–$\frac{1}{2}$ in. (6–12 mm) thick; nut 4–6-ribbed, strongly compressed	Shaggy, but plates more often straight
C. tomentosa mockernut hickory	Usually 7–9 leaflets; pubescent beneath and on rachis	$1\frac{1}{2}$–2 in. (4–5 cm) diameter; husk $\frac{1}{8}$–$\frac{1}{4}$ in. (3–6 mm) thick; nut 4-ribbed	Dark with interlacing ridges; shallowly furrowed
C. glabra pignut hickory	5 or 7 leaflets; pubescent and with many scales	1–2 in. (2.5–5 cm) diameter, pyriform to oval; husk thin and dehiscent or only partly so; nut rounded or 4-ribbed	Dark and deeply furrowed or light gray and scaly

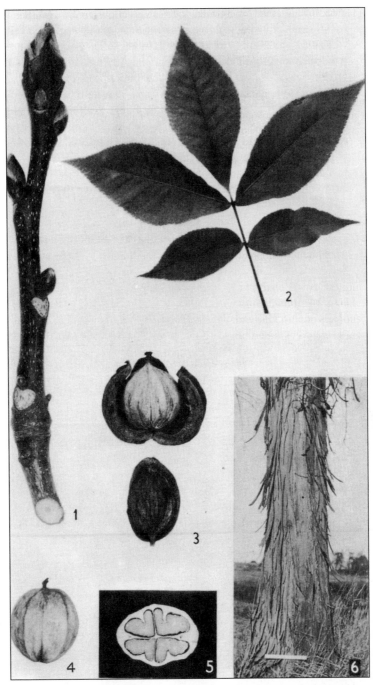

FIGURE 9-23 *Carya ovata,* shagbark hickory. (1) Twig ×1$\frac{1}{4}$. (2) Leaf ×$\frac{1}{4}$. (3) Nut with dehiscing husk ×$\frac{1}{3}$. (4) Nut ×$\frac{1}{3}$. (5) Cross section of nut ×$\frac{1}{3}$. (6) Bark.

dark yellow-green, glabrous above; pale yellow-green, glabrous, or puberulent below; *rachis* stout, grooved, glabrous or puberulent; *fall color* golden yellow.

Fruit $1\frac{1}{4}$ to $2\frac{1}{2}$ in. (3 to 6 cm) in diameter, solitary or paired, subglobose, depressed at the apex; nut subglobose, brownish white to pinkish white, 4-ribbed, with a reddish brown to nearly black, readily dehiscent husk, $\frac{1}{4}$ to $\frac{1}{2}$ in. (6 to 12 mm) thick; *seed* sweet.

Twigs stout, gray-brown to reddish brown, usually somewhat pubescent, lenticellate; *terminal buds* $\frac{1}{2}$ to $\frac{3}{4}$ in. (12 to 18 mm) long, broadly ovoid to ellipsoidal, obtuse, covered with 3 or 4 visible dark brown, loosely fitting pubescent scales; *lateral buds* somewhat smaller, ovoid, divergent; *leaf scars* slightly elevated, somewhat obdeltoid to semicircular; *bundle scars* scattered or in 3 clusters.

Bark smooth and gray on young stems, soon breaking up and becoming shaggy with large vertical plates that curve away from the trunk at the ends.

Range

Eastern (Map 9-8). Elevation: sea level to 2000 ft northward, to 3000 ft in the southern Appalachians. Dry to mesic woods, floodplains, valleys, upland slopes, and limestone flatwoods and glades.

General Description

Shagbark hickory is a medium-sized tree, 70 to 80 ft tall and 12 to 24 in. dbh (max. 125 by 4 ft), and in the forest develops a clear, straight, cylindrical bole with a small, open crown; if given sufficient space, however, the crown is typically oblong, and this shape characterizes many of the hickories when grown in the open. Seedlings of shagbark hickory develop a large and remarkably deep taproot, which may penetrate downward 1 to 3 ft the first season with a corresponding top growth of only a few inches; this feature is typical of most of the hickories except those found on wet soils.

This species varies greatly over its range in the kind of site occupied; in the North, it is often found on upland slopes and flatwoods with other hickories and oaks, but farther south it is more

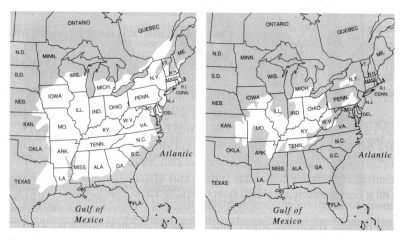

MAP 9-8   **(Left)** *Carya ovata.* **(Right)** *Carya laciniosa.*

common in deep, moist soils of alluvial origin, with the hardwoods of that region. This shift in common-site conditions occurs in other hickories and other genera. Shagbark hickory, at least in youth, will stand considerable forest competition and may be rated as moderately tolerant.

Large crops of seed are produced nearly every other year, especially by trees in the open, and it is not unusual to obtain 2 or more bushels of nuts from a single large tree. Nuts are distributed by squirrels and other rodents. Squirrels fail to claim some of their buried nuts, and the following year young hickory seedlings spring up, often in the most unlooked-for places. Many other wildlife species eat hickory nuts. Young trees produce vigorous sprouts when cut, and root suckers are also common, especially from older trees.

Although one of the fastest-growing hickories, shagbark is still a slow-growing tree and attains an age of 250 to 300 years. The shaggy bark makes it picturesque as a specimen tree, but the litter of husks and nuts does not recommend it for city planting. Like most hickories, this tree is highly susceptible to damage by fire. Although numerous insects and diseases affect shagbark hickory, none are life threatening.

Shagbark hickory hybridizes with *C. cordiformis, C. illinoinensis, C. laciniosa,* and *C. carolinae-septentrionalis.*

Cárya carolìnae-septentrionàlis (Ashe) Engl. and Graebn. (*Carya ovata* var. *australis* [Ashe] Little) Carolina hickory, southern shagbark hickory

This species resembles *C. ovata* except for its smaller leaves and fruit, slender twigs, and smaller, nearly glabrous, black or brownish black cylindrical buds. It reaches commercial proportions and is fairly abundant in the Piedmont and ranges from central North Carolina to northern Georgia and northeastern Mississippi, and west through eastern and central Tennessee. It occurs most frequently in basic soils or around limestone glades. Although treated as a variety by Little (1979) and Stone (1997), it is considered a distinct species by many. It hybridizes with *C. ovata* on the cedar glades where they are sympatric.

Cárya laciniòsa (Michx. f.) Loud.   shellbark hickory, bigleaf shagbark hickory

Distinguishing Characteristics (Fig. 9-24)

Leaves 12 to 20 in. (30 to 50 cm) long, with 5 to 9 (usually 7) ovate, oblong-lanceolate, or obovate, sessile or nearly sessile leaflets; *terminal leaflets* 5 to 9 in. (12.5 to 23 cm) long, 3 to 4 in. (7.5 to 10 cm) wide, larger than the laterals; *margin* finely serrate, but the teeth lack the subterminal tuft of hairs; *apex* acute to acuminate; *base* cuneate or inequilaterally rounded; *surfaces* dark green, lustrous above, pale yellow-green to yellowish brown, velvety pubescent below; *rachis* stout, grooved, glabrous or pubescent; *fall color* yellow-brown.

Fruit $1\frac{3}{4}$ to $2\frac{1}{2}$ in. (4.5 to 6 cm) long, solitary or paired, ellipsoidal to subglobose, depressed at the apex, with an orange to chestnut-brown husk, $\frac{1}{4}$ to $\frac{1}{2}$ in. (6 to 12 mm) thick; *nut* subglobose to obovoid, strongly compressed laterally, yellowish brown to reddish brown, 4- to 6-ribbed; *seed* sweet.

Twigs stout, orange-brown, prominently orange lenticellate; *terminal buds* $\frac{3}{4}$ to 1 in. (18 to 25 mm) long, ovoid, obtuse, covered with 6 to 8 visible dark brown, loosely fitting scales; *lateral buds* smaller, ovoid-oblong, divergent; *leaf scars* obcordate.

Bark similar to that of *C. ovata* but usually with straighter plates.

FIGURE 9-24 *Carya laciniosa,* shellbark hickory. (1)Twig ×1$\frac{1}{4}$. (2) Side view of section of husk after dehiscence ×$\frac{1}{2}$. (3) Nut ×$\frac{1}{2}$.

Range

Eastcentral (Map 9-8). Disjuncts in several outlying states. Elevation: sea level to 1000 ft. Floodplains, mesic woodlands.

General Description

This hickory, as its common name indicates, is very similar to shagbark hickory. It is, however, much more frequently found on wet alluvial bottoms in neutral or alkaline soils where it attains greater size and age. It often occurs in nearly pure groves, or mixed with other bottomland hardwoods, on areas that are inundated for several weeks during high water. Some common associates include bur, swamp chestnut, cherrybark and pin oaks, red maple, American elm, and sweetgum. It is very tolerant of low-light intensities. The growth rate is perhaps slower than that of other hickories. The maximum measured size of shellbark is 145 ft tall and 4 ft dbh. It has the largest nut of all hickories, which is sweet and delicious.

It hybridizes with *C. cordiformis, C. illinoinensis,* and *C. ovata.*

Cárya tomentòsa (Poir.) Nutt.   mockernut hickory

Distinguishing Characteristics (Fig. 9-25)

Leaves 8 to 20 in. (20 to 50 cm) long, with 7 or 9 (rarely 5) sessile or nearly sessile, lanceolate to obovate-oblanceolate, glandular-resinous, fragrant leaflets; *terminal leaflets* 4 to 8 in. (10 to 20 cm) long, 2 to 3 in. (5 to 8 cm) wide, larger than the lateral leaflets; *margin* finely to coarsely serrate; *apex* acute to acuminate; *base* inequilaterally rounded or cuneate; *surfaces* dark yellow-green, lustrous above, pale yellow-green to orange-brown, densely pubescent below with fascicled hairs; *rachis* stout, grooved, pubescent; *fall color* golden brown.

Fruit 1$\frac{1}{2}$ to 2 in. (4 to 5 cm) long, solitary or paired, obovoid to ellipsoidal, deeply 4-channeled from the base to the depressed apex, with a dark red-brown husk, $\frac{1}{8}$ to $\frac{1}{4}$ in. (3 to 6 mm) thick; *nut* light reddish brown, 4-ribbed, somewhat laterally compressed; *seed* sweet.

Twigs stout, reddish brown to brownish gray, pubescent; *terminal buds* $\frac{1}{2}$ to $\frac{3}{4}$ in. (12 to 18 mm) long, subglobose, reddish brown, tomentose, the outer scales soon deciduous, showing the paler, silky ones beneath; *lateral buds* smaller, somewhat divergent; *leaf scars* 3-lobed to obcordate.

Bark dark gray, firm, close, with low, rounded interlacing ridges and shallow furrows.

Range

Eastern (Map 9-9). Elevation: sea level to 3000 ft in the southern Appalachians. Well-drained, fertile uplands, dry to mesic slopes, and floodplains.

FIGURE 9-25 *Carya tomentosa,* mockernut hickory. (1) Leaf $\times\frac{1}{4}$. (2) Portion of tomentose petiole $\times2$. (3) Twig $\times1\frac{1}{4}$. (4) Nut $\times\frac{1}{3}$. (5) Cross section of nut $\times\frac{1}{3}$. (6) Nut with dehiscing husk $\times\frac{1}{3}$. (7) Bark.

General Description

Mockernut hickory is a small- to medium-sized tree, 40 to 60 ft tall and 10 to 20 in. dbh, but it may attain a maximum size of 146 ft by 4 ft. It is found on many soil types with varying amounts of moisture. This tree is the common hickory on dry, sandy soils in the pine forests of the southern Coastal Plains. Mockernut is common on terraces in the second bottoms along the Mississippi and reaches its best development in Arkansas, Missouri, and the lower Ohio River Valley. Most of the commercial mockernut is found on fertile uplands. Hardwood associates include

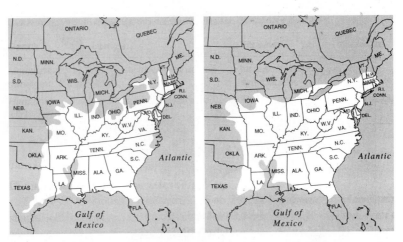

MAP 9-9 **(Left)** *Carya tomentosa.* **(Right)** *Carya glabra,* including var. *odorata.*

oaks and hickories, sweetgum, maples, American beech, and many others depending upon the site. Mockernut's growth rate is slow, and although intolerant, it responds quickly to release. It can live for 500 years.

The wood is used extensively for furniture, flooring, barrels, tool handles, baseball bats, skis, veneer, charcoal, and pulp. It is also excellent for firewood. The nuts are a preferred mast for wildlife.

Mockernut hickory hybridizes with *C. illinoinensis* and *C. texana.*

Cárya glàbra (Mill.) Sweet var. ***glabra***   pignut hickory

Distinguishing Characteristics (Fig. 9-26)

Leaves 6 to 10 in. (15 to 25 cm) long, 5 or 7 (mostly 5 but variable from tree to tree) leaflets sessile or nearly sessile, lanceolate to oblanceolate or obovate; *terminal leaflets* 4 to 6 in. (10 to 15 cm) long, $1\frac{1}{2}$ to 2 in. (4 to 5 cm) wide, somewhat larger than the lateral leaflets; *margin* finely serrate; *apex* acute to acuminate; *base* cuneate or inequilaterally rounded; *surfaces* dark yellow-green, glabrous above, more or less pubescent to glabrous below; *rachis* slender, glabrous, green; *fall color* golden yellow.

Fruit 1 to 2 in. (2.5 to 5 cm) long, obovoid or pyriform, the smooth, somewhat shiny, thin husk tardily dehiscent about halfway to the base, often not releasing the nut until after it falls to the ground; *nut* smooth, not ribbed; *seed* bitter.

Twigs slender, lustrous reddish brown, glabrous; *terminal buds* $\frac{1}{4}$ to $\frac{1}{2}$ in. (6 to 12 mm) long, ovoid to subglobose, obtuse to acuminate, covered by a few reddish brown, glabrous scales, the outer usually deciduous; *lateral buds* smaller, somewhat divergent; *leaf scars* obdeltoid to obcordate.

Bark on young trees smooth, eventually developing close, rounded interlacing ridges, medium to dark gray.

FIGURE 9-26   *Carya glabra* var. *glabra,* pignut hickory. (Left) Pear-shaped nut showing incomplete dehiscence of husk (compare with var. *odorata*) $\times \frac{1}{2}$. (Right) Bark.

Cárya glàbra var. **odoràta** (Marsh.) Little (*Carya ovalis* [Wangenh.] Sarg.)   red hickory

Distinguishing Characteristics (Fig. 9-27)

Leaves similar to var. *glabra* but with 7 leaflets, less often also 5 on the same tree, petiole reddish.

Fruit 1 to $1\frac{1}{2}$ in. (2.5 to 4 cm) long, oval to subglobose, with a dull (not shiny), granular-textured husk, thin, splitting freely to the base; *nut* brownish white, thin-shelled, 4-ribbed above the center; *seed* usually sweet.

Bark light gray, at first smooth, later somewhat platy or shaggy; eventually closely and deeply furrowed with interlacing ridges, scaly or ragged on the surface.

Range of the species

Eastern (Map 9-9). Elevation: sea level to 4800 ft in the southern Appalachians. Xeric to mesic forests, uplands, slopes, and ridges. The two varieties are sympatric.

General Description of the species

Pignut hickory and red hickory are usually medium-sized trees 50 to 60 ft tall and 12 to 24 in. dbh (max. 150 by 4 ft). They are important upland trees occurring in mixture with black, red, and white oaks and mockernut hickory. Nuts are an important food to many species of wildlife. Shade tolerance varies greatly.

The treatment of these pignuts has been problematical and variable for many years—from recognition as two separate species, *C. glabra* and *C. ovàlis,* or varieties as treated here, to no distinction at all between the two (Stone, 1997). Although the extremes are distinct morphologically and ecologically, intergradation is so extensive throughout the range that many trees can be identified only as *C. glabra. Carya glabra* s.l. hybridizes with *C. cordiformis, C. floridana, C. pallida,* and *C. texana.*

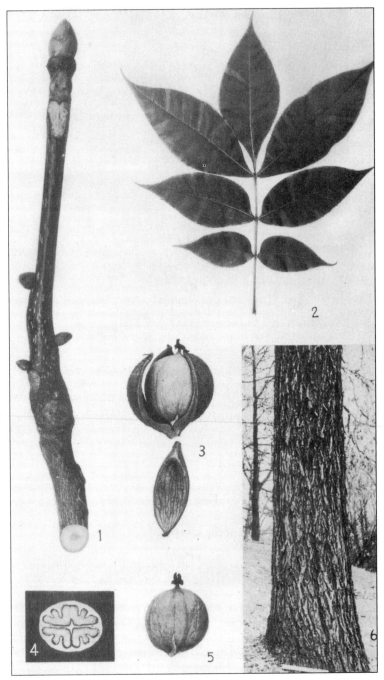

FIGURE 9-27   *Carya glabra* var. *odorata,* red hickory. (1) Twig $\times 1\frac{1}{4}$. (2) Leaf $\times\frac{1}{3}$. (3) Subglobose nut showing full dehiscence of husk (compare with var. *glabra*) $\times\frac{3}{4}$. (4) Cross section of nut $\times\frac{3}{4}$. (5) Nut $\times\frac{3}{4}$. (6) Bark.

Cárya pállida (Ashe) Engl. and Graebn.   sand hickory, pale hickory

Sand hickory is similar to *C. glabra.* It is found on dry soils from southern New Jersey through the Piedmont and Coastal Plains to western Florida and Louisiana. The leaflets are narrow, conspicuously silvery below, with fascicled hairs on the petiole and rachis but not on the back of the leaflets, and the crown is denser than that of the other associated hickories; the nut is pear-shaped like *C. glabra* var. *glabra,* but the husk is dehiscent to the base. It hybridizes with *C. glabra* and *C. texana.*

Cárya texàna Buckl.   black hickory

The black hickory is found on dry, often rocky hillsides, sandy uplands, and along creeks mainly west of the Mississippi River from Missouri to Texas. This is a pignut type like *C. glabra* but with usually 7 leaflets, and the buds and leaflet backs are covered with dense rusty-brown hairs and yellow-brown scales. It hybridizes with *C. glabra, C. pallida, C. tomentosa,* and *C. aquatica.*

Cárya floridàna Sarg.   scrub hickory, Florida hickory

This is similar to the black hickory but confined largely to the scrub vegetation of coastal dunes and sand ridges of northcentral peninsular Florida. It has 3 to 7 leaflets and many brownish scales on the leaflet backs. It hybridizes with *C. glabra.*

Cárya myristicifórmis (Michx. f.) Nutt.   nutmeg hickory

Nutmeg hickory has (5)7 or 9 leaflets with numerous scales on the back. The thin husk is prominently winged along the sutures from apex to base more like the pecan hickories, and the shape of the broadly ellipsoidal to obovoid thick nut suggests that of a nutmeg; the seed is sweet. The bark is gray to brownish, thin, fissured and scaly into long strips or broad plates. This hickory is found in swamp forests, along stream banks, or elsewhere on rich, moist soils, often on calcareous soils or marl ridges. Although reaching 80 to 100 ft by 2 ft dbh, it is of minor importance because of its rare, scattered occurrence. It is the rarest of the hickories. Small, widely separated populations are found in North Carolina, South Carolina, Alabama, Mississippi, Arkansas, Louisiana, Oklahoma, and Texas; also in Mexico.

Pecan Hickories (Table 9-11)

Cárya illinoinénsis (Wangenh.) K. Koch   pecan

Distinguishing Characteristics (Fig. 9-28)

Leaves 12 to 20 in. (30 to 50 cm) long, with 9 to 17 sessile or nearly sessile, lanceolate to oblanceolate, usually falcate leaflets; *leaflets* 2 to 7 in. (5 to 18 cm) long, 1 to 3 in. (2.5 to 7.5 cm) wide; *margin* serrate or doubly serrate; *apex* acute to acuminate; *base*

TABLE 9-11   COMPARISON OF IMPORTANT PECAN HICKORIES

Species	Leaves	Nuts
C. illinoinensis pecan	9 to 17 leaflets, lanceolate and often falcate	Ellipsoidal; husk 4-winged base to apex; seed sweet
C. cordiformis bitternut hickory	7 to 11 leaflets, oblong-lanceolate	Subglobose; husk 4-winged above the middle; seed bitter

FIGURE 9-28 *Carya illinoinensis,* pecan. (1) Twigs ×1$\frac{1}{4}$, fast-growing (left) and slow-growing (right). (2) Nut with husk ×$\frac{3}{4}$. (3) Nut ×$\frac{3}{4}$. (4) Cross section of nut ×$\frac{3}{4}$. (5) Leaf ×$\frac{1}{4}$.

cuneate or inequilaterally rounded; *surfaces* dark yellow-green, glabrous or nearly so above, pale yellow-green, glabrous or pubescent below; *rachis* slender, glabrous or pubescent; *fall color* yellow.

Fruit usually in clusters of 3 to 12, ellipsoidal, 1$\frac{1}{4}$ to 2 in. (3 to 5 cm) long; with a dark-brown husk, 4-winged from apex to base; *nut* ellipsoidal, bright reddish brown, smooth or slightly 4-ridged; *seed* sweet.

Twigs moderately stout, reddish brown, dotted with orange-brown lenticels, usually more or less pubescent; *terminal* buds $\frac{1}{4}$ to $\frac{1}{2}$ in. (6 to 12 mm) long, acute, somewhat 4-angled, valvate, yellowish brown, scurfy and appearing naked; *lateral buds* smaller, often divergent; *leaf scars* obovate.

Bark on mature trees light brown to brownish gray, divided into interlacing, somewhat scaly ridges separated by narrow fissures (Fig. 9-29).

Range

Southeastern (Map 9-10 illustrating the native range). The present-day distribution extends east to the Atlantic coast, being naturalized from pecan groves. Elevation: sea level to 1600 ft. Bottomlands.

General Description

Pecan is said to be the largest of the native hickories and varies from 110 to 140 ft in height and 2 to 4 or more ft dbh (max. 130 by 7 ft). This species is found as a scattered tree on moist but

FIGURE 9-29   Bark of pecan.

well-drained ridges in river bottoms in company with sycamore, sweetgum, American elm, willow and water oaks, persimmon, poplars, hackberries, black willow, and other bottomland hardwoods. It is the least tolerant and the fastest growing of the hickories.

The timber produced from the pecan hickories is not as strong or heavy as that of the "true hickories," but it is suitable for flooring, paneling, veneer, and furniture. Pecan is valuable chiefly for its fruit, which is produced in large amounts and used widely in the United States. Efforts to improve the size and quality of the nuts have been successful (hence, "papershell pecans"), and fruit from large planted groves of these trees is harvested on a commercial scale; pecan is now found planted and naturalized through all the southern states as far north as Virginia.

The spelling of the specific epithet has varied over the years: *illinoensis, illinoiensis,* and *illinoinensis.* The last of these is the original spelling and has been designated as the correct one.

Pecan is the state tree of Texas. It hybridizes with *C. aquatica, C. cordiformis, C. laciniata, C. ovata,* and *C. tomentosa.*

Cárya cordifórmis (Wangenh.) K. Koch   bitternut hickory

Distinguishing Characteristics (Fig. 9-30)

Leaves 6 to 10 in. (15 to 25 cm) long, with 7 to 11 sessile or nearly sessile, lanceolate or ovate-lanceolate to oblong-lanceolate leaflets; *terminal leaflets* 3 to 6 in. (7.5 to 15

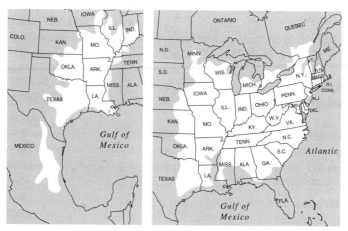

MAP 9-10   **(Left)** *Carya illinoinensis*; now naturalized eastward to eastern North Carolina. **(Right)** *Carya cordiformis.*

FIGURE 9-30 *Carya cordiformis*, bitternut hickory. (1) Twig $\times 1\frac{1}{4}$. (2) Leaf $\times\frac{1}{3}$. (3) Nut with husk $\times\frac{3}{4}$. (4) Nut $\times\frac{3}{4}$. (5) Cross section of nut $\times\frac{3}{4}$. (6) Bark.

cm) long, 1 to $1\frac{1}{4}$ in. (2.5 to 3 cm) wide, slightly larger than the lateral leaflets; *margin* finely to coarsely serrate; *apex* acuminate; *base* cuneate to unequally rounded; *surfaces* bright green, lustrous, glabrous above, pale green, pubescent or glabrous below; *rachis* slender, slightly grooved, pubescent; *fall color* golden yellow.

Fruit $\frac{3}{4}$ to $1\frac{1}{4}$ in. (2 to 3 cm) long, solitary or paired, subglobose, with a yellowish green, often minutely scurfy husk, 4-winged above the middle; *nut* subglobose, light reddish brown or gray-brown, thin-shelled; *seed* bitter.

Twigs moderately stout, greenish brown to gray-brown, more or less pubescent, lenticellate; *terminal buds* $\frac{1}{3}$ to $\frac{3}{4}$ in. (8 to 18 mm) long, cylindrical or somewhat 4-angled, valvate, sulfur-yellow, and scurfy pubescent; *lateral buds* smaller, often short-stalked, divergent; *leaf scars* 3-lobed, obcordate; *pith* brownish white.

Bark close and firm, remaining smooth for many years, eventually with shallow furrows and low, narrow, interlacing ridges, sometimes slightly scaly on the surface.

Range

Eastern (Map 9-10). Elevation: sea level to 2000 ft. Mesic forests, alluvial bottomlands, valleys, and stream banks, also dry uplands northward.

General Description

Bitternut hickory, the only member of the pecan group in the north, is a medium-sized tree and commonly attains a height of 60 to 80 ft and 12 to 24 in. dbh (max. 150 by 4 ft). Like mockernut hickory, it is found on many soil types, from those of dry, gravelly uplands to rich, moist bottomlands. Bitternut hickory is associated chiefly with other hickories and oaks, but many other hardwoods may also be in the mixture. It is usually more tolerant than its associates, and it is the shortest-lived of the hickories (to 200 years old). Bitternut hickory can produce numerous stump and root sprouts.

Probably the most abundant and uniformly distributed of the hickories, it reaches its largest size in the bottomlands of the lower Ohio basin; it is also the common hickory of Kansas, Nebraska, and Iowa, where it occurs on gravelly ridges bordering streams. It is an important source of lumber, pulp, and firewood.

Bitternut hickory hybridizes with *C. glabra, C. illinoinensis, C. laciniosa,* and *C. ovata.*

Cárya aquática (Michx. f.) Nutt.   water hickory, bitter pecan

This species is common in swamps and on other low wet sites through the Coastal Plains from Virginia to Florida, west to eastern Texas, and north in the Mississippi Valley to southern Illinois. It is most similar to *C. illinoinensis* and hybridizes with it but differs in its fruit, which is a shorter, flattened, obovoid, 4-ribbed nut with a bitter seed, 5 to 13 narrow leaflets, and shaggy bark. In the absence of the nut, it is often difficult to distinguish the two species. Water hickory, although of minor importance, will grow on sites too poor for the better-grade hardwoods, and hence merits some consideration. Common associates include overcup and Nuttall oaks, green ash, American and cedar elms, waterlocust, and hackberries. It hybridizes with *C. texana.*

Myricaceae: The Wax-Myrtle Family

This family includes three genera and ca. 50 species of evergreen shrubs and small trees found usually in swamps or on dry sandy soils in the temperate and warmer parts of the world. Two genera and eight species occur in North America. The leaves are alternate, simple, and aromatic and bear minute, yellow-orange or black resin dots, visible with a hand lens. The flowers are anemophilous, unisexual, and in rather short, inconspicuous aments, appearing in early spring. *Cómptonia peregrìna* (L.) Coult. with pinnatifid, fern-like, linear-lanceolate leaves is the common, shrubby "sweet-fern" of the Northeast. The myricas, bayberries, or wax-myrtles have leaves with entire or toothed margins. *Mỳrica gàle* L., sweet gale, has the broadest distribution of all myricas. It occurs in bogs and lake borders from northeastern United States and eastern Canada into Alaska south to Oregon. The drupes lack a waxy coating but are surrounded by glabrous, spongy bracteoles that aid in flotation during water dispersal. The waxy covering of the fruit of *Mỳrica pensylvánica* Loisel., northern bayberry, furnishes most of the bayberry wax used for candles. For a description of candle making, see Dengler (1967). Both of the following species are cultivated as ornamentals. The fruit is eaten by small birds. The pollen is allergenic.

The roots of *Myrica* have a symbiotic relationship with actinomycetes (moldlike bacteria), which form root nodules and have the capacity to convert atmospheric nitrogen into a form usable by plants. This capacity for nitrogen fixation is generally associated with legumes, but *Myrica, Alnus, Elaeagnus, Casuarina,* and *Ceanothus* are notable additions. This process is of major ecological significance because great quantities of nitrogen compounds are added to the soil.

Mỳrica cerífera L.   wax myrtle, southern bayberry

This is a common, large shrub or small tree of the Coastal Plain from New Jersey south to southern Florida, west to eastern Texas and southeastern Oklahoma. It has escaped cultivation into the Piedmont. It occurs in moist habitats and often forms dense thickets. The leaves are oblanceolate with yellow-orange glands mainly on the backs and coarse teeth only above the middle.

Mỳrica califórnica Cham.   Pacific bayberry

This shrub or small tree occurs quite abundantly on low hills, sand dunes, and riverbanks near the coast from British Columbia to southern California, from sea level to 500 ft elevation. The leaves have black glands below.

Fagaceae: The Beech Family

The beech family includes nine genera and ca. 800 species of trees and shrubs scattered throughout both hemispheres, but is most characteristic of the forests in the North Temperate Zone. The importance of this group to American forestry can hardly be overemphasized because in this country it takes first place among the hardwoods in the production of lumber and other forest products. It is also important for ornamentals and as a source of mast, and many species form the dominant trees in many associations. The most important genus is *Quercus* (oak).

Five genera (Table 9-12) with ca. 97 species are found in North America. Two genera, *Lithocarpus* and *Chrysolepis,* are confined to the Pacific coast region; *Castanea* and *Fagus* are restricted to the East, and *Quercus* is widely distributed over a large portion of the United States and southern Canada, with the exception of the Great Plains.

The pollen of many species is a common cause of allergies in susceptible individuals.

TABLE 9-12   COMPARISON OF NATIVE GENERA IN THE FAGACEAE

Genus	Leaves	Flowers	Fruit
Fagus beech	Deciduous; thin	Staminate in heads; pistillate in short 2- to 4-flowered spikes	Nut sharply 3-angular in cross section; 2 within a bur with weak, unbranched spines; maturing first year
Castanea chestnut	Deciduous; thin, somewhat leathery	In erect, unisexual and bisexual spikes, rigid or flexible	Nut rounded or smooth, angular in cross section; 1–3 within a bur with stiff, sharp, branched spines with simple hairs; maturing first year
Chrysolepis western chinkapin	Persistent; thick, leathery	In erect, unisexual or bisexual spikes, rigid or flexible	Nut rounded or smooth, angular in cross section; 1–several within a bur with stiff, sharp, branched spines with yellowish glands; maturing second year
Lithocarpus tanoak	Persistent; leathery	In erect or ascending, unisexual or bisexual spikes, rigid or flexible	Acorn round in cross section; solitary; enclosed at base by a cup with long, slender, strongly reflexed scales hooked at tip; maturing second year
Quercus oak	Deciduous or persistent; thin or leathery	Staminate spikes (aments) pendent; pistillate ament usually stiff	Acorn round in cross section; solitary; in a scaly cup, variously shaped and covering base or more; scales imbricate; maturing first or second year

Botanical Features of the Family

Leaves persistent, or deciduous and often marcescent; alternate, simple, stipulate, usually pinnately veined, and short-petioled.

Flowers primarily anemophilous but sometimes aided by insects in *Chrysolepis,* *Lithocarpus,* and *Castanea,* imperfect (plants monoecious) and borne in one of several ways. *Staminate* flowers usually in aments, but in *Fagus* they form a globose head; the aments pendent or erect, the flowers variously grouped on the axis; *staminate* flowers have a 4- to 7-lobed calyx and 4 to 8 (rarely more) stamens. *Pistillate* flowers borne in short few-flowered spikes on new growth (*Fagus* and *Quercus*) or in clusters near the base of the smaller staminate aments (*Castanea, Chrysolepis, Lithocarpus*). When such bisexual aments occur, there are also others of the purely staminate type present; *pistillate* flower consists of a 4- to 8-lobed calyx adnate to a 3- (rarely 6-) celled ovary with a style for each cell, containing 1 to 2 ovules, only one of which matures.

Fruit a nut with an outer cartilaginous coat, and partially or wholly encased in a cupule (bur or cup); *nut* one-seeded by abortion, seed lacking endosperm; cotyledons large and fleshy, and as in *Juglans, Carya,* and certain other genera with large seeds, always remaining (with the exception of *Fagus*) within the seed coats upon germination (i.e., hypogeal).

FAGUS L.   beech

There are ca. 10 species of *Fagus,* all in the Northern Hemisphere. Besides the one North American species (eastern United States, Canada, and Mexico), one is European, another is in the Caucasus, and the others range through temperate eastern Asia.

European beech, *Fàgus sylvática* L., is an important timber tree of Europe, where it is grown in pure or mixed stands under forest management. It is distinguished by the leaves, which have 5 to 9 pairs of secondary veins in contrast to 9 to 14 pairs in *F. grandifolia,* the crenate rather than serrate leaf margins, and softer fruit spines. The leaves, buds, and bark of *F. sylvatica* are also darker than those of the American species. The fruit, known as *beech mast,* is used in Europe for fattening hogs and as a source of vegetable oil. Numerous cultivars of this species are very widely cultivated as beautiful ornamental lawn, street, or park trees.

The name *beech* has a very ancient origin and signifies "book." It is said that the early writings of the Germanic peoples were inscribed upon tablets made of this wood. Gutenberg printed the first Bible from movable type carved from beech wood.

The bluish gray, smooth bark of the beech is a most distinctive feature, but often, and almost characteristically, made very unsightly by initials, hearts, and dates carved upon its surface.

Young trees of some oaks, and particularly beech, characteristically hold their brown or tan leaves until spring. Trees with this very conspicuous feature during the winter are called *marcescent.*

Fàgus grandifòlia Ehrh.   American beech

Distinguishing Characteristics (Fig. 9-31)

Leaves deciduous, $2\frac{1}{2}$ to 6 in. (6 to 15 cm) long, 1 to 3 in. (2.5 to 7.5 cm) wide; *shape* elliptical to oblong-ovate; *margin* remotely serrate with sharp, incurved teeth; *apex* acuminate; *base* broadly cuneate; *surfaces* silky at first, becoming glabrous above and various degrees of puberulent below plus tufts of hairs in the vein axils, secondary veins parallel to each other; texture of the leaf somewhat papery to the touch; *petiole* short; *stipules* $\frac{3}{4}$ to $1\frac{3}{4}$ in. (2 to 4 cm) long, narrow, and straplike, membranous to papery; *fall color* yellow to bronze.

Flowers appearing after the unfolding of the leaves in the spring; *staminate* in globose heads, each flower with a 4- to 8-lobed calyx and 8 to 16 stamens; *pistillate* in 2- to 4-flowered spikes, surrounded by a cupule, each flower with a 4- to 5-lobed calyx attached to a 3-celled ovary.

Fruit an edible nut, $\frac{1}{2}$ to $\frac{3}{4}$ in. (12 to 18 mm) long, triangular in cross section; paired or in threes within a woody, 4-parted cupule or bur covered with weak, unbranched spines.

Twigs slender, at times slightly zigzag; *pith* homogeneous, terete; *pseudoterminal buds* $\frac{3}{4}$ to 1 in. (18 to 25 mm) long, slender, lance-shaped, sharp-pointed, often appearing and interpreted as a true terminal; scales numerous and imbricated in 4 ranks; *lateral buds* similar; *leaf scars* small, inconspicuous; *stipular scars* narrow, nearly encircling the twig.

Bark thin, smooth, light blue-gray in color, often mottled and not changing appreciably as the tree grows older.

FIGURE 9-31 *Fagus grandifolia,* American beech. (1) Twig $\times 1\frac{1}{4}$. (2) Staminate flowers $\times\frac{1}{2}$. (3) Leaf $\times\frac{1}{2}$. (4) Fruit (nuts) in a prickly husk $\times\frac{3}{4}$. (5) Nuts in partly opened husk $\times\frac{3}{4}$. (6) Bark.

Range

Eastern (Map 9-11). Elevation: sea level to 3000 ft northward, to 6000 ft in the southern Appalachians. Moist, rich uplands and lowlands, and cool north-facing slopes.

General Description

American beech is one of the most distinctive and common trees of the eastern hardwood forest, and in winter its smooth blue-gray trunk and marcescent condition are easily recognized even at some distance. The tree averages 70 to 80 ft in height and 2 to 3 ft dbh (max. 140 by 6 ft). In the

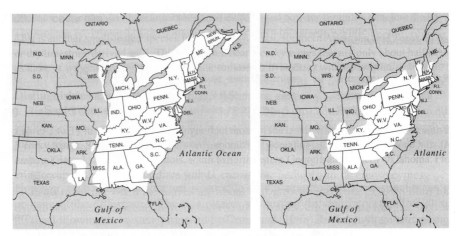

MAP 9-11 **(Left)** *Fagus grandifolia.* **(Right)** *Castanea dentata.*

open it has a short, stocky bole with a wide-spreading crown, whereas under forest conditions a clear, straight, massive trunk and smaller crown develop. The root system is shallow and extensive and tends to produce numerous sprouts. Such root suckering is very characteristic of beech, and old trees are often surrounded by thickets of young growth, mostly of this origin rather than from seed. Stump and root collar sprouts, except from stumps under 4 in. in diameter, are short-lived.

American beech is found on a wide variety of soils from the sandy loams of the Northeast to the alluvial bottoms of the Ohio and Mississippi Valleys. Beech requires much more moisture for growth than many other hardwoods. Trees are not usually found on soils where the surface layers dry out quickly or are subject to prolonged flooding. This species may be the most tolerant of northern hardwoods, and with sugar maple and yellow birch it forms the typical maple-beech-birch forest of that region. Other associates include red spruce, white pine, hemlock, black cherry, northern red oak, and white ash. Several of these are also southern associates, and to them may be added such species as southern magnolia, basswood, sweetgum, oaks, and hickories.

American beech reaches its largest size on the better-drained sites of alluvial bottoms in the Ohio and Mississippi Valleys and in the western half of the southern Appalachians. However, even under the best conditions, growth is slow.

Large crops of seed may be produced at 2- to 3-year or longer intervals. This species may attain an age of 300 to 400 years, but old trees are usually affected by butt rot. The thin bark and shallow root system make beech very susceptible to damage by fire. Beech bark disease, initiated by scale insects and followed by *Nectria* bark canker fungi, has caused severe mortality in the Northeast (Houston, 1994). This disease is spreading westward and southward.

American beech is a variable species and shows at least three different races (Camp, 1950; Cooper and Mercer, 1977; Hardin and Johnson, 1985). The northern form is "gray beech" found from Nova Scotia to the Great Lakes, and on the higher mountains of North Carolina and Tennessee mainly on neutral to alkaline soils. "White beech" is found on the southern Coastal Plains and northward on poorly drained acid sites. Between these two forms, mainly on well-drained acid sites and mixing with them, is "red beech." The beech of much of the glaciated area is a mixture derived in various combinations from the three basic forms. Careful studies are needed to clarify this variation on a local basis because each stand is not only variable but has a "norm" often differing from that of adjacent sites.

The wood is used for flooring, furniture, veneer, butcher's blocks and cutting boards, plywood, and fuel. The nuts are an important source of mast for wildlife.

CASTANEA Mill.   chestnut, chinkapin

This is a small genus with ca. 10 species found in southern Europe, northern Africa, southwestern and eastern Asia, and eastern United States. Because the superior flavored nuts of the blight-stricken American chestnut are no longer available, three other species are used. Spanish or European chestnut, *C. satìva* Mill.; Japanese chestnut, *C. crenàta* Sieb. and Zucc.; and Chinese chestnut, *C. mollíssima* Bl., are all important in their native lands. The United States imports the nuts but also has a few groves of the better varieties for local production, especially in Georgia and Maryland. These are also used as ornamental yard or park trees.

Pollination in *Castanea* has been described as anemophilous, entomophilous, or a combination of both. The anemophilous syndrome is generally expressed by a monoecious condition, flowering before the leaves, large quantities of smooth, lightweight pollen, and pistillate flowers inconspicuous and lacking both odor and nectar. The entomophilous syndrome normally includes flowering after the leaves appear, less amounts of heavier, sticky pollen, and flowers showy and aromatic. Johnson (1988) found that the American species of *Castanea* are primarily wind pollinated. Any role played by insects is a passive, indirect one by dislodging pollen from the anthers, but the insects do not visit the pistillate flowers. The combination of anemophilous and entomophilous features in these species may reflect an intermediacy between ancestral entomophily and the derived anemophily characteristic of other Fagaceae.

Botanical Features of the Genus

Leaves deciduous, serrate, stipulate.

Flowers borne in ascending, staminate, and bisexual aments, appearing in late spring when the leaves are nearly full grown; *staminate* in clusters of 3 to 7 along the elongated and conspicuous ament axis, each with a 6-lobed calyx and 10 to 20 long-stalked stamens; *pistillate* solitary, or in clusters of 2 to 3, in a cupule at the base of special, short, staminate aments (bisexual aments); calyx 6-lobed, adnate to a 6-celled ovary bearing occasionally a few abortive stamens.

Fruit a lustrous-brown, rounded, edible nut borne singly or in clusters of 2 or 3 in a 2- to 4-valved bur covered with needle-sharp branched spines.

Twigs slender to moderately stout, usually brown, glabrous or tomentose; *pith* stellate; *terminal buds* lacking; *lateral buds* with 2 or 3 visible scales; *leaf scars* half round; *bundle scars* 3 or more; *stipular scars* present.

Castànea dentàta (Marsh.) Borkh.   American chestnut

Distinguishing Characteristics (Fig. 9-32)

Leaves 5 to 9 in. (13 to 23 cm) long, $1\frac{1}{2}$ to 3 in. (4 to 7.5 cm) wide; *shape* oblong-lanceolate; *margin* coarsely and sharply serrate, with bristle-tipped or hair-tipped teeth;

FIGURE 9-32 *Castanea dentata,* American chestnut. (1) Twig ×1$\frac{1}{4}$. (2) Leaf ×$\frac{1}{2}$. (3) Inflorescence ×$\frac{1}{3}$. (4) Portion of staminate inflorescence showing clusters of flowers ×1. (5) Base of inflorescence showing pistillate flower ×1. (6) Nut ×1. (7) Nut in partly opened prickly husk ×$\frac{3}{4}$. (8) Chestnut sprouts.

apex acuminate; *base* acute; *surfaces* dull yellow-green and glabrous; phyllotaxy variable, usually $\frac{1}{2}$ or $\frac{2}{5}$; *fall color* yellow to orange-yellow.

Fruit 2 to 2$\frac{1}{2}$ in. (5 to 6 cm) in diameter, covered with sharp, branched spines; *nut* $\frac{1}{2}$ to 1 in. (12 to 25 mm) wide, nearly flat on one or two sides, borne 2 or 3 in each bur.

Twigs somewhat lustrous, chestnut-brown; *lateral buds* ca. $\frac{1}{4}$ in. (6 mm) long, ovoid, brown.

Bark dark brown, shallowly fissured into broad, flat ridges.

Range

Eastern (Map 9-11). Elevation: sea level to 4400 ft in the southern Appalachians.

General Description

The American chestnut was a highly valued tree; its growth was rapid, the wood durable, and the flavorful roasted nuts contributed to the food supply. In the forest, the nuts were an important wildlife food. Tannin was extracted from both bark and wood. All things considered, chestnut may well have been the most important hardwood of North America. However, in 1904, a chestnut blight disease, caused by the fungus *Cryphonectria parasitica,* presumably brought in from eastern Asia, was first noticed on a few trees at the Bronx Zoo. Within 50 years the blight had spread over the entire range of the chestnut, causing wholesale destruction. The standing-dead trunks were important for the highly prized "worm-eaten chestnut" lumber until the 1960s. Its ability to sprout vigorously after the trunk is dead helps the species to persist in spite of continued attacks by the fungus dispersed from cankers present on other trees. These sprouts occasionally have fruit, but no resistant form has been found.

With the beginning of the chestnut blight, scientists began a vigorous program of hybridization and, in the 1970s, experimented with a strain of the fungus with reduced virulence. Thus far, however, there is no prospect of the American chestnut again becoming a significant forest tree. Certain hybrids and selections may be available as ornamentals and for the nuts. Its original dominance on many sites has been replaced by various oak and hickory species and yellow-poplar.

The American chestnut is more closely related to the European and Asian chestnuts than it is to chinkapin. Over 1800 references on American chestnut are listed by Forest et al. (1990).

Castànea pùmila Mill.   chinkapin

Chinkapin (or chinquapin) differs from the American chestnut in its smaller, tomentose leaves; smaller, single-seeded burs; and usually tomentose twigs. It is a small tree or shrub of dry soils from Pennsylvania to Florida and west to Oklahoma and Texas. The nuts are eaten by people and wildlife. Although quite variable in habit and other features, Johnson (1988) found that the most realistic classification was to recognize only a single species (including *C. alnifòlia* Nutt. and *C. ozarkénsis* Ashe). Little (1979) and Nixon (1997) treat the Ozark chinkapin as a distinct species.

CHRYSOLEPIS Hjelmquist   western chinkapin, evergreen chinkapin

This genus includes two species of evergreen trees and shrubs, both native to the Pacific coast. *Chrysolepis sempervìrens* (Kell.) Hjelm., Sierra chinkapin or bush golden chinkapin, is a small, low, timberline, rhizomatous shrub common in the Coast Ranges and the high Sierra Nevada of California.

These species have been previously included in *Castanópsis* (D. Don) Spach (Little, 1979). *Castanopsis* is a larger genus, native of Asia, with a very different cupule structure (Nixon, 1997).

Botanical Features of the Genus

Leaves persistent, coriaceous, entire or obscurely toothed, stipulate.

Flowers in erect, unisexual or bisexual aments; *staminate* in cymose clusters of three along the amentiferous stalk, the individual flowers similar to those of chestnut; *pistillate* solitary or in clusters of two or three per cupule, the ovary 3-celled; appearing in early summer.

Fruit a large, lustrous, brown to yellowish brown nut borne solitary or in clusters of two or three and encased in a 2- or 4-valved husk with stiff, sharp, branched spines, maturing at the end of the second season.

Chrysolèpis chrysophýlla (Dougl. ex Hooker) Hjelm.   giant golden chinkapin

Distinguishing Characteristics (Fig. 9-33)

Leaves lanceolate to oblong-ovate, 2 to 5 in. (5 to 13 cm) long, $\frac{5}{8}$ to $1\frac{1}{2}$ in. (1.5 to 4 cm) wide, entire, often slightly revolute, dark green above, covered with dense, minute, golden-yellow hairs below; stipules prominent and often persistent.

Flowers like those of chestnut.

Fruit densely covered with branched spines, the husk containing one or two edible nuts that mature at the end of the second season.

Twigs slender, at first covered with golden-yellow scales, becoming reddish brown and somewhat scurfy, terminal bud present, with imbricate scales; stipular scar present; pith stellate.

Bark initially dark gray and smooth, maturing to very wide reddish brown plates, bright red within.

Range

Southwestern Washington, south from the Columbia River along the west slopes of the Cascades through Oregon to central California, along the Coast Ranges to the San Jacinto Mountains. Elevation: sea level to 6000 ft. Gravelly and rocky slopes, canyons, and dry ridges.

General Description

The giant golden chinkapin is either a large, erect shrub with leaves folded upward along the midrib or a tree with leaves typically flat (Nixon, 1997). Under average conditions for growth, the tree will reach a height of 60 to 80 ft and 3 ft dbh, although much larger trees (max. 127 ft by 5 ft) have been observed in the moist valleys of northern coastal California. Under forest conditions, this species develops a clear bole for one-half to two-thirds of its length, and also a dense, ovoid to conical crown; the crown of open-grown trees is generally large and wide-spreading. A taproot is developed by juvenile trees, but this gives way in later life to a well-developed lateral system of anchorage. It is very tolerant when young but requires full sunlight for adequate mature growth.

Considerable seed is produced at frequent intervals, and natural regeneration by this means is generally satisfactory. Some renewal is also traceable to stump sprouts. Growth is moderately rapid, and maturity is reached in about 200 to 500 years.

This species is quite gregarious on poor, dry soils and often forms pure stands over wide areas. The tree also appears as an understory species in stands of redwood and Douglas-fir, or mixed with western juniper, Jeffrey pine, canyon live oak, and as a component of chaparral.

The wood is used for furniture, paneling, veneer, and cabinetwork. The nuts are an important source of mast for birds and small and large mammals. They are sweet and edible but difficult to remove from the cupules until completely ripe.

FIGURE 9-33   *Chrysolepis chrysophylla,* giant golden chinkapin. (1) Leaf ×$\frac{3}{4}$. (2) Nut with prickly husk ×1. (3) Nut ×1. (4) Bark (*Photo by E. Fritz*).

LITHOCARPUS Blume   tanoak

Lithocarpus numbers 100 to 200 species of evergreen trees and shrubs. The genus is widespread through southeastern Asia and Indonesia but is not found elsewhere except for a single species indigenous to western North America. The Asiatic forms will not endure cold winters; hence, their use as ornamentals is not possible in the cooler regions of the temperate zone. A few species have been successfully introduced into southern United States, but they are rarely used for decorative purposes. The fruits of *L. édulis* (Mak.) Nakai and *L. cornèus* (Lour.) Rehd. of Japan and south China, respectively, are rather sweet and edible. Botanically, this genus is very interesting, because its flowers are nearly identical with those of either *Castanea* or *Chrysolepis* although its fruit resembles that of *Quercus* (Nixon, 1997).

Botanical Features of the Genus

Leaves persistent, leathery, stipulate.

Flowers similar to those of *Castanea* or *Chrysolepis;* appearing in summer.

Fruit a nut (acorn), partially or rarely wholly enveloped by the cuplike cupule with reflexed scales; maturing at the end of the second season.

Lithocárpus densiflòrus (Hook. and Arn.) Rehd.   tanoak, tanbark-oak

Distinguishing Characteristics (Fig. 9-34)

Leaves oblong to oblong-lanceolate, 2 to 5 in. (5 to 13 cm) long, $\frac{3}{4}$ to $2\frac{1}{4}$ in. (2 to 6 cm) wide, repand-dentate to entire-revolute, rounded-truncate at the base, light green above, brownish to golden woolly pubescent in early summer below, ultimately becoming bluish white and sparsely pubescent toward the end of the season; *petiole* tomentose.

Flowers similar to those of *Chrysolepis.*

Fruit a bitter acorn, $\frac{3}{4}$ to $1\frac{1}{4}$ in. (1.8 to 3 cm) long, the cup shallow, with narrow, spreading scales, tomentose, lined with lustrous red pubescence.

Twigs at first densely pubescent but later dark reddish brown and often covered with a glaucous bloom.

Bark broken into heavy rounded ridges separated by deep narrow fissures.

Range

Southwestern Oregon, south along the Coast Ranges to southern California; south in the Sierra Nevada to Mariposa County. Elevation: sea level to nearly 5000 ft. Moist valleys and mountain slopes in mixed evergreen and redwood forests.

FIGURE 9-34   *Lithocarpus densiflorus,* tanoak. (1) Leaf $\times\frac{3}{4}$. (2) Acorn ×1.

General Description

Tanoak includes two varieties (Nixon, 1997), either a shrub with small leaves or a moderately large tree 70 to 90 ft in height and 24 to 36 in. dbh, but under the most favorable conditions reaching somewhat larger proportions (max. 150 by 9 ft). The boles of forest trees are clear, but often asymmetrical, and support narrow pyramidal crowns of ascending branches; open-grown trunks are short and often disappear in a mass of widespread limbs that form broad, nearly globose crowns. Anchorage consists of a deep, well-developed taproot and a number of short laterals. Best development is reached on deep, rich, moist, but well-drained, gravelly and sandy soils with redwood, Douglas-fir, Oregon white oak, giant chinkapin, and California black oak. At elevations of 3,500 to 4,500 ft along the middle-west slopes of the Sierra Nevada, tanoak occurs in mixture with black oak, Pacific madrone, bigleaf maple, white fir, sugar pine, ponderosa pine, and Douglas-fir. Fairly extensive, nearly pure stands of coppice growth appear on cutover areas.

Viable seed is borne in abundance after 30 to 40 years. The growth of seedlings proceeds at a moderate rate, and the trees reach maturity between 200 and 300 years. Tanoak can exist under forest cover throughout its life, and the rapidity with which suppressed trees often develop upon being released from dense shade is remarkable.

The bark of tanoak contains appreciable amounts of tannin and at one time was an important commercial source of this material for the western states. It is now used for fuel and pulp. Acorns are a very important mast for a wide variety of wildlife and were a principal food for Native Americans.

QUERCUS L.   oak

The oaks are widely distributed throughout the temperate regions of the Northern Hemisphere and extend southward at higher elevations to the tropics (Colombia in the Western Hemisphere; northern Africa and Indonesia in the Eastern Hemisphere). The total number of species is ca. 400, with 250 in the new world. There are 90 species native to North America north of Mexico (Nixon, 1997), and according to Little (1979), 58 of these reach tree size.

The oak genus with its many species is the most important aggregation of hardwoods found on the North American continent, if not in the entire Northern Hemisphere; in fact, the central and southern hardwood forests of the United States may be thought of as oak forests, with the other broad-leaved species playing a secondary role. An interesting discussion of the development of oak forests is given by Abrams (1992). Oaks furnish more native timber annually than any other group of broad-leaved trees, and the amount is surpassed only by the conifers, which occupy first place in this respect. Other products of lesser importance but widespread use include cork from two southern European species,* tannin from the bark of certain others, and ink derived from the insect leaf galls of a few old world species; the sweet acorns of several North American oaks were used extensively for food by Native Americans. Indeed, throughout its vast range, this genus has historically been an important food source for primitive people. It continues to be of major importance to many forms of wildlife.

The sturdy qualities and appearance of many of the oaks, together with their longevity in comparison with other hardwoods, have made them since very ancient times the objects of admiration and worship among the early people of the old world. This feeling still persists in a modified form, and many of the historic as well as ornamental trees of the United States and other countries are oaks of one species or another.

Several Asian and European species have become popular ornamentals in the United States and Canada. Sawtooth oak, *Q. acutíssima* Carruth.; Chinese evergreen oak, *Q. myrsinifòlia* Bl.; blue Japanese oak, *Q. glaúca* Thunb.; and English oak, *Q. ròbur* L., are frequently seen. English oak has become naturalized and hybridizes with *Q. alba*.

From the botanical standpoint, oaks are of great interest due to their intrinsic variation as well as hybridization and introgression. For an example of variation, see Baranski (1975), and for a discussion of oak hybrids see Cottam et al. (1982), Hardin (1975), Little

---

*The world's cork supply is obtained chiefly from *Q. sùber* L. of southern Europe and northern Africa. See Cooke (1946) and Fowells (1949) for reports on attempts at growing these trees in the United States, and de Oliveira and de Oliveira (1994) for more information on the tree and the cork industry.

(1979), Palmer (1948), Rushton (1993), and Tucker (1993). Due to extensive hybridization within sections of the genus, the hybrids are not indicated in the general descriptions. See Nixon (1997) for the hybrids and binomials given to them.

Because of the size and complexity of this genus, only the most important or characteristic species are fully described in this text. Other species of ecological interest are mentioned briefly. Identification of local species can be accomplished with the aid of local or regional field guides to trees or in FNA vol. 3 (1997).

Several insects, including gypsy moth, and diseases, especially oak wilt, attack eastern oaks. Oak wilt is more serious on red than white oaks. Wood borers cause much defect in the lumber of many oak species.

Oak is the state tree of Iowa.

Botanical Features of the Genus

Leaves deciduous or persistent; lobed or not; *margin* crenate, serrate, or entire; *shape* and *size* often very variable even on the same tree; *stipules* usually deciduous or at times persistent at the upper nodes; phyllotaxy $\frac{2}{5}$.

Flowers (Fig. 9-35) imperfect (plants monoecious), appearing on the old or new growth, before, with, or after the unfolding of the new leaves; *staminate* aments pendent, clustered; individual flowers with a 4- to 7-lobed calyx, which encloses 6 stamens (rarely 4 to 12); *pistillate* flowers solitary or in few- to many-flowered spikes from the axils of the new leaves; individual flowers consist of a 6-lobed calyx surrounding a 3- (rarely 4- to 5-) celled ovary, the whole partly enclosed in a cupule.

Fruit an acorn maturing in 1 or 2 seasons; *nut* 1 per cup, round in cross section, enclosed at base by a scaly cupule, variously saucer- to cup- to bowl-shaped, the scales imbricate; usually 1-seeded by abortion.

Twigs stout to slender, straight, commonly angled; *pith* homogeneous, stellate; *buds* clustered at the ends of the twig; *terminal* bud present, with many scales imbricated in 5 ranks; *lateral buds* similar but smaller; *leaf scars* semicircular; *bundle scars* scattered, numerous; *stipular scars* minute.

Infrageneric Classification

On the basis of similarities in leaf, flower, pollen, and fruit characteristics, morphology of foliar trichomes (Hardin, 1979), molecular data, and interspecific hybridization, oaks are classified into two subgenera and three sections. The classification of subgenera and sections used here follows Nixon (1993, 1997). Much more research is needed to determine species relationships among the oaks. What follows is an outline of the groupings used here.

FIGURE 9-35 *Quercus,* oak. (1) Spring twig ×$\frac{1}{2}$ showing flowers and young leaves. (2) Pistillate flowers ×4. (3) Staminate flower ×4.

Outline of Subgenera, Sections, and Smaller Groups

I. Subgenus *Cyclobalanopsis.* The cycle-cup oaks. Asian.

II. Subgenus *Quercus.* The scale-cup oaks. Northern Hemisphere.

    A. Section *Quercus* (=*Leucobalanus*). The white oaks. (Leaves lacking spinose teeth or bristle-tipped lobes; often with stellate hairs and lacking multiradiate hairs; acorns maturing in one season; acorn cup scales basally thick; inner pericarp glabrous; seed sweet)

        1. True white oaks. (Leaves deciduous; more or less deeply lobed) *Q. alba, macrocarpa, lyrata, stellata, margaretta, douglasii, garryana, lobata, gambelii*

        2. Chestnut oaks. (Leaves deciduous; crenate, coarsely toothed, or sinuate) *Q. michauxii, prinus, bicolor, muehlenbergii*

        3. Live oaks. (Leaves persistent; usually unlobed) *Q. virginiana, geminata, arizonica, oblongifolia*

    B. Section *Lobatae* (=*Erythrobalanus*). The red and black oaks. (Leaves mostly with bristle-tipped lobes; if unlobed, the margins, apices, or both, often with spines or bristles; often with multiradiate hairs and lacking stellate hairs; acorns usually maturing in two seasons; acorn cup scales rarely thick; inner pericarp tomentose; seed bitter)

        1. True red and black oaks. (Leaves deciduous; lobed) *Q. rubra, velutina, shumardii, falcata, pagoda, coccinea, palustris, ellipsoidalis, texana, marilandica, laevis, kelloggii*

        2. Willow, laurel, and water oaks. (Leaves deciduous or tardily so; unlobed) *Q. phellos, nigra, hemisphaerica, laurifolia, imbricaria, incana*

        3. Western live oaks. (Leaves persistent; unlobed) *Q. wislizeni, agrifolia, emoryi*

    C. Section *Protobalanus.* The intermediate oaks. (Leaves persistent, often with aristate teeth; stellate or multiradiate hairs present; acorns maturing in two seasons; acorn cup scales more or less thick, tomentose; inner pericarp densely to sparsely tomentose appearing glabrous) *Q. chrysolepis, palmeri, cedrosensis, tomentella, vaccinifolia*

True White Oaks

Deciduous; leaves more or less deeply lobed; eastern and western species (Table 9-13).

Quércus álba L.   white oak

Distinguishing Characteristics (Fig. 9-36)

Leaves deciduous, 4 to 9 in. (10 to 23 cm) long, 2 to 4 in. (5 to 10 cm) wide; *shape* oblong-obovate; *margin* entire, 7- to 9-lobed, with deep to shallow sinuses extending evenly toward the midrib, sometimes with secondary lobes; *apex* rounded; *base* cuneate; *surfaces* glabrous or puberulent, bright green above, slightly paler below; *fall color* yellow, red, or purplish brown.

TABLE 9-13   COMPARISON OF IMPORTANT TRUE WHITE OAKS

Species	Leaves	Acorns
Q. alba white oak	Deeply to shallowly lobed, sinuses narrow; pale green and glabrous or puberulent below	$\frac{3}{8}$–$1\frac{1}{4}$ in. (1–3 cm) long, ovoid-oblong; cup bowl-like with warty scales
Q. macrocarpa bur oak	Irregularly lobed, the center pair of sinuses the deepest; pale and pubescent below	$\frac{3}{4}$–2 in. (2–5 cm) long, broadly ovoid; cup deep, conspicuously fringed on the margin
Q. lyrata overcup oak	Deeply lobed, with broad irregular sinuses, the lowest ones the deepest; green and puberulent or pale and pubescent below	$\frac{1}{2}$–1 in. (1.2–2.5 cm) long, subglobose; almost wholly enclosed in a deep, thin, unfringed cup
Q. stellata post oak	Cruciform, the middle lobes the largest; tawny tomentose below	$\frac{1}{2}$–1 in. (1.2–2.5 cm) long, ovoid-oblong; cup bowl-like with thin scales
Q. garryana Oregon white oak	Evenly and deeply lobed; pale with orange-brown pubescence below	1–$1\frac{1}{4}$ in. (2.5–3 cm) long, ovoid to obovoid; cup shallow with pubescent, free-tipped scales

Fruit solitary or paired, sessile or short-stalked; *nut* $\frac{3}{8}$ to $1\frac{1}{4}$ in. (1 to 3 cm) long, ovoid-oblong, enclosed for one-quarter of its length in a light chestnut-brown, bowl-like cup with thickened, warty scales.

Twigs moderately stout, purplish gray to greenish red; *terminal buds* $\frac{1}{8}$ to $\frac{3}{16}$ in. (3 to 5 mm) long, globose to broadly ovoid, reddish brown, glabrous.

Bark light ashy gray, very variable in appearance; on young to medium-sized trees often broken up into small, vertically aligned blocks, scaly on the surface; later irregularly plated, with the plates attached on one side, or deeply fissured, with narrow rounded ridges. (Figure 9-36, pts. 5 and 6, shows the amount of variation to be expected, with pt. 5 the more common except on very old trees.)

Range

Eastern (Map 9-12). Elevation: sea level to 5500 ft in the southern Appalachians. Dry to mesic woods.

General Description

This tree is the most important species of the white oak group and is said to furnish nearly three-fourths of the timber harvested as white oak. Any estimate, however, is subject to error because only white oak and red oak are recognized in the trade; each type includes the wood of a number of species that lose their identity when the logs are manufactured into lumber or other forest products.

White oak at its best is a large tree 80 to 100 ft tall and 3 to 4 ft dbh (max. 150 by 9 ft). In the open it is characterized by a short stocky bole and a wide-spreading crown of rugged appearance; this outline with some variation is a feature of most oaks. Under forest conditions, white oak develops a tall straight trunk with a small crown.

The oaks as a group are noted for their deep root systems, and at the end of the first season, white oak seedlings with but 3 to 4 in. of slender top growth terminate below in a prodigious taproot $\frac{1}{4}$ to $\frac{1}{2}$ in. in diameter at the surface and extending 1 ft or more into the ground (Fig. 7-17, pt. 5).

FIGURE 9-36 *Quercus alba,* white oak. (1) Twig ×1$\frac{1}{4}$. (2, 3) Variation in leaf lobing ×$\frac{1}{2}$, see also Figure 5-1. (4) Acorn ×1. (5, 6) Barks.

Although white oak is found on many types of soil, it reaches its best development in coves or on the higher bottomlands where the soil is deep and moist, with good internal drainage. Common associates include hickories, other oaks, basswood, white ash, black cherry, yellow-poplar, and sweetgum; a complete list would include a much larger number of species (see Burns and Honkala, 1990).

Although often found in uneven-aged mixed hardwood forests, this oak, after clear-cutting, may form even-aged stands. The ability to complete many cycles of growth, declining vigor, and death

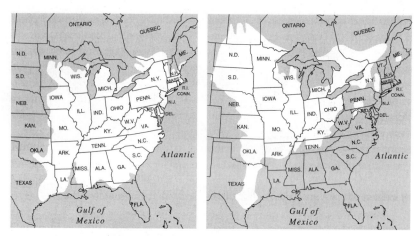

MAP 9-12 **(Left)** *Quercus alba.* **(Right)** *Quercus macrocarpa.*

of the aerial portions of a seedling while its roots increase both in size and stored energy are characteristics of this and most other oaks. This is in part responsible for much successful oak regeneration and the maintenance of these only moderately shade tolerant species in so many climax forest types. These stool and small-stump sprouts can respond dramatically to release and produce trees of excellent form and quality. Sprout growth from high stumps is often affected by heart rot.

White oak is often a prolific seeder, but good seed years do not occur regularly, and locally several years (4 to 10) may pass without any crop whatever. Like many of the white oak group, the acorns of this species germinate in the fall soon after they are released. This is a disadvantage northward because the roots often do not have time to penetrate the soil before they are frozen. Nearly 200 species of wildlife eat the acorns. White oak is highly preferred by gypsy moth.

The growth of white oak is not fast, and like many slow-growing trees it may attain a considerable age, reaching a maximum of about 450 years. This species, like a number of other oaks, is intermediate in tolerance, with a tendency to become more intolerant with age.

White oak is a majestic, large, shade tree but is seldom planted due to its slow growth and difficulty of transplanting. The wood is used for furniture, flooring, cabinets, veneer, shakes, railroad ties, and tight cooperage. White oak casks for ageing whiskey and many wines are well known, and the discarded ones are often cut in half and used as planters.

White oak is a wide-ranging species in terms of latitude and elevation and is quite variable in leaf shape (see Fig. 5-1), acorn size, and other features. Much of this variation is intrinsic (Baranski, 1975), but added variation is due to introgression from the other associated species of the white oak section (Hardin, 1975).

White oak is the state tree of Connecticut, Illinois, and Maryland.

Quércus macrocárpa Michx.   bur oak, mossycup oak

Distinguishing Characteristics (Fig. 9-37)

Leaves deciduous, 4 to 12 in. (10 to 30 cm) long, 3 to 6 in. (8 to 15 cm) wide; *shape* obovate to oblong; *margin* entire, 5- to 9-lobed, the two center sinuses usually reaching nearly to the midrib; *apex* rounded; *base* cuneate; *surfaces* dark green and lustrous above, pale and pubescent below; *fall color* yellow or brown.

Fruit solitary, usually stalked, variable in size; *nut* $\frac{3}{4}$ to 2 in. (2 to 5 cm) long, broadly ovoid, downy at the apex, enclosed one-half or more in a deep cup which is conspicuously fringed on the margin.

Twigs stout, yellowish brown, usually pubescent; after the second year often with conspicuous corky ridges; *terminal buds* $\frac{1}{8}$ to $\frac{1}{4}$ in. (3 to 6 mm) long, mostly obtuse, usually tawny pubescent.

Bark similar to that of *Q. alba* but darker and more definitely ridged vertically.

Range

Northeastern and central (Map 9-12). Local disjuncts in several states. Elevation: 300 to 3000 ft. Dry to mesic ravines, ridges, flatwoods, savannas, bottomlands, or upland on calcareous soils.

FIGURE 9-37 *Quercus macrocarpa,* bur oak. (1) Twig ×1$\frac{1}{4}$. (2) Leaf ×$\frac{1}{2}$. (3) Acorn ×$\frac{3}{4}$. (4) Bark.

General Description

Bur oak is a moderately shade tolerant, medium-size to large tree 70 to 80 ft tall and 2 to 3 ft dbh (max. 125 by 7 ft) with a massive trunk and broad crown of stout branches. On dry upland soils, the root system is remarkably deep; the taproots of 8-year-old saplings may be more than 14 ft in length. Over its very extensive range, this oak is found mixed with numerous hardwoods on many types of soil from sandy plains to moist alluvial bottoms. Bur oak is noted for its resistance to drought; invading the grasslands of the prairies and Great Plains, it may form extensive open, parklike stands. It can live to 440 years old.

Besides its value as a timber producer, the broad crown, distinctive leaves, fringed acorns, and corky twigs make it a popular ornamental; it is also more resistant to city smoke and gas injury than most other oaks.

Quércus lyràta Walt.   overcup oak

Distinguishing Characteristics (Figs. 9-38 and 9-39)

Leaves deciduous, 5 to 8 in. (13 to 20 cm) long, $1\frac{1}{2}$ to 4 in. (4 to 10 cm) wide; *shape* oblong-obovate; *margin* entire and very variable, mostly with 5 to 9 lobes separated by broad irregular sinuses; *apex* acute; *base* cuneate; *surfaces* dark green and glabrous above, green and nearly glabrous, or silvery white and downy below; *fall color* yellow, red, or brown.

Fruit solitary or paired, sessile or stalked; *nut* $\frac{1}{2}$ to 1 in. (1.2 to 2.5 cm) long, the diameter usually greater than the length, subglobose to ovoid, two-thirds to almost entirely enclosed in a deep, thin, unfringed cup.

FIGURE 9-38   *Quercus lyrata,* overcup oak. (1) Bark. (2) Twig ×1$\frac{1}{4}$. (3) Acorn ×1. (4) Leaf ×$\frac{1}{2}$.

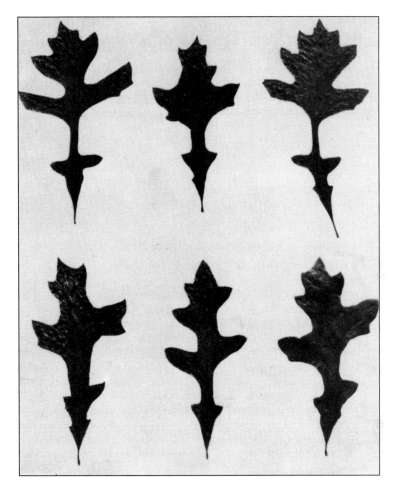

FIGURE 9-39   Leaf variation in one tree of overcup oak.

Twigs slender, gray; *terminal buds* about $\frac{1}{8}$ in. (3 mm) long; ovoid to globose, and covered with light chestnut-brown, somewhat tomentose scales; stipules often persistent near the tip of the twig.

Bark somewhat similar to that of *Q. alba,* but brownish gray and rough, with large irregular plates or ridges; trunk frequently with a twisted appearance.

Range

Southeastern (Map 9-13). Disjuncts in central Tennessee and northwestern Georgia. Elevation: sea level to 500 ft. River banks, floodplains, adjacent lower slopes, and swamp edges.

General Description

Overcup oak perhaps more nearly resembles bur oak than any other associated species because of its deeply cupped acorns and somewhat similar leaves. It is usually of poor form with a short, often

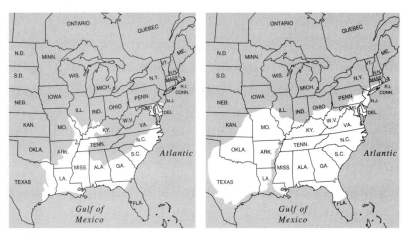

MAP 9-13   **(Left)** *Quercus lyrata.* **(Right)** *Quercus stellata.*

crooked or twisted bole and a large, open crown with crooked branches and relatively few smaller branchlets. Overcup oak may attain a height of 126 ft and a dbh of 7 ft, but it is commonly much smaller. It can reach 400 years of age.

This species is not abundant or important except in the main bottomlands of the lower Mississippi and the lower valleys of its tributaries. Here overcup oak is one of the most common swampland trees growing on wet, poorly drained clay soils, and it is better adapted to withstand prolonged inundation than most of the accompanying species. Common associates are those species common to these wet, southern bottomlands. It is moderately shade tolerant.

Overcup acorns are often disseminated by flood waters and, unlike most others of the white oaks, may remain dormant until spring when the waters recede. Its seedlings, however, may be killed by prolonged inundation. Although overcup oak usually produces timber of poor quality, it merits consideration because of its abundance over large areas and its ability to grow on the poorest of the bottomland sites.

Quércus stellàta Wangenh.   post oak

Distinguishing Characteristics (Fig. 9-40)

Leaves deciduous, thick, somewhat leathery, 4 to 6 in. (10 to 15 cm) long, 3 to 4 in. (7.5 to 10 cm) wide; *shape* oblong-obovate; *margin* entire and usually deeply 5-lobed, the two middle lobes squarish, nearly opposite and at right angles to the midrib, giving the leaf a cruciform appearance; *apex* rounded; *base* cuneate; surfaces dark green with scattered stellate-pubescence above, tawny tomentose below; *fall color* brown, yellow, or red.

Fruit solitary or paired, sessile or short-stalked; *nut* $\frac{1}{2}$ to 1 in. (1.2 to 2.5 cm) long, ovoid-oblong, sometimes slightly striped, smooth or pubescent at the apex; enclosed for about one-third of its length in a bowl-shaped cup with thin scales (less often warty like those of white oak).

Twigs stout, somewhat tawny-tomentose; *terminal buds* about $\frac{1}{8}$ in. (3 mm) long, subglobose to broadly ovoid, covered with chestnut-brown pubescent scales.

FIGURE 9-40 *Quercus stellata,* post oak. (1) Twig ×1¼. (2) Acorn ×1. (3) Leaf ×½. (4) Bark.

Bark similar to that of *Q. alba* but darker often reddish brown and with more definite longitudinal ridges and less scaly.

Range
Southeastern (Map 9-13). Elevation: sea level to 3000 ft. Upland dry, sandy, gravelly, or rocky ridges to more mesic flatwoods.

General Description
This is a small- to medium-sized tree 40 to 50 ft tall and 12 to 24 in. dbh (max. 105 by 5 ft) but may attain only shrubby proportions under adverse conditions. The crown usually consists of fewer large branches than in white oak, and they are characteristically gnarled in appearance. The associates of post oak on the so-called barrens or in sand-hill country include black, blackjack, and scarlet oaks, upland hickories, pitch and shortleaf pines, and eastern redcedar. Post oak is an intolerant species of very slow height growth that survives and reproduces in part because of its great drought resistance. It can live to nearly 400 years old. Post oak is susceptible to the chestnut blight fungus (*Cryphonectria parasitica*), which has caused considerable damage in some localities. Acorns are a very important source of food for wildlife. Post oak is often planted as an ornamental shade tree.

The common name *scrub oak* is frequently used collectively for several species of oaks (*Q. stellata, Q. margaretta, Q. laevis, Q. marilandica, Q. incana*) that form an often dense, scrubby layer under a canopy of longleaf or slash pine, particularly when fire is excluded for many years.

On the driest, deepest sands or occasionally loamy soil is another species, closely related to post oak, ***Q. margarétta*** Ashe ex Small or sometimes considered as *Q. stellata* var. *margaretta* (Ashe ex Small) Sarg. (Little, 1979). Sand post oak is distinguished by glabrous or puberulent twigs, smaller leaves with lobes more rounded and less cruciform, acorn cup scales generally thinner, and winter buds larger and more acute. It is found from southeastern Virginia south to Florida, west to central Texas, and north to Missouri.

In the lower Mississippi Valley on the silty loam soils of the second bottoms or terraces is the closely related delta post oak, *Q. símilis* Ashe (*Q. stellata* var. *paludòsa* Sarg.) with puberulent twigs and less cruciform leaves. The usual associates of this bottomland form include several other oaks, sweetgum, elms, ashes, hickories, and occasionally loblolly pine or spruce pine.

Quércus douglásii Hook. and Arn.   blue oak, mountain white oak

Distinguishing Characteristics. *Leaves* deciduous, blue-green, oblong to oval, $1\frac{1}{4}$ to 4 in. (3 to 10 cm) long, $\frac{3}{4}$ to $1\frac{3}{4}$ in. (2 to 4.5 cm) wide, with entire, sinuate and/or toothed to lobed margins. *Fruits* $\frac{3}{4}$ to $1\frac{1}{4}$ in. (2 to 3 cm) with shallow, bowl-like cups (Fig. 9-41).

Range. Northern to southern California in the Coast Ranges and Sierra Nevada. Elevation: 3000 to 3500 ft. Low valleys and foothills where it forms pure groves or mixed forests with Digger pine and several other oaks.

General Description. Mature trees may exceed a height of 80 ft and dbh of 30 in. (max. 94 ft by $6\frac{1}{3}$ ft), but they are usually much smaller. The geographical range has been reduced by clearing in recent years, and the remaining woodlands show poor reproduction. It is used locally for fuel and fenceposts.

Quércus garryàna Dougl. ex Hook.   Oregon white oak

Distinguishing Characteristics (Fig. 9-42)

Leaves deciduous, 3 to 6 in. (7.5 to 15 cm) long, 2 to 4 in. (5 to 10 cm) wide; *shape* oblong to obovate; *margin* entire, with 5 to 7 rounded lobes separated by moderately deep sinuses, often more or less revolute; *surfaces* lustrous dark green and glabrous above, pale green and with orange-brown pubescence below; *petiole* ca. $\frac{3}{4}$ in. (18 mm) long, stout, pubescent; *fall color* red.

Fruit solitary or paired, sessile or short-stalked; *nut* 1 to $1\frac{1}{4}$ in. (2.5 to 3 cm) long, ovoid to obovoid, light brown, enclosed only at the base in a shallow bowl-like cup with pubescent or tomentose free-tipped scales.

Twigs stout, orange-red, sparsely to densely pubescent during the first winter; *buds* $\frac{1}{4}$ to $\frac{1}{2}$ in. (6 to 13 mm) long, covered with rusty-brown tomentum or sparsely glandular-pubescent.

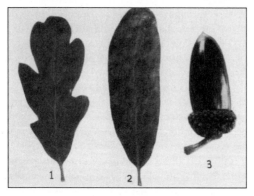

FIGURE 9-41   *Quercus douglasii,* blue oak. (1, 2) Leaf variation ×1. (3) Acorn ×1.

FIGURE 9-42   *Quercus garryana,* Oregon white oak. (1) Twig ×1$\frac{1}{4}$.
(2) Leaf ×$\frac{1}{2}$. (3) Acorn ×1.

FIGURE 9-43   Bark of Oregon white oak.

Bark (Fig. 9-43) smooth on young stems; on old trunks broken up into thin, gray-brown or gray scaly ridges, separated by narrow, shallow fissures.

Range

Northwestern (Map 9-14). Elevation: sea level to 3000 ft northward, 1000 to 5000 ft southward. Dry, rocky slopes or alluvial soils. It extends farther south in the Sierra Nevada than indicated on the map (Nixon, 1997).

MAP 9-14   *Quercus garryana.*

General Description

Oregon white oak, indigenous to the Pacific Northwest, is a shrub or medium-sized tree seldom more than 50 to 70 ft tall and 2 to 3 ft dbh (max. 107 by 8 ft). The bole, even under forest conditions, is usually short and rather crooked. It rises above a well-developed lateral root system and merges into a wide-spreading, round-topped crown. Oregon white oak grows on almost any kind of soil, commonly on hot, dry, grassy but rocky slopes in a region where rainfall averages 35 in. annually. Best development is attained on the alluvial soils of Washington and Oregon. Here, limited pure stands are found, but the tree usually occurs as an individual or in small groves mixed with such species as Douglas-fir, ponderosa and Digger pines, California black oak, and Pacific madrone.

Large crops of seeds are produced every few years, but reproduction by this means is usually poor, owing to the inability of the germinating seeds to penetrate the heavy sods upon which they are so frequently disseminated. Stump sprouts and root suckers are vigorous, and these afford the best means of natural regeneration. Growth is slow. Trees 3 ft dbh may be 250 years old, but a life span of 500 years may be attained. This oak is usually rated as intolerant, and eventually dies when overtopped by Douglas-fir.

It is the most important of the western oaks and used for furniture, construction lumber, cabi-network, interior trim, fence posts, locally for fuel, and as an ornamental.

Three varieties have been recognized (Nixon, 1997) based on habit and pubescence.

Quércus lobàta Née   valley oak, California white oak

Distinguishing Characteristics.   *Leaves* deciduous, 2 to 4 in. (5 to 10 cm) long, $1\frac{1}{4}$ to $2\frac{1}{2}$ in. (3 to 6 cm) wide, oblong, elliptical, or obovate, 7- to 11-lobed, both surfaces finely pubescent. *Fruit* $1\frac{1}{4}$ to $2\frac{1}{4}$ in. (3 to 6 cm) long, elongated-conic, one-third enclosed in a deep bowl-like cup, the cup scales free at their tips and forming a fringe about the cup. *Twigs* slender, pubescent. *Bark* with the appearance of alligator leather.

Range.   Northern to southern California, from the Sierra Nevada Mountains to the Coast Ranges. Elevation: mostly below 2000 ft, but occasionally to 5000 ft. Valleys and slopes.

General Description.   This white oak is largely restricted to low valley sites in western California. It is the largest of California oaks and attains a height of 90 to 125 ft or more and a dbh of 3 to 5 ft (max. 160 by 12 ft). Through the central portion of the state, it forms nearly pure open groves. The bole is usually short, but massive, and it merges into a large widespread hemispherical to subglobose crown with the ends of the branches drooping. Best development is reached on deep rich loams, although the tree also occurs on poor, dry, sandy, and gravelly sites.

This tree grows with greater rapidity than any other California oak and can attain large size in a comparatively short time. The oldest trees are commonly defective, heart rot being quite prevalent, and hence are of little value except for fuel. Ring counts of large stumps indicate that an age of more than 200 to 250 years is rarely attained. Due to large-scale clearing, the range has been dramatically reduced in recent years. Nearly 800 references to this and other species of California oaks are listed by Griffin et al. (1987).

Quércus gambélii Nutt.   Gambel oak, Rocky Mountain white oak

Distinguishing Characteristics.   *Leaves* deciduous, elliptical or oblong, rounded at apex, acute at base, deeply 5- to 9-lobed, 2 to 6 in. (5 to 15 cm) long, $1\frac{1}{4}$ to $3\frac{1}{4}$ in. (3 to 8 cm) wide; shiny dark green above, paler below with soft pubescence; fall color yellow or reddish. *Fruit* short-stalked, $\frac{1}{2}$ to $\frac{7}{8}$ in. (1.2 to 2.1 cm) long, oval, with thick, scaly cup enclosing $\frac{1}{4}$ to $\frac{1}{3}$ of the nut.

Range.   Northern Utah, east to southern Wyoming and western Oklahoma, south to western Texas, west to southern Arizona; local disjunct in southern Nevada. Elevation: 5000 to 8000 ft. Slopes and valleys in the foothills.

General Description.   Tree 20 to 70 ft tall with rounded crown forming dense groves, or a thicket-forming shrub on dry slopes and valleys, in mountains, foothills, and plateaus. It is the common oak in the Rockies and a frequent associate of ponderosa pine. It was named for its discoverer, William Gambel, a naturalist from Philadelphia who made important collections in the southern Rockies in the 1840s.

Chestnut Oaks

Deciduous; leaves crenate, coarsely toothed, or sinuate; eastern species (Table 9-14).

Quércus michaùxii Nutt. (*Quercus prìnus* L. sensu some authors)   swamp chestnut oak, cow oak, basket oak

Distinguishing Characteristics (Fig. 9-44)

Leaves deciduous, 4 to 9 in. (10 to 23 cm) long, 2 to $5\frac{1}{2}$ in. (5 to 14 cm) wide; *shape* broadly obovate to oblong-obovate; *margin* deeply crenate to coarsely dentate, often with large glandular-tipped teeth; *apex* acute or slightly obtuse; *base* cuneate; *surfaces* dark green and lustrous above, pale green to silvery white and felty pubescent below with erect hairs; *fall color* dark red or brown.

TABLE 9-14   COMPARISON OF IMPORTANT CHESTNUT OAKS

Species	Leaves	Acorns
Q. michauxii swamp chestnut oak	Obovate, crenate-dentate; felty pubescent below	Cup thick, the scales wedge-shaped, and distinct, tips thick and silky-tomentose
Q. montana chestnut oak	Elliptical to obovate, crenate; finely pubescent below	Cup thin, the scales more or less fused, tips thin, reddish, and glabrous

FIGURE 9-44   *Quercus michauxii,* swamp chestnut oak. (1) Twig ×1$\frac{1}{4}$. (2) Acorn ×1. (3) Leaf ×$\frac{1}{2}$. (4) Bark.

Fruit solitary or paired, subsessile or usually stalked; *nut* 1 to $1\frac{1}{2}$ in. (2.5 to 4 cm) long, ovoid to oblong, enclosed for not more than a third of its length in a thick bowl-shaped cup, with distinct, somewhat wedge-shaped scales with the tips thickened and silky tomentose.

Twigs moderately stout, reddish brown to orange-brown; *terminal buds* $\frac{1}{4}$ in. (6 mm) long, acute, covered with thin, red, puberulous scales, pale on the margin.

Bark irregularly furrowed or scaly, ashy gray, tinged with red; freshly cut surfaces reddish brown.

Range

Southeastern (Map 9-15). Disjuncts in northeastern Tennessee and southern Kentucky. Elevation: sea level to 1000 ft. Bottomlands, ravine slopes, flatwoods over limestone.

General Description

Swamp chestnut oak is probably the most important species of this group. It is a well-formed tree with a straight massive trunk and narrow crown, averaging 60 to 80 ft tall and 2 to 3 ft dbh (max. 152 by 9 ft). This oak, at least in the lower Mississippi Valley, is somewhat irregular in occurrence, but when present, it is frequently abundant.

Swamp chestnut oak is intolerant and is found on the best, relatively well-drained, loamy ridges and silty clay and loamy terraces in bottomlands. Principal associates are cherrybark, Shumard, white, and delta post oaks and blackgum; other bottomland hardwoods are also present. Early growth is limited more by available moisture than by light.

The quality of the wood is second only to that of the best white oaks, and for this reason, the tree has been intensively cut in many localities; the name "basket oak" undoubtedly originated from its use for basket splits.

See the discussion of the name *Q. prinus* under *Q. montana*.

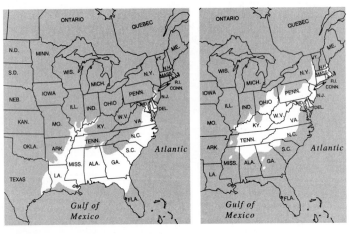

MAP 9-15   **(Left)** *Quercus michauxii*. **(Right)** *Quercus montana*.

Quercus montána Willd. (*Quércus prìnus* L.)   chestnut oak, rock oak

Distinguishing Characteristics (Fig. 9-45)

Leaves deciduous, 4 to 8 in. (10 to 20 cm) long, $1\frac{1}{2}$ to 4 in. (4 to 10 cm) wide; *shape* obovate to elliptical, or nearly lanceolate; *margin* coarsely crenate, or repand-crenate; *apex* rounded; *base* cuneate; *surfaces* yellowish green and lustrous above, paler and often finely pubescent below with appressed stellate hairs throughout and erect hairs on veins; *fall color* yellow to orange.

Fruit solitary or paired, subsessile or stalked; *nut* $\frac{3}{4}$ to $1\frac{3}{8}$ in. (2 to 3.5 cm) long, ovoid to ellipsoidal, very lustrous, enclosed at the base, or more often from one-third to one-half of its length, in a thin cup with somewhat fused scales with thin, reddish, and glabrous tips.

Twigs orange to reddish brown; *terminal buds* $\frac{1}{4}$ in. (6 mm) long, usually acute or acuminate, covered with bright chestnut-brown scales.

Bark brownish gray to nearly black, on older trees very deeply and coarsely furrowed.

Range

Northeastern (Map 9-15). Elevation: 1500 to 5000 ft. Xeric, rocky ridges and upper slopes, also more mesic sites.

FIGURE 9-45   *Quercus montana* (*Q. prinus*), chestnut oak. (1) Twig ×$1\frac{1}{4}$. (2) Leaf ×$\frac{1}{2}$. (3) Acorn cup showing thin wall ×1. (4) Acorn ×1. (5) Bark.

General Description

Chestnut oak is a moderately tolerant, medium-sized tree 50 to 60 ft tall and 2 ft dbh (max. 150 by 7 ft). It is found typically on poor dry upland rocky sites where it may form pure stands. On somewhat better soils, common associates include white and pitch pines, eastern redcedar, white, scarlet, and black oaks, sweetgum, red maple, and numerous other hardwoods. It can reach 400 years of age.

Chestnut oak can produce many stump and root collar sprouts that are initially capable of very rapid growth, and much of its regeneration on cutover or burned land is of sprout origin. Best growth is made in well-drained coves and other moist sites. The bark of chestnut oak is relatively high in tannin. Its foliage is highly preferred by gypsy moth.

The name *Q. prinus* L. is now completely confused in the botanical and ecological literature because it has been used for both chestnut oak and swamp chestnut oak for many years by various authors. It is safer to use *Q. montana* and *Q. michauxii* for these, or at least give the common name when using *Q. prinus* (Hardin, 1979; Nixon, 1997).

Quércus bicolor Willd.   swamp white oak

Distinguishing Characteristics (Fig. 9-46). *Leaves* deciduous, 4 to 7 in. (10 to 18 cm) long, 2 to $4\frac{1}{2}$ in. (5 to 11 cm) wide, obovate, shallowly lobed or coarsely toothed; fall color red. *Fruit* $\frac{3}{4}$ to $1\frac{1}{4}$ in. (2 to 3 cm) long, usually paired, borne on a slender peduncle. *Twigs* straw-brown, dull, terminal buds $\frac{1}{16}$ to $\frac{1}{8}$ in. (2 to 3 mm) long, orange-brown, essentially glabrous. *Bark* on the upper limbs peeling off in ragged, papery scales; deeply furrowed below into flat scaly ridges or blocky, dark brown.

FIGURE 9-46   *Quercus bicolor,* swamp white oak. (1) Tip of twig ×1$\frac{1}{4}$. (2) Leaf ×$\frac{1}{2}$. (3) Acorn ×1. (4) Bark.

Range.   Southern Ontario, east to southern Quebec and southern Maine, south to northern Virginia, west to Missouri, north to southeastern Minnesota; local disjuncts in North and South Carolina, Tennessee, southern Missouri, and northeastern Kansas. Elevation: sea level to 1000 ft, locally to 2000 ft. Low, wet bottomlands, edge of swamps and streams.

General Description.   Swamp white oak is a moderately shade tolerant, medium-sized tree 60 to 70 ft tall and 2 to 3 ft dbh (max. 100 by 7 ft), often with a poorly pruned bole and irregular crown. This species is not abundant except locally where it occurs along stream banks or on moist or peaty flats and along swamp margins. It is highly tolerant of poorly aerated soils. When present at all, the supply may be large, and even though the quality of the timber is often inferior, it is harvested and used with that of the other white oaks. It is an excellent, large shade tree, tolerating very wet and dry soils, and park, home, and urban settings.

Quércus muehlenbérgii Engelm.   chinkapin oak

Distinguishing Characteristics (Fig. 9-47).   *Leaves* deciduous, 4 to 6 in. (10 to 15 cm) long, $1\frac{1}{2}$ to 3 in. (4 to 7.5 cm) wide, obovate to oblong-lanceolate, coarsely serrate with glandular-tipped teeth; fall color reddish. *Fruit* sessile or short stalked, $\frac{1}{2}$ to 1 in. (1.2 to 2.5 cm) long, with a thin bowl-shaped cup. *Twigs* slender, orange-brown, terminal buds $\frac{1}{8}$ in. (3 mm) long, orange-brown. *Bark* ashy gray, more or less rough and flaky.

Range.   Southern Michigan and southern Ontario, east to western Vermont, south to northwestern Florida, west to central Texas, north to Iowa and southern Wisconsin; local disjuncts in western Texas and southeastern New Mexico. Elevation: 400 to 3000 ft. Limestone outcrops and calcareous soils, dry slopes or bluffs, moist bottomlands, and river banks.

General Description.   This shade-tolerant oak is rare over most of its range and hence is of little commercial value. It is usually a small- to medium-sized tree found on dry limestone out-

FIGURE 9-47   *Quercus muehlenbergii*, chinkapin oak. (1) Twig ×1$\frac{1}{4}$. (2) Acorn ×1. (3) Leaf ×$\frac{1}{2}$. (4) Bark.

crops and alkaline soils. In the Ohio and Mississippi Valleys, it is found on moist soils in the higher bottoms; here it averages 60 to 80 ft tall and 2 to 3 ft dbh (max. 160 by 6 ft). It can reach 400 years of age.

Live Oaks

Evergreen; leaves usually unlobed; eastern and western species of the white oak section.

Quércus virginiàna Mill.   live oak

Distinguishing Characteristics (Fig. 9-48)

Leaves persistent until the following spring; $\frac{3}{4}$ to 5 in. (2 to 13 cm) long, $\frac{3}{8}$ to 2 in. (1 to 5 cm) wide; *shape* oblong, elliptical or rarely obovate; *margin* usually entire (sometimes sparingly toothed) and very slightly revolute ; *apex* obtuse; *base* cuneate or acute; *surfaces* dark glossy green above, paler below and densely covered with fused-stellate hairs that are closely appressed to the surface and barely obvious even with a hand lens.

Fruit usually in clusters of 2 to 5 on a peduncle of varying length; maturing in one season; *nut* $\frac{5}{8}$ to 1 in. (1.5 to 2.5 cm) long, dark brown to nearly black, ellipsoidal, one-third enclosed in a turbinate cup; seed sweet.

Bark medium to dark gray, deeply furrowed, with short, rounded, rough ridges, becoming blocky.

Range

Southeastern (Map 9-16). Local disjuncts in southern Oklahoma. Elevation: sea level to 300 ft and to 2000 ft in Texas. Mixed low woodlands, hammocks, borders of salt marshes, coastal dunes, and maritime forests.

FIGURE 9-48   *Quercus virginiana,* live oak. (1) Leaves $\times \frac{1}{2}$. (2) Acorns ×1. (3) Bark.

MAP 9-16   *Quercus virginiana*

General Description

This species is generally considered to be the only important "live oak" of eastern North America, although laurel oak holds its leaves so late in the spring that they are almost evergreen. Live oak is usually a medium-sized tree 40 to 50 ft tall and 3 to 4 ft dbh. The short, enlarged, buttressed trunk, sometimes 6 to 8 ft dbh (max. 78 by 11 ft), divides into three or four wide-spreading, horizontal branches; the crown is close, round-topped, and in the largest open-grown trees may be upward of 125 to 150 ft across (Fig. 9-49).

Live oak is a moderately tolerant, dry sand-plain species, but it is found on a variety of sites including hammocks. On moist locations, some common associates are southern magnolia, redbay, sweetgum, American holly, laurel and water oaks, and hickories.

Live oak is extremely difficult to kill. When a tree is cut or girdled, numerous sprouts spring up from the root collar and surface roots. When these sprouts are cut, many more appear. Live oak decline, a wilt disease, can be a serious problem.

When grown ornamentally, especially on rich soils, growth is fast. Although very large trees are supposed to be many centuries old, studies have shown that the largest specimens now standing generally are about 200 to 300 years of age.

Certain individual trees yield acorns much sweeter and more edible than others, and it is said that the Native Americans produced an oil somewhat comparable to olive oil from them.

The wood is exceedingly strong and one of the heaviest of native woods. During the era of "wooden ships and iron men" it was utilized in building frigates and other naval vessels. The

FIGURE 9-49   Live oak showing typical spreading form of an old, open-grown tree. (*Courtesy of U.S. Forest Service.*)

earliest planting and intensive management of forest trees for timber in North America centered around this naval demand for ships' knees from live oaks.

Live oak is the state tree of Georgia.

Quércus geminàta Small   sand live oak

The sand live oak, sometimes considered *Q. virginiana* var. *geminàta* (Small) Sarg., is distinguished from *Q. virginiana* by a more revolute leaf margin, the lower surface with more raised and conspicuous fused-stellate hairs, reticulate venation conspicuous above, and flowering 2 to 3 weeks later than *Q. virginiana* in the same area. It is the common form of live oak on coastal dunes along the Atlantic and Gulf coasts and inland areas in central Florida on light-colored sands. It occurs from southeastern North Carolina to central Florida and west to southeastern Louisiana.

Quércus arizónica Sarg.   Arizona white oak, Arizona oak

This tree with persistent, spinose, blue-green leaves (Fig. 9-50), $1\frac{1}{2}$ to 3 in. (4 to 7.5 cm) long, and plated bark is common to moist benches and canyon walls in the Southwest. Under favorable growing conditions, it attains a height of 40 to 60 ft and a dbh of 24 to 36 in., and develops a short trunk that supports a massive round-topped crown. It is usually dwarfed at timberline. Arizona oak ranges through southern New Mexico, Arizona, western Texas, and northern Mexico between elevations of 5000 to 10,000 ft.

Quércus oblongifòlia Torr.   Mexican blue oak

This is a small tree rarely attaining a height of more than 30 ft and a dbh of 30 in. It occurs in association with *Q. arizonica* and *Q. emoryi,* and its foliage (Fig. 9-50), 1 to 2 in. (2.5 to 5 cm) long, usually entire and blue-green, is a source of forage for range stock. It is somewhat more restricted in elevation, however, and rarely occurs over 6000 ft. This southwestern tree is found in extreme western Texas, southwestern New Mexico, southeastern Arizona, and into northern Mexico.

FIGURE 9-50   Leaves of southwestern evergreen oaks $\times\frac{1}{2}$. (1) *Q. emoryi,* Emory oak. (2) *Q. arizonica,* Arizona white oak. (3) *Q. oblongifolia,* Mexican blue oak.

True Red and Black Oaks

Leaves deciduous or tardily so, lobed with bristle tips; eastern and western species (Table 9-15).

Quércus rùbra L.   northern red oak

Distinguishing Characteristics (Fig. 9-51)

Leaves deciduous; 4 to 9 in. (10 to 23 cm) long, 3 to 6 in. (7.5 to 15 cm) wide; *shape* oblong to obovate; *margin* with 7 to 11 often toothed lobes separated by regular U-shaped sinuses extending about halfway to the midrib; *apex* acute to acuminate; *base* obtuse to broadly cuneate;

TABLE 9-15 COMPARISON OF IMPORTANT TRUE RED AND BLACK OAKS

Species	Leaves	Acorns
Q. rubra northern red oak	Oblong-obovate, 7- to 11-lobed, with narrow, regular U-shaped sinuses; glabrous except for axillary tufts below	Variable, but mostly subglobose, $\frac{5}{8}$–$1\frac{1}{8}$ in. (1.5–2.8 cm) long; usually enclosed only at base by a flat, thick, saucerlike cup
Q. velutina black oak	Obovate to ovate, 5- to 7-lobed, with sinuses of varying depth; more or less scurfy and yellowish to copper-colored below, axillary tufts white to rufous	Ovoid-oblong, $\frac{5}{8}$–$\frac{3}{4}$ in. (1.5–1.9 cm) long; nut $\frac{1}{3}$ to $\frac{1}{4}$ enclosed in a bowl-like cup; cup scales dull (not shiny), free at tips
Q. shumardii Shumard oak	Obovate to oval, 7- to 9-lobed, with many bristle tips and moderately deep sinuses; glabrous except for axillary tufts below	Oblong-ovoid, tapering to apex, $\frac{5}{8}$–$1\frac{1}{8}$ in. (1.5–2.8 cm) long; nut enclosed only at base by a thick, saucerlike cup
Q. falcata southern red oak	Heteromorphic; lower crown leaves usually only 3-lobed at apex; middle and upper crown leaves 5- to 7-lobed; base U-shaped; grayish or rusty pubescent below	Subglobose, $\frac{1}{2}$–$\frac{5}{8}$ in. (1.2–1.5 cm) long; nut enclosed only at base by a flat, saucerlike cup
Q. pagoda cherrybark oak	More or less uniformly 5- to 11-lobed, the lobes pointing outward; base cuneate	Same as above
Q. coccinea scarlet oak	Ovate to oval, deeply 5- to 9-lobed, the sinuses nearly circular; glabrous except for axillary tufts below	Depressed-globose, $\frac{1}{2}$–1 in. (1.2–2.5 cm) long, usually with several concentric rings at the apex; nut $\frac{1}{2}$ enclosed in a bowl-like cup with shiny, appressed scales
Q. palustris pin oak	Obovate to oval, usually 5-lobed, the sinuses deep and broadly U-shaped or squarish, the central lobes nearly perpendicular to midrib; pale and glabrous except for axillary tufts below	Hemispherical, $\frac{3}{8}$–$\frac{5}{8}$ in. (9–15 mm) long; nut enclosed only at base by a flat, saucerlike cup with free-tipped scales
Q. texana Texas red oak	Obovate, usually 5- to 7-lobed, the sinuses deep; pale and glabrous except for axillary tufts below	Oblong-ovoid, $\frac{3}{4}$–$1\frac{1}{4}$ in. (2–3 cm) long; nut $\frac{1}{4}$–$\frac{1}{2}$ enclosed in a bowl-like cup

surfaces somewhat lustrous, or dull above, glabrous below except for occasional often inconspicuous axillary tufts; *fall color* dark red.

Fruit solitary or paired; *nut* $\frac{5}{8}$ to $1\frac{1}{8}$ in. (1.5 to 2.8 cm) long; variable in shape, but usually subglobose, enclosed at the base in a flat, thick, saucerlike cup, the kernel white; *cup scales* pubescent, their tips appreciably darkened.*

Twigs moderately stout, greenish brown to reddish brown, glabrous; *terminal buds* $\frac{1}{4}$ in. (6 mm) long, ovoid, pointed, terete, covered by numerous reddish brown, more or less hairy scales.

*Figure 9-51, pts. 3 and 4, show the variation to be expected in acorns from different trees. Previously, trees bearing the shallow-cup type were designated as a separate variety. Because intergradations are found, and the two forms do not appear to breed true, there seems to be no good reason for continuing to keep them separate.

FIGURE 9-51   *Quercus rubra,* northern red oak. (1) Twig ×1$\frac{1}{4}$. (2) Acorns at end of first season ×2. (3, 4) Variation in mature acorns ×1. (5) Leaf ×$\frac{1}{2}$. (6) Bark of young tree. (7) Bark of old tree.

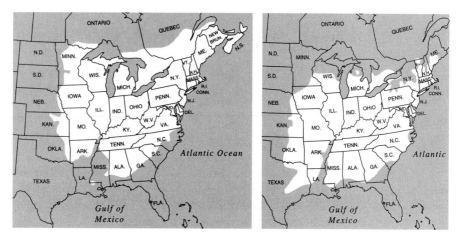

MAP 9-17   **(Left)** *Quercus rubra.* **(Right)** *Quercus velutina.*

Bark smooth on young stems, greenish brown; ultimately brown to nearly black and broken up into wide, flat-topped ridges, separated by shallow fissures; on very old trees more narrowly ridged or corrugated; inner bark light red or brownish red.

Range
Eastern (Map 9-17). Elevation: sea level to 5500 ft. Mesic, well-drained slopes, coves, and drier ridges.

General Description
This species is probably the most important ornamental and timber tree of the genus. It is also one of the most widespread of northern oaks, and is a medium-sized to large tree 60 to 80 ft tall and 2 to 3 ft dbh (max. 140 by 8 ft). Although developing a short massive trunk with an extensive crown in the open, this tree when grown in the forest produces a tall, straight, columnar bole that supports a small rounded crown. The root system, although deep like that of many other oaks, is often lacking in a well-developed and persistent taproot that develops less rapidly than the laterals after the seedling stage.

Northern red oak may be found on sandy loam soils in mixture with other northern hardwoods and white pine. Best development, however, is usually made on fine-textured, well-watered soils with good surface drainage, associated with such species as basswood, black cherry, ashes, hickories, other oaks, yellow-poplar, and sweetgum. Northern red oak forms nearly pure stands at high altitudes in the southern Appalachians, but it is also listed as a component in more than 20 forest-cover types, and therefore the number of associated species is very large. On a tolerance scale, this oak, like many others, is intermediate.

Good seed crops occur at 2- to 5-year intervals. Germination of red oak acorns does not take place until the following spring, and many of the acorns are destroyed by insects or eaten by many different animals. Growth of seedlings is fast, and on the best soils nursery stock sometimes attains a height of nearly 5 ft the first season. This oak is also a prolific sprouter, both as a seedling and as a young tree, and quickly restocks cutover areas by this means. Although not so long-lived as white oak, this species may live up to two to three centuries.

Northern red oak is widely used as an ornamental because of comparative ease in transplanting, the autumnal coloration of the leaves, and the symmetrical form of the tree. The wood is used for flooring, furniture, railroad ties, posts, and pulp.

Northern red oak is the state tree of New Jersey and the provincial tree of Prince Edward Island.

Quércus velùtina Lam.   black oak

Distinguishing Characteristics (Fig. 9-52)

Leaves deciduous; 4 to 9 in. (10 to 23 cm) long, 3 to 6 in. (7.5 to 15 cm) wide; *shape* obovate to ovate; *margin* with 5 to 7 often toothed lobes separated by oval or U-shaped sinuses of variable depth*; *apex* acute to acuminate; *base* broadly cuneate to nearly truncate, *surfaces* exceedingly lustrous and dark green above, yellow-green to coppery-colored below, and more or less scurfy pubescent with conspicuous axillary tufts; *fall color* dark red.

Fruit solitary or paired; *nut* $\frac{5}{8}$ to $\frac{3}{4}$ in. (1.5 to 1.9 cm) long, ovoid to hemispherical, often striated, light red-brown, one-fourth to one-third enclosed in a deep bowl-like cup, the kernel yellow; *cup scales* loosely imbricated above the middle, dull chestnut-brown, tomentose.

Twigs stout, reddish brown, glabrous; *terminal buds* $\frac{1}{4}$ to $\frac{1}{2}$ in. (6 to 12 mm) long, ovoid, sharp-pointed, angular, covered with numerous, hoary tomentose scales.

Bark thick, nearly black on old stems, deeply furrowed vertically, and with many horizontal breaks; inner bark bright orange-yellow and bitter to the taste.

Range

Eastern (Map 9-17). Elevation: sea level to 5000 ft. Xeric, upland ridges and slopes, also submesic woodlands.

General Description

Black oak, a tree 50 to 60 ft tall and 2 to 3 ft dbh (max. 150 by 8 ft), is one of the most common of eastern upland oaks. Although making its best growth on moist, rich, well-drained soils, where it approaches red oak in stature, it is sensitive to competition on such sites and is found more often on poor, dry, sandy, or heavy glacial clay hillsides. Black oak is featured by a deep taproot, a somewhat tapering, often limby bole, and an irregularly rounded crown. It is intermediate in tolerance (less than white and chestnut oaks, about the same as northern red and scarlet). Although pure stands may occur, this oak is usually found in mixture with a large number of other species. In the South, it is commonly found in the shortleaf pine belt and occasionally in the bottoms, where it occupies the highest and best-drained areas.

Black oak is a persistent sprouter; good seed years may occur at 2- to 3-year intervals, although as with all oaks this periodicity can be very irregular. Growth is slower than in red oak but faster than in most other associated oaks; trees over 200 years old are rarely found. Repeated defoliation by gypsy moth can kill black and numerous other oaks. Oak wilt can be a serious disease.

---

*This species illustrates a characteristic of several of the red oak group: heteromorphic foliage. The narrow-lobed form (Fig. 9-52, pt. 3) is common in the upper crown of mature trees, and the broad-lobed form (Fig. 9-52, pt. 2) is common in the lower crown or on immature trees.

FIGURE 9-52   *Quercus velutina,* black oak. (1) Twig ×1¼. (2, 3) Variation in leaf lobing ×½. (4) Variation in acorns ×1. (5) Bark.

Quércus shumárdii Buckl.   Shumard oak

Distinguishing Characteristics (Fig. 9-53)

Leaves deciduous; 3 to 8 in. (8 to 20 cm) long, 2½ to 5 in. (6 to 13 cm) wide; *shape* obovate to oval; *margin* with 5 to 9 lobes again divided near their ends and with many bristle tips, the lobes separated by moderately deep oval or U-shaped sinuses; *apex* acute to acuminate; *base* obtuse to nearly truncate; *surfaces* dark green and lustrous above, paler and glabrous below except for axillary tufts; *fall color* red or brown.

FIGURE 9-53 *Quercus shumardii,* Shumard oak. (1) Twig ×1$\frac{1}{4}$. (2) Acorn ×1. (3) Leaf ×$\frac{1}{2}$. (4) Bark (*Courtesy of U.S. Forest Service*).

Fruit solitary or in pairs; *nut* $\frac{5}{8}$ to 1$\frac{1}{4}$ in. (1.5 to 3 cm) long, oblong-ovoid, slightly tapering toward the apex, enclosed at the base in a thick, shallow, saucer-shaped cup, the kernel whitish; *cup scales* appressed, pale pubescent or nearly glabrous.

Twigs slender to moderately stout, gray to grayish or greenish brown, glabrous; *terminal buds* $\frac{1}{4}$ in. (6 mm) long, ovoid, sharp pointed, usually angled, covered by gray to gray-brown, downy, or nearly glabrous scales.

Bark on old trees very thick, broken into pale to whitish scaly ridges by deep, much darker-colored furrows.

Range

Eastern (Map 9-18). Disjuncts in southern Pennsylvania, western West Virginia, northeastern Tennessee, and northern Virginia. Elevation: sea level to 2500 ft. Mesic woodlands, hammocks, bluffs, ravines, river banks, and limestone hills.

General Description

This species, one of the largest of the southern red oaks, may attain a height of 100 to 125 ft and a dbh of 4 to 5 ft (max. 130 by 8 ft). A long, clear, symmetrical bole rises above a slightly buttressed base and moderately shallow root system, whereas the crown is usually open and wide-spreading. The tree never occurs in pure stands but rather as an occasional individual or in small groves in mixture with other hardwoods, particularly swamp chestnut, cherrybark, and delta post oaks, black

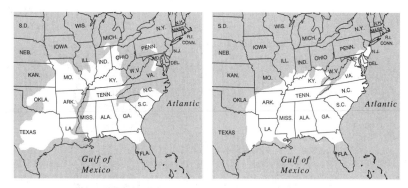

MAP 9-18   **(Left)** *Quercus shumardii.* **(Right)** *Quercus falcata.*

tupelo, green ash, and hickories. This oak will grow on a number of soil types, but it is mostly a bottomland species and develops best on river terraces and small-stream floodplains. On such sites, it may make very rapid radial growth. It is intolerant, and full light is needed for satisfactory reproduction. Its sparse distribution is somewhat puzzling but may be due to its shade-intolerance and erratic regeneration. It grows to a large size in the Coastal Plain and Mississippi Valley regions, but is nowhere very abundant.

It has been stated that the wood of this species is mechanically superior to that of many other red oaks. However, it is mixed with other red oak timbers and is not sold as Shumard oak.

Quércus falcàta Michx.   southern red oak

Distinguishing Characteristics (Figs. 9-54 and 9-55)

Leaves deciduous; 4 to 9 in. (10 to 23 cm) long, 2 to 6 in. (5 to 15 cm) wide; *shape* obovate to ovate; heteromorphic: those of saplings and the lower crown of older trees shallowly 3-lobed near the apex (Fig. 9-55); those of the upper crown or throughout older trees are 5- to 7-lobed, the lateral lobes often falcate and the terminal lobe prolonged (Fig. 9-54); *apex* acuminate or falcate; *base* usually U-shaped; *surfaces* dark lustrous green above, grayish green to tan and tomentose below, turning rusty upon drying; *fall color* red to brown.

Fruit solitary or paired; *nut* $\frac{1}{2}$ to $\frac{5}{8}$ in. (1.2 to 1.5 cm) long, subglobose, orange-brown, sometimes striated, more or less stellate pubescent, enclosed one-third or less in a thin, shallow cup, the kernel yellow; *cup scales* reddish brown, appressed, pubescent except on the margins.

Twigs dark red, pubescent, or nearly glabrous; *terminal buds* $\frac{1}{8}$ to $\frac{1}{4}$ in. (3 to 6 mm) long, ovoid, acute with reddish brown, puberulent scales.

Bark dark brown to nearly black, thick, the rough ridges separated by deep, narrow fissures; inner bark only slightly yellow (otherwise similar to that of *Q. velutina*).

Range

Eastern (Map 9-18). Elevation: sea level to 2500 ft. Well-drained, fertile, dry to submesic uplands.

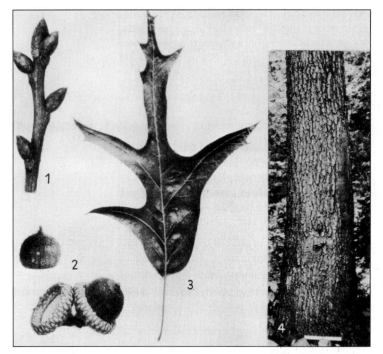

FIGURE 9-54 *Quercus falcata,* southern red oak. (1) Twig ×1¼. (2) Acorns ×1. (3) Leaf from upper part of large tree ×½. (4) Bark.

FIGURE 9-55 Trilobed leaf of young southern red oak ×½.

General Description

This species is one of the most common of upland southern oaks and is particularly characteristic of the drier, poorer soils of the Piedmont region; it is rare in the bottomlands of the Mississippi Delta. However, throughout its range, the largest trees are found on fertile stream bottoms. Its occurrence in the various pine types of the Coastal Plains is sporadic. It is a medium-sized tree 50 to 80 ft tall and 1 to $2\frac{1}{2}$ ft dbh (max. 128 by 9 ft) with a deep root system, a short trunk, and an extensive rounded crown that even under forest conditions is composed of large branches. This oak is similar in its silvical features to black oak, and it is found with it and other dry-soil oaks and hickories, black tupelo, sweetgum, southern magnolia, and shortleaf and Virginia pines. It is used for lumber and fuel.

Quércus pagòda Raf. (*Q. falcata* var. *pagodifòlia* Ell.)   cherrybark oak

Distinguishing Characteristics (Fig. 9-56)

   Leaves more uniformly 5- to 11- and shallow-lobed than those of *Q. falcata,* with the upper margins of the lobes more nearly at right angles to the midrib; *base* broadly cuneate; *fall color* red.

   Bark smooth, soon with narrow, flaky, or scaly ridges; otherwise similar to that of *Q. falcata.*

FIGURE 9-56   *Quercus pagoda,* cherrybark oak. (1) Leaf ×$\frac{1}{2}$. (2) Acorn ×1. (3) Bark.

Range

Southeastern (Map 9-19). Elevation: sea level to ca. 800 ft. River banks, poorly drained bottomlands, terraces, and mesic slopes.

General Description

This oak is a more massive, better-formed tree than the southern red oak and often reaches heights of 100 to 130 ft and dbh of 3 to 5 ft (max. 150 by 9 ft), which classes it among the largest of southern oaks. The tree is found in a number of different bottomland types, but it develops best on poorly drained, loamy, bottomland soils, where its associates include swamp chestnut oak, wetland hickories, green ash, and sweetgum. On higher ground, including the rich well-drained soil of old fields, to green ash and sweetgum may be added American beech, southern magnolia, black tupelo, yellow-poplar, and others.

Cherrybark oak has intermediate tolerance and is usually dominant or codominant in the stand. It is especially plentiful in the lower Mississippi Valley, and on account of its fast growth, clear bole, and the superior quality of the wood it has been rated as the best red oak of this region, and it is often used for furniture and trim. It is often planted for shade.

Quércus coccínea Muenchh.   scarlet oak

Distinguishing Characteristics (Fig. 9-57)

Leaves deciduous; 3 to 7 in. (7.5 to 18 cm) long, 2 to 5 in. (5 to 13 cm) wide; *shape* ovate, obovate, or oval; *margin* deeply 5- to 9-lobed, with wide nearly circular sinuses;

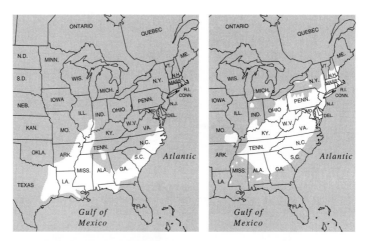

MAP 9-19 **(Left)** *Quercus pagoda.* **(Right)** *Quercus coccinea.*

FIGURE 9-57 *Quercus coccinea*, scarlet oak. (1) Twig ×1$\frac{1}{4}$. (2) Leaf ×$\frac{1}{2}$. (3) Acorn ×1. (4) Top of acorn showing concentric rings ×3. (5) Bark.

apex acute to acuminate; *base* truncate to broadly cuneate; *surfaces* bright lustrous green above, paler and glabrous below except for axillary tufts; *fall color* scarlet.

Fruit solitary or paired; *nut* $\frac{1}{2}$ to 1 in. (1.2 to 2.5 cm) long, oval to hemispherical, or depressed-globose, reddish brown, rarely striated, usually with concentric rings near the apex, one-third to one-half enclosed in a deep bowl-like cup, the kernel whitish; *cup scales* appressed, reddish brown, lustrous.

Twigs slender, reddish brown, glabrous; *terminal buds* $\frac{1}{8}$ to $\frac{1}{4}$ in. (3 to 6 mm) long, ovoid, often angled, covered by dark reddish brown scales often pale pubescent above the center of the bud.

Bark on mature trees dark brown to nearly black, broken up into irregular ridges separated by shallow fissures of varying width; often flaky on the upper branches; otherwise similar to that of *Q. velutina,* except inner bark red.

Range
Eastern (Map 9-19). Disjuncts in southwestern Georgia, southern Alabama, southwestern Mississippi, northern Indiana, southern Michigan, western New York. Elevation: sea level to 5000 ft. Xeric uplands and slopes.

General Description
Scarlet oak is a medium-sized tree 70 to 80 ft tall and 2 to 3 ft dbh (max. 102 by 5 ft). Its ability to make rapid growth on dry soils of ridges and upper slopes makes it an important component of the climax forests of the Appalachians and eastward. In this region, it may grow in nearly pure stands or in mixture with other oaks and hickories, and pitch, shortleaf, and Virginia pines.

Scarlet oak is quite intolerant, and although its growth on good sites may rival that of yellow-poplar and red oak, it is usually excluded from such sites by more tolerant species. Though quite sensitive to the frequent fires on upland sites, it sprouts vigorously following fire. This oak is often used as an ornamental because of its general hardiness, rapid growth, drought tolerance, and brilliant autumnal leaf coloration. It may reach an age of nearly 400 years.

Quércus palústris Muenchh.   pin oak

Distinguishing Characteristics (Fig. 9-58)
Leaves deciduous; 3 to 5 in. (7.5 to 13 cm) long, 2 to 4 in. (5 to 10 cm) wide; *shape* obovate to broadly oval; *margin* 5-lobed, less commonly 7- to 9-lobed, with irregular wide to narrow often angled sinuses extending two-thirds or more to the midrib, the median lobes nearly perpendicular to the midrib; *apex* acute to acuminate; *base* truncate to broadly cuneate; *surfaces* bright green and lustrous above, paler below and glabrous except for axillary tufts; *fall color* red.

Fruit solitary or clustered; *nut* $\frac{3}{8}$ to $\frac{5}{8}$ in. (0.9 to 1.5 cm) long, nearly hemispherical, light brown, often striated, enclosed only at the base in a thin, saucerlike cup, the kernel yellow; *cup scales* appressed, red-brown, puberulent, dark-margined.

Twigs slender, lustrous, reddish brown; *terminal buds* $\frac{1}{8}$ in. (3 mm) long, ovoid, acute, with reddish brown scales.

Bark grayish brown, smooth for many years, eventually with low, scaly ridges separated by shallow fissures.

FIGURE 9-58   *Quercus palustris*, pin oak. (1) Twig ×1¼. (2) Leaf ×½. (3) Acorn ×1. (4) Bark.

Range

Northeastern and northcentral (Map 9-20). Disjuncts in southwestern Ontario, western Vermont, and the Piedmont of North Carolina and South Carolina. Elevation: sea level to 1000 ft. Wet to mesic bottomlands or upland flats.

General Description

Pin oak is a medium-sized tree 70 to 80 ft tall and 2 to 3 ft dbh (max. 135 by 7 ft). The root system is shallower than in many of the other oaks and supports a trunk more or less studded with small tough branchlets that do not prune readily; these are also characteristic of the larger limbs (hence, perhaps the common name "pin oak"). The crown of open-grown trees, initially strongly excurrent becoming broadly pyramidal with slightly drooping lower branches, together with the deeply cut leaves, presents a very pleasing appearance, especially in the fall when its red autumnal coloration is at its peak.

Pin oak makes good growth on wet clay flats, where water may stand for several weeks during late winter and spring. The prolific mast produced on such sites is an important game food, particularly for waterfowl. It is also found on better-drained soils in the bottoms, but here its

MAP 9-20   *Quercus palustris.*

intolerance gives other species an advantage. Common associates include sweetgum, overcup and bur oaks, elms, green ash, and red maple.

Pin oak is one of the most commonly planted ornamental trees in urban landscapes in eastern United States and Canada. Its rapid growth, range of habitat tolerance, lack of serious diseases or insect damage, attractive shape, and fall coloration are ideal for streets and yards. However, pin oak will develop foliar chlorosis on alkaline soils.

Quércus ellipsoidàlis E. J. Hill   northern pin oak, jack oak

Distinguishing Characteristics. Very similar to pin oak and scarlet oak except for the ellipsoidal acorn one-third to one-half enclosed in a deep bowl-shaped cup; the kernel is yellow.

Range. Southern Manitoba and Minnesota, east to northcentral Wisconsin, central Michigan and southern Ontario, south into northwestern Ohio, northern Indiana, northern Illinois, and Iowa. Elevation: 600 to 1300 ft. Dry, upland woods, near ponds, and along streams, rarely on mesic slopes.

General Description. This is a medium-sized tree reaching 65 ft tall and 3 ft dbh. The wood is hard, heavy, and strong and used for flooring and furniture.

Quércus texàna Buckley (*Q. nuttallii* Palmer)   Texas red oak, Nuttall's oak

Distinguishing Characteristics

Leaves deciduous; 4 to 8 in. (10 to 20 cm) long, 2 to 5 in. (5 to 13 cm) wide; *shape* obovate; *margin* usually 5- to 7- (rarely 9-) lobed, lobes rather broad, separated by deep sinuses; *apex* acute to acuminate; *base* truncate to broadly cuneate; *surfaces* dull, dark green above, paler below, glabrous except for axillary tufts; *fall color* brown.

Fruit (Fig. 9-59) solitary or clustered; *nut* $\frac{3}{4}$ to $1\frac{1}{4}$ in. (2 to 3 cm) long, oblong-ovoid, reddish brown, often striated, one-fourth to one-half enclosed in a stalked, deep, thick cup; *cup scales* appressed, or free at the rim.

Twigs moderately slender, gray-brown to reddish brown, glabrous; *terminal buds* nearly $\frac{1}{4}$ in. (6 mm) long, ovoid, slightly angled, with numerous gray-brown, glabrous or slightly downy scales (similar to those of *Q. shumardii*).

Bark dark gray-brown, smooth; on older trees broken into broad flat ridges divided by narrow lighter-colored fissures (similar to that of *Q. palustris*).

Range

Southcentral (Map 9-21). Disjuncts in southeastern Oklahoma and eastern Texas. Elevation: sea level to 500 ft. Poorly drained alluvial woodlands of river and stream bottoms.

General Description

When Thomas Nuttall explored the Arkansas bottomlands more than a century ago, he listed among other species, *Q. coccinea*, scarlet oak. Not until 1927 was it discovered that what he presumed to be this tree was probably the species that now

FIGURE 9-59
Acorn of *Quercus texana*, Texas red oak ×$\frac{3}{4}$.

MAP 9-21   *Quercus texana.*

is often called Nuttall's oak. The species, long called *Q. nuttallii,* is now considered the same as *Q. texana* and when combined, the latter name has priority over the former (Dorr and Nixon, 1985).

Texas red oak (most often called Nuttall's oak) is a medium-sized tree attaining, however, in virgin stands a height of 100 to 120 ft and a dbh of 5 ft. It is commercially important for railroad ties, fuel, and charcoal, and it is one of the few noteworthy species found on poorly drained clay flats in the first river bottoms. Its very rapid growth on such sites and the ability of its seedlings to tolerate some shade recommend its intensive management for lumber production. Reforestation of bottomland fields often involves planting this oak. Because of its tolerance to flooding, it is important in green-tree reservoirs where it provides an abundance of mast for ducks and other animals. Reproduction is consistently good, and second-growth trees reach merchantable size (2 ft dbh) in about 70 years. The common associates include sweetgum, American elm, green ash, red maple, and overcup, willow, and water oaks.

It is often used as an ornamental.

Quércus marilándica Muenchh.   blackjack oak

Distinguishing Characteristics (Fig. 9-60).   *Leaves* deciduous, coriaceous, $2\frac{1}{2}$ to 5 in. (6 to 13 cm) long, 2 to 4 in. (5 to 10 cm) wide, variable but typically broadly obovate, and sometimes very shallowly 3-lobed at the apex; lustrous above, tawny pubescent below; fall color yellow or brown. *Fruit* $\frac{5}{8}$ to $\frac{3}{4}$ in. (15 to 19 mm) long, oblong, one-half enclosed in a thick bowl-shaped cup with loose, reddish brown scales. *Twigs* stout, somewhat scurfy-pubescent; buds often more than $\frac{1}{4}$ in. (6 mm) long, angled, similar to those of *Q. velutina* but usually more reddish brown and more rounded in cross section. *Bark* black, very rough and "blocky."

FIGURE 9-60   *Quercus marilandica,* blackjack oak. (1) Twig ×1$\frac{1}{4}$. (2) Acorns ×1. (3) Leaf ×$\frac{1}{2}$. (4) Bark.

Range. Long Island, New Jersey, and central Pennsylvania, south to northwestern Florida, west to southeastern and central Texas, north to southern Iowa. Elevation: near sea level to 3000 ft. Open, dry woodlands, ridges, slopes, and sandy flatwoods.

General Description. Blackjack oak is a small, poorly formed tree characteristic of dry, sterile soils where it is associated in open mixtures with such species as post, black, and southern red oaks, mockernut hickory, and eastern redcedar. It is not common in the North, but very abundant southward, and in the Southeast composes a large part of the forest growth on the poorest of sites. Although of value chiefly as a cover crop, this oak is utilized locally as mine props, fuel, charcoal, and railroad ties.

Quércus laèvis Walt.   turkey oak

Distinguishing Characteristics (Fig. 9-61). *Leaves* deciduous, 4 to 8 in. (10 to 20 cm) long, 3 to 6 in. (7.5 to 15 cm) wide, very variable in size and shape, commonly much dissected and with 3 to 5 (7) spreading, slender, falcate lobes; lustrous above, glabrous below except for axillary tufts; fall color red. *Fruit* $\frac{3}{4}$ to 1 in. (2 to 2.5 cm) long, oval, one-third enclosed in a thin bowl-shaped and top-shaped cup, with the scales extending slightly over the rim, the kernel yellow. *Twigs* stout, usually glabrous; buds to $\frac{1}{2}$ in. (13 mm) long, narrow and tapering, rusty pubescent above the middle. *Bark* rough, on old trees nearly black.

Range. Coastal Plain from southeastern Virginia to central Florida and west to southeastern Louisiana. Elevation: near sea level to 500 ft. Well-drained sandy ridges and flatwoods.

General Description. Turkey oak is an intolerant, small tree found in the coastal pine belt and the sand-hill country of the South. The leaves are very distinctive and are not readily mistaken for those of other associated species. The plane surfaces of the leaves are oriented more or less vertically, an adaptation to the hot sun and highly reflective, white, sandy soil. It is an excellent fuelwood species, and its acorns are a major food source for many animals.

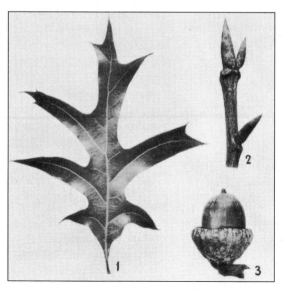

FIGURE 9-61   *Quercus laevis,* turkey oak. (1) Leaf $\times\frac{1}{2}$. (2) Twig $\times 1\frac{1}{4}$. (3) Acorn $\times 1$.

Quércus kellóggii Newb.   California black oak

Distinguishing Characteristics (Fig. 9-62). *Leaves* deciduous, 5- to 7-lobed, 3 to 8 in. (7.5 to 20 cm) long, 2 to 5 in. (5 to 13 cm) wide, elliptical, dark green above and pale pubescent below; fall color yellow. *Acorn* oblong-ellipsoidal, 1 to $1\frac{1}{2}$ in. (2.5 to 4 cm) long, with deep bowl-like cups. *Bark* gray to black and deeply fissured.

Range. McKenzie River basin in central Oregon, south over the Cascade and Siskiyou mountains, along the Coast Ranges and western slopes of the Sierra Nevada to southern California near the Mexican border. Elevation: 1000 to 8000 ft. Dry sandy and gravelly soils of canyon floors, benches, and moist mountain slopes.

General Description. This tree is typical of the transition zone in California. It is one of the most abundant and frequent oaks in California but does not tolerate poor soil drainage. Its most common associates include Douglas-fir, Digger and ponderosa pines, Pacific dogwood, and canyon live oak. At elevations above 4000 ft, it is also found with sugar pine, incense-cedar, giant sequoia, and white fir. It is generally a tree of moderate proportions sometimes attaining a height of 80 ft with a broad, rounded crown (max. 124 ft tall and $11\frac{1}{2}$ ft dbh); the bole is usually short, often crooked, and in old trees, highly defective. It is an important fuelwood species, and its wood has many desirable properties for commercial use.

FIGURE 9-62 *Quercus kelloggii*, California black oak. (1) Leaf $\times\frac{1}{2}$. (2) Acorn $\times 1$.

Willow, Water, and Laurel Oaks

Deciduous or tardily so; mature leaf form unlobed; eastern species (Table 9-16).

Quércus phéllos L.   willow oak

Distinguishing Characteristics (Fig. 9-63)

Leaves deciduous, often tardily so; 2 to 5 in. (5 to 13 cm) long, $\frac{3}{8}$ to $\frac{3}{4}$ in. (1 to 2 cm) wide; *shape* linear to linear-lanceolate, occasionally oblanceolate; *margin* entire; *apex* and *base* acute; *surfaces* glabrous above, glabrous or rarely pubescent below; *fall color* yellow-brown.

Fruit solitary or paired; *nut* $\frac{3}{8}$ to $\frac{1}{2}$ in. (10 to 12 mm) long, subglobose, glabrate, light yellowish or greenish brown, enclosed at the base in a saucerlike cup, the kernel yellow; *cup scales* thin, pubescent, with a somewhat greenish red tinge.

Twigs slender, glabrous, red to reddish brown; *terminal buds* $\frac{1}{8}$ in. (3 mm) long, ovoid, sharp-pointed, covered by chestnut-brown scales paler on the margin.

Bark smooth and steel-gray on young stems, ultimately breaking up into thick rough ridges separated by deep, irregular fissures and becoming nearly black on mature trees.

TABLE 9-16 COMPARISON OF WILLOW, WATER, AND LAUREL OAKS

Species	Leaves	Acorns
Q. phellos willow oak	Readily deciduous; mostly linear-lanceolate	$\frac{3}{8}-\frac{1}{2}$ in. (10–12 mm) long; nut greenish brown, hemispherical; nut enclosed at base by a saucerlike cup; kernel yellow
Q. nigra water oak	Deciduous; spatulate or narrowly obovate, sometimes shallowly 3-lobed at apex, the juvenile leaves variously lobed	$\frac{3}{8}-\frac{5}{8}$ in. (10–15 mm) long; nut black or nearly so, subglobose; nut enclosed at base by a saucerlike cup; kernel orange
Q. hemisphaerica laurel oak	Semievergreen or tardily deciduous; oblanceolate or oblong-obovate to elliptical, apex often apiculate or aristate	$\frac{1}{4}-\frac{1}{2}$ in. (7–11 mm) long; tan, ovoid to hemispherical; nut enclosed at base by a saucerlike cup; kernel yellow

Range

Southeastern (Map 9-22). Disjuncts in northeastern Florida and southeastern Georgia, eastern Missouri, and western Illinois. Elevation: sea level to 1000 ft. Bottomlands or low flatwoods and upland old fields.

FIGURE 9-63 *Quercus phellos,* willow oak. (1) Twig ×1$\frac{1}{4}$. (2) Leaf ×$\frac{1}{2}$. (3) Acorns ×1. (4) Bark.

General Description

Willow oak is an intolerant, medium-sized to large tree, 80 to 100 ft tall and 3 to 4 ft dbh (max. 130 by 7 ft), although much smaller in the northern part of its range. Open-grown trees are very distinctive with their dense oblong to oval crowns and bright green willowlike leaves. In the forest, the crown, although less developed, is still more or less full or rounded, and the slender lower branches do not prune readily. Numerous spurlike branchlets throughout make this tree similar in appearance to pin oak, which it is often called in some sections of the South.

Willow oak is a bottomland tree. Growth is best on well-drained loams and poorest on poorly drained heavy clays. Common associates are laurel, cherrybark, Nuttall, water, overcup, and swamp chestnut oaks, sweetgum, American elm, cedar elm, green ash, and water hickory. As an ornamental, willow oak has few if any superiors throughout the southern part of its range, and in many southern cities it is the most widely planted shade and street tree. This use has probably led to the natural spread of the tree to a number of more upland sites. Willow oak is also important for lumber and pulpwood, and it produces a heavy acorn crop eaten by many animals. Willow, water, and laurel oaks are important alternate hosts for fusiform rust.

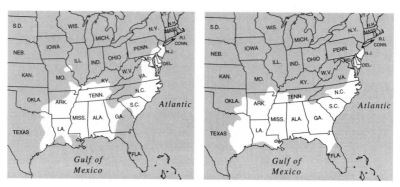

MAP 9-22 **(Left)** *Quercus phellos.* **(Right)** *Quercus nigra.*

Quércus nìgra L.   water oak

Distinguishing Characteristics (Fig. 9-64)

Leaves deciduous, but many remain green and attached to the tree until late winter; *size* variable, but usually 2 to 5 in. (5 to 13 cm) long, and $\frac{3}{4}$ to 2 in. (2 to 5 cm) wide; *shape* spatulate, obovate, and oblong; *margin* very variable, usually entire and not lobed or slightly 3-lobed at the apex; or, on sprouts, saplings, or vigorous growth, deeply 3-lobed at the apex, or variously pinnately lobed; *apex* acute to obtuse; *base* mostly acute to cuneate; *surfaces* glabrous except for occasional axillary tufts below; *fall color* yellow-brown.

Fruit solitary or paired; *nut* $\frac{3}{8}$ to $\frac{5}{8}$ in. (10 to 15 mm) long, hemispherical, black or nearly so, often striated, minutely tomentose or glabrous, enclosed at the base by a thin, saucerlike cup, the kernel orange; *cup scales* appressed, tomentose except on the often darker margins.

Twigs slender, dull red, glabrous; *terminal buds* $\frac{1}{8}$ to $\frac{1}{4}$ in. (3 to 6 mm) long, ovoid, sharp-pointed, angular, with reddish brown scales.

Bark gray-black, relatively close, often with irregular patches, eventually with rough, wide, scaly ridges.

Range

Southeastern (Map 9-22). Elevation: sea level to 1000 ft. Mixed upland woods, bottomlands, hammocks, and old fields.

General Description

In the forest, water oak is a medium-sized tree 60 to 70 ft tall and 2 to 3 ft dbh (max. 125 by 6 ft) with a tall, slender bole and a somewhat symmetrical, rounded crown of ascending branches.

This intolerant tree is a bottomland species although also found on moist uplands; between these limits, it is exceedingly cosmopolitan, occurring in most bottomland types with the exception of permanent swamps. Best development is made on the better-drained loamy or clay ridges in the bottoms with sweetgum, although, because of the diversity in site, the number of associates is very large. On old fields in the bottomlands, especially in the lower Mississippi Delta, second-growth

FIGURE 9-64   *Quercus nigra,* water oak. (1) Twig ×1¼. (2) Leaf variation ×½. (3) Acorns ×1. (4) Bark (*Photo by R. A. Cockrell*).

water oak is one of the most common trees, and together with cherrybark oak often composes nearly 80 to 90 percent of the stand; growth is fast, and trees 50 to 70 years old are merchantable. This species is a favorite ornamental in many southern cities. It is also sometimes used for timber, fuel, and plywood.

Quércus hemisphaèrica Bartr. ex Willd.   laurel oak, Darlington oak

Distinguishing Characteristics (Fig. 9-65)

Leaves rather coriaceous and opaque, semievergreen or tardily deciduous and falling during early spring just before the new leaves appear, or earlier in severe winters; 1 to 3 in. (2.5 to 8 cm) long, ⅝ to 1¼ in. (1.5 to 3 cm) wide; *shape* predominantly oblanceolate although variably elliptical to oblong-obovate but usually widest above the middle; *margin* entire to repand, sometimes undulate or irregularly lobed on saplings or vigorous shoots; *apex* acute or acuminate, often obtuse but apiculate, or occasionally rounded and aristate; *base* narrowly rounded or broadly cuneate; *surfaces* lustrous green above, paler and only minutely puberulent below and with a more or less conspicuous yellow midrib.

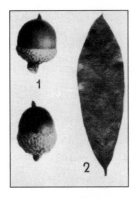

Fruit sessile or subsessile, usually solitary; *nut* ovoid to hemispherical, tan and short pubescent, $\frac{1}{4}$ to $\frac{1}{2}$ in. (7 to 11 mm) long, one-fourth or less enclosed in a thin, saucerlike cup, the kernel yellow; *cup scales* red-brown, pale pubescent, appressed.

Twigs slender, brown to gray, glabrous; *terminal buds* $\frac{1}{8}$ to $\frac{3}{16}$ in. (2 to 4 mm) long, ovoid to oval, acute, covered by lustrous, red-brown scales.

Bark medium gray and moderately smooth on young stems but ultimately darker with shallow, narrow furrows and long, broad, flattened or concave ridges.

FIGURE 9-65 *Quercus hemisphaerica,* laurel oak. (1) Upper acorn ×1. (2) Leaf ×$\frac{1}{2}$. *Quercus laurifolia,* swamp laurel oak. (1) Lower acorn ×1.

Range

Southeastern (Map 9-23). Disjuncts more inland in several states including southern Arkansas, southcentral Tennessee, and northwestern Mississippi. Elevation: sea level to 500 ft. Well-drained, sandy woodlands, terraces along rivers and streams, hammocks, maritime forests.

General Description

Laurel oak may be 60 to 70 ft tall and 2 to 3 ft dbh (max. 148 by 7 ft), but it is usually much smaller. The crown is pyramidal-rounded. Common associates include willow, live, and Nuttall oaks, sweetgum, and loblolly, slash, and longleaf pines. Laurel oak usually occurs as a scattered tree and is not very abundant except in Florida. This species is widely used in the South as an ornamental street and yard tree but is shorter-lived than its associates, the willow oak and live oak. Though tolerant of shade, young laurel oak grows rapidly in full sun.

In contrast with *Q. laurifolia,* this is a tree of woodlands in well-drained sandy soils (see the discussion of the next species). In contrast with *Q. phellos,* which it closely resembles, the leaves of *Q. phellos* are predominantly lanceolate and generally more narrow, acute at both ends, sometimes more pubescent below, and never undulate or lobed. Ecologically, *Q. phellos* overlaps both *Q. hemisphaerica* and *Q. laurifolia.*

Quércus laurifòlia Michx.   swamp laurel oak, diamond leaf oak

Distinguishing Characteristics.   This species is distinguished from laurel oak by its ecology and its *leaves,* which are more membranaceous and with conspicuous reticulate veins as seen with transmitted light, promptly or tardily deciduous or nearly overwintering in the South; spatulate-oblanceolate, or widest at or near the middle (rhombic), occasionally lobed; apex acute to rounded and rarely aristate; base narrowly cuneate; upper surface dull, lower surface slightly puberulent to glabrate; *flowering* with new leaf growth about 2 weeks before *Q. hemisphaerica* in the same general area.

Range (Map 9-23).   Swamp laurel oak is found in much wetter habitats (bottomlands, poorly drained silty soils, wet hammocks, and swamp margins) than laurel oak. The geographical ranges of the two are generally the same.

General Description.   The names of the laurel oaks have been confused for many years. Several authors have not distinguished the laurel oak and swamp laurel oak (Little, 1979). In

MAP 9-23 *Quercus hemisphaerica* and *Q. laurifolia*

the combined concept, the name is *Q. laurifolia* s. lat., but when separated, the laurel oak is *Q. hemisphaerica*, and the swamp laurel oak is *Q. laurifolia.*

This species is occasionally used as an ornamental and the wood used mostly for pulp and firewood.

Quércus imbricària Michx.   shingle oak

Distinguishing Characteristics (Fig. 9-66).   *Leaves* deciduous, 3 to 6 in. (7.5 to 15 cm) long, $\frac{3}{4}$ to 2 in. (2 to 5 cm) wide, oblong to elliptical, margin entire to slightly repand, dark green and shiny above, pale to brown and pubescent below; fall color yellow or reddish. *Fruit* typically subglobose, $\frac{1}{2}$ to $\frac{5}{8}$ in. (12 to 15 mm) long, the nut enclosed one-third to one-half in a thin bowl-shaped cup with appressed red-brown scales. *Twigs* slender, dark-green to greenish brown, buds ovoid, pointed. *Bark* gray-brown, close, eventually with broad low ridges separated by shallow furrows.

Range.   Pennsylvania south to western North Carolina, west to Arkansas, north to southern Iowa and southern Michigan; disjuncts in Louisiana and Alabama. Elevation: to 3500 ft in the southern Appalachians. Stream banks and mountain slopes and valleys.

General Description.   The scientific name, given to this tree by the explorer-botanist Michaux, indicates its use for the production of split shingles or shakes. Shingle oak, or, as it is sometimes called, "northern laurel oak," is a medium-sized tree 50 to 60 ft tall and 2 to 3 ft dbh (max. 80 by 5 ft). It is especially characteristic of the lower Ohio Valley, where it reaches its best development on moist soils along streams or on hillsides in mixture with such species as pin and overcup oaks, elms, and hickories.

A tree occasionally found in the southern Piedmont and Coastal Plain and frequently mistaken for *Q. imbricaria* is *Q. ×rudkinii* Britt. (*Q. marilandica × phellos*). It has leaves the size and shape of shingle oak but lacks the pubescence on the abaxial surface.

Quércus incàna Bartr.   bluejack oak

This is another of the entire-margined, unlobed, willow and laurel oak group which is a common, small tree in the sandhill scrub oak community in the Coastal Plain from southeastern Virginia to central Florida and west to Louisiana and eastern and central Texas, north to southwestern Arkansas and southeastern Oklahoma. Its frequent associates are *Q. laevis, Q. margaretta, Q. marilandica,* and longleaf or slash pine. It is recognized by its small leaves, 2 to 4 in. (5 to 10 cm) long, and $\frac{1}{2}$ to 1 in. (1.2 to 2.5 cm) wide, which are grayish white pubescent beneath and pale blue- or gray-green above; fall color reddish.

FIGURE 9-66 *Quercus imbricaria,* shingle oak. (1) Twig ×1$\frac{1}{4}$. (2) Leaf ×$\frac{1}{2}$. (3) Acorn ×1.

Western Live Oaks

Evergreen; leaves unlobed; western species of the red oak section.

FIGURE 9-67 *Quercus wislizeni,* interior live oak. Juvenile leaf ×1.

Quércus wislizèni A.DC.   interior live oak, Sierra live oak

This tree is commonly shrubby but may reach some 50 to 70 ft in height and 12 to 24 in. dbh (max. 89 by 7 ft) (Fig. 9-67). It has short leaves 1 to $2\frac{3}{4}$ in. (2.5 to 7 cm) long, $\frac{1}{2}$ to 2 in. (1.2 to 5 cm) wide, oblong-elliptical, glabrous above, with hollylike juvenile leaves; those on larger trees are entire, similar to those of canyon live oak. Interior live oak leaves persist for two seasons. The *acorn* is $\frac{3}{4}$ to $1\frac{1}{2}$ in. (2 to 4 cm) long, ovoid, deeply encased in a deep, thin, bowl-like cup. The largest trees develop short, thick boles and irregular but wide-spreading crowns. It is most common on dry river bottoms, washes, and slopes where the soil is dry but rich. It is sometimes found in pure groves, but it is much more commonly noted in association with Digger pine. This tree occurs from Mt. Shasta south along the Sierra Nevada and Coast Ranges to Baja California at elevations from sea level to 5000 ft. It is quite variable and has been divided into several varieties (Nixon, 1997).

Quércus agrifòlia Nee   coast live oak, holly-leafed oak

This closely resembles the previous species in both form and general appearance. Its *leaves,* however, are somewhat larger, $\frac{3}{4}$ to 3 in. (2 to 7.5 cm) long, $\frac{1}{2}$ to $1\frac{1}{2}$ in. (1.2 to 4 cm) wide, convex, rugose above, with axillary tufts of hairs below, and persisting for only 1 year. Trees are found on dry loams and gravelly soils in open valleys and shallow canyons and on moist slopes in the coastal regions of southern California and Baja California from sea level to 4500 ft. Heights of 60 to 90 ft and dbh of 24 to 36 in. or more are fairly common (max. 12 ft dbh). This species is quite gregarious and forms pure, open, but extensive stands. In mixed forests, California sycamore, white alder, numerous other oaks, and bigcone Douglas-fir are common associates. When fully exposed to the ocean, it seldom develops beyond shrubby proportions.

Quércus émoryi Torr.   Emory oak

This is probably the most common arborescent oak of southern Arizona and New Mexico. Its oblong-lanceolate, repand-sinuate, 1 to $3\frac{1}{2}$ in. (2.5 to 9 cm) long *leaves* (Fig. 9-50), oblong-ovoid *acorn* maturing in 1 year, and alligator-leatherlike *bark* are its principal identifying features. Under the most favorable conditions for growth, this species attains a height of from 40 to 60 ft and a dbh of as many inches. It is commonly encountered as a massive shrub, however, and as such is often browsed by both horses and cattle. Geographically, Emory oak is restricted to western Texas, southern Arizona, New Mexico, and northern Mexico.

Intermediate Oaks

Evergreen shrubs and small- to medium-sized trees; leaves entire or spiny toothed; western species combining characters of the other two sections.

Quércus chrysolèpis Liebm.   canyon live oak

This is the second-largest oak in California, at least in terms of diameter (max. $11\frac{1}{2}$ ft dbh) and crown spread. At the time of its measurement some years ago, one famous old tree in the Hupa

FIGURE 9-68   *Quercus chrysolepis,* canyon live oak. Acorn and leaves ×1.

Valley was 95 ft tall, had a dbh of more than 6 ft and a crown with a spread of more than 125 ft. The leaves of the canyon live oak, which persist for 3 or 4 years, are thick and turn a dark blue-green in their second season. They are polymorphic, those on sprouts and young trees aculeate (hollylike), those on older trees entire (Fig. 9-68), 1 to 3 in. (2.5 to 7.5 cm) long. Large ovoid *acorns,* $\frac{3}{4}$ to 2 in. (2 to 5 cm) in length, with thick, shallow cups covered with golden tomentum, are most distinctive. This tree frequents cool, moist, canyon walls and dry, mountain slopes. Growth is never rapid, maturity being reached in about 300 years. Its principal forest associates include several other oaks, incense-cedar, and ponderosa pine. Canyon live oak extends from southwestern Oregon south through the Coast Ranges and Sierra Nevada to Baja California, east to central and southern Arizona and southwestern New Mexico. Elevation: 1000 to 6500 ft in the north; 5500 to 7500 ft in the south. It is important for fuelwood, food for wildlife, and urban plantings.

Canyon live oak hybridizes with the related Palmer or Dunn oak, *Q. pálmeri* Engelm. *(Q. dunnii* Kellogg), which is a small, often shrubby, southwestern oak of the open pinyon-juniper woodland or canyon flats at 3500 to 6000 ft elevation in California and Arizona. It has evergreen, rounded leaves, $\frac{1}{2}$ to $1\frac{1}{2}$ in. (1.2 to 3.5 cm) long, coarsely dentate with bristle tips longer than in *Q. chrysolepis,* and with secondary veins 12 or fewer and not parallel (rather than more than 12 and parallel in *Q. chrysolepis*). See Tucker and Haskell (1960) for a thorough discussion of these species and the past confusion.

There are three other species in this section. See Manos (1993) and Nixon (1997) for a distribution map and discussion of their characteristics, variation, and hybridization.

Betulaceae: The Birch Family

Although somewhat larger during previous geological periods, the Betulaceae now number only six genera and ca. 120 species of deciduous trees and shrubs that, with few exceptions, are restricted to the cooler regions of the Northern Hemisphere.

The introduced European hornbeam, *Carpinus bétulus* L., is often used ornamentally in the United States. Another European species, *Corylus avellàna* L., produces the filbert nuts of commerce, and in recent years most of the local supply has come from American-grown trees of this species.

Five genera with 33 species are represented in North America (Table 9-17). The remaining genus *Ostryópsis* (monotypic) is restricted to eastern Asia.

The early spring pollen is a common cause of allergies in susceptible individuals. All genera are equally allergenic.

Botanical Features of the Family

Leaves deciduous, alternate, simple, stipulate.

Flowers imperfect (plant monoecious), anemophilous, appearing before or with the leaves, or rarely autumnal; *staminate* aments preformed (except in *Carpinus*), pendulous, the individual flowers borne in clusters of 1 to 6 in the axils of bracts, each consisting of 2 to 20 stamens with or without a calyx and no corolla; *pistillate* flowers in short, spikelike or capitate aments, the individual flowers borne at the base of bracts, solitary or

TABLE 9-17   COMPARISON OF NATIVE GENERA IN THE BETULACEAE

Genus	Fruit
Betula birch	Samaras very small, laterally winged; borne in a compact ament with 3-lobed, leathery bracts dehiscent with fruit
Alnus alder	Samaras small, laterally winged; borne in compact ament with thickened, woody bracts persistent long after release of fruits
Carpinus hornbeam	Nutlets small, longitudinally ribbed, not winged; each borne in the axil of a green, 3-lobed, thin, leafy bract; several bracts in a loose, pendent, racemose cluster
Ostrya hophornbeam	Nutlets small, not winged; each enclosed in a tan, inflated, bladderlike sac; several forming a loosely imbricate, pendent, racemose cluster
Corylus hazel	Nuts large and not winged; each surrounded by 2 coarsely toothed, leathery, leafy bracts, either ± distinct and broad or fused into a tubelike beak

in clusters of two or three, comprising a 2-celled, superior ovary surmounted by a short style and a 2-lobed stigma, calyx present or absent, corolla absent.

Fruit a small- to medium-sized, 1-celled, 1-seeded nut (or nutlet), or samara, subtended by a papery, leafy, or somewhat coriaceous bract, or in a semiwoody and persistent ament.

BETULA L.   birch

The genus *Betula* numbers ca. 35 species of trees and shrubs widely scattered through the Northern Hemisphere from the Arctic Circle to southern Europe, the Himalayas, China, and Japan in the old world, and in North America from the Arctic regions to the southeastern United States. Several species form vast forests in areas of the far North, whereas others, which are extremely dwarfed in habit, are found on the slopes of mountains at or near timberline. The dwarf or resin birch, *B. glandulòsa* Michx., a small shrub common to the peat bogs and high mountains of the West and North, is a suitable browse plant for both sheep and cattle, and in Alaska it is one of the important summer grazing plants for reindeer.

Because of their handsome foliage and showy bark, many of the birches are used for decorative purposes. Cultivars of the European white birch, *B. pubéscens* Ehrh. and *B. péndula* Roth (*B. alba* L. in part), and river birch, *B. nigra,* are exceedingly popular. There are cultivars of several other species in the nursery trade.

Eighteen species of *Betula* are endemic to or naturalized in the temperate regions of North America. Of the seven or eight arborescent species, three (Table 9-18) attain commercial size, abundance, and quality, and the wood has many uses including furniture, cabinets, and plywood. Interspecific hybridization is common (Barnes et al., 1974; DeHond and Campbell, 1989; Furlow, 1997; Little, 1979).

Botanical Features of the Genus

Leaves deciduous, 2-ranked, mostly ovate, oval, or triangular; serrate (often doubly), dentate, or lobed.

TABLE 9-18   COMPARISON OF IMPORTANT NATIVE BIRCHES

Species	Leaves	Fruiting Aments	Twigs	Young Bark
B. alleghaniensis yellow birch	Doubly serrate; base rounded to remotely cordate	Ovoid, erect; bracts pubescent	Aromatic	Bronze
B. lenta sweet birch	Singly or doubly serrate; base cordate	Oblong, erect; bracts essentially glabrous	Aromatic	Black
B. papyrifera paper birch	Doubly serrate; base obtuse, rounded, or truncate	Cylindrical, pendent; bracts puberulent	Nonaromatic	Chalky white

Flowers *staminate* aments preformed, borne in clusters of two or three, or solitary, the individual flowers consisting of 4 stamens adnate to a 4-parted calyx, each bract bearing 3 flowers; *pistillate* aments solitary, the individual flowers naked, borne in clusters of three and subtended by a 3-lobed bract.

Fruit a small samara borne in an erect or pendent ament, the apically 3-lobed bracts usually deciduous from the persistent axis at maturity, releasing the samaras; samaras compressed, laterally winged. A key to the samaras of northeastern birches has been published by Cunningham (1957).

Twigs slender, in certain species aromatic; usually greenish to reddish brown, glabrous or pubescent; spur shoots commonly present on old growth; *terminal buds* absent; *lateral buds* distichous, with several imbricated scales, only three of which are usually visible; *leaf scars* lunate to semioval, with 3 nearly equidistant bundle scars; *pith* small, homogeneous, terete, or remotely triangular; *stipular scars* narrow, slitlike, unequal.

Bétula alleghaniénsis Britton (*B. lutea* Michx. f.)   yellow birch

Distinguishing Characteristics (Fig. 9-69)

Leaves 3 to 5 in. (7.5 to 13 cm) long, $1\frac{1}{2}$ to 2 in. (4 to 5 cm) wide; *shape* ovate to oblong-ovate; *margin* sharply doubly serrate; *apex* acute to acuminate; *base* rounded or remotely cordate and inequilateral; *surfaces* dull, dark green, glabrous above, pale yellow-green and with tufts of hairs in the axils of the principal veins below; *petiole* slender, pubescent, $\frac{1}{2}$ to $\frac{3}{4}$ in. (12 to 18 mm) long; *fall color* bright yellow.

Fruit in an ovoid, sessile, or short-stalked, erect ament, at maturity $\frac{3}{4}$ to $1\frac{1}{4}$ in. (2 to 3 cm) long; *bracts* pubescent on the back, longer than broad, often tardily deciduous; *samara* pubescent toward the apex, oval to ovate, about as wide as its lateral wings.

Twigs slender, yellowish brown to brown, aromatic, with a slight wintergreen taste; *buds* ovate, acute, with chestnut-brown scales, ciliate on the margin.

Bark on young stems and branches yellowish, golden gray, or bronze-colored, separating at the surface and peeling horizontally into thin, curly, papery strips; eventually breaking up into reddish-brown fissures and plates on mature trunks.

FIGURE 9-69 *Betula alleghaniensis,* yellow birch. (1) Twig ×1$\frac{1}{4}$. (2) Spur shoot ×1$\frac{1}{4}$. (3) Fruiting ament ×$\frac{3}{4}$. (4) Samara ×3. (5) Pistillate bract ×3. (6) Leaf ×$\frac{1}{2}$. (7) Bark of young tree. (8) Bark of old tree.

Range

Northeastern and southern Appalachians (Map 9-24). Disjunct in southwestern Kentucky. Elevation: sea level to 2500 ft northward, 3000 to 6300 ft in the southern Appalachians. Northern swamps, lakes and stream banks; upland slopes and ridges in the mountains.

General Description

This species, the most important of the native birches, is a medium-sized tree 60 to 70 ft tall and 2 to $2\frac{1}{2}$ ft dbh (max. 110 by $4\frac{1}{2}$ ft). In the forest, the crown is irregularly rounded, and the long, usually well-formed bole terminates below in a shallow, wide-spreading root system.

Yellow birch is probably the most typical of northeastern hardwoods. In the mountainous sections, it develops best on sandy loam soils in mixture with sugar maple and American beech (beech-birch-maple mixture), which grow under the light shade of such pioneer species as the aspens and pin cherry; the associated conifers include eastern hemlock, red spruce, balsam fir, and eastern white pine. Farther south, yellow birch occurs on a variety of soils but always in moist, cool locations such as steep northerly slopes and the edges of sphagnum bogs. Temperature seems to be an important factor influencing distribution, and it may be more than coincidental that the southern limit for the species follows roughly the 70°F summer isotherm.

Yellow birch is a prolific seeder, usually producing large crops at 2- or 3-year intervals after about age 40. The winged nutlets are light and are carried a considerable distance by the wind; this probably accounts for the excellent reproduction that often takes place after fires. Seeds germinate on almost any dependably moist place, including moss-covered boulders and old partially rotted stumps or logs. The young seedlings not only thrive in such situations but commonly persist for many years, and eventually may even send their roots downward over the surface of the boulder or stump until they reach the ground. Eventually, the stump may rot away completely, and the tree may be left on "stilts." Among the conifers, hemlock is the chief competitor of yellow birch in this respect, and it is common to find several small trees of each species competing for the supremacy of a single stump. Elsewhere, after a good seed year, there are countless thousands of small birch seedlings scattered over the forest floor. The roots have difficulty in growing through the layer of leaves and other litter. Disturbance of this layer by logging or fire greatly improves subsequent seedling reproduction.

Yellow birch is moderately tolerant, less so than hemlock, sugar maple, and American beech, but more so than the aspens and pin cherry. Growth is not rapid, and trees 50 years old may be 50

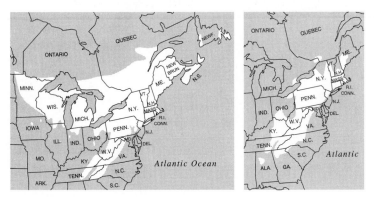

MAP 9-24 **(Left)** *Betula alleghaniensis.* **(Right)** *Betula lenta.*

ft tall and 6 in. in diameter. Maturity is reached in about 150 years, and old trees may reach 350 years old. It has a large number of damaging agents.

The papery bark curls are highly flammable and burn even when wet, a useful feature to know when starting a campfire in rain-soaked woods. Oil of wintergreen may be obtained by distilling the young twigs or inner bark, but the amount is much less than in sweet birch. Yellow birch is widely harvested for furniture, paneling, cabinets, and veneer. It is also one of the main hardwoods used in the distillation of wood alcohol and other products. Yellow birch is a preferred browse for deer and moose. It is the provincial tree of Quebec.

Yellow birch hybridizes with *P. papyrifera* and *P. pumila*.

Bétula lénta L.   sweet birch, black birch

Distinguishing Characteristics (Fig. 9-70)

Leaves $2\frac{1}{2}$ to 5 in. (5 to 13 cm) long, $1\frac{1}{2}$ to 3 in. (4 to 7.5 cm) wide; *shape* ovate to oblong-ovate; *margin* sharply and singly or irregularly doubly serrate; *apex* acute; *base* cordate, or unequally rounded; *surfaces* dull, dark green, glabrous above, pale yellow-green, with tufts of white hairs in the axils of the principal veins below; *petiole* stout, puberulent, $\frac{1}{2}$ to $\frac{3}{4}$ in. (12 to 18 mm) long; *fall color* bright yellow.

Fruit in an oblong-ovoid, short-stalked or sessile, erect ament $\frac{3}{4}$ to $1\frac{1}{2}$ in. (2 to 4 cm) long; *bracts* glabrous, or sparingly ciliate and slightly pubescent near the base, usually somewhat longer than broad, tardily deciduous; *samara* glabrous, obovoid, slightly broader than its lateral wings.

FIGURE 9-70   *Betula lenta,* sweet birch. (1) Bark (*Photo by H.P. Brown*). (2) Pseudoterminal bud ×$1\frac{1}{4}$. (3) Spur shoot ×$1\frac{1}{4}$. (4) Leaf ×$\frac{1}{2}$. (5) Fruiting ament ×$\frac{3}{4}$. (6) Samara ×3. (7) Pistillate bract ×3.

Twigs slender, reddish brown, lenticellate, with a strong wintergreen odor and taste; *buds* lustrous, glabrous, sharply pointed, chestnut-brown, divergent.

Bark on young trees smooth, reddish brown to nearly black (Fig. 9-71), with prominent horizontal lenticels; on mature trees brownish black and breaking up into large thin irregular scaly plates.

Range

Northeastern and southern Appalachians (Map 9-24). Local disjuncts in northcentral Alabama, southwestern Kentucky, and southern Quebec. Elevation: sea level northward, 1700 to 6000 ft in the southern Appalachians. Mesic uplands, coves, and slopes.

General Description

Sweet birch is a medium-sized tree 50 to 60 ft tall and 12 to 24 in. dbh (max. 100 by 5 ft), and in the forest develops a long, clear bole that terminates below in a rather deep, wide-spreading root system. This species reaches its best development in Tennessee and Kentucky but is often more abundant northward. Deep, rich, and moist but well-drained soils are preferred, yet this tree is also found on rocky sites where the roots often grow over and around boulders and rocky ledges, a habit much more typical, however, of the yellow birch. Never occurring in pure stands, sweet birch is found as a scattered tree with such species as white pine, hemlock, yellow birch, sugar maple, American beech, black cherry, white ash, basswood, and yellow-poplar.

Sweet birch and yellow birch differ from most other native birches by the presence of wintergreen oil in the twigs and inner bark of stems and roots. The twigs are a favorite for chewing.

Although oil of wintergreen may be extracted from the twigs and bark of sweet birch, it is now mostly produced synthetically. The uses of the wood are the same as that of yellow birch.

The seeding habits and other silvical features of the sweet and yellow birches are similar, except that the former species is not so aggressive, is more exacting in its seedbed requirements, and is intolerant.

Bétula papyrífera Marsh.   paper birch, white birch

Distinguishing Characteristics (Fig. 9-72)

Leaves 2 to 4 in. (5 to 10 cm) long, $1\frac{1}{2}$ to 2 in. (4 to 5 cm) wide; *shape* ovate to oval; *margin* coarsely doubly serrate; *apex* acute to acuminate; *base* rounded or obtuse; *surfaces* dull, dark green, glabrous above, pale yellowish green, glabrous or pubescent, and minutely glandular below; *petiole* black-glandular, slender, $\frac{3}{4}$ to 1 in. (18 to 25 mm) long; *fall color* light yellow.

Fruit in a pendent, cylindrical, stalked ament, $1\frac{1}{2}$ to 2 in. (4 to 5 cm) long; bracts puberulent on the back, about as long as broad, more or less deciduous at maturity; *samara* glabrous, elliptical to oval, narrower than its lateral wings.

FIGURE 9-71   Young sweet birch, with yellow birch in the right background.

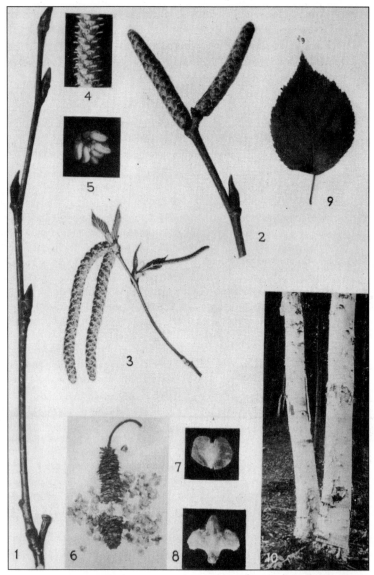

FIGURE 9-72 *Betula papyrifera,* paper birch. (1) Twig ×1 $\frac{1}{4}$. (2) Immature staminate aments ×1 $\frac{1}{4}$. (3) Flowering staminate aments ×$\frac{1}{2}$. (4) Portion of staminate ament ×5. (5) Staminate flower ×5. (6) Fruiting ament shedding bracts and fruits ×$\frac{3}{4}$. (7) Samara ×3. (8) Pistillate bract ×3. (9) Leaf ×$\frac{1}{2}$. (10) Bark of young trees.

Twigs slender, dull reddish brown to orange-brown, lenticellate; *buds* ovoid, acute, gummy, covered with chestnut-brown scales.

Bark at first dark brown, soon turning chalky to creamy white, separating into thin, papery strips; at the base of old trees, nearly black and deeply fissured.

Range

Northern (Map 9-25). Local disjuncts scattered south of the normal range. Elevation: sea level to 4000 ft. Margins of lakes, streams, and swamps; also uplands.

General Description

Paper birch, the most widely distributed of the native birches, is primarily a Canadian species and extends southward only through the northern United States where it is found in mountainous regions or in other localities where a cool, moist site is available.

This species is a medium-sized tree 50 to 70 ft tall and 12 to 24 in. dbh (max. 120 by 6 ft). The crown is pyramidal or later irregularly rounded, relatively open, and the long, cylindrical, often curved bole terminates below in a shallow root system.

Paper birch occurs as a scattered tree in the mixed conifer-hardwood forests of the North; here its associates include white, red, and jack pines, red, black, and white spruces, balsam fir, maples, aspens, American beech, yellow birch, and black ash. In some localities this birch, together with white spruce and balsam fir, composes a large portion of the landscape. After fire, paper birch often

MAP 9-25   *Betula papyrifera* and *B. cordifolia.*

seeds in over large areas where mineral soil has been exposed, and especially on moist sites it may form nearly pure stands; in this respect, paper birch is similar to the aspens and is commonly found in mixture with them on old burns. These species are all intolerant and often serve as a cover under which the slower-growing, more tolerant trees (especially spruce) become established. Paper birch is a fast-growing, short-lived tree and rarely attains an age of more than 80 years (max. 140). Young stems can produce an abundance of basal sprouts when cut or burned.

The bark was used by Native Americans for utensils, canoes, and wigwam covers and is now utilized for novelties of various sorts; it is also flammable and useful in starting a fire. Such uses by increasing numbers of campers have resulted in extremely unsightly peeled birches at northern campsites. The wood is excellent for fuel.

The wood has, in recent years, become more valued for veneer and pulp, and interest in the management of the species is increasing. Its buds, foliage, flowers, and fruits are eaten by many wildlife species.

Although valued along with the European white birch as an ornamental, both trees may be damaged or killed by the bronze birch borer. Birch dieback, a poorly understood environmental disease, can cause heavy losses to both paper and yellow birch.

Paper birch is the state tree of New Hampshire and the provincial tree of Saskatchewan. It hybridizes with *B. alleghaniensis, B. populifolia,* and *B. pumila.*

Bétula cordifòlia Regel   mountain paper birch

This species, considered as *B. papyrifera* var. *cordifolia* (Regel) Fern. by Little (1979), is distinguished from paper birch by its usually cordate leaf base and other morphological differences, seedling growth habits, chromosome number (Grant and Thompson, 1975), and betulin content of the outer bark (O'Connell et al., 1988). Mountain paper birch occurs on moist, rocky slopes and in rich, open forest in eastern Canada south to Minnesota, Wisconsin, northern Michigan, northern New York and New England, and local in the southern Appalachians of Virginia, West Virginia, North Carolina, and Tennessee (Map 9-25). It hybridizes with *B. alleghaniensis, B. occidentalis, B. papyrifera,* and *B. populifolia.*

Bétula populifòlia Marsh.   gray birch

Distinguishing Characteristics (Fig. 9-73).   *Leaves* triangular, with a narrow, pointed apex, doubly serrate; fall color pale yellow. *Staminate aments* usually solitary. *Fruit* in a spreading or ascending, cylindrical, glabrous or puberulent ament with deciduous bracts; bracts about the same in length as in width; samara somewhat narrower than its lateral wings. *Twigs* slender, lenticels warty-glandular, buds ovoid, gummy. *Bark* at first brownish, soon grayish white, exfoliating very little in comparison with that of paper birch; with black triangular patches, or inverted Vs, usually present on the trunk below the branch insertions.

Range.   Southern Ontario, east to Cape Breton, south to Pennsylvania and New Jersey; local disjuncts in northern Virginia, Ohio, and Indiana.

General Description.   Gray birch is the smallest of the northeastern tree birches and commonly attains a height of only 20 to 30 ft and a dbh of 15 in. or less (max. 60 by 2 ft). The root system is shallow, and the bole usually poorly shaped and limby, with an irregular, open, pyramidal crown. This birch is now very plentiful, especially through New England, and covers large areas on abandoned farms and burned-over land; it will grow on the poorest of sterile soils, and by prolific seeding has gained a foothold in advance of other species. In this respect, it is similar to both paper birch and the aspens. On sterile sites, gray birch is often associated with pitch pine

FIGURE 9-73 *Betula populifolia,* gray birch. (1) Leaf ×$\frac{1}{2}$. (2) Pistillate bract ×3. (3) Samara ×3. (4) Fruiting ament ×$\frac{3}{4}$. (5) Twig with lateral bud ×1$\frac{1}{4}$.

and bear oak (*Q. ilicifolia* Wangenh.), these three species forming a very monotonous vegetation type covering wide areas. Farther south, or on better soils, other oaks occur with gray birch, and among the conifers white pine is most frequently seen. Gray birch serves at first as a nurse tree for various other tree species but soon causes it considerable damage by excessive crowding. In its silvical features, gray birch is similar to paper birch. It is used as an ornamental, for fuel, and for cabinetwork.

Gray birch hybridizes with *B. papyrifera, B. cordifolia* and *B. pumila.*

Bétula nìgra L.   river birch, red birch

Distinguishing Characteristics (Fig. 9-74).   *Leaves* rhombic-ovate, conspicuously and often deeply doubly serrate, with a broadly wedge-shaped base; fall color yellow. *Fruit* in a cylindrical, erect, pubescent ament with deciduous bracts, maturing in late spring; bracts longer than broad; samara somewhat broader than its lateral wings. *Twigs* reddish brown, slender, usually puberulent and with rough lenticels, buds acute. *Bark* salmon-pink, papery, later becoming coarsely scaly.

Range.   Southern New Hampshire, south to northern Florida, west to eastern Texas, north to southeastern Minnesota; local disjuncts in New York, Vermont, and Massachusetts. Elevation: sea level to 1000 ft northward; to 3000 ft in the southern Appalachians. Alluvial banks of rivers and streams and in floodplains.

General Description.   River birch, the only species of *Betula* at low elevations in the Southeast, is a medium-sized tree 70 to 80 ft tall and 2 to 3 ft dbh, usually much smaller (max. 100 by 5 ft), with a trunk that often divides 15 to 20 ft from the ground into several arching branches. It occurs as a scattered tree with such species as sycamore, elm, silver and red maples, willows, green ash, boxelder, and cottonwoods.

Besides the distinction of being the only typically southeastern birch, it is also the only one that matures its fruit in late spring. At this time, high water is receding, and silty shore lines are exposed, which offer the best possible place for the windblown or waterborne seeds to germinate.

The silvical features are similar to those of the other birches. This species is seldom cut, except for pulpwood, because of its poor form and the high percentage of undesirable tension wood. Because of its beauty and relative lack of serious pests, its use as an ornamental is increasing, particularly the cultivar 'Heritage', which is more hardy and with more colorful bark. River birch has much potential in plantings to restore degraded land, wet or dry. Even north of the natural range,

FIGURE 9-74 *Betula nigra*, river birch. (1) Fruiting ament ×$\frac{3}{4}$. (2) Pistillate bract ×3. (3) Samara ×3. (4) Leaf ×$\frac{1}{2}$. (5) Bark.

river birch makes a desirable specimen tree for yards and parks where it often makes good growth even on dry soils. It is occasionally used as a street tree in the Pacific Northwest.

Bétula occidentàlis Hooker (*B. fontinàlis* Sarg.)   water birch

Water birch is a large shrub or small tree of lakeshores, streamsides, slopes, ridges, and moist open woods. It is common in the Rocky Mountains and is one of the most widespread of the western birches. It occurs from Alaska east to eastern Manitoba and south to New Mexico, Arizona, and southern California. Elevation: 330 to 9840 ft. The leaves are small, and the bark is smooth, dark red-brown to bronze, and not readily exfoliating. Water birch hybridizes with *B. papyrifera*.

ALNUS Mill.   alder

Alnus includes ca. 25 species of trees and large shrubs mostly distributed through the cooler regions of the Northern Hemisphere. The Eurasian species extend southward to northern Africa, Iran, the Himalayas, and Taiwan. In the new world, alders are found from Alaska and Canada to Guatemala and on mountains to Argentina.

The European or black alder, *A. glutinòsa* (L.) Gaertn., originally introduced for charcoal manufacture, is now used ornamentally, with several cultivars, in the eastern United States and Canada and has become naturalized in some areas. It is also widely planted on spoil banks and is being tested for intensive management for pulp. Several cultivars of *A. incàna* (L.) Moench., white alder, are commonly used as ornamentals in northern United States and Canada. About eight species of *Alnus,* seven of which attain tree size, are native to the United States and Canada. Only one western species, however, reaches commercial proportions and abundance. The common stream-bank alders of the eastern states and Canada are *A. rugòsa* (DuRoi) Spreng., speckled alder, of the Northeast (treated as *A. incana* subsp. *rugosa* [DuRoi] Clausen by Furlow [1997]), and *A. serrulàta* (Ait.) Willd., hazel alder, of the Southeast. Several other common shrubby alders are described by Furlow (1997). The twigs and leaves of alders are important as deer and moose browse, and the buds and seeds are eaten by birds. Native Americans used alders medicinally.

Alnus roots have nodules of nitrogen-fixing actinomycetes, and large quantities of nitrogen compounds are added to the soil from the roots and decaying leaves. See also under *Myrica*.

Botanical Features of the Genus

Leaves mostly ovate, oval, or obovate; usually irregularly serrate or dentate; 3-ranked.

Flowers *staminate* aments preformed, in racemose clusters of 3 to 5, the individual flowers consisting of 1 to 4 stamens attached to a 4-parted calyx and subtended by 3 to 5 bractlets, with 3 to 6 flowers for each scale; *pistillate* aments often preformed, in clusters of 2 or 3, the individual flowers composed of a naked ovary surmounted by 2 stigmas and subtended by 2 to 4 bractlets, in clusters of 2 to 4 at the base of bracts; early spring flowering.

Fruit a small samara borne in a persistent semiwoody ament; *samara* compressed, laterally winged, chestnut-brown.

Twigs slender to moderately stout, reddish or tinged with red; *terminal buds* absent; *lateral buds* stalked (in most species), covered by 2 or 3 valvate or several imbricate scales; *leaf scars* raised, more or less triangular to semicircular, generally with 3 bundle scars; *stipular scars* narrow and triangular; *pith* homogeneous, triangular in cross section.

Álnus rùbra Bong.   red alder

Distinguishing Characteristics (Fig. 9-75)

Leaves 3 to 6 in. (7.5 to 15 cm) long, $1\frac{1}{2}$ to 3 in. (4 to 7.5 cm) wide; *shape* ovate to elliptical; *margin* doubly serrate-dentate, the teeth glandular, strongly revolute; *apex* acute; *base* obtuse or rounded; *surfaces* dark green, glabrous or glabrate above, paler and rusty pubescent on the midrib and principal veins below; *petiole* grooved, $\frac{1}{4}$ to $\frac{1}{2}$ in. (6 to 12 mm) long; *fall color* yellow-brown.

Fruit in an oblong to ovoid, long-stalked or rarely sessile ament, $\frac{1}{2}$ to $1\frac{1}{4}$ in. (1.2 to 3 cm) long; *bracts* truncate, with greatly thickened, often rugose apices; *samaras* orbicular to obovoid, with membranous lateral wings or a single encircling wing.

Twigs slender to moderately stout, bright red to reddish brown, terete or somewhat 3-angled on vigorous shoots; *pseudoterminal buds* $\frac{1}{3}$ to $\frac{2}{3}$ in. (8 to 17 mm) long, stalked, covered by 2 or 3 red, scurfy-pubescent scales; *lateral buds* slightly smaller, stalked, somewhat divergent, usually heavily resinous; *leaf scars* raised, triangular to semicircular, with 3 bundle scars; *pith* remotely triangular in cross section, greenish white.

FIGURE 9-75 *Alnus rubra,* red alder, (1) Twig ×1$\frac{1}{4}$. (2) Young staminate (left) and pistillate (right) aments ×$\frac{3}{4}$. (3) Fruiting aments ×1$\frac{1}{4}$. (4) Samara ×2. (5) Leaf ×$\frac{1}{2}$. (6) Bark.

MAP 9-26
Alnus rubra.

Bark grayish white, pale gray, or blue-gray, smooth or covered with small warty excrescences; on large trees breaking up into large flat plates of irregular contour; inner bark bright reddish brown.

Range

Northwestern (Map 9-26). Elevation: sea level to 2500 ft. Along streams and lower slopes.

General Description

Red alder is the largest alder of North America and the most important hardwood in the Pacific Northwest, a region supporting few commercial broadleaved trees. The mature tree commonly varies from 80 to 120 ft in height and 10 to 35 in. dbh (max. 135 by $7\frac{1}{2}$ ft). In dense stands, it develops a clear, symmetrical, slightly tapered bole, rising from a shallow, spreading root system, and supporting a narrow, domelike crown. The crown of open-grown trees, however, is broadly conical and often extends nearly to the ground. Red alder is essentially a coastal species, and except in northern Washington and Idaho, where it occurs sporadically, it is rarely found more than 100 miles from the ocean. It makes its best growth on moist rich loamy bottomlands, slopes, and benches, although it also attains tree size on gravelly or rocky soils. This species occurs in pure stands or in mixture with the coastal mountain species of the Northwest.

Red alder is one of the first trees to appear following wildfires and certain kinds of logging. It never forms a permanent forest but improves the soil for subsequent species, first by building up a mull humus layer on the forest floor, and, second, by increasing the nitrogen content of the soil through the agency of nitrogen-fixing nodules on its roots. The nitrogen content of the leaves is also high. This tree is an annual seeder, with exceptional crops of light, widely dispersed seeds at about 4-year intervals. The seeds germinate well on either organic or mineral soils. Partial shade may be tolerated for the first 2 or 3 years, but after that, full light is needed. Growth is rapid, and 10-year-old trees may be 35 to 40 ft tall. Sawlogs are produced in 40 to 60 years. Red alder is rated as intolerant in comparison with its associates, except black cottonwood. Maturity is reached in 60 to 90 years. Sprouts from the stumps are often vigorous, but seldom reach large size. Whole stands may be defoliated by the tent caterpillar, but the attacks are cyclic, and little permanent damage results.

The common name "red" is for the color of the inner bark and heartwood. Its wood is used for furniture, veneers, pulpwood, fuel, and wooden novelties.

Álnus rhombifòlia Nutt.   white alder, Sierra alder

Distinguishing Characteristics.   *Leaves* narrowly elliptic to rhombic, rarely ovate, pale green, singly (rarely doubly) serrate or serrate-dentate, the teeth glandular; midrib often glandular above; fall color yellow. *Fruit* in an oblong ament, $\frac{3}{8}$ to $\frac{3}{4}$ in. (1 to 2 cm) long; samara ovoid, remotely winged. *Twigs* slender, orange-red; buds stalked with 2 scales; pith triangular in cross section. *Bark* gray, divided into irregular plates covered by thin, appressed scales.

Range.   Washington and western Idaho, south through the Coast Ranges of California and on the west slopes of the Sierra Nevada to southern California. Disjunct in western Montana. Elevation: near sea level to 8000 ft, but generally below 5000 ft. Foothill woodlands and stream banks.

General Description.   Under favorable conditions, this tree attains a height of 50 to 80 ft and a dbh of 18 to 36 in. (max. 115 by 4 ft). Pure stands occur along the banks of streams and in canyon bottoms when moisture is plentiful, but white alder is seldom observed along water courses where the stream flow is intermittent. California sycamore, bigleaf maple, Pacific dogwood, and Oregon ash are the principal associates in mixed forests. Few species compare in altitudinal distribution with white alder. Along the northcentral California coast, it often occurs only a few feet above sea level, whereas in the southern Sierra Nevada it ascends to elevations of 8000 ft.

Like most alders, this species is a prolific seeder. Young trees endure considerable competition and often form dense thickets when provided with plenty of moisture. Growth is fairly rapid, and maturity is reached in about 50 to 70 years.

White alder produces low-grade lumber, and it is doubtful if the wood will ever be of value except for fuel. It is also used as an ornamental.

CARPINUS L.   hornbeam

About 25 species are included in this genus, which is restricted to the Northern Hemisphere. One species, ***Carpìnus caroliniàna*** Walt., American hornbeam (bluebeech, ironwood, water beech), is found in the United States east of the Great Plains. It is a small, usually poorly formed, understory tree with elliptical, doubly serrate, distichous leaves and a twisted, fluted trunk, with dark bluish gray, smooth bark (max. 65 ft tall, 2 ft dbh). The fruit is easily identified by the three-lobed, leafy bract that subtends the nutlet (Fig. 9-76). The species is very tolerant and commonly occurs as an understory in hardwood mixtures. The wood is extremely hard and tough, hence the common name "ironwood" often applied to this species. The twigs and leaves are browsed by deer, and the nuts are eaten by birds and small mammals.

Carpinus bétulus L., European hornbeam, with many different cultivars, is cultivated extensively in North America. It has not become naturalized, although it persists around old homesites.

FIGURE 9-76   (Left) *Carpinus caroliniana,* American hornbeam. Nutlets and subtending leafy bracts ×2. *(Photo by Dr. T. L. Mellichamp.)* (Right) *Ostrya virginiana,* hophornbeam. Nutlets enclosed in papery saclike bracts ×$\frac{3}{4}$. *(Courtesy of U.S. Forest Service.)*

OSTRYA Scop.   hophornbeam

Of some five known species of this genus, three are native to North America. Eastern hophornbeam, *Óstrya virginiàna* (Mill.) K. Koch, is tolerant and has about the same range as American hornbeam but extends farther west in the plains region. The habits of both species are somewhat similar, but the American hornbeam is perhaps more common on moist sites. Eastern hophornbeam is easily distinguished from other trees by its bark, which has a "shreddy" appearance (i.e., broken into small, thin, narrow, vertical strips that curve away from the trunk). The leaves are similar to those of *Carpinus* but with some of the secondary veins with a prominent, submarginal, descending branch; the fruit, however, is very different and consists of nutlets enclosed in oval, flattened, papery sacs borne in hoplike clusters (Fig. 9-76). Its maximum height is 76 ft and dbh of 3 ft. The wood is used for fuel, fence posts, tool handles, and other hard wooden objects. It is an attractive, small shade tree, and its buds, catkins, and fruits are important to many wildlife species.

Óstrya knowltònii Cov.   Knowlton hophornbean, western hophornbeam

This tree of streamsides and rocky slopes in moist canyons is restricted to small isolated areas in Arizona, Utah, New Mexico, and western Texas. The petiole and young twig have stipitate glands. A related species, *Óstrya chisosénsis* Correll, Chisos hophornbeam, lacking the stipitate glands, is an endemic to the Big Bend National Park, Texas, and is of conservation concern.

These are often browsed by deer, and the nutlets are eaten by various wildlife.

CORYLUS L.   hazel, filbert

There are ca. 15 species of this genus scattered throughout the Northern Hemisphere, and two are native in North America. The leaves are distichous, oval to broadly elliptical, doubly serrate, and more or less pubescent on both surfaces; the nuts are fairly large (as the European filbert of commerce) and enclosed by leafy bracts. A western one, *Còrylus cornùta* var. *califórnica* (A.DC.) Sharp, California hazel, is a shrub or small tree; the two eastern ones, *C. cornuta* Marsh. var. *cornuta* and *C. americàna* Marsh, are both shrubs. *Corylus cornuta* , known as beaked hazel, is distinguished by the bracts that form a tubelike beak beyond the nut and lack stalked glands on the twig. The American hazel (*C. americana*) has broad, leafy bracts around the nut and long-stalked glands on the twig. Both have edible nuts, usually eaten quickly by squirrels. The twigs and leaves are also good deer browse.

Casuarinaceae: The Beefwood Family

The beefwood family includes four genera and 95 species in Indomalaysia, Australia, and the western Pacific. They are very distinctive, monoecious or dioecious, evergreen trees and shrubs with drooping, green, slender, jointed twigs appearing as long needles from a distance. The leaves are minute, scalelike, in whorls and more or less connate, forming toothed sheaths at each node. Flowers are anemophilous and in spikes (staminate) or heads (pistillate) in spring and summer. The pollen is allergenic. The small samaras are borne in woody, conelike multiples.

A single genus with three species has become naturalized in North America (Wilson, 1997). All are shade intolerant and fix atmospheric nitrogen (see under *Myrica*). Hybridization occurs between all three species.

Casuarìna equisetifòlia L. ex J.R. & G. Forst.   horsetail casuarina, Australian-pine, beefwood tree

Branchlets with angular ridges, densely pubescent; scale leaves persistent at apex of twig, 6 to 8 per whorl; cone cylindric to subglobose, ca. $\frac{3}{4}$ in. (19 mm) long and $\frac{3}{8}$ in. (10 mm) in diameter. It reaches 130 ft tall and is important as a hedge or windbreak around citrus groves or cultivated fields, along seashores, inland as an ornamental, and for erosion control. It has become naturalized and weedy and is of much concern in some areas of southern Florida, reproducing by seed. It does not spread by root sprouts. It is least tolerant of wet sites.

Casuarìna cunninghamiàna Miq.   river-oak casuarina

Branchlets with angular ridges, glabrous or only sparsely pubescent; scale leaves withering at apex of twig, 8 to 10 per whorl; cone cylindric to subglobose, ca. $\frac{1}{2}$ in. (13 mm) long and $\frac{1}{4}$ in. (5 mm) in diameter. It is commonly planted in Arizona and southern California and less frequently in Florida. It infrequently spreads by root sprouts.

Casuarìna glaùca Sieb. ex Spreng.   Australian-pine, scaly-bark beefwood

Branchlets with flat or rounded ridges, glabrous and often glaucous; scale leaves marcescent and long-recurved, 12 to 17 per whorl; cone subglobose, $\frac{3}{8}$ to $\frac{3}{4}$ in. (9 to 18 cm) long and ca. $\frac{3}{8}$ in. (7 to 9 cm) in diameter. It is widely cultivated and now considered a pest in Florida because of extensive root sprouting. It is the most cold hardy of the three listed here.

CARYOPHYLLIDAE

Cactaceae: The Cactus Family

Even though the Cactaceae are not important as trees, several attain treelike proportions, and their massive candelabra or organ-pipe appearance adds a unique aspect to landscapes of the southwestern United States and Florida Keys. They are an important source of shelter, water, and food for wildlife. Cacti are often the only woody plants in many parts of the arid West and form a characteristic vegetation including a tremendous variety of sizes and shapes.

The family includes nearly 100 genera and ca. 1400 species of trees, shrubs, herbs, and vines. All are xerophytic stem succulents and indigenous to the Western Hemisphere. In terms of numbers, best development is reached in Central America and Mexico, although several are found in South America, and others occur in the southern United States. There are three genera of native cactus trees in North America, some reaching 50 to 60 ft tall and $2\frac{1}{2}$ ft dbh.

The saguaro (pronounced *sah-WAH-ro*) is the best known, and it forms one of the most striking features of the desert landscape and is the trademark of the Sonoran Desert. The largest individuals may be 150 to 200 years old.

Botanical Features of the Family

Leaves of most cacti are reduced to spines, and even the spines may be lacking in some, so that photosynthesis takes place in the green, succulent stems; spines arise from swollen areas called *areoles*, and many cacti also have numerous, minute, barbed bristles (*glochids*) on the areoles.

Flowers usually perfect, insect/bird/bat pollinated, actinomorphic, solitary, with many petaloid tepals that are white, yellow, or red; stamens many; ovary inferior.

Fruit a many-seeded berry, edible.

Carnègiea gigantèa (Engelm.) Britt. & Rose (*Cereus giganteus* Engelm.)   saguaro, giant cactus

Columnar and unbranched for many years, but later with 2 to 10 erect branches, usually shorter than the main stem; stems spiny, fluted with 12 to 30 vertical ribs, yellow-green, lacking glochids; 20 to 35 (50) ft tall, 1 to $2\frac{1}{2}$ ft dbh (Fig. 9-77). Flowers 2 to 3 in. (5 to 7.5 cm) across, the tepals white; pollination in May by insects, bats, and birds. Fruit 2 to $3\frac{1}{2}$ in. (5 to 9 cm) long, a red or purple berry, edible, usually spineless. This occurs on rocky and gravelly soils of desert foothills, canyons, and outwashes in Arizona and southwestern California at 600 to 3600 ft elevation. This is the state flower of Arizona.

DILLENIIDAE

Theaceae: The Tea Family

The tea family, including 22 genera and 610 species, is mainly tropical in distribution. In the southeastern United States, there are three genera and four species that are native trees and shrubs. The family is best known for the tea of commerce, *Caméllia sinénsis* (L.) Kuntze, the various highly ornamental cultivars of *Caméllia japónica* L. and *C. sasánqua* Thunb. ex J.A.Murr., and *Franklínia alatamáha* Bartr. ex Marsh. Franklinia is a tree of cultivation now, having last been seen in the wild in 1790 along the Altamaha River in coastal Georgia and now apparently extinct from its native habitat. The other natives are shrubs of *Stewártia,* silky camellia, and the loblolly-bay, described below.

Botanical Features of the Family

Leaves nonaromatic, simple, alternate, entire or serrulate, estipulate, deciduous or evergreen.

Flowers solitary, large, perfect, entomophilous; 5 sepals, 5 petals, many stamens, pistil single and superior.

Fruit a capsule or berrylike.

FIGURE 9-77 *Carnegiea gigantea,* saguaro cactus. (*Photo by J. W. Hardin.*)

Gordònia lasiánthus (L.) Ellis   loblolly-bay

Distinguishing Characteristics.   *Leaves* persistent, elliptical to oblanceolate, 4 to 6 in. (10 to 15 cm) long, 1 to 2 in. (2.5 to 5 cm) wide, with small, blunt serrulations. *Flowers* solitary, 2 to 4 in. (5 to 10 cm) across, the 5 white petals fringed on the margin, the many golden-yellow, erect stamens conspicuous; appearing early to late summer. *Fruit* a capsule.

Range.   Coastal Plain, eastern North Carolina to central Florida and west to southern Mississippi. Elevation: sea level to 500 ft. It is a common tree of wet habitats, pocosins, bays, swamp margins, and low flatwoods.

General Description.   Trees to 60 ft tall with a narrow, compact crown. It is tolerant of shade. The leaf margins are frequently eaten by insects. It is of little commercial value, although the wood is occasionally used in cabinetwork. Although it has attractive foliage and is striking in flower, it is difficult to grow under cultivation and is seldom used as an ornamental. See the discussion of "bay trees" under *Persea.*

Tiliaceae: The Linden Family

The Tiliaceae comprise ca. 50 genera and 680 species of trees, shrubs, and herbs, widely scattered throughout the world, but most abundant in the Southern Hemisphere. Jute, which consists of the bast fibers of several herbaceous species of the genus *Corchòrus,* finds use in the production of papers, twine and rope, and coarse fabrics such as burlap. *Elaeocárpus, Pentàce, Gréwia,* and *Tilia* are notable timber-producing genera. Although three genera of this family are found on the North American continent, only one, *Tilia,* is arborescent.

Botanical Features of the Family

Leaves deciduous, alternate, distichous, simple, stipulate.

Flowers perfect, actinomorphic, entomophilous, borne in cymes or corymbs; *sepals* 5; *petals* 5; *stamens* numerous, in multiples of 5; *pistil* 1, with a 2- to 10-celled, superior ovary.

Fruit a capsule, drupe, berry, or nutlike; seeds without endosperm.

TILIA L.   basswood, linden, lime

Tília includes 45 species widely distributed in eastern North America, Mexico, Europe, central China, and southern Japan. Several exotics are popular ornamentals in both eastern and western United States and Canada, particularly the European cut-leaf linden, *T. platyphýllos* Scop. 'Laciniata'; little-leaf linden, *T. cordàta* Mill.; silver linden, *T. tomentòsa* Moench.; and the Japanese linden, *T. japónica* (Mig.) Simonkai. None have become naturalized in North America. The American basswoods are also suitable for decorative use, and basswood honey is widely known in the eastern part of the country. The name *basswood* may have originated on account of the strong, tough bark or bast fibers, which were used by primitive peoples for cordage of various sorts. Basswood or linden wood is lightweight, white, and soft and is used for cabinetwork, by wood-carvers and wood-turners, and for charcoal used by artists and for making gun powder.

Although primarily entomophilous, the pollen does become dry and windblown and is allergenic.

The classification of the North American basswoods has been problematical, with as many as 20 species in some earlier treatments to only one highly polymorphic species without varietal recognition (Godfrey, 1988). Jones (1968) recognized three species; however, these intergrade, are highly variable, and have only a few distinguishing features, so it seems most realistic to recognize a single species, *T. americana,* with three varieties that have fairly distinct geographical ranges but with broad zones of secondary intergradation (Hardin, 1990).

Vegetatively, basswood is often confused with red mulberry. Basswood can be distinguished by its 2 or 3 green or reddish bud scales, the twig somewhat zigzag and lacking milky juice, the teeth of the leaf margin sharp with a thickened tip on each tooth, and the upper leaf surface not scabrous.

Tília americàna L.   American basswood, American linden

Distinguishing Characteristics (Fig. 9-78)

Leaves $2\frac{1}{2}$ to 7 in. (6 to 18 cm) long, 2 to $6\frac{1}{4}$ in. (5 to 16 cm) wide; *shape* broadly ovate to suborbicular; *venation* pinnipalmate; *margin* coarsely and sharply serrate, the teeth with thickened tips; *apex* acuminate; *base* unequally cordate to almost truncate; *surfaces* nearly glabrous except for axillary tufts of hairs below to densely tomentose; cotyledons palmately 5-lobed; *fall color* yellow to orange.

Flowers appearing in late spring when the leaves are nearly full grown, $\frac{1}{2}$ to $\frac{5}{8}$ in. (12 to 15 mm) wide, borne in few-flowered cymes, the peduncle long and slender, pendant, attached to a narrow, leaflike bract 4 to 5 in. (10 to 13 cm) long.

Fruit $\frac{1}{4}$ to $\frac{3}{8}$ in. (6 to 10 mm) in diameter, short-ovoid or subglobose, nutlike, grayish tomentose; clustered at the end of a long, pendant peduncle attached to the persistent, leaflike bract.

Twigs green to red; zigzag; *pith* homogeneous, terete; *terminal buds* lacking; *lateral buds* inequilateral, mucilaginous, usually with 2 or 3 visible, green or reddish scales; *leaf scars* half-elliptical; *bundle scars* numerous, scattered, not always distinct; *stipular scars* prominent.

Bark on young trees green or grayish green, later gray to brown, breaking up into narrow ridges, somewhat scaly on the surface.

General Description

Basswood is highly regarded as a timber tree. It varies from 70 to 80 ft in height and 2 to 3 ft dbh (max. 116 by 7 ft) with a long, clear, cylindrical, sometimes buttressed bole and deep but wide-spreading root system. Best development is reached on moist, deep, loamy soils. Because this tree is found in 16 forest types, the list of associates is lengthy. Some of the more common are eastern hemlock, northern red oak, red and sugar maples, white ash, yellow buckeye, American elm, and black cherry. American basswood is a very prolific stump sprouter and quickly regenerates by this means; in fact, the occurrence of a circle of sprouts around a stump or an old decadent tree is so characteristic that basswood can often be identified at some distance by this feature.

Seeds are borne almost yearly but require 2 years or longer for germination unless given special treatment. The young seedlings are easily recognized in the forest by their palmately 5-lobed cotyledons (Fig. 7-17), which look quite different from the mature basswood leaves. Subsequent growth is rather fast, and the tree matures in 90 to 140 years. Basswood is tolerant and will grow

FIGURE 9-78   *Tilia americana*, basswood. (1) Leaf $\times\frac{1}{2}$. (2) Bark of old tree and young sprout. (3) Inflorescence with bract $\times\frac{1}{2}$. (4) Fruits with persistent bract $\times\frac{1}{2}$. (5) Twig $\times1\frac{1}{4}$.

under a considerable cover, especially in youth. It is an important soil improver; in a study of the mineral content of the leaves of 24 hardwoods and softwoods, basswood was highest in calcium and magnesium and also yielded significant amounts of nitrogen, phosphorus, and potassium.

In addition to its valuable timber, basswood is highly prized in certain localities for its honey. The wood is also prized for cabinetmaking, hand carving, and turning. The Iroquois and other Native Americans made rope from the bark by soaking it in water for several weeks or months to allow the nonfibrous portions to ret, after which it was twisted into the desired form. The leaves and twigs are browsed by deer, and the fruits are eaten by various birds and small mammals. This species and its cultivars are popular street, park, and yard trees. The recent decline in this species in North America as well as the European species is unexplained.

Tília americàna var. *americàna*   American basswood

Distinguishing Characteristics.   *Leaves* nearly glabrous or puberulent with simple hairs except for tufts of hairs in the axils of the major veins below; green on both surfaces.

Range.   Northeastern (Map 9-27). Elevation: sea level to 3200 ft. Moist bottomlands, rich coves, and slopes.

General Description.   This is the northern variety. It occurs as a codominant with sugar maple in the northwestern portion of the eastern deciduous forest (Braun, 1950).

Tília americàna var. *heterophýlla* (Vent.) Loud. (*T. heterophylla* Vent.)   white basswood, beetree

Distinguishing Characteristics.   *Leaves* densely whitish or tan stellate-tomentose on the backs. The stellate hairs are closely matted, and the surface is often glaucous. An occasional tree or shade leaves may become glabrate during the season.

Range.   Southwestern Pennsylvania, south to northwestern Florida, west to northeastern Mississippi, north to southern Illinois; disjuncts in western New York, Missouri, and northwestern Arkansas. Elevation: 200 to 5000 ft. Rich coves and slopes, and mesic forests.

General Description.   This is the basswood of the southern Appalachians. It is a large tree (to 125 ft tall) and one of the frequent codominants, with yellow buckeye and sugar maple, in the cove hardwoods of the mountains (Braun, 1950). It can live for 200 years. Populations north of North Carolina and west into Missouri are generally introgressants with var. *americana*.

MAP 9-27   *Tilia americana* var. *americana*; distribution in southern Appalachians is var. *heterophylla*.

Tília americàna var. ***caroliniàna*** (Miller) Castigl. (*T. caroliniana* Miller and *T. floridana* Small)
Carolina basswood

Distinguishing Characteristics. *Leaves* with the lower leaf surfaces tomentose with gray-ish or brownish, loose, easily detached, stellate and fasciculate hairs that either persist or are lost during the season; or, leaves puberulent from the beginning with a few fasciculate hairs along the main veins; pale or glaucous.

Range. North Carolina, south to central Florida, west to central Texas, north to southeastern Oklahoma. Elevation: near sea level to 2000 ft. Mesic to drier flatwoods, edge of ponds, streams, and Florida sinks.

General Description. This is a small tree of mixed hardwoods and pines of the southeastern Coastal Plain and outer Piedmont. Introgressants with var. *heterophylla* are common south into western Florida.

Tamaricaceae: The Tamarisk Family

This family of trees and shrubs includes four genera and 78 species and is native to Eurasia and Africa. The genus *Tamarix* is cultivated, and several species have become naturalized.

Botanical Features of the Family

Plants halophytes and xerophytes with slender, upright, spreading, or drooping branches resembling a juniper.

Leaves simple, alternate, small and scalelike and appressed against the slender twig, estipulate, persistent or deciduous, with salt-excreting glands. Many of the branches fall with the leaves in autumn.

Flowers minute but showy, perfect with 4 or 5 sepals, petals, and stamens; ovary superior; white to pink, crowded in slender racemes, spikes, or panicles; appearing in the spring and summer; entomophilous.

Fruit a small capsule with hairy, windblown seeds.

TAMARIX L.   tamarisk

This genus includes 54 Eurasian species. Several have been introduced into the United States for windbreaks, hedges, shade, erosion control, and as attractive ornamental shrubs or trees. They often grow in alkaline soils and are able, by their salt-excreting glands, to get rid of excess salts. They are also used for firewood and, when forming thickets, produce excellent cover for wildlife. The chief impact is ecological.

The genus is distinctive with its slender branches and tiny pink to white flowers in showy, crowded clusters at the ends of branches. The species, however, are difficult to distinguish, and the names found in the literature are often confused—particularly *T. chinensis, T. gallica, T. pentandra,* and *T. ramosissima.* Duncan and Duncan (1988) have a key to six of the species.

Támarix chinénsis Lour.   tamarisk, salt-cedar

This is one of the most common of several naturalized species in the southern and western United States and is recognized by having 5 sepals and 5 petals.

It was introduced in the sixteenth century as an attractive ornamental, as a windbreak, and for erosion control. It is also an important honey tree. It is a conspicuous and ecologically important tree of stream and river banks, lake shores, and marine shorelines. However, it is now weedy and of much concern, particularly in the riparian wetlands of western United States, because of its deep roots, which can draw up groundwater from 15 feet below the surface, lowering water tables and drying up surface water holes. It has been estimated that these trees can use 5 million acre-feet of water per year in the arid Southwest. It generally grows to ca. 16 ft tall and 4 in. dbh (rarely to 44 ft and 2 ft dbh) and reproduces by seeds and root sprouts often forming dense thickets. Once established, it is difficult to eradicate and is displacing small desert animals that depend on streamside habitats.

Támarix parviflòra DC.   small-flower tamarisk

This species is the most easily identified with its 4 sepals and 4 petals rather than 5 as in the other species. It is common in California.

Salicaceae: The Willow or Poplar Family

The Salicaceae include two genera and 435 species of trees and shrubs widely distributed throughout the world but are most abundant in the cooler regions of the Northern Hemisphere. The tropical species are restricted to mountainous regions, where they usually occur at or near timberline. The willow family, except for the genus *Populus,* is not currently a timber-contributing group of major importance, but many of its species are important in the natural regeneration of our forests and in conservation of soil and water, are important to wildlife, and are popular as ornamentals.

The seeds of poplars and willows are short-lived and require abundant and continuous moisture for germination. Because of this, they are, in nature, restricted to sites that are quite moist during the period of seed dispersal. In many species, propagation by cuttings or root sprouts is exceptionally good, and this feature is of value in obtaining a network of fast-growing roots for erosion control or rapid propagation of clones.

Certain species in this family are often difficult to separate from each other, and numerous natural hybrids have also been described. In contrast, some forms are very distinctive in appearance and are widely used as ornamentals. *Salix* and *Populus* (Table 9-19) are represented in North America by ca. 40 tree species.

TABLE 9-19   COMPARISON OF GENERA IN THE SALICACEAE

Genus	Leaves	Flowers	Capsules	Buds
Salix willow	Usually several times longer than wide; short-petioled	With nectar glands; bract margins entire	Not inserted upon a disk	Covered by a single caplike scale; terminal bud absent
Populus poplar	Usually about as long as wide; long-petioled	Without nectar glands; bract margins laciniate	Inserted upon a cup-shaped disk	Covered by several imbricated scales; terminal bud present

Botanical Features of the Family

Leaves deciduous, alternate, simple, stipulate; the petioles often glandular.

Flowers imperfect (plants dioecious) or in some cases also androgynous; both sexes in aments, usually appearing before the leaves; individual flowers solitary, each subtended by a bract; *staminate* flowers with 1 to many stamens; *pistillate* flowers consisting of a 1-celled pistil with 2 to 4 parietal placentae bearing many ovules, the styles with 2 to 4, often 2-lobed, stigmas; anemophilous and entomophilous.

Fruit a 1-celled, 2- to 4-valved capsule containing numerous small, comose seeds which are shed in late spring or early summer. These are extremely light in weight and are often carried considerable distances by wind. They rapidly lose their vitality, however, and unless a moist location is available soon after their release, they dry out and die. Given proper conditions, germination takes place rapidly, often within 24 to 48 hours.

SALIX L.   willow

The genus *Salix* numbers ca. 400 species that are largely scattered through the cooler regions of the Northern Hemisphere. A few forms are tropical (Indonesia), and some are found in South Africa and southern South America. Others extend beyond the Arctic Circle to the northern limits of tree growth, where they become greatly dwarfed and in some instances are actually reduced to a creeping or matlike habit. A number of shrubby forms serve as browse in the western cattle country, and the honeybee and bumblebee obtains much of its nectar and nutritious pollen for rearing spring broods from the early-flowering species. In the chemical history of aspirin, extract of willow bark was one of the precursors (Collier, 1963). Among the exotics commonly used for ornamental purposes are the graceful and stately weeping willow (*S. babylònica* L.) of eastern Asia, laurel willow (*S. pentándra* L.), white willow (*S. álba* L.), and crack willow (*S. frágilis* L.), all of European origin. The last two are common escapes and are often found naturalized in the eastern United States and Canada. Although beautiful, all willows are short-lived.

About 100 species of *Salix* occur in North America, but only about 30 of them attain tree size. Most of the commercial willow timber is produced by *S. nigra.* Hybridization and introgression occur among related, sympatric species.

Willow pollen is weakly allergenic. Some species are a source of honey.

Botanical Features of the Genus

Leaves alternate or rarely subopposite, distichous, mostly narrowly lanceolate to elliptical; margins entire or finely to coarsely toothed, usually short-petioled, or sessile; stipules often persisting for several weeks or longer.

Flowers (Fig. 9-79) entomophilous and anemophilous, appearing before the leaves; aments ascending, the individual flowers (both sexes) with a basal nectar gland and a densely pubescent, entire-margined bract; *staminate* flowers with 1 to 12 (mostly 2) stamens; *pistillate* flowers with a single pistil composed of 2 carpels.

Fruit (Fig. 9-79) a 2-valved, 1-celled capsule containing a number of cottony or silky-haired seeds.

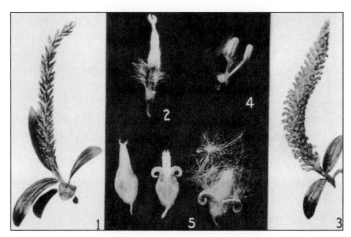

FIGURE 9-79 *Salix,* willow. (1) Pistillate ament $\times\frac{3}{4}$. (2) Pistillate flower ×6. (3) Staminate ament $\times\frac{3}{4}$. (4) Staminate flower ×6. (5) Capsule with stages of dehiscence and comose seeds ×5.

Twigs slender to stout, often brittle and easily broken at the base; glabrous, pubescent or hoary-tomentose; ranging through shades of red, orange, yellow, green, purple, or brown; lenticellate; *terminal buds* absent; *lateral buds* usually appressed, covered by a single scale either fused into a cap or with free overlapping margins; *leaf scars* V-shaped, usually with 3 bundle scars; *stipular scars* often inconspicuous; *pith* homogeneous, terete.

Sàlix nìgra Marsh.　　black willow

Distinguishing Characteristics (Fig. 9-80)
Leaves 3 to 6 in. (7.5 to 15 cm) long, $\frac{3}{8}$ to $\frac{3}{4}$ in. (1 to 2 cm) wide; *shape* lanceolate, often falcate; *margin* finely serrate; *apex* acuminate; *base* obtuse or rounded; *surfaces* light green and somewhat lustrous above, glabrous (except on the veins) and green below; *petiole* short, terete; *fall color* yellowish.

Flowers terminal in aments, appearing on short leafy twigs; *staminate* flowers with 3 to 5 long, yellow stamens; *pistillate* with a solitary pistil and 2 nearly sessile stigmas.

Fruit a capsule ca. $\frac{1}{4}$ in. (6 mm) long, slender, ovoid-conic, short-stalked, glabrous.

Twigs slender, purplish green to pale orange-brown or yellow; *leaf scar* inconspicuous; *buds* reddish brown or yellowish, $\frac{1}{16}$ in. (1 to 2 mm) long.

Bark brown to nearly black, divided into deep fissures separating thick, interlacing, sometimes scaly ridges.

Range
Eastern (Map 9-28). Elevation: sea level to 5000 ft. River and stream banks, lake shores, wet depressions, ditches, and canals.

FIGURE 9-80 *Salix nigra,* black willow. (1) Bark. (2) Leaf with stipules on twig $\times\frac{1}{2}$. (3) Fruiting ament $\times\frac{3}{4}$. (4) Twigs with lateral buds $\times 1\frac{1}{4}$.

MAP 9-28 *Salix nigra.*

General Description

Black willow is a small- to medium-sized tree 30 to 60 ft tall and about 14 in. dbh (max. 140 by 4 ft) with a broad, irregular crown and a superficial root system. Reproduction by seeds may be somewhat restricted because they must germinate on moist mineral soil soon after being shed; however, propagation by natural sprouts and root suckers is excellent.

Growth is rapid, and maturity is reached in 50 to 70 years. Like most willows, this species is very intolerant and is usually found as a dominant tree in hardwood mixtures or in pure stands. In the South, especially along the Mississippi River, pioneer well-stocked stands of black willow develop. On the best fine-textured alluvial soils, trees 40 years old may be 100 ft tall and nearly 20 in. dbh, with straight boles clear of limbs for as much as 40 ft.

Black willow is the largest and only commercially important of the willows and used for furniture, cabinetwork, pulp, and charcoal, and as an ornamental.

POPULUS L.   aspens, balsam poplars, cottonwoods

The genus *Populus* consists of 35 species widely distributed throughout the Northern Hemisphere, ranging from the forests of northern Africa, the Himalayas, China, and Japan northward beyond the Arctic Circle. In the new world, poplars are found from Alaska and Canada to northern Mexico.

Throughout much of the temperate world, races, ecotypes, clones, and hybrids are common. Within this rich variety, there are many whose form, foliage, and fast growth have attracted use as ornamental yard and street trees. Among the more important exotics in North America are the old world white poplar (*P. álba* L.), introduced in 1784, and the gray poplar (P. ×*canéscens* [Ait.] Sm.), which have become naturalized and weedy in places; the Simon poplar (*P. simònii* Carr.), with its nearly rhombic leaves; the clones of "Lombardy poplar" (*P. nìgra* L. 'Italica'), introduced in 1784, with its fastigiate branches; and the widely cultivated Carolina poplar, P. ×*canadénsis* Moench = *P. deltoides* × *P. nigra,* with only the staminate form known. The leaves of Carolina poplar may be distinguished from those of eastern cottonwood by their obtuse or cuneate, usually eglandular bases and the finer serrations of the margin. The disadvantages of all these poplars as ornamentals include their early decline and breakup, insect pests and diseases, and aggressive root invasion, which can clog sewers and drain pipes.

Another hybrid of interest is *P.* ×*jackii* Sarg. (*P. balsamifera* × *P. deltoides*). The cultivar called balm-of-Gilead (*P.* ×*gileadénsis* Rouleau; *P. candicans* Ait.) is a sterile female clone frequently planted for landscape purposes and for medicinal use. The large, aromatic, resinous buds, known locally as "balm buds," are boiled and the steam inhaled or steeped in water and the liquid used as a folk medicine for pain, snakebite, broken bones, pulmonary, and many other ailments. The buds are rich in salicin, a glucoside that breaks down to salicylic acid. The effects of the buds may be similar to taking aspirin. Salicin is also readily hydrolyzed to saligenin, which is of use as a pain reliever. The bark and flowers have also been used medicinally. "Balm-gily" trees are still seen around old homesites in the northeastern United States and southern Appalachians.

Throughout the genus, the high value of the wood as a source of pulp, veneer, and lumber coupled with some extremely rapid growth rates have prompted much activity by forest-tree geneticists. An enormous and highly successful project over the last half century has been the development of poplar hybrids. For historical perspective, see

Schreiner (1949, 1950). Certain varieties (in the form of rooted cuttings) grow to be remarkable trees, even on spoil banks from coal mining. Average 10-year-old trees may be 30 ft tall and $5\frac{1}{2}$ in. dbh. One 12-year-old tree was 56 ft by $10\frac{1}{2}$ in. Pulpwood in 8 years, small logs in 15 years, and veneer stock in 20 years are entirely possible. To cover barren ground with a forest so quickly is a great environmental achievement.

About 10 species of *Populus* are native to North America, but only five or six of them occur in commercial size and quantity. Natural hybridization and introgression occur among related, sympatric species. In addition, a number of cultivars are frequently planted throughout the United States and Canada. A thorough discussion of *Populus* biology is presented by Stettler et al. (1996).

Poplar pollen is mildly allergenic.

Botanical Features of the Genus

Leaves ovate to deltate (lanceolate or nearly so in *P. angustifolia* James and *P. ×acuminata* Rydb. of the Rocky Mountains); crenate-serrate, dentate, or lobed; usually with long, terete, or laterally compressed petioles.

Flowers anemophilous, in drooping aments appearing before the leaves, individual flowers (both sexes) solitary, inserted on a disk and subtended by a pubescent, fimbriate-margined bract; *staminate* flowers with either 6 to 12, or 12 to many stamens; *pistillate* flowers with a single pistil composed of 2 or 3 (rarely 4) carpels.

Fruit a 2- to 4-valved capsule containing a number of tufted seeds. The seeds are shed in large numbers, and at the height of their dispersal, windblown masses of white "cotton" may accumulate to a depth of 6 in. or more in neighboring depressions.

Twigs stout to slender, mostly olive-brown to lustrous reddish brown, glabrous, pubescent, or hoary-pubescent; *terminal bud* present, resinous or nonresinous, covered by several imbricated scales; *lateral buds* of nearly the same size as the terminal bud, divergent or appressed, the first or lowest scale directly above the leaf scar; flower buds conspicuously larger; *leaf scars* nearly deltate to elliptical, with 3 bundle scars, either single or divided; *pith* homogeneous, stellate in cross section.

The genus includes three major groups of species, the aspens, balsam poplars, and cottonwoods (Table 9-20). In addition to the morphological differences indicated in the table, they also differ in range and habitat. Aspens are northern species adapted to cold climates in either moist or dry soils. They are fire adapted and readily sprout from roots when the trunks are burned or killed. The resulting clones can cover extensive areas. Balsam poplars are also northern in distribution and occur in swamps and along river

TABLE 9-20   COMPARISON OF ASPENS, BALSAM POPLARS, AND COTTONWOODS

Group	Flowers	Capsules	Buds
Aspens	With 6 to 12 stamens	Thin-walled	Essentially nonresinous
Cottonwoods and balsam poplars	With 12 to 60 stamens	Thick-walled	Resinous and aromatic

banks. They also sprout from roots following fires and form clones. Cottonwoods are more southern in distribution and grow along creek or river banks and in river bottomlands where they can withstand prolonged flooding.

Aspens

Pópulus tremuloìdes Michx.   quaking aspen, trembling aspen, popple

Distinguishing Characteristics (Fig. 9-81)

Leaves with blades $1\frac{1}{2}$ to 3 in. (4 to 7.5 cm) long and wide; *shape* suborbicular to broadly ovate; *margin* finely crenate-serrate; *apex* acute to acuminate; *base* cuneate, rounded, to subcordate; *surfaces* somewhat lustrous, green and glabrous above, duller and glabrous below; *petioles* laterally flattened, $1\frac{1}{2}$ to 3 in. (3.8 to 7.5 cm) long; *fall color* bright yellow to golden (Fig. 9-82).

Flowers imperfect, trees dioecious; variable in sexual expression due to elevation and habitat with staminate trees at higher and drier sites, pistillate at lower and more moist sites.

Fruit ca. $\frac{1}{4}$ in. (6 mm) long, narrowly conical, curved, 2-valved.

Twigs slender, lustrous, reddish brown; *terminal buds* conical, sharp-pointed, sometimes very slightly resinous, covered by 6 to 7 visible, reddish brown, imbricated scales; *lateral buds* incurved, similar to the terminal buds but smaller.

Bark smooth, greenish white to cream-colored, becoming furrowed with long, flat-topped ridges, dark brown or gray.

Range

Northern (Map 9-29). Elevation: sea level to 10,000 ft. Lowlands along streams and swamp margins, or upland slopes and valleys following fire.

General Description

Quaking aspen is the most widely distributed native tree species of North America and one of the most variable (Mitton and Grant, 1996). Massive and very old clones (reaching ages of 1 million years) exist, and it is the combination of asexual and sexual reproduction that underlies its capacity for wide range and high genetic variability. It is fast-growing, relatively short-lived, and may be a twisted shrub less than 3 ft tall but commonly attains heights of 50 to 60 ft and dbh of 1 to 2 ft (max. 120 by $4\frac{1}{2}$ ft). This tree is very intolerant and under competition develops a long, clear bole and small rounded crown. The root system is wide-spreading, and sometimes surface roots may extend nearly 80 ft from the base of the tree. Depending upon soil depth and texture, other roots may go down 3 to 5 ft or more. Over its vast range, quaking aspen is found on many types of soil from moist loamy sands to shallow rocky soils and clay.

In old-growth forests, this species occurred as a scattered dominant tree, or along the edges of openings or water courses where there was sufficient light. Logging and subsequent fires usually destroyed the organic litter over wide areas and exposed mineral soil. Such sites are favorable to aspen, an aggressive pioneer, and extensive pure stands of this species spring up and serve as a cover for the more tolerant northern conifers and hardwoods, which develop slowly beneath this semi-open canopy. After 30 years or more, competition becomes excessive, much of the aspen dies, and the relatively few remnant trees maintain a position of dominance over the dense growth of other species below. In the lake states, these include the northern pines, spruces, and balsam fir,

FIGURE 9-81 (1–3) *Populus tremuloides,* quaking aspen. (1) Twig ×1¼. (2) Leaf ×½. (3) Bark. (4–6) *Populus grandidentata,* bigtooth aspen. (4) Leaf ×½. (5) Bark of older tree. (6) Twig ×1¼.

which in turn are finally replaced by hemlock and maples. Other aspen associates are many, including bigtooth aspen, paper birch, and balsam poplar, and in the West, Douglas-fir, lodgepole pine, and white fir.

Unless managed, a pioneer stand of aspen is transient, and although some individual trees may reach an age of 150 years (200 in the West), most lake states aspens over 60 years old are not worth harvesting because of their rapid deterioration. At least 500 organisms from deer and beaver to

FIGURE 9-82   Grove of quaking aspen in the fall. (*Courtesy of U.S. Forest Service.*)

insects, fungi, and viruses feed upon aspen (Graham et al., 1963). Hypoxylon canker is the most serious disease of quaking and bigtooth aspens. Aspen stands are important wildlife habitat. Beaver especially prefer the bark, leaves, twigs, and branches of aspens. Trees are thin-barked, and until mature, they are easily killed by fire.

When aspen stands are cut or burned, innumerable root suckers grow quickly from those roots less than 3 or 4 in. deep and form a dense new forest. In Michigan, Minnesota, and Wisconsin, there are extensive areas of aspen forests that are valuable for pulpwood production and are easily regenerated by coppice systems. The wood is especially useful for waferboard, plywood, and paper. Further details about aspen can be found in Adams (1990) and DeByle and Winokur (1985).

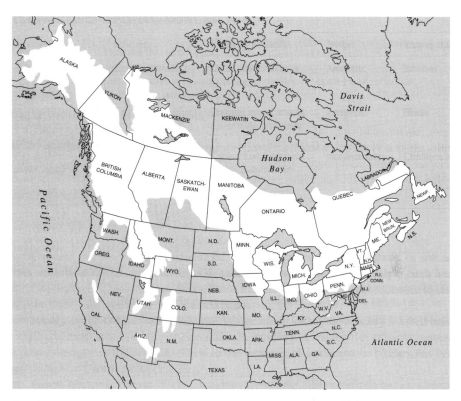

MAP 9-29 *Populus tremuloides.*

Pópulus grandidentàta Michx.   bigtooth aspen

Distinguishing Characteristics (Fig. 9-81). *Leaves* with blades suborbicular to ovate, $2\frac{3}{4}$ to 5 in. (7 to 13 cm) long, usually larger than those of the preceding species, coarsely toothed, young leaves whitish tomentose; petioles $1\frac{1}{2}$ to $2\frac{1}{2}$ in. (3.8 to 6.3 cm) long, slender, laterally flattened; fall color pale yellow to orange. *Fruit* similar to that of quaking aspen. *Twigs* often stouter than those of quaking aspen, dull, brownish gray; lateral buds gray, puberulous, divergent. *Bark* yellowish gray to tan and some with a green or orange cast, often not readily separated from that of the preceding species, later brown-gray and furrowed.

Range.   Southeastern Manitoba, east to Cape Breton, south to Virginia, west to northeastern Missouri; disjuncts in western North Carolina and central Tennessee. Elevation: sea level to 2000 ft northward, to 3000 ft in the southern mountains. Mesic to drier forests, floodplains, and uplands.

General Description.   Bigtooth aspen is a medium-sized tree, 60 to 70 ft tall and 2 ft dbh (max. 121 by 5 ft). It is similar in its silvical features to the preceding species. However, in the lake states, bigtooth aspen is less adaptable to site variation than quaking aspen and predominates on well-drained uplands of medium to good quality. With the cessation or control of forest fires, conditions are increasingly unfavorable for the development of new extensive stands of the highly intolerant aspens. Because these two species, especially bigtooth aspen, will produce under management in Michigan more wood per acre per year than any other tree native to the state (Graham et al., 1963), much effort has been made to develop effective silvicultural procedures. It is a valuable source of pulp.

Balsam Poplars and Cottonwoods

This group includes about six native species. Three are of primary importance (Table 9-21).

Pópulus balsamífera L.   balsam poplar, tacamahac poplar

Distinguishing Characteristics (Fig. 9-83)

Leaves with blades 3 to 6 in. (7.5 to 15 cm) long and about half as wide; *shape* broadly ovate to ovate-lanceolate; *margin* finely crenate-serrate; *apex* abruptly acute to acuminate; *base* rounded or cordate; *surfaces* lustrous dark green, glabrous above, pale green or commonly with rusty brown blotches below; *petiole* terete, 2 to $3\frac{1}{2}$ in. (5 to 9 cm) long; *fall color* yellow.

Fruit $\frac{1}{4}$ to $\frac{1}{3}$ in. (6 to 8 mm) long, ovoid, glabrous, 2-valved.

Twigs moderately stout, reddish brown to dark brown, lustrous, lenticellate; *terminal buds* ovate to narrowly conical, covered by 5 imbricated scales sealed by a fragrant, amber-colored, sticky resin; *lateral buds* smaller, appressed, or divergent near the apex.

Bark on young stems and limbs greenish brown to reddish brown, on older trunks eventually becoming gray to grayish black and dividing into flat, scaly or shaggy ridges separated by narrow, V-shaped fissures.

Range

Northern (Map 9-30). Elevation: sea level to 5500 ft in the Rockies. Lowland swamps, stream banks, floodplains, sandbars, and lower slopes.

General Description

Balsam poplar is a fast-growing, medium-sized tree 60 to 85 ft tall and 12 to 24 in. dbh (max. 128 by 5 ft) with a long cylindrical bole, narrow, open, pyramidal crown, and shallow root system. It reaches its greatest size in the far Northwest (Mackenzie River valley of northwestern Canada).

TABLE 9-21   COMPARISON OF IMPORTANT POPLARS AND COTTONWOODS

Species	Leaves	Capsules
P. balsamifera balsam poplar	Ovate-lanceolate; petiole terete, often with 2 glands at apex	2-valved, glabrous
P. deltoides eastern cottonwood	Deltoid; petiole laterally compressed, with 2 glands at apex	3- or 4-valved, glabrous
P. trichocarpa black cottonwood	Ovate-lanceolate; petiole terete, without glands at apex	3-valved, pubescent

FIGURE 9-83   *Populus balsamifera,* balsam poplar. (1) Bark. (2) Leaf $\times\frac{1}{2}$. (3) Capsule $\times 2$. (4) Twig $\times 1\frac{1}{4}$.

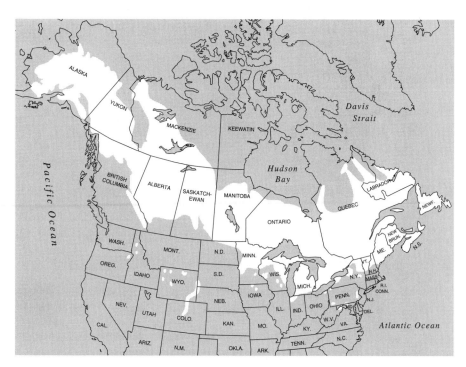

MAP 9-30   *Populus balsamifera.*

In this region, considered too cold for most trees, it is the largest, most characteristic species and occurs in open pure stands or mixed with other northern trees. Balsam poplar grows farther north than any other tree species in North America. Throughout its wide range, the associates of balsam poplar include balsam fir, white spruce, the aspens, and paper birch. It reproduces by seed, root sprouts, stump sprouts, and broken branches. It can reach 230 years of age. Like other poplars, this species is intolerant. Besides the timber and pulpwood that it furnishes, the wood is used for excelsior, boxes, and crates. Moose, deer, elk, and other animals browse the stems.

Pópulus deltoìdes Bartr. ex Marsh.   eastern cottonwood, eastern poplar

Distinguishing Characteristics (Fig. 9-84)

Leaves with blades 3 to 7 in. (7.5 to 18 cm) long, 3 to 5 in. (7.5 to 13 cm) wide; *shape* deltate to ovate-deltate; *margin* crenate-serrate, the teeth glandular; *apex* acuminate to acute; *base* truncate to slightly cordate; *surfaces* lustrous green, glabrous above, somewhat paler and glabrous below; *petiole* laterally flattened, $1\frac{1}{2}$ to 3 in. (4 to 7.5 cm) long, glandular; *fall color* yellow to yellow-orange.

Fruit ca. $\frac{3}{8}$ in. (10 mm) long, ovoid, 2- to 4-valved.

Twigs stout, angular, yellowish brown, glabrous; *terminal buds* ca. $\frac{3}{4}$ in. (18 mm) long, narrowly ellipsoidal to conical, lustrous brown, resinous, covered by 6 or 7 imbricated scales; *lateral buds* somewhat smaller, divergent.

Bark light greenish yellow or yellowish gray on young stems, eventually becoming ashy gray and dividing into thick, broad, flattened or rounded ridges separated by deep fissures.

Range

Eastern (Map 9-31). This map includes the plains cottonwood, var. *occidentalis* Rydb. The typical variety occurs from southeastern North Dakota and central Minnesota south through eastern Texas and eastward to the Atlantic coast. Elevation: sea level to 1000 ft. Floodplains, stream and riverbanks, and valleys.

General Description

This species, the most important and largest of the eastern poplars or cottonwoods, is a medium-sized to large tree 80 to 100 ft tall and 3 to 4 ft dbh (max. 175 by 12 ft). Open-grown trees have a spreading crown supported by a massive trunk that is often divided near the ground and terminates below in an extensive superficial root system; in the forest, the bole is long, clear, and cylindrical, and the crown is much smaller.

Not common in the Northeast and Appalachian regions, typical eastern cottonwood covers a wide range from the central plains to the southern Atlantic coast. On the best alluvial soils in the Mississippi Valley, growth is exceedingly fast, and young trees may grow 5 ft or more in height and 1 in. dbh yearly for the first 25 to 30 years. Trees 100 ft tall but only 9 years old have been measured. Cottonwood is very intolerant and occurs in pure stands or open mixtures with such species as black willow, sycamore, American elm, and some of the bottomland oaks. In the South, cottonwood may seed in heavily on old fields in mixture with sweetgum, by which it is eventually replaced.

Like other poplars, eastern cottonwood sheds large quantities of silky-haired seeds, which may travel by air or on the surface of water for many miles. The small, tender seedlings initially grow

FIGURE 9-84 *Populus deltoides,* eastern cottonwood. (1) Twig $\times 1\frac{1}{4}$. (2) Leaf $\times\frac{1}{2}$. (3) Staminate aments $\times\frac{1}{2}$. (4) Staminate flowers, front and rear views $\times 3$. (5) Pistillate flower and fimbriate-margined bract $\times 3$. (6) Infructescences of capsules $\times\frac{2}{3}$. (7) Bark.

MAP 9-31 *Populus deltoides.*

very slowly and may be killed by high temperatures, brief moisture stress, or pelting rains. Propagation by unrooted cuttings is good, and young trees produced in this way make rapid growth. Cottonwood is a short-lived species, and growth peaks at about 45 years. Trees over 70 years old rapidly deteriorate, and the maximum life span is probably not greater than two centuries. Numerous insects and diseases cause damage, especially in cultivated plantations. This tree is a favored browse plant of many animals. The wood is used for boxes, furniture, plywood, and pulp. Although commonly planted near homes for "instant shade," its brittleness, pest problems, massive size, and short longevity suggest its use should be greatly tempered.

Cottonwood is the state tree of Kansas and Nebraska. A broad zone of intergradation occurs along the east side of the Great Plains, from South Dakota to northern Texas, where the eastern cottonwood and plains cottonwood overlap in distribution.

Populus deltoides var. ***occidentalis*** Rydb. (*P. deltoides* ssp. *monilifera* [Ait.] Eckenw.; *P. sargentii* Dode)
plains cottonwood

Distinguishing Characteristics. The plains cottonwood differs from the eastern cottonwood by the *leaves,* which have fewer teeth, 5 to 15 per side (20 to 25 in eastern cottonwood), a more rounded base (truncate to slightly cordate in eastern cottonwood), a more prolonged apex, and only 1 or 2 glands at base of the blade (3 to 5 in eastern cottonwood); *twigs* light yellow (yellow-brown in eastern cottonwood) and the buds pubescent (glabrous in eastern cottonwood).

Range. The plains cottonwood extends from southern Alberta, southern Saskatchewan, and southwestern Manitoba southward to northern Texas and northwestern New Mexico. Elevation: 1000 to 6900 ft. It grows in deep, well-drained loam soils along rivers and streams, sandbars, and bottomlands.

General Description. This is one of the largest trees of the Great Plains. It attains a height of 65 to 100 (130) ft and a dbh of 4 to 7 ft. It is fast growing but short lived. It is especially common on moist alluvial soils through the plains and prairie states, where a winding belt of green cottonwood indicates the presence of a stream or water course. These "gallery forests" form a very characteristic aspect of the plains landscape and form the only natural shade and windbreak for hundreds of square miles. Although not found naturally on dry soils, this species was extensively planted for shelterbelts around homesteads by the early settlers and when once established has proved to be relatively drought-resistant and has become naturalized outside of its natural range.

The tender parts of the trees are browsed by various wildlife, and the seeds are eaten by birds. The wood is used for fuel, posts, and baskets, and the tree is widely planted for quick, but temporary, shade. This is the state tree of Wyoming. It intergrades with the eastern cottonwood on the southeastern side of the Great Plains and hybridizes with *P. angustifolia* and *P. balsamifera*.

Pópulus trichocárpa Torr. and Gray   black cottonwood, California poplar

Distinguishing Characteristics (Fig. 9-85)

Leaves with blades 3 to 6 in. (7.5 to 15 cm) long, 2 to 4 in. (5 to 10 cm) wide; *shape* ovate to ovate-lanceolate; *margin* finely crenate to crenate-serrate; *apex* acute to long-acuminate; *base* rounded or slightly cordate; *surfaces* dark green, glabrous above, rusty brown to silvery white, or occasionally pale green below; *petioles* $1\frac{1}{2}$ to 3 in. (4 to 7.5 cm) long, terete in cross section; *fall color* yellow.

Fruit $\frac{1}{3}$ to $\frac{1}{4}$ in. (8 to 13 mm) long, 3-valved, pubescent.

Twigs slender to moderately stout, orange-brown to light yellow-brown or greenish brown, slightly angular, lenticellate; *terminal buds* $\frac{3}{4}$ in. (18 mm) long, ovoid-conical, with 6 or 7 visible imbricated scales, resinous and giving a fragrant odor when crushed; *lateral buds* smaller, often divergent and falcate.

Bark tawny yellow to gray, and smooth on young stems; later, dark gray to grayish brown and separated by deep furrows into narrow, flat-topped ridges.

Range

Western (Map 9-32). Disjunct in southwestern North Dakota. Elevation: sea level to 2000 ft northward, to 9000 ft southward. Valleys, stream banks, floodplains, and upland slopes.

General Description

Black cottonwood, the largest of the American cottonwoods, is also the tallest, native broad-leafed tree of the Pacific Northwest. In the Puget Sound basin and vicinity, trees sometimes attain heights of 125 to 150 ft and dbh of 48 to 60 in. (max. 188 by 10 ft). Forest trees develop long, clear boles with narrow, cylindrical, round-topped crowns. This species occurs most frequently on moist, sandy, gravelly, or deep well-aerated alluvial soils. Dwarfed trees may be seen occasionally on poor, dry, sterile sites.

Black cottonwood forms limited pure stands and groves, especially on newly formed river bars. It also occurs with other moist site species throughout its range.

Large seed crops are released annually. The seed exhibits a high degree of viability, but its capacity to germinate is of very short

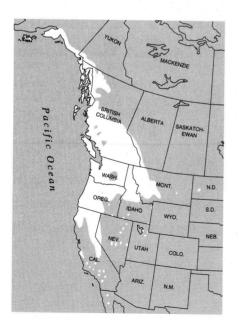

MAP 9-32   *Populus trichocarpa.*

FIGURE 9-85   *Populus trichocarpa,* black cottonwood. (1) Twig ×1$\frac{1}{4}$. (2) Pistillate ament ×$\frac{1}{2}$. (3) Staminate ament ×$\frac{1}{2}$. (4) Staminate flower, bract removed ×3. (5) Pistillate flower, bract removed ×3. (6) Capsule ×2. (7) Dehiscing capsule with comose seeds ×2. (8) Leaf ×$\frac{1}{2}$. (9) Bark (*Photo by G.B. Sudworth, U.S. Forest Service*).

duration. A moist sandy soil provides a suitable seedbed, and vigorous seedling development usually occurs on such sites. This species is very intolerant, and trees lacking vigor are soon suppressed by their more rapidly growing associates. Growth is quite rapid, particularly through the seedling and sapling stages. Trees on good sites grow fairly rapidly throughout their life span, with maturity being attained in about 150 to 200 years. Recently in the hardwood-deficient forests of the Pacific Northwest, much interest has developed in the cultivation of this species. It is used for boxes, crates, pulp, and light woodenware. Numerous mammals seriously damage this species in young plantations. It is planted for shade in Alaska and elsewhere.

Black cottonwood is treated as a subspecies of *P. balsamifera* by some authors.

Pópulus angustifòlia James   narrowleaf cottonwood

Distinguishing Characteristics.   *Leaves* with blades 2 to 5 in. (5 to 13 cm) long, $\frac{1}{2}$ to 1 in. (1.2 to 2.5 cm) wide, lanceolate, apex acuminate, base rounded, margin serrulate, glabrous or nearly so, shiny green above, paler below; *petiole* short, laterally flattened at the base; fall color yellow. *Fruit* 2-valved. *Buds* sticky, long-pointed, and fragrant.

Range.   In the West from southern Alberta and Saskatchewan south into northern Mexico and western Texas and from eastcentral California east to western South Dakota and northwestern Nebraska; generally at 3000 to 9000 ft elevation. Along stream banks and on moist, upland flats.

General Description.   Tree to 50 ft tall and $1\frac{1}{2}$ ft dbh with slender, upright branches (Fig. 9-86). This is a common cottonwood in the northern Rocky Mountains and the southwestern United States. As the common name indicates, this has the narrowest leaf of all the poplars and is more willowlike. It is commonly cultivated as a yard tree and used for fuel and fence posts.

FIGURE 9-86   Narrowleaf cottonwood in Utah. *(Courtesy of U.S. Forest Service.)*

Pópulus fremóntii Wats.   Fremont cottonwood

Distinguishing Characteristics. *Leaves* with blades 2 to 3 in. (5 to 7.5 cm) long and wide, broadly deltate, short-pointed at apex and nearly truncate at base, the margin coarsely crenate-serrate, glabrous; petiole $1\frac{1}{2}$ to 3 in. (3.7 to 5 cm) long, yellow, laterally flattened; fall color bright yellow.

Range. It is found from western Texas west to southern California, north to western Nevada, and east to southern Colorado. It is the common cottonwood at low elevations (to 6500 ft) in the Southwest, growing along stream and river bottoms with sycamore and willow, and in grasslands and woodlands.

General Description. Tree 40 to 100 ft tall and 2 to 4 ft dbh with broad, flattened, open crown and light yellowish green leaves. It is a variable species with three recognized varieties (see Little, 1979). It is used mainly for firewood and fence posts and planted for shade.

Pópulus heterophýlla L.   swamp cottonwood, swamp poplar

Distinguishing Characteristics. *Leaves* with blades 4 to 7 in. (10 to 18 cm) long, 3 to 6 in. (7.5 to 15 cm) wide, ovate, crenate-serrate, cordate or rounded at the base; petiole $2\frac{1}{2}$ to $3\frac{1}{2}$ in. (6 to 8.5 cm) long, terete; fall color yellow. *Fruit* ovoid, 2- or 3-valved, long pedicelled. *Twigs* moderately slender, brownish gray, with an orange pith and slightly resinous, stout buds. *Bark* furrowed, somewhat scaly.

Range. Along the Coastal Plain from southern Connecticut to northern Florida, west in the Gulf states to western Louisiana, and north in the Mississippi Valley to southern Illinois, Indiana, Ohio, and Michigan.

General Description. Swamp cottonwood is a medium-sized tree, 65 to 100 ft tall and 2 to 3 ft dbh (max. 130 ft by 6 ft), scattered through the bottomland forests of the southern Coastal Plain and Mississippi Valley. It is similar in its habits to the other cottonwoods and, although the wood is utilized locally, it is of secondary importance among the southern hardwoods. Swamp cottonwood can grow on soils too poorly drained for eastern cottonwood.

Ericaceae: The Heath Family

The Heath family includes 107 genera and some 3400 species of trees, shrubs, rarely vines, and herbs widely scattered through the cooler regions of the world, but particularly abundant in southeastern Asia and South Africa. The family is of little consequence as a timber-contributing group, but several genera, particularly *Rhododéndron**, *Kálmia, Enkiánthus, Leucóthoe, Piéris, Arctostáphylos,* and *Callùna,* include species that are highly prized for ornamental purposes. *Kalmia* (mountain laurel) is the state flower of Connecticut and Pennsylvania.

The burls and root wood of briar, *Érica arbórea* L., a small southern European tree, are used in making the bowls of tobacco pipes. Huckleberries (*Gaylussàcia* spp.), as well as cranberries and blueberries (*Vaccínium* spp.), are important members of this family. The leaves of several shrubby genera, notably *Kalmia, Rhododendron, Leucothoe, Lèdum,* and *Menzièsia,* contain appreciable amounts of andromedotoxin, a substance

---

*This genus includes the "rhododendrons," which are mostly evergreen with campanulate flowers, and the "azaleas," which are deciduous or evergreen with more funnelform flowers. However, there is no constant distinction between the two groups.

that is very poisonous to livestock, particularly sheep. They are, however, browsed by deer in the winter without danger if eaten in small quantity along with other plants.

Certain of the above plants, as well as many others in this family, are of interest to the forester and ecologist on account of their widespread occurrence in acidic bogs, where they often dominate the shrubby flora. When they are taken from their natural habitat, it is usually necessary to keep the soil acid; otherwise, these shrubs lose their vigor and often die. Two important tree genera in the United States are *Arbutus* and *Oxydendrum.*

Botanical Features of the Family

Leaves deciduous or persistent, alternate (rarely opposite or whorled), simple, estipulate.

Flowers perfect, mostly sympetalous and mainly 5-parted, actinomorphic or zygomorphic, entomophilous; ovary superior or inferior.

Fruit a capsule, berry, or drupe.

ARBUTUS L.   madrone

This genus includes ca. 14 species scattered through the forests of the Mediterranean basin and North and Central America. The Pacific madrone, an evergreen tree native to western North America, is described here. Two other species are found in the Southwest, the Arizona madrone and Texas madrone, both of very restricted distributions. *Arbutus ùnedo* L., strawberry tree, from Eurasia, is cultivated for its rich, reddish brown bark.

Arbùtus menzièsii Pursh   Pacific madrone

Distinguishing Characteristics.   *Leaves* persistent until the new leaves are fully grown, 2 to 5 in. (5 to 13 cm) long, oval to oblong, coriaceous, entire, or finely to coarsely serrate on vigorous growth; dark, shiny green above, pale to white below, shed in summer of second year, turning red before dehiscing. *Flowers* white, urn-shaped, in drooping terminal panicles; appearing in early spring. *Fruit* an orange-red, berrylike drupe. *Twigs* slender, green, red, or brown; terminal buds ovoid, with numerous imbricated scales; lateral buds minute; leaf scars semicircular, with a single bundle scar. *Bark* very distinctive, dark reddish brown, dividing into thin, scaly plates that are deciduous during summer and fall.

Range.   Coastal British Columbia south through western Washington and Oregon to southern California. Elevation: sea level to 5900 ft. Upland slopes and canyons, and low woods near coast.

General Description.   Pacific madrone is a medium-sized tree 80 to 100 ft tall and 2 to 4 ft dbh (max. 125 ft by 9 ft). In dense stands it forms a clear, symmetrical bole but is apt to produce a short, crooked trunk in rather open situations. The tree is found on a variety of soils but becomes shrubby on very poor sites. Best development is attained on well-drained soils near sea level. This species is quite gregarious and often forms nearly pure stands, although it occurs in much greater abundance as an understory species in Douglas-fir and redwood forests or in association with Digger pine, ponderosa pine, California black oak, bigleaf maple, tanoak, red alder, and golden chinkapin.

The tree is a dependable annual seeder; the seeds exhibit a high percentage of viability, and germination is especially good in loose moist soils. Madrone appears able to endure dense shade throughout its life, although top light is conducive to more rapid growth.

The wood is used for furniture, interior trim, veneer, weaving shuttles, charcoal, and fuel. The tree is also a source of honey and used as an ornamental. Deer browse the leaves and branchlets, and the fruits are eaten by birds and small mammals. Fire is a major damaging agent to thin-barked madrone. The chief cause of dieback is due to several canker diseases (Burnes and Honkala, 1990).

OXYDENDRUM DC.   sourwood

This is a monotypic genus and the only tree of the Ericaceae in eastern United States.

Oxydéndrum arbòreum (L.) DC.   sourwood

Distinguishing Characteristics.   *Leaves* deciduous, 3 to 8 in. (8 to 20 cm) long, 1 to $2\frac{3}{4}$ in. (2.5 to 7 cm) wide, sour tasting, lanceolate or elliptic, narrowly serrulate at least toward the apex, essentially glabrous except strigose with stiff, straight hairs on the midrib below and petiole, apex acute, base rounded to cuneate; fall color dark red to scarlet or yellow. *Flowers* ca. $\frac{1}{4}$ in. (6 mm) long, white, pendent, and urn-shaped on elongated branches of a terminal, drooping panicle; appearing in midsummer. *Fruit* a gray capsule, ca. $\frac{3}{8}$ in. (10 mm) long, held erect on the slightly drooping branches of the infructescence. *Twig* slender, yellow-green to red, buds partly embedded, leaf scars rounded with a single bundle scar, terminal bud lacking, stipular scars lacking, pith solid and homogeneous. *Bark* longitudinally furrowed and often blocky on large trees.

Range.   Southwestern Pennsylvania, central Virginia, and southeastern Maryland, south to northwestern Florida, west to Louisiana, north to southern Indiana. Elevation: near sea level to 5000 ft in the southern Appalachians. Mesic woods, bluffs, ravines, coves, slopes, and valleys.

General Description.   A small- to medium-sized tree, to 50 ft tall (max. 110 ft) and 1 ft dbh, moderately intolerant and commonly found with mixed hardwoods or pine-hardwoods. It is very conspicuous in flower and fall color and is an exceptional ornamental tree. The primary commercial importance is in the honey, which brings a higher price than most other types. The foliage is browsed by deer.

RHODODENDRON L.   rhododendron

This genus of rhododendrons and azaleas is one of the best known in horticulture and includes some 850 species found throughout the cooler and temperate portions of the Northern Hemisphere. About 21 species are native to North America, and many others have been introduced, especially from the Orient, for ornamental use. *Rhododéndron macrophýllum* D. Don ex G. Don is the Pacific rhododendron of the Northwest and the state flower of Washington.

Rhododéndron máximum L.   rosebay rhododendron, great rhododendron, great-laurel

Distinguishing Characteristics (Fig. 9-87).   *Leaves* persistent, leathery, oblong, revolute. *Flowers* in large, showy terminal clusters, white to pink; appearing in late spring. *Fruit* an elongated sticky capsule with many extremely small seeds.

Range.   Mostly in the Appalachians and adjacent territory from southern New York to northern Georgia; also in Nova Scotia, Ontario, and Maine.

FIGURE 9-87   *Rhododendron maximum*, rosebay rhododendron. (1) Flower bud ×1. (2) Infructescence ×$\frac{1}{2}$. (3) Inflorescence and leaves ×$\frac{1}{2}$. (*Photos by D.M. Brown.*)

General Description.   This species is a very common and beautiful, large shrub or small tree and often forms extensive, almost impenetrable thickets, known locally as "rhododendron hells." It is the state flower of West Virginia. A very common associate is mountain-laurel, *Kálmia latifòlia* L., with somewhat smaller leaves and abundant white to pink flowers. Together, these two species cover some 3 million acres of the southern Appalachians.

Sapotaceae: The Sapodilla Family

This family of 53 genera and 975 species is pantropical in distribution with only a few temperate species. They are trees and shrubs with milky sap. Sapodilla (*Manilkàra zapòta* [L.] Royen), native of southern Mexico and Central America, has edible fruits, and the latex, chicle, is used in chewing gum. It is widely cultivated in tropical regions. There are four species of *Bumelia* in the southern United States that reach tree size; only one is described here.

Botanical Features of the Family

Leaves alternate, simple, mostly estipulate, entire, coriaceous, often with dense T-shaped hairs, evergreen or deciduous.

Flowers small, perfect, white, in cymes; entomophilous; *sepals* and *petals* 4 to 12, each connate; *stamens* 3 to many, epipetalous; *ovary* superior, 2- to many-carpellate.

Fruit a berry with 1 to several large seeds.

Bumèlia lanuginòsa (Michx.) Pers.   gum bumelia, woolly buckthorn

Distinguishing Characteristics.   *Leaves* tardily deciduous, on spur branches, each 1 to 3 in. (2.5 to 7.5 cm) long, $\frac{3}{8}$ to 1 in. (1 to 2.5 cm) wide, elliptical or obovate, apex round, base cuneate, margin entire, shiny dark green above, densely gray- or rusty-pubescent below, with milky sap. *Flowers* bell-shaped, 5-lobed, white, on clusters of long axillary pedicels; appearing in summer. *Fruit* $\frac{3}{8}$ to $\frac{1}{2}$ in. (10 to 12 mm) long, an elliptical black berry with one seed. *Twig* ending in a simple thorn to $\frac{3}{4}$ in. (19 mm) long, terminal bud lacking, pith homogeneous, leaf scars with 3 bundle scars. *Bark* dark gray, furrowed with narrow, scaly ridges.

Range. Central Florida west to southern and western Texas, north to Kansas and central Missouri; disjunct in southwestern New Mexico and southeastern Arizona. Elevation: mainly to 2500 ft, but a southwestern variety to 5000 ft. Open, sandy woods of valleys and upland slopes.

General Description. Tree to 50 ft tall and 1 ft dbh, with straight trunk and narrow crown of short, stiff branches, often thorny. The wood is used locally for cabinets and tool handles. The fruits are eaten by birds and small mammals.

Ebenaceae: The Ebony Family

The Ebenaceae include two genera and about 485 species of trees and shrubs widely scattered through the tropical and warmer forested regions of both hemispheres. However, previous to the Ice Age, ebenaceous plants flourished in Greenland and near the Arctic Circle in both Asia and North America. The present floras of tropical Africa and those of the Indo-Malayan region are featured by many members of this family. Although true ebony wood is produced by *Dióspyros ébenum* J. König ex Retz., a number of other species also produce black or brownish black timbers that are marketed under the name of ebony.

Botanical Features of the Family

Leaves mostly deciduous, alternate, simple, entire, estipulate.

Flowers perfect and imperfect (plants usually dioecious or polygamous), small, actinomorphic, perianth 3- to 7-parted, each connate, stamens 3–many, ovary superior; entomophilous.

Fruit a berry with large seeds, edible, often with enlarged calyx attached.

DIOSPYROS L.   persimmon, ebony

This is the largest genus of the Ebenaceae, with ca. 475 species, widely scattered through eastern and southwestern Asia, the Mediterranean region, Indo-Malaya, and North America.

Two species, Texas persimmon, *D. texàna* Scheele, and common persimmon described here, are the only native members of this family found in the United States. Only the common persimmon is of any importance as a timber species. *Diospyros kàki* L.f., Japanese persimmon, is cultivated in the South for its very large, yellowish, edible fruits.

Botanical Features of the Genus

Leaves deciduous or persistent, entire.

Flowers regular (plant dioecious or polygamous).

Fruit globose, oblong, or pyriform, with 1 to 10 seeds, and subtended by the enlarged woody calyx, maturing after the leaves drop in the fall.

Dióspyros virginiàna L.   common persimmon

Distinguishing Characteristics. (Fig. 9-88). *Leaves* deciduous, $2\frac{1}{2}$ to 6 in. (6 to 15 cm) long, $1\frac{1}{2}$ to 3 in. (3.8 to 7.5 cm) wide, oblong-ovate to elliptical or oval, with a somewhat metallic luster, entire, venation slightly arcuate, and even the smallest veins are evident on the gray-green lower surface; fall color yellow to reddish. *Flowers* imperfect, tree dioecious; greenish yellow, somewhat urn-shaped or tubular, $\frac{3}{8}$ to $\frac{5}{8}$ in. (10 to 15 mm) long, appearing in spring after the leaves. *Fruit* a subglobose berry, $\frac{3}{4}$ to $1\frac{1}{2}$ in. (2 to 4 cm) in diameter, orange-colored, tinged with purple when ripe, with large seeds and subtended by the persistent woody calyx. *Twigs* moderately slender, grayish brown, glabrous or pubescent; terminal buds lacking, the laterals dark gray, covered by two overlapping scales; leaf scars lunate with a single bundle scar; pith intermittently homogenous or chambered in the same twig; stipular scars lacking. *Bark* blackish, broken up into small, conspicuous, square, scaly blocks.

Range. Southern Connecticut and Long Island, south to southern Florida, west to central Texas, north to southeastern Iowa. Elevation: sea level to 3500 ft. Dry flatwoods, upland woods, valleys, and bottomlands.

General Description. Common persimmon is ordinarily a small- to medium-sized tree, averaging 30 to 50 ft in height and 12 in. dbh (max. 130 by 4 ft). It is drought tolerant. The bole is usually short with a shallow, or in youth moderately deep, root system and a characteristic round-topped crown. The tree is shade tolerant and is often found on moist bottomlands, where it occurs as scattered trees with other hardwoods. It is a prolific root sprouter, and on old fields may form thickets. It is also very common along roadsides and fence rows where wildlife help plant the seeds.

FIGURE 9-88 *Diospyros virginiana,* common persimmon. (1) Bark. (2) Seed $\times\frac{3}{4}$. (3) Berry $\times\frac{3}{4}$. (4) Leaf $\times\frac{1}{2}$. (5) Tip of twig $\times 3$.

The fruit is exceedingly astringent when green, but edible when ripe, and is consumed locally. The distribution of persimmon depends largely upon the widespread use of the fruit for food by a wide variety of animals, because the seeds are large and not readily disseminated by other means. Several varieties have been developed for fruit production, and the tree is an attractive ornamental. The wood is very hard and used for furniture, veneer, weaving shuttles, and golf club heads. Persimmon is a source of honey.

See under *Nyssa biflora* for ways of distinguishing sterile *Diospyros* from *Nyssa.*

Styracaceae: The Storax Family

This family has 11 genera and 160 species mainly in warm temperate and tropical areas of the Mediterranean region, southeast Asia, and North and South America. There are two genera native in the United States. They are trees and shrubs with resinous bark and usually stellate hairs on the leaves.

Botanical Features of the Family

Leaves alternate, simple, estipulate.

Flowers white, mainly perfect, regular; sepals and petals 4 to 8, distinct or connate; stamens 4 to many, usually epipetalous; carpels 3 to 5, connate, ovary superior or inferior; entomophilous.

Fruit a capsule, rarely samara, or dry or fleshy drupe.

Halèsia tetráptera Ellis (*H. carolina* L. and *H. monticola* [Rehd.] Sarg. of earlier authors)
Carolina silverbell

Distinguishing Characteristics. *Leaves* 3 to 6 in. (7.5 to 15 cm) long, $1\frac{1}{2}$ to $2\frac{1}{2}$ in. (4 to 6 cm) wide, elliptical, acuminate at apex, margin serrulate, some stellate pubescence on the lower side; fall color yellow. *Flowers* white, drooping in clusters of 2 to 5, bell-shaped, $\frac{1}{2}$ to 1 in. (1.2 to 2.5 cm) long, petals 4, connate; appearing before or with the developing leaves. *Fruit* oblong, 4-winged, $1\frac{1}{4}$ to 2 in. (3 to 5 cm) long, dry. *Twigs* brown, slender, stellate pubescent, bundle scar 1, stipular scars lacking, pith chambered, terminal bud lacking. *Bark* reddish brown, furrowed into loose, broad, scaly ridges.

Range. West Virginia to southern Alabama and west to eastern Oklahoma. Elevation: ca. 1200 to 5500 ft. Moist, rich woods along streams, in coves, and on slopes of the mountains.

General Description. Shade-tolerant tree 30 to 80 ft tall, but in the southern Appalachians reaching 100 ft tall and 3 ft dbh. In flower in early to midspring, it is one of the very showy trees of the southern Appalachian slopes and coves and a beautiful native ornamental for the landscape.

The once commonly used name for this tree, *H. carolina,* should be applied instead to the little silverbell of the southeastern Coastal Plain (Duncan and Duncan, 1988; Godfrey, 1988; Reveal and Seldin, 1976).

Symplocaceae: The Sweetleaf Family

This is a small, monotypic family with ca. 250 species in moist tropics and subtropics of Asia and the Americas. There is a single species in the United States.

Symplòcos tinctòria (L.) L'Her.   sweetleaf, horse-sugar

Distinguishing Characteristics. *Leaves* alternate or appearing subopposite, simple, estipulate, deciduous or tardily so in the northern part of its range and evergreen farther south, semicoriaceous, 3 to 5 in. (7.5 to 13 cm) long, 1 to 2 in. (2.5 to 5 cm) wide, elliptical or lanceolate, slightly serrulate or entire; fall color purplish green to bronze. *Flowers* small, 6 to 12 in dense, pale yellow clusters on the twigs before the new leaves appear, each perfect, actinomorphic, epigynous, with small, 5-parted calyx and corolla, but these nearly hidden by the many long stamens; entomophilous. *Fruit* an elliptical berry, ca. $\frac{3}{8}$ to $\frac{1}{2}$ in. (8 to 12 mm) long, greenish. *Twigs* glabrous to pubescent, light green, slender, pith chambered, bundle scar 1, estipulate. *Bark* gray, smooth or slightly furrowed.

Range.   Coastal Plain and Piedmont from southern Delaware to central Florida, west to eastern Texas, north into Oklahoma, Arkansas, and west Tennessee; in the southern Appalachians of western North Carolina and southeastern Tennessee southward. Elevation: below 1000 ft in the Coastal Plain and Piedmont; to 4000 ft in the Appalachians. Found on mountain slopes and ridges with mixed hardwoods and pine, or in moist flatwoods and maritime forests.

General Description.   Shrub or small tree to 20 ft tall. The common names refer to the "sweetish" initial taste of the leaves, which are eaten by livestock and wildlife. The "tinctoria" refers to the yellow dye once obtained from the leaves and bark.

ROSIDAE

Rosaceae: The Rose Family

The Rosaceae include ca. 95 genera and some 2825 species of trees, shrubs, vines, and subshrubs, widely scattered throughout the world but more numerous in temperate climates. From the standpoint of forestry, the family is relatively unimportant except for cherry wood, but it is extremely valuable, agriculturally, for apple, pear, plum, cherry, apricot, almond, peach, strawberry, loganberry, raspberry, blackberry, quince, and loquat. This family also includes a number of highly prized, ornamental genera such as *Spiraèa, Crataègus, Physocárpus, Amelánchier, Sórbus, Ròsa, Kérria, Rhodotỳpos, Pyracántha, Cydònia, Pỳrus, Màlus,* and *Prùnus.*

Ten genera are represented by arborescent forms in the United States, but *Prunus* is the only timber-producing one of importance.

Botanical Features of the Family

Leaves deciduous or persistent; alternate (rarely opposite); simple, unifoliate, or compound; mostly stipulate.

Flowers perfect, actinomorphic, 5- (rarely 4-) parted; stamens many; hypanthium present or absent; ovaries inferior or superior; entomophilous.

Fruit a pome, drupe, capsule, follicle, or achene, rarely a capsule.

PRUNUS L.   cherry, plum

This genus includes over 200 species of trees and shrubs that are widely distributed through the cooler regions of the Northern Hemisphere. Peach and nectarine, *P. pérsica* (L.) Batsch; garden plum, *P. doméstica* L.; almond, *P. dúlcis* (Mill.) D. A. Webb

(*P. amýgdalus* Batsch); apricot, *P. armeníaca* L.; sweet cherry (mazzard), *P. àvium* (L.) L.; and the sour cherry, *P. cérasus* L., are all well-known members of this genus. There are also numerous cultivars of various species that are popular ornamentals for their flowers rather than fruits. Peach is the state flower of Delaware.

About 30 species of *Prunus* are included in the flora of the United States and Canada. Of the some 18 arborescent forms, only *P. serotina* is important for timber.

Botanical Features of the Genus

Leaves deciduous or persistent, alternate, simple; usually with 2 glands on petiole just below the blade or on the base of the blade.

Flowers in terminal or axillary racemes, umbels, or corymbs, appearing before, with, or after the leaves; *perianth* perigynous; *calyx* 5-lobed; *petals* 5, separate, white or pink; *stamens* 15 to 20; *pistil* 1 (ovary 1-celled), superior but surrounded by the deciduous hypanthium.

Fruit a thin dry, or thick fleshy, 1-seeded drupe; pit bony, smooth, or rugose.

Twigs slender or stout, usually bitter to the taste, red to brown, often conspicuously lenticellate; *terminal buds* present, the scales imbricate; *lateral buds* of nearly the same size as the terminals; *leaf scars* semicircular for the most part, with scattered bundle scars; *stipular scars* present; *pith* homogeneous; spur growth is common in this genus.

Prùnus serótina Ehrh.   black cherry

Distinguishing Characteristics (Fig. 9-89)

Leaves deciduous, 2 to 6 in. (5 to 15 cm) long, $1\frac{1}{4}$ to 2 in. (3 to 5 cm) wide; *shape* narrowly oval to oblong-lanceolate; *margin* finely serrate with callous incurved teeth; *base* cuneate; *apex* acuminate or abruptly pointed; *surfaces* dark green and very lustrous above, paler below, and at maturity usually with dense, pale or reddish brown hairs along the midrib near the base of the blade; *petiole* glandular just below the blade; *fall color* yellow to orange.

Flowers white, borne in racemes, appearing when the leaves are from half- to nearly full-grown.

Fruit $\frac{3}{8}$ to $\frac{1}{2}$ in. (8 to 12 mm) in diameter, depressed globose, almost black when ripe, ripening from June to October, flesh dark purple.

Twigs with a bitter almond taste, slender, reddish brown, or gray; usually with short spur shoots on older growth; *terminal buds* ca. $\frac{3}{16}$ in. (5 mm) long, ovate, chestnut brown, with several visible scales; *lateral buds* similar but smaller; *leaf scars* small, semicircular; *bundle scars* 3, often not distinct; *stipular scars* minute.

Bark on young stems smooth, reddish brown to nearly black, with conspicuous, narrow, horizontal lenticels; on older trunks (Fig. 9-90) exfoliating in small, persistent, shiny, platy scales with upturned edges.

Range

Eastern (Map 9-33). Elevation: sea level to 5000 ft in the southern Appalachians; 4500 to 7500 ft in the Southwest. Dry to mesic forests, fence rows, old fields, and edges of woods.

FIGURE 9-89   *Prunus serotina*, black cherry. (1) Twig $\times 1\frac{1}{4}$. (2) Inflorescence $\times \frac{1}{2}$. (3) Infructescence of drupes $\times \frac{1}{2}$. (4) Leaf $\times \frac{1}{2}$. (5) Bark of young tree. (6) Bark of old tree.

General Description

Black cherry, largest of the native cherries, is a medium-sized tree 60 to 80 ft tall and 2 to 3 ft dbh (max. 146 by 7 ft). In the forest, it develops a long, straight, clear, cylindrical bole, a narrow oblong crown, and a relatively shallow root system. Best growth is made on rich, deep, moist soils where the tree may occur in pure stands of limited extent, or more commonly mixed with such species as northern red oak, white ash, sugar maple, basswood, white pine, and hemlock. In the Northeast, this

FIGURE 9-90   Bark of black cherry. (*Photo by J. W. Hardin.*)

species often reaches commercial proportions in virgin forests of the red spruce–white pine–northern hardwood association on sandy soils. In these and other dense stands, old-growth black cherry is usually a dominant tree because of its intolerance and early rapid growth. Open-grown cherry on good sites may grow 3 ft in height and $\frac{1}{2}$ in. dbh a year for the first 20 years. In some studies, this tree at age 60 was 80 to 100 ft tall and 20 to 24 in. dbh. Although it has a tremendous geographical range, much of the finest black cherry lumber and veneer come from the Allegheny Plateau of New York, Pennsylvania, and West Virginia.

Some seed is produced yearly with large crops at 1- to 5-year intervals; they are spread by birds, and under favorable conditions show a 90% germination. Regeneration by sprouts is also good. Black cherry attains an age of 150 to 200 years. Because of its thin scaly bark, it is severely damaged by surface fires. The defoliating eastern tent caterpillars and cherry scallop shell moth are the major insect pests. The black knot fungus and many other fungous pathogens often restrict the development and utility of cherry. Deer often prevent black cherry from regenerating in the Allegheny Plateau.

The wood is prized for furniture, paneling, handles, and turned bowls; the inner bark for medicine; the twigs and leaves are a good deer browse, and rabbits feed on seedlings. Wilted foliage, if eaten, is poisonous to livestock due to a cyanogenic glycoside.

MAP 9-33   *Prunus serotina.*

Prùnus pensylvánica L.f.   pin cherry, fire cherry

Distinguishing Characteristics.   *Leaves* thin, yellow-green, usually narrower, more lanceolate, prolonged acuminate, and less shiny than those of black cherry; fall color bright yellow or red. *Flowers* borne in umbels. *Twig* with lens-shaped lenticels. *Bark* more reddish on young stems.

Range.   British Columbia to Newfoundland, south to northern Georgia, west to Colorado. Elevation: sea level to 6000 ft in the southern Appalachians. Clearings, often after a fire.

General Description.   Pin cherry is a common species occurring in and quickly invading disturbed or burned areas. It is a very intolerant, small tree (but can reach 80 ft tall and $1\frac{1}{2}$ ft dbh) of more ecological than commercial importance. Its lifespan is generally about 30 years. The fruits are important for birds and small mammals, and the leaves and branchlets are important deer browse. Viable seeds accumulate in the soil to nearly one-half million seeds per acre and can remain viable for 50 or more years. It also reproduces readily from numerous root sprouts.

Prùnus virginiàna L.   chokecherry

Distinguishing Characteristics.   Shrub or small tree with sharply serrate, obovate leaves and racemose flowers; fall color yellow; drupes dark red or blackish.

Range.   Together with its several varieties, it covers nearly the entire northern continental United States, and in Canada it is transcontinental except for coastal British Columbia; sea level to 8000 ft elevation. Found in moist soils of open sites.

General Description.   This species is very common along hedgerows where the seeds are left by birds. In sections where rabbit damage to other hardwoods and even conifers is extreme, it flourishes untouched, whereas in certain northern forest plantations it is practically the only species to survive in frost pockets where exceedingly low winter temperatures prevail. Chokecherry sprouts prolifically, and this, together with its general hardiness, seem to ensure its success even under the most extreme conditions. Although usually classed as a forest weed, this species may be useful in erosion control and is used in the plains states in shelterbelts. The fruits are an important food for many wildlife.

Prùnus subcordàta Benth.   Klamath plum

A thicket-forming shrub or small tree from Oregon and California, it is the only wild plum in the Pacific states. It is readily identified by the nearly rounded leaves and the dark red or yellow, edible drupe.

Prùnus emarginàta Dougl. ex Eaton   bitter cherry

This is a tree of widespread occurrence in both the northern Rocky Mountains and Pacific regions from British Columbia to southern California, and eastward in Arizona and western New Mexico. It has oblong-obovate leaves and bright red drupes turning black at maturity.

MALUS Mill.   apple

This genus numbers ca. 55 species of trees and shrubs scattered throughout the forests of the Northern Hemisphere. The most important species is *Màlus sylvéstris* var. *doméstica* (Borkh.) Mansf. (*M. pumila* Mill.), the common apple. This tree was introduced in colonial times and is now naturalized over a large portion of the United States. The leaves are elliptical to ovate, whitish pubescent below and along the petiole; the young twigs and buds are somewhat white tomentose, and the former have a characteristic semisweet taste. There are a number of native species of little or no importance as timber producers. Apple blossom is the state flower of Arkansas and Michigan. It is also a source of honey.

PYRUS L.　pear

This genus includes ca. 20 species distributed through Europe, Asia, and northern Africa. The introduced common pear, **Pỳrus commùnis** L., escapes cultivation, and the tree bears a superficial resemblance to the apple. The leaves of the pear, however, are thinner, more lustrous, and nearly glabrous; the twigs are less pubescent and often have a thorn terminating a spur branch. Pear trees are a source of honey.

An extremely common ornamental, introduced from China in 1908 for its showy white flowers, glossy dark green leaves, and scarlet to purple fall color is **P. calleryàna** Decne., the callery pear, with several cultivars. It has become naturalized and is becoming an invasive weed in cleared areas.

SORBUS L.　mountain-ash

Of ca. 100 species widespread mainly in temperate areas, four native species are found in the United States and Canada, two in the East, and two in the West. The common eastern species are **Sórbus americàna** Marsh., American mountain-ash, and **S. decòra** (Sarg.) Schneid., showy mountain-ash. They are small trees or shrubs common to rocky hillsides or shores and easily recognized by alternate, glabrous, odd-pinnately compound leaves with 13 to 17 lanceolate, acuminate, sharply serrate leaflets, fall color red; clusters of bright red fruits; twig with resinous buds and an odor of almond when broken.

The European mountain-ash or rowan tree, **Sórbus aucupària** L., with numerous cultivars, has been introduced for ornamental purposes. It differs from the native species in its white pubescent twigs and leaflets, nonresinous white woolly buds, and acute to obtuse leaflets. It has escaped from cultivation and is often found growing wild in Alaska, across southern Canada, and northern United States.

AMELANCHIER Medic.　serviceberry, shadbush, Juneberry

This genus includes ca. 33 species, and one or more of them is found in every state and province of the United States and Canada. They are important deer browse, and the fruits are eaten by humans and wildlife. Of some 10 species native to the United States and Canada, seven reach small tree size. The name "shadbush" often used for these trees, is based on the flowering time, which is the same as when the shad "run the rivers" to spawn. Although the genus is easily recognized, the species are often difficult to distinguish.

Amelánchier arbòrea (Michx. f.) Fern.　downy serviceberry

This generally small tree (but reaches 100 ft tall by 2 ft dbh) occurs in moist woods from New Brunswick south to northern Florida, west to eastern Texas, and north to eastern Minnesota. It is most common at low elevations. There are also many cultivars planted as ornamentals. *Leaves* singly serrate, oblong to elliptical, white tomentose at flowering but becoming smooth, the petioles pubescent; fall color yellow to red. *Flowers* white in small racemes, appearing just before or with the developing leaves; petals long and straplike. *Fruits* like very small, reddish to purple apples and quite edible. *Bark* gray, often streaked with darker lines. *Twigs* with a faint bitter almond taste, the terminal buds are slender, $\frac{1}{4}$ to $\frac{1}{2}$ in. (6 to 12 cm) long; stipular scars lacking.

Amelánchier laèvis Wieg. (*A. arborea* var. *laevis* [Wieg.] Ahles)   Allegheny serviceberry

This species differs from *A. arborea* by having young leaves bronze colored, leaves glabrous or puberulent and at least half-grown at flowering, petioles and pedicels glabrous, and pomes purple-black and sweet. It is found from Minnesota to Newfoundland, south to eastern Kansas, Missouri, Indiana, Ohio, Delaware, and in the Appalachians to Georgia and Alabama, generally at 3000 to 6000 ft elevation. It is very showy with its white flowers in the spring.

Amelánchier alnifòlia (Nutt.) Nutt.   western serviceberry, western shadbush

This shrub or small tree, 20 to 40 ft tall, extends along the coast from southeastern Alaska to northern California, and eastward to southwestern Ontario and Wisconsin. It has purple to blue-black pomes; smooth, gray, and often muscular-appearing stems; and leaves that are usually entire below the middle.

CRATAEGUS L.   hawthorn, thornapple

Crataègus is a very complex genus due to interspecific hybridization and apomixis (Dickinson, 1985). Approximately 35 aborescent species may be recognized in the United States (Little, 1979, 1980a and b) and 30 in Canada (Farrar, 1995). The group is of considerable interest to the plant taxonomist, but the identification of closely related forms is seldom attempted. All are shrubs or small trees with simple serrate or lobed leaves and branches that are armed with thorny modified branchlets and fruits like small apples (Fig. 9-91). The hawthorns are typical of open pastures, where they are often aggressive and difficult to eradicate. Several species are used as ornamentals. Hawthorn is the state flower of Missouri. These are important for deer and cattle browse, and Figure 9-91, pt. 1, shows the typical effects of overgrazing on pasture trees.

THE LEGUMES

The legumes consist of 642 genera and 17,275 species of trees, shrubs, lianas, and herbs widely scattered throughout the world. They are second only to the grasses in their economic importance. Among the herbaceous species are such important forage and food plants as clover, vetch, alfalfa, bean, soybean, peanut, and pea. They are also important for forest products such as lumber, tannins, gums, resins, dyes, and drugs. Several are important as crop weeds or poisonous to livestock. Legumes often have root nodules containing nitrogen-fixing bacteria of great value in enriching the soil.

FIGURE 9-91 *Crataegus,* hawthorn. (1) Pasture trees after grazing. (2) Inflorescence $\times \frac{3}{4}$. (3) Pomes and seed $\times \frac{3}{4}$. (4) Thorn $\times 1$. (5) Twig $\times 1\frac{1}{2}$.

Legumes have long been classified in one family, Leguminosae or Fabaceae s. lat., on the basis of the entomophilous flower, single superior carpel, generally compound (rarely simple) leaves, and the unique fruit type (the *legume*), which normally splits along two sutures, or it may remain indehiscent. As a single family, it was then divided into three subfamilies based mainly on floral morphology. It is felt by many that these three groups are more consistent with the customary concepts of *families* of flowering plants, and they are treated that way here (Table 9-22). However, the phylogenetic relationships are still not well resolved, and with additional investigation, it may be appropriate to again combine the three groups into one family (Judd et al., 1994, 1999).

Mimosaceae: The Mimosa Family

This is a family of 64 genera and 2950 species. A common native genus is *Prosòpis,* which includes the mesquites of the Southwest. These are spiny trees and shrubs characteristic of desert and dry prairie landscapes. The leaves and pods are edible and are most important to the existence of many species of wildlife. The wood is hard and durable and is used in outdoor cooking for its aromatic smoke. Although mainly entomophilous and an important source of honey, the windblown pollen is allergenic. *Acàcia* is another tree of the southern United States. This is a very large genus with 1200 species widely distributed in the tropics and subtropics. Many species are used extensively for such products as gums, tannings, wood, fuel, food, forage, dyes, and perfumes, and as cultivated ornamentals.

TABLE 9-22   COMPARISON OF THE THREE LEGUME FAMILIES

Family	Leaves	Floral symmetry	Petals	Stamens
Mimosaceae	Mostly bipinnate	Actinomorphic	Usually valvate, connate	Much longer than petals; 4–10 to many
Caesalpiniaceae	1-pinnate, bipinnate, or unifoliolate	Slightly to strongly zygomorphic	Imbricate; upper petal inside laterals, 2 lower ones separate	Usually 10 and same length or shorter than petals; separate
Fabaceae	1-pinnate or unifoliolate	Strongly zygomorphic	Imbricate; upper petal (standard) outside laterals (wings), 2 lower ones usually fused (keel)	Usually 10 and same length or shorter than petals; usually enclosed within keel; 10 separate or fused into filament tube, or 9 fused and 1 separate

Albízia julibríssin Durazzini   silktree, mimosa

> **Distinguishing Characteristics.**   *Leaves* bipinnate, 6 to 15 in. (15 to 38 cm) long, leaflets $\frac{3}{8}$ to $\frac{5}{8}$ in. (10 to 15 mm) long, with the major vein near the leading margin, the margins entire. *Flowers* many in a round head, each flower small but with long stamens whitish at base and pink to red toward the ends; appearing after the leaves. *Fruit* a flat legume, 5 to 8 in. (13 to 20 cm) long. *Twig* with 3-lobed leaf scars and 3 bundle scars; stipular scars present; terminal bud lacking; pith solid.
>
> **Range.**   Mimosa is a native of Asia from Iran to China and Japan and is widely planted as an ornamental. It has escaped and become naturalized in the midwestern and southeastern United States, mainly along roadsides, disturbed areas, and edges of woods.
>
> **General Description.**   This is a tree with a short trunk and broad, flattened crown to 20 ft tall. It is a showy tree, flowering over a long period during the summer. The leaflets fold up at night. It was introduced in 1745.

Caesalpiniaceae: The Caesalpinia Family

> This family has 153 genera and 2175 species of trees, shrubs, lianas, and herbs. Seven aborescent genera are represented in North America, with *Gleditsia* being the most important timber tree.

GLEDITSIA L.   honeylocust

> This genus includes ca. 14 species of trees scattered through the forests of North and South America, eastern and central Asia, and tropical Africa. They produce little timber, although several species are highly prized for ornamental purposes.
>
> The genus is represented in the United States by two species. *Gledítsia aquática* Marsh., waterlocust, is an unimportant bottomland species of the South, and *G. triacanthos,* the honeylocust, is a tree of secondary commercial importance. *Gleditsia* ×*texàna* Sarg. is a commonly occurring hybrid between these two.

Botanical Features of the Genus

Leaves deciduous, alternate, 1- to 2-pinnately compound, often clustered on 2-year-old twigs, stipules inconspicuous.

Flowers perfect and imperfect (polygamous and often functionally dioecious), regular, greenish yellow, borne in axillary spikelike racemes; *calyx* 3- to 5-lobed; *petals* 3 to 5; *stamens* 6 to 8, distinct; *pistils* with 2 to many ovules.

Fruit either an elongated, compressed, many-seeded, usually indehiscent legume, or a short, few seeded, tardily dehiscent legume; seeds suborbicular, with hard endosperm.

Twigs stout, conspicuously zigzaged, armed with heavy, 3-branched (rarely 2- to many-branched) thorns; *pith* homogeneous; *terminal buds* lacking; *lateral buds* partly submerged, superposed, scaly; *leaf scars* U-shaped; *bundle scars* 3; *stipular* scars lacking.

Gledítsia triacánthos L.   honeylocust

Distinguishing Characteristics (Figs. 9-92 and 9-93)

Leaves both 1-pinnately and bipinnately compound, the pinnate with 15 to 30 nearly sessile leaflets, the bipinnate leaves with 4 to 7 pairs of pinnae, 4 to 8 in. (10 to 20 cm) long; *leaflets* $\frac{3}{8}$ to $1\frac{1}{4}$ in. (1 to 3 cm) long, $\frac{1}{2}$ to $\frac{7}{8}$ in. (1.3 to 2.2 cm) wide on pinnate leaves, smaller on bipinnate ones; *shape* of leaflets ovate, ovate-lanceolate, or elliptical; *margin* crenulate; *apex* acute or rounded; *base* acute or inequilateral; *surfaces* glabrous, dark green, lustrous above; dull yellow-green, glabrous or nearly so below; *rachis* grooved above, pubescent, swollen at base; *fall color* yellow.

Flowers small, perfect and imperfect, appearing before the leaves; *petals* greenish white; *pistil* with many ovules.

Fruit a reddish brown to purplish brown, strap-shaped, usually twisted legume, 6 to 18 in. (15 to 45 cm) long and about 1 in. (25 mm) wide; *seeds* oval, dark brown, $\frac{1}{3}$ in. (8 mm) long, surrounded by sweet-tasting, edible, yellow-green pulp.

Twigs greenish brown to reddish brown, lustrous, with 3-branched (rarely simple or 1-branched) thorns, 2 to 3 in. (5 to 8 cm) long; *lateral buds* minute, superposed.

Bark on mature trees grayish brown to nearly black, often conspicuously lenticellate, broken up into long, narrow, longitudinal, and superficially scaly ridges that are separated by deep fissures; often with clusters of large, branched thorns.

Range

Eastern (Map 9-34). The map shows its original range, much extended now through the eastern United States, from New England to South Carolina, by cultivation and naturalization; rare in southwestern Ontario. Elevation: sea level to 2000 ft. Floodplains of streams and rivers, lake shores, hammocks, old fields, fence rows, and limestone upland woods.

General Description

Honeylocust is a medium-sized tree 70 to 80 ft tall and 2 to 3 ft dbh (max. 140 by 6 ft), with a rather short bole and open, narrow or spreading crown; the root system is wide and deep. Root sprouts are frequent. Although commonly found on rich, moist bottomlands or on soils of a limestone origin, this species is very hardy and drought-resistant when planted elsewhere, especially in the plains and prairie states, where it has been widely used as windbreaks. When trimmed,

FIGURE 9-92   *Gleditsia triacanthos,* honeylocust. (1) Twig $\times 1\frac{1}{4}$. (2) Branched thorn $\times 1\frac{1}{4}$. (3) Bipinnate leaf $\times\frac{1}{2}$. (4) Once-pinnate leaf $\times\frac{1}{2}$. (5) Bark.

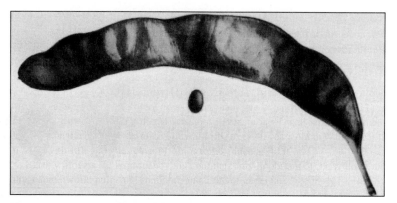

FIGURE 9-93 Legume and seed of honeylocust ×$\frac{1}{2}$.

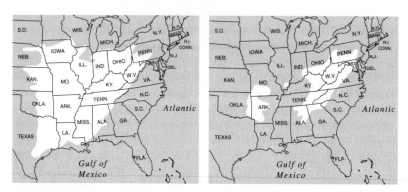

MAP 9-34 **(Left)** *Gleditsia triacanthos* (original range; now naturalized east to the coast and planted in the western states). **(Right)** *Robinia pseudoacacia* (original range; now naturalized throughout much of the eastern United States, southern Canada, Pacific states, and Europe).

honeylocust makes a desirable tall hedge, which soon becomes impassable because of the many forbidding thorns.

Good seed crops occur nearly every year. Growth is rapid, and maturity is reached in about 120 years, although older trees are sometimes found. Honeylocust is intolerant; hence, it occurs in the open or as a dominant tree in the forest, or in mixture with other bottomland hardwoods. Honeylocust, though a legume, does not appear to have root nodules.

This species owes its name to the sweetish pulp between the seeds. When green, these pods are eagerly eaten by cattle and wildlife, and the propagation of thornless honeylocust to be planted as a pasture tree has been recommended. Seed dispersal is by stream waters, wind, deer, cattle, and rodents. The tree is also a source of honey.

The thornless *G. triacanthos* f. *inermis* (Pursh) Schneid. is sometimes found growing wild, and in recent years clones of it and its many cultivars have been planted extensively as street or yard trees in cities. Certain cultivars bear no fruit. Many American elm shade trees killed by the Dutch elm disease are being replaced by the thornless, podless honeylocust; thus, a tree previously rare in cities has become exceedingly common. It is now overused because it is attacked by several insects and fungous diseases. However, it is a beautiful shade tree, and it tolerates a great variety of soil conditions that limit so many other species.

Strong, durable, honeylocust wood has been used for bows, construction lumber, and fence posts; the thorns and pods have been used for various items by local people.

Cércis canadénsis L.   eastern redbud

Distinguishing Characteristics. *Leaves* alternate, unifoliolate (appearing simple), blade nearly orbicular or reniform, $2\frac{1}{2}$ to 5 in. (6 to 12.5 cm) long and wide, entire margined, membranaceous, and glabrous; *petioles* $1\frac{1}{2}$ to 4 in. (4 to 10 cm) long; fall color yellow. *Flowers* bright pink, clustered and appearing before the leaves in early spring. *Fruit* a flat legume, $2\frac{1}{2}$ to $3\frac{1}{4}$ in. (6 to 8 cm) long, turning black. *Twig* slender, brown to black with small, black, rounded buds often superposed, leaf scar with a fringe at the top, 2 or 3 bundle scars, stipular scars lacking, and terminal bud lacking. *Bark* smooth, dark gray to black.

Range. New Jersey south to central Florida, west to southern Texas, north to southeastern Nebraska. Elevation: sea level to 2200 ft. Rich, moist, mixed deciduous or pine woods.

General Desciption. This is a shade-tolerant small tree to about 45 ft tall and commonly used as an ornamental. The pink flowers are very showy in the spring, often contrasting with the white of flowering dogwood. Cultivars with white flowers and darker foliage are available. It is the state tree of Oklahoma.

California redbud, **C. occidentàlis** Torr. ex Gray, is similar except for a more leathery leaf and shorter petioles, and occurs along mountain streams, dry hillsides, or in canyons in California, southern Nevada, southern Utah, and Arizona.

Chinese redbud, **C. chinénsis** Bunge, is a large, upright shrub that is becoming popular as an ornamental.

Gymnócladus dioìcus (L.) K. Koch   Kentucky coffeetree

Distinguishing Characteristics. *Leaves* alternate, deciduous, bipinnate, 12 to 32 in. (30 to 80 cm) long; leaflets 2 to $2\frac{1}{2}$ in. (5 to 6 cm) long, $\frac{3}{4}$ to $1\frac{1}{2}$ in. (2 to 4 cm) wide, with acute apex. *Flowers* imperfect (trees dioecious), white to purplish in large panicles, appearing in spring after the leaves.

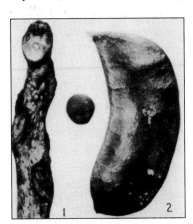

Fruit (Fig. 9-94) a hard, thick pod, 4 to 7 in. (10 to 18 cm) long, $1\frac{1}{2}$ to 2 in. (4 to 5 cm) wide, dark reddish brown, with a thick, inedible pulp surrounding the large seeds. *Twig* (Fig. 9-94) stout, brown, with large leaf scars and thick salmon-colored to brown pith, bundle scars 3 to 5, terminal bud lacking, lateral buds partly concealed, stipular scars minute, fringed. *Bark* gray, deeply furrowed into narrow scaly ridges.

Range. Southern Ontario, east to central New York, southwestward to Oklahoma, north to southern Minnesota; cultivated and naturalized eastward. Elevation: 300 to 2000 ft. Valleys and mixed forests, never common.

General Description. This is a medium to large (max. 102 ft tall) tree. The roasted seeds were once used as a substitute for coffee; however, raw seeds and the pulp between them are poisonous. It is a prolific root sprouter.

FIGURE 9-94   *Gymnocladus dioicus,* Kentucky coffeetree. (1) Twig ×$1\frac{1}{4}$. (2) Legume and seed ×$\frac{1}{2}$.

Fabaceae s. str. (Papilionaceae): The Bean or Pea Family

This family, with 425 genera and 12,150 species, is the largest, most advanced, and economically the most important of the three legume families. There are eight genera and 11 species reaching tree size in the United States.

ROBINIA L.   locust

Robinia includes four species of trees and shrubs, all of which are found in the United States and Mexico, and one in Canada. Several of these have ornamental value, and *R. pseudoacacia* is a timber tree of some importance.

Botanical Features of the Genus

Leaves deciduous, alternate, 1-pinnately compound, mostly spinose-stipulate.

Flowers perfect, papilionate, borne in racemes that appear just after the leaves; *calyx* 5-lobed; *corolla* consisting of a large obcordate standard, 2 obtuse wings, and an incurved keel; *stamens* 10, diadelphous; *pistil* with many ovules.

Fruit a many-seeded, nearly sessile legume; *seeds* reniform, without endosperm.

Twigs moderately stout, angular, somewhat zigzag, reddish brown, usually with stipular spines at least on vigorous branches; *pith* homogeneous; *terminal buds* lacking; *lateral buds* naked, submerged beneath the leaf scar, often superposed; *leaf scars* broadly ovate to somewhat reniform; *bundle scars* 3.

Robínia pseudoacàcia L.   black locust

Distinguishing Characteristics (Fig. 9-95)

Leaves 6 to 14 in. (15 to 35 cm) long, pinnately compound with 7 to 19 subopposite or alternate leaflets; *leaflets* $1\frac{1}{2}$ to 2 in. (4 to 5 cm) long, $\frac{1}{2}$ to $\frac{3}{4}$ in. (12 to 18 mm) wide; *shape* elliptical, ovate-oblong, or ovate; *margin* entire; *apex* mucronate or notched; *base* rounded; *surfaces* dull, dark blue-green, glabrous above, paler and glabrous, except for slight pubescence on the midrib below; *fall color* yellow.

Flowers ca. $\frac{3}{4}$ in. (18 mm) long, white, fragrant, in a drooping raceme.

Fruit a flat, brown, oblong-linear, glabrous legume, 2 to 4 in. (5 to 10 cm) long and ca. $\frac{1}{2}$ in. (12 mm) wide; *seeds* 4 to 8, ca. $\frac{3}{16}$ in. (5 mm) long, reniform.

Twigs (see generic description).

Bark on mature trees reddish brown to nearly black, deeply furrowed into rounded, interlacing, fibrous, superficially scaly ridges. The inner bark contains a poisonous principle; stock have died from browsing the bark or young shoots, and children have become ill by chewing them.

Range

Eastcentral (Map 9-34). Elevation: 500 to 5000 ft. Now naturalized throughout southern Canada, United States, and parts of Europe and Asia. Cove forests and open, upland slopes, fence rows, disturbed ground, and limestone soils.

FIGURE 9-95   *Robinia pseudoacacia,* black locust. (1) Twig with stipular spines and buried buds $\times 1\frac{1}{4}$. (2) Leaf $\times\frac{1}{2}$. (3) Inflorescence $\times\frac{1}{2}$. (4) Legume $\times\frac{3}{4}$. (5) Dehisced legume with seeds $\times\frac{3}{4}$. (6) Bark.

General Description

Black locust is a medium-sized tree 40 to 60 ft tall and 1 to 2 ft dbh (max. 100 by 7 ft), and on good sites it may develop a clear straight bole. However, there are several growth forms of this species, and certain trees are spreading in habit with poorly developed trunks. The root system, although often shallow and wide-spreading, may develop several very deep roots, and in the arid Southwest, such vertical roots may be from 20 to 25 ft long. The crown is open and irregular. This tree does best on moist, rich, loamy soils or those of limestone origin, but is very cosmopolitan and is found on a wide variety of sites, especially old fields and similar cleared areas.

Abundant seed is produced almost every year. Black locust is fast-growing, especially in youth, and on the best soils it will average 2 to 4 ft in height a year. It is intolerant and is not found in dense woods except as a dominant tree. Some associates are dry-soil oaks, hickories, yellow-poplar, white ash, black walnut, eastern redcedar, and shortleaf and Virginia pines. When cut, this locust sprouts vigorously from stump and roots; in this way, it spreads across abandoned fields.

Black locust was once planted extensively for railroad ties and fence posts, but on certain sites the locust borer has severely damaged—even destroyed—whole plantations. Tree vigor is the important factor, and fast-growing trees on the best soils exhibit the greatest resistance. Black locust is commonly used for erosion control and spoil-bank stabilization. It is naturally restocking large areas in the southern mountains. Like most legumes, this species improves the soil through the nitrogen-fixing bacteria in root nodules; the leaf litter decomposes very rapidly, releasing much nitrogen, calcium, and potassium. It is a source of honey.

The wood swells or contracts very little with changes in moisture content and therefore finds a special use for such items as insulator pins and "tree nails," the latter once used in wooden-ship construction. Deer browse the foliage, and birds and small mammals eat the seeds. The bark, how-ever, is toxic to livestock and humans. The wood is exceptional for fuel.

There are several forms of this species; one of these (a clone or mixture of clones), shipmast locust, *R. pseudoacacia* var. *rectíssima* Raber, is especially valuable because of its tall, straight bole, and fewer branches. Numerous cultivars also exist.

Black locust was introduced into Germany very early (about 1601) and has now become in Europe one of the most widely distributed of exotic North American trees. Its use should be tem-pered with the knowledge that it has many pest problems and can become quite invasive.

Robínia neomexicàna Gray   New Mexico locust

Distinguishing Characteristics. *Leaves* pinnate, 4 to 10 in. (10 to 25 cm) long, with 13 to 21 leaflets, each $\frac{1}{2}$ to $1\frac{1}{2}$ in. (1.2 to 4 cm) long, $\frac{1}{4}$ to 1 in. (6 to 25 mm) wide, elliptical, with bristle tip, entire; fall color yellow. *Flowers* purplish pink, in drooping racemes, appearing in late spring and early summer after the leaves. *Fruit* a flat legume, $2\frac{1}{2}$ to $4\frac{1}{2}$ in. (6 to 11 cm) long. *Bark* light gray, furrowed into scaly ridges.

Range. Southeastern Nevada, east to southern and central Colorado, south to western Texas, west to southeastern Arizona. Elevation: 400 to 8500 ft. Canyons and moist slopes.

General Description. Spiny tree to 25 ft tall and 8 in. dbh. It is planted as an ornamental and for erosion control. Foliage and flowers are browsed by livestock and wildlife; Native Americans ate the flowers and pods.

Cladrástis kentúkea (Dum.-Cours.) Rudd (*Cladrastis lutea* [Michx. f.] K. Koch)   yellowwood

Distinguishing Characteristics. *Leaves* 8 to 12 in. (20 to 30 cm) long, pinnate with (5) 7–9 (11) leaflets, alternate or subopposite on the rachis, each $2\frac{1}{2}$ to 8 in. (6 to 20.5 cm) long, $1\frac{1}{4}$ to 5 in.

(3 to 12.5 cm) wide, broadly elliptical, abruptly acuminate; fall color yellow. *Flowers* appearing with the leaves, white, fragrant, in drooping panicles ca. 1 ft (30 cm) long; the lower keel petals separate. *Legume* thin, flat, papery, indehiscent, pendent, 1–3 in. (3 to 8 cm) long, $\frac{3}{8}$ to $\frac{1}{2}$ in. (8 to 11 mm) wide. *Bark* smooth and gray. *Twig* with leaf scars nearly encircling the bud, which is short, naked, and golden pubescent; bundle scars 3–9; terminal bud absent; lateral buds often superposed; stipular scars lacking; pith homogeneous.

Range.   This is a rare tree; Kentucky, southwestern Virginia, eastern Tennessee, and western North Carolina, south to Alabama, west to eastern Oklahoma; disjunct in southern Indiana and southern Ohio. Elevation: 300 to 3500 ft. Stream banks, moist mountain coves and slopes, and limestone cliffs.

General Description.   Yellowwood is a tree to 50 ft tall and $1\frac{1}{2}$ ft dbh and very showy in flower. It is cultivated as an ornamental tree, and the yellow heartwood is used for paneling, gunstocks, turning into platters or bowls, and for a yellow dye. The nomenclature is somewhat debatable, and both names are found in the current literature. There are five other species in eastern Asia. For additional discussion regarding this tree, see Spongberg and Ma (1997).

Elaeagnaceae: The Oleaster Family

This is a small family of three genera and 45 species in warm temperate to subtropical regions. They are deciduous or evergreen trees and shrubs with silvery or brown stellate hairs or scales (lepidote) on the twigs and leaves. The roots have nodules with nitrogen-fixing bacteria; see under *Myrica*. Four to six species of *Elaeagnus* are found in North America. Silverberry, *E. commutàta* Bernh., is a native to Alaska, western Canada, and the northern Great Plains. Two Asiatic shrubs, autumn elaeagnus, *E. púngens* Thunb., and thorny elaeagnus, *E. umbellàta* Thunb., have become naturalized and weedy in many areas. Russian-olive deserves more description here.

Botanical Features of the Family

Leaves deciduous or persistent, alternate (rarely opposite or whorled), simple, entire, estipulate.

Flowers in late spring and early summer, solitary or in small umbels, perfect or imperfect (plant dioecious), regular, with 2 to 4 connate sepals, no petals, stamens 2, 4, or 8, carpel 1 superior to appearing inferior, entomophilous.

Fruit an achene surrounded by a fleshy calyx tube and drupelike in appearance.

Elaeágnus angustifòlia L.   Russian-olive

This is a deciduous shrub or small tree to 25 ft tall, sometimes with spiny twigs, and covered with silvery or brownish scalelike, stellate trichomes on twigs, leaves, flowers, and fruits. *Leaves* alternate, $1\frac{1}{2}$ to $3\frac{1}{4}$ in. (4 to 8 cm) long, $\frac{3}{8}$ to $\frac{3}{4}$ in. (1 to 2 cm) wide, lanceolate or oblong, entire, gray-green. *Flowers* bell-shaped, ca. $\frac{3}{8}$ in. (1 cm) long, fragrant, axillary. *Fruit* drupelike, $\frac{3}{8}$ to $\frac{1}{2}$ in. (1 to 1.2 cm) long, elliptical, yellow to pinkish, sweet. It is a native of Europe and Asia and has been planted as an ornamental, for erosion along rivers, and as windbreaks in the plains. It has become naturalized and locally weedy in moist woods, along streams and rivers, especially in the West, and on slopes throughout the United States and southern Canada. The silvery lepidote leaves and twigs make this particularly attractive as a yard, park, or street tree, and the fruits are eaten by birds and small mammals.

Myrtaceae: The Myrtle Family

A tropical and subtropical family with 129 genera and 4620 species, it is found mainly in the Southern Hemisphere, particularly South America and Australia. None are native in North America, but several trees are widely cultivated in the southern United States and California, and some have become naturalized, weedy, and conspicuous components of many areas.

They are evergreen trees or shrubs with alternate (*Eucalýptus*), opposite or whorled, simple leaves, which are entire, punctate-leathery, and aromatic when crushed; the flowers are showy, with a thick hypanthium and 4 or 5 petals and many stamens; the fruit is a berry or capsule; pollination by insects, birds, or bats.

Some trees are extensively cultivated for the fruit (guava, surinam cherry), spices (cloves, allspice), eucalyptus oil used in medicines, or as ornamentals (bottlebrush, eucalyptus). Two genera have become important in the United States. ***Eucalýptus*** spp. from Australia have been introduced and widely planted as ornamentals in the southern United States, mainly California and Florida. Some have become naturalized in California and form extensive forests. They have been planted for their attractive leaves and bark, for rapid growth, and for windbreaks, soil stabilization, and fuel. The highly flammable leaves and twigs, however, are a dangerous and unwanted fuel during wildfires in California. These are tall trees to 120 ft tall and 3 ft dbh, evergreen, with narrow, leathery leaves on drooping branches and with an odor of camphor when crushed, and mottled green, yellow, and brown bark. Some species are grown in plantations, in various parts of the world, for pulpwood, fuel, and lumber (Hora, 1981; Zobel et al., 1987), and form possibly the world's most productive industrial forests (Laarman and Sedjo, 1992). It is mainly entomophilous, but the pollen is suspected to be allergenic.

Melaleùca quinquenérvia (Cav.) S. T. Blake   cajeput-tree, punktree

This evergreen tree is a native of Australia but has become naturalized in southern Florida, particularly in the Everglades. It was originally planted in southern Florida to dry up this region. It is so prolific and widespread by small, windblown seeds, as well as root sprouts, that it is now detrimental and of great environmental concern. It is recognized by its alternate, leathery, slender, aromatic leaves with five veins, showy white flowers, and multilayerd, spongy bark.

Cornaceae: The Dogwood Family

The Cornaceae s. lat., including the Nyssaceae, have 14 genera and 120 species of trees, shrubs, and few herbs scattered through the forests of the Northern Hemisphere. See Eyde (1988) for the reasons for maintaining one family. Two genera are of importance (Table 9-23).

TABLE 9-23   COMPARISON OF THE NATIVE GENERA OF THE CORNACEAE S. LAT.

Genus	Leaves	Twigs	Drupes
Cornus dogwood	Usually opposite; veins arcuate	Pith homogeneous; bud scales 2, valvate	Blue, white, or red; 1 to 2 seeds; pit barely grooved
Nyssa tupelo	Alternate; veins not arcuate	Pith diaphragmed; bud scales 3 to 5, imbricate	Blue or purple; 1 seed; pit winged or longitudinally ribbed

Botanical Features of the Family

Leaves deciduous, alternate or opposite, simple, usually entire, estipulate.

Flowers perfect or imperfect, 4- or 5-parted, ovary inferior, entomophilous.

Fruit a drupe or berry.

CORNUS L.   dogwood, cornel

Córnus is a genus of small trees and shrubs (rarely herbs) numbering about 65 species. Except for a single Peruvian species, the dogwoods are restricted to the Northern Hemisphere. Probably the best known of the exotic ornamentals in North America are the cornelian cherry, *Cornus más* L., and kousa dogwood, *C. koùsa* Hance.

Sixteen species of *Cornus* are native to the United States, 12 native in Canada, and 11 attain tree size. Most of these are of but minor importance, and only two of them will be considered in detail. *Cornus alternifòlia* L., alternate-leaf dogwood or pagoda dogwood, with alternate leaves, greenish brown twigs, and blue fruit, is a small tree of the East; *Cornus drummóndii* C. A. Meyer, rough-leaf dogwood, is a shrub or small tree of central United States with leaves having scattered, bristly hairs on the upper surface; *Cornus occidentàlis* (Torr. and Gray) Cov., western dogwood, is a large shrub or small tree native to the Pacific Northwest with hairy inflorescence branches.

Botanical Features of the Genus

Leaves opposite, or rarely alternate, with entire or finely toothed margins, and arcuate venation.

Flowers perfect, usually small, in terminal cymes, panicles, or large-bracted heads; flower parts in fours, calyx and corolla present.

Fruit a red, white, blue, or green 2-celled, 2-seeded drupe.

Córnus flórida L.   flowering dogwood

Distinguishing Characteristics (Fig. 9-96). *Leaves* opposite, $2\frac{1}{2}$ to 5 in. (6 to 13 cm) long, $1\frac{1}{2}$ to $2\frac{1}{2}$ in. (4 to 6 cm) wide, oval or elliptical, arcuately veined, fall color dark red. *Flowers* in heads, with four very conspicuous white or pink, notched bracts formed by the enlarged bud scales; appearing before the leaves. *Fruit* a bright red drupe. *Twigs* slender, somewhat angled, purplish or greenish with a glaucous bloom; terminal flower buds subglobose and stalked, leaf buds acute, and covered with two valvate scales; leaf scars narrow with 3 bundle scars; pith homogeneous; stipular scars lacking. *Bark* broken up into small blocks.

FIGURE 9-96   *Cornus florida*, flowering dogwood. (1) Twig ×$1\frac{1}{4}$. (2) Flower bud ×$1\frac{1}{4}$. (3) Head with 4 white bracts and small central flowers ×$\frac{1}{2}$. (4) Flower ×3. (5) Cluster of drupes ×1. (6) Leaf ×$\frac{1}{2}$. (7) Branch in flower (*Photo by C.A. Brown*). (8) Bark.

Range.   Southern Michigan, southern Ontario, and southern Maine, south to central Florida, west to eastern Texas, and north to central Missouri. Elevation: sea level to 5000 ft in the southern Appalachians. Hardwood and mixed conifer forests, mountain slopes, and coves.

General Description.   Flowering dogwood is a small, bushy, tolerant tree that rarely attains a height of more than 40 ft (max. 55 ft) and dbh of 12 to 18 in. The trunk is short with little taper; from 6 to 10 ft above the ground, it usually breaks up into several large, wide-spreading limbs resulting in a low, dense crown. Best development is reached as an understory species in association with other hardwoods. This tree is usually found on well-drained friable soils. Its leaves are exceptionally rich in calcium. Because it is shallow-rooted, dogwood is very drought-sensitive.

Flowering dogwood is a widely used ornamental, with numerous cultivars, and makes a striking display when it is in full bloom in the early spring, and then with the red fruit and red autumn leaves. It is the state flower of North Carolina and Virginia and the state tree of Missouri and Virginia. All parts are important deer browse, particularly in the spring, and numerous animals, including three dozen bird species, eat the fruits. The wood is white and hard and used for spools and small pulleys, and although once used for shuttles for weaving, these have now been replaced by plastic.

Dogwood anthracnose, a damaging fungous disease, introduced from Asia, was first reported on *C. nuttallii* and *C. florida* in the mid to late 1970s. It has spread through the Northeast south to South Carolina on flowering dogwood and in the West on Pacific dogwood. It has the potential to cause major destruction to the native stands, particularly in forest situations. A drastic decline is noticeable in the southern Appalachians.

Córnus nuttállii Audubon   Pacific dogwood

Distinguishing Characteristics (Figs. 9-97, 9-98). *Leaves* opposite, $2\frac{1}{2}$ to $4\frac{1}{2}$ in. (6 to 11 cm) long, $1\frac{1}{4}$ to $2\frac{3}{4}$ in. (3 to 7 cm) wide, ovate to obovate. *Flowers* in heads, with 4 to 7 white, acute or truncate bracts, appearing in spring and early summer or sporadically until fall. *Fruit* a bright red to orange-red drupe. *Twigs* similar to those of the preceding species. *Bark* smooth or with thin scaly plates near the base of the tree.

Range. Southwestern British Columbia, south to western Oregon, and in mountains to southern California. Elevation: sea level to 6000 ft. Moist slopes and valleys as an understory of conifers.

General Description. This species is a small tree, rarely ever attaining a height of more than 60 ft or a dbh of 12 to 20 in. (max. 80 ft by 4.5 ft). Best development is reached in the Puget Sound basin and in the redwood belt of California. Under forest conditions, it develops a slightly tapered stem that extends through the crown, but in more open situations the bole is commonly short and supports a number of spreading limbs forming an ovoid to rounded-conical crown, or it may break up at the ground into several nearly erect stems resulting in a bushy habit. Pacific dogwood is one of North America's finest ornamental trees. The tree is peculiar in that it frequently flowers a second time during the late summer while the fruits of the first flowering are turning red. The snowy-white bracts and the brilliant red fruits against a background of lustrous green

FIGURE 9-97   *Cornus nuttallii,* Pacific dogwood. (1) Leaf $\times\frac{1}{2}$. (2) Head of flowers with 5 bracts $\times\frac{3}{4}$.

FIGURE 9-98 Bark of Pacific dogwood.

foliage produce an extremely beautiful effect. This is the provincial flower of British Columbia.

NYSSA L.   tupelo

This genus consists of seven arborescent or shrubby species: two in Asia, one in Central America, and four in southern North America (Table 9-24). Fossil leaves, pollen, fruit, and wood of many Tertiary *Nyssa* species are widely distributed across Europe, Asia, and North America. Large numbers of fruits are found in the brown coal of Brandon, Vermont (Eyde, 1963).

Botanical Features of the Genus

Leaves deciduous, alternate, simple, petiolate, often crowded toward the tips of the twigs, especially on side branchlets.

Flowers perfect or imperfect (plant polygamo-dioecious), appearing with or before the leaves, small, greenish white, borne in capitate clusters, racemes, or solitary.

Fruit an ovoid or oblong drupe, with a 1-celled, 1-seeded, winged or ribbed pit.

Twigs slender to stout, glabrous, greenish brown, olive-brown, tan, or reddish brown; *pith* mostly diaphragmed; *terminal buds* present, with several imbricated scales; *lateral buds* similar but smaller; spur growth often abundant; *leaf scars* reniform to semicircular; *bundle scars* 3; *stipular scars* lacking.

Nyssa aquática L.   water tupelo

Distinguishing Characteristics (Figs. 9-99, 9-100, 9-101, and 9-102)

Leaves 4 to 8 in. (10 to 20 cm) long, 2 to 4 in. (5 to 10 cm) wide; *shape* oblong-obovate; *margin* entire to repand-toothed near apex; *apex* acute to acuminate; *base*

TABLE 9-24   COMPARISON OF IMPORTANT SPECIES OF TUPELO

Species	Leaves	Twigs	Drupes
N. aquatica water tupelo	4–8 in. (13–20 cm) long, oblong-obovate, sometimes irregularly toothed above	Stout; terminal buds globose, obtuse, ca. $\frac{1}{8}$ in. (3 mm) long	Ca. 1 in. (2.5 cm) long, reddish purple; pit prominently ribbed
N. sylvatica black tupelo	2–5 in. (5–13 cm) long, obovate, rarely toothed	Slender; terminal buds ovoid, acute, ca. $\frac{1}{4}$ in. (6 mm) long	$\frac{3}{8}-\frac{1}{2}$ in. (9–12 mm) long, blue-black; pit indistinctly ribbed

FIGURE 9-99   *Nyssa aquatica,* water tupelo. (1) Twig ×1¼. (2) Staminate inflorescence ×¾. (3) Leaf ×½. (4) Drupes ×¾. (5) Pit showing ridges ×1½.

cuneate to rounded; *surfaces* dark green and lustrous above, paler and more or less downy below; *fall color* red.

FIGURE 9-100   Fruit pits of *Nyssa* ×1.
(1) *N. aquatica.* (2) *N. sylvatica.*
(3) *N. biflora.* (4) *N. ogeche.*

Flowers appearing in March and April; *staminate* in dense clusters; *pistillate* solitary on long slender peduncles; stamens present, often functioning.

Fruit pendent on slender stalks; ca. 1 in. (2.5 cm) long, oblong, dark reddish purple; *pit* light brown to whitish and conspicuously ribbed with about 10 thin ridges.

Twigs rather stout, reddish brown; *pith* diaphragmed; *terminal buds* yellowish, usually small and somewhat globose or rounded, 1/16 to 1/8 in. (1.5 to 3 mm) in diameter; *lateral buds* small and inconspicuous; *leaf scars* rounded; *bundle scars* 3, conspicuous.

Bark thin, brownish gray with scaly ridges.

FIGURE 9-101   Water tupelo swamp. Note buttresses and high water marks. (*Courtesy of U.S. Forest Service.*)

FIGURE 9-102   Bark of water tupelo. (*Photo by J.C. Th. Uphof.*)

Range

Southeastern (Map 9-35). Elevation: sea level to 500 ft. Floodplain forests, swamps, pond and lake margins.

General Description

Water tupelo is a medium-sized to large tree 80 to 90 ft tall and 3 to 4 ft dbh (max. 110 by 8 ft). The trunk is buttressed at its base but tapers rapidly to a long, clear bole; the crown is rather narrow and usually open. Water tupelo is one of the most characteristic of southern swamp trees and is found on sites that are periodically under water. For a short time during the summer, this tree may be found on dry ground, but in the fall, winter, and spring the water is often 3 to 6 ft deep over the same area (Fig. 9-101). Water tupelo occurs in almost pure and very dense stands or mixed with baldcypress, which is its almost constant associate. Others include overcup and water oaks, black willow, swamp cottonwood, red maple, sweetgum, and slash pine.

This tupelo is a prolific annual seeder, and the heavy seeds are distributed by water, birds, and rodents. Following the recession of high water, they become lodged in the mud and may germinate even after more than a year of submergence. The growth of trees on moist but well-drained bottomlands is rapid, whereas that of trees standing on extremely wet sites is much slower. Water tupelo is intolerant, and on the better sites is suppressed by other species.

The timber is commercially important for lumber, veneer, and pulp, and the root wood, which is especially light and spongy, is used locally for bottle corks and fishnet floats. The fruit is important to wood ducks and other wildlife. Tupelo honey is an important item in a number of

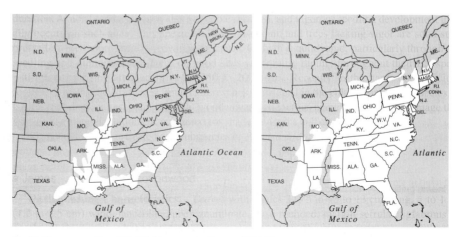

MAP 9-35   **(Left)** *Nyssa aquatica.* **(Right)** *Nyssa sylvatica* and *N. biflora.*

localities where the trees are abundant; however, the famous gourmet tupelo honey is from the Ogeechee tupelo mentioned later.

Nyssa sylvática Marsh.   black tupelo, blackgum, sourgum

Distinguishing Characteristics (Figs. 9-100, 9-103). Black tupelo differs from the preceding species as follows: *leaves* smaller, 2 to 5 in. (5 to 13 cm) long, 1 to 3 in. (2.5 to 7.5 cm) wide, obovate and usually entire or with 1 or 2 large teeth on the upper third of the blade, the small veins

FIGURE 9-103   *Nyssa sylvatica,* black tupelo. (1) Twig $\times 1\frac{1}{4}$. (2) Leaf $\times \frac{1}{2}$. (3) Drupes $\times \frac{3}{4}$. (4) Bark.

inconspicuous on the lower surface; fall color orange-yellow to brilliant red to dark red; *fruit* in clusters of 4 or 5, smaller, $\frac{3}{8}$ to $\frac{1}{2}$ in. (9 to 12 mm) long, with an indistinctly ribbed pit; *twigs* more slender, and the ovoid terminal buds are ca. $\frac{1}{4}$ in. (6 mm) long; *bark* "blocky," or with the appearance of alligator hide; and the base of the trunk is less swollen.

Range.   Eastern (Map 9-35). Elevation: sea level to 4000 ft. Mesic forests, swamp borders, also dry slopes.

General Description.   Black tupelo is usually a medium-sized tree, 50 to 60 ft tall and 2 to 3 ft dbh (max. 121 by 5 ft). Best growth is made on moist alluvial soils with some drainage, but the tree is very cosmopolitan and grows on relatively dry upland sites. It is found in 29 forest-cover types, and its associates include both wet-soil and dry-soil hardwoods and conifers. Black tupelo can live over 500 years. Although black tupelo is recorded as tolerant, swamp tupelo is intolerant.

Large trees are often hollow all the way to the ground, due to various decay-producing fungi and many wood-boring insects. However, the tree may continue to grow for many decades, adding new wood and increasing its girth each year while the hollow cavern in its trunk gets larger as well. Such trees are of no value for lumber but is a boon to wildlife because the hollow center can provide shelter for several generations of wildlife species from insects and nesting birds to rabbits, squirrels, 'possums, and even hibernating bears. The tree can still bear leaves, be browsed by deer, and its berries eaten by birds and various mammals. In earlier days, hollow gum logs were often cut and used for "bee gums," "rabbit gums," or storage containers for various other items (Elliott, 1994).

The wood is soft and light, but the fibers interlock so it is very tough. It is used for crates and boxes, and if not hollow, the logs can be used for railroad ties and wharf pilings, gun stocks, chopping bowls, ox yokes, and mauls. With its glossy leaves and outstanding fall color, black tupelo is a fine, although relatively slow growing, ornamental shade tree.

Nyssa biflòra Walter (*N. sylvatica* var. *biflora* [Walt.] Sarg.)   swamp tupelo

Much of the tupelo harvested from southern Coastal Plain swamps is from swamp tupelo. This may be distinguished from black tupelo by its narrower ($\frac{5}{8}$ to $1\frac{1}{2}$ in. [1.5 to 4 cm] wide), often oblanceolate leaves, and fruit pit, which has ridges slightly more prominent than those of *N. sylvatica* (Fig. 9-100). Swamp tupelo, as its name indicates, is typical of very wet sites; the more upland and northern *N. sylvatica* commonly inhabits higher ground and moist sites. Swamp tupelo is considered to be a distinct species according to a study by Burckhalter (1992).

This and other species of *Nyssa* may produce root sprouts that grow vigorously and form thickets around the base of the tree (Eyde, 1963). They are important deer browse. Flowers are an important source of nectar for honey.

Black and swamp tupelo are easily confused with persimmon (*Diospyros virginiana*) without flowers or fruits. *Diospyros* differs by having leaves gray-green below with evident darker venation, even the ultimate veins obvious without magnification, rather than green below with obscure ultimate veins; twig with a pseudoterminal rather than true terminal bud; one bundle scar rather than three; year-old twigs with chambered rather than diaphragmed pith; and the adaxial petiole and proximal midrib surfaces with minute glandular hairs rather than eglandular.

The remaining tree species, *N. ogèche* Bartr. ex Marsh, Ogeechee tupelo, with tomentose flowers and more or less hairy leaves, is found in coastal Georgia, adjacent South Carolina, and northern Florida. Thousands of seedlings have been planted in western Florida for honey production. It is related to *N. aquatica* and has the largest fruits (Fig. 9-100), which are used to make preserves and beverages.

Aquifoliaceae: The Holly Family

The Aquifoliaceae include four genera and 420 species of trees and shrubs of wide distribution through the temperate and tropical forests of both hemispheres.

Two genera and about 15 species are found in the United States and Canada. *Nemopánthus mucronàtus* (L.) Trel., the mountain-holly, is a large shrub or small tree of southern Canada and eastern United States. *Ilex* is the important genus of trees.

Botanical Features of the Family

Leaves often evergreen, but in some species deciduous, alternate, simple, stipulate.

Flowers usually imperfect (most species dioecious), superior ovary, small, 5-parted, entomophilous.

Fruit a drupe with several stones (pyrenes).

ILEX L.   holly

The genus *Ilex* consists of 400 species of evergreen or deciduous trees and shrubs, widely scattered throughout the world and appearing in the floras of every continent with the exception of Australia. *Ilex aquifòlium* L., English holly, was successfully introduced into North America many years ago and bears fruit more plentifully than the American species, *I. opaca.* The leafy sprays of both species are used for Christmas decoration. In the West, the climate of the Puget Sound region and southwestern British Columbia (where it has become naturalized) is favorable to holly culture, and holly farms have paid good dividends. The English holly and many of its cultivars are those chiefly favored, although other species and hybrids are very popular ornamentals. Holly culture is also practiced to a certain extent in the Southeast, where the native species is used almost exclusively.

The flora of eastern United States includes about 16 species of *Ilex,* 14 of which are usually arborescent. One of these, *Ilex opaca,* is a timber tree of secondary importance.

Botanical Features of the Genus

Leaves deciduous or persistent; entire, serrate, or aculeate (spiny-toothed); stipulate with very small, nearly black, triangular, persistent stipules.

Flowers perfect and imperfect (dioecious or polygamous), corolla white, axillary, small.

Fruit a red or black (or rarely yellow) drupe.

Ìlex opàca Ait.   American holly

Distinguishing Characteristics (Fig. 9-104). *Leaves* persistent, leathery, 2 to 4 in. (5 to 10 cm) long, $\frac{3}{4}$ to $1\frac{1}{2}$ in. (2 to 4 cm) wide, elliptical, entire or spiny-toothed. *Flowers* imperfect (plant dioecious), solitary or cymose, small, greenish white. *Fruit* red (reputedly poisonous if eaten), containing a few ribbed stones. *Bark* light gray or sometimes rough and warty, and usually with variously colored crustose lichens.

Range. Eastern Massachusetts, south to central Florida, west to southcentral Texas, north to southeastern Missouri. Elevation: sea level to 5000 ft in the southern Appalachians. Floodplains and mixed hardwood forests.

FIGURE 9-104   *Ilex opaca,* American holly. Branch with leaves and drupes ×$\frac{1}{2}$.

General Description.   This species is the largest of the native hollies and varies from 40 to 50 ft in height and 1 to 2 ft dbh (max. 100 by 4 ft). The bole is straight and regular but usually short. The tree does best on deep, moist bottomlands but will persist, especially in the North, on dry, gravelly soil. American holly is very tolerant of forest competition and is quite resistant to saltwater spray; it is used along the coast in exposed places as an ornamental.

Fruit dispersal is facilitated by birds. Growth is slow, and the tree matures in 100 to 150 years. It may be propagated by cuttings and is now raised in this way for decorative purposes. In this connection, the production of fruit can be assured by taking cuttings from pistillate trees only. The branches and leaves are in demand during the Christmas season, and in certain areas the natural stands of this tree have been nearly exterminated. Holly seems to have been a typically southern tree, but during the last few centuries it has moved steadily northward (seeds carried by birds and landscapers), now reaching New England. It is also cultivated, and over 1000 cultivars have been named.

Since the days of the Greeks, Romans, and Druids, and presumably much earlier, hollies with their evergreen leaves and bright red fruits attracted attention and played an important part in primitive magic, medicine, and folklore (Dengler, 1966). It is the state tree of Delaware.

The wood is hard and white and used for veneer, cabinetwork, handles, carvings, and various specialty items. The fruit is an important food for birds but should not be eaten by children. It is a source of honey.

Euphorbiaceae: The Spurge Family

This is a very large family with 313 genera and 8100 species mainly tropical but also represented in temperate regions. It includes trees, shrubs, and herbs. Several are of great economic importance for timber, rubber, casterbean oil, tung oil, tapioca, vegetable tallow, dyes, and the showy and popular poinsettia.

The family is not important in North American forestry, yet two Asian trees have become naturalized and so invasive in the southeastern United States that they merit brief mention. These trees are deciduous, have alternate, simple leaves with 2 glands on the upper petiole just below the blade, milky juice, and a capsule.

Aleurites fórdii Hemsl.   tung-tree

> This is a small tree cultivated along the Gulf Coast for its seeds, which yield an important oil. *Leaves* long-petioled with 2 red glands at the top, stipulate, the blades large, ovate, with cordate base and palmate veins, the margins entire, sometimes shallowly lobed at the top; *flowers* large and showy, the petals 5–8, white with red veins inside near base; *capsule* large, green to red-brown, dehiscing into 3–5 sections with large, angled seeds. The seeds are extremely poisonous.
>
> This tree has escaped cultivation along borders of woods, roadsides, and hedgerows from Georgia to Louisiana.

Sàpium sebíferum (L.) Roxb.   Chinese tallow-tree, popcorn-tree

> This is a small- to medium-sized tree, cultivated since 1850 for its attractive leaves, "popcorn"-like seeds, and fall coloration. *Leaves* long-petioled, $1\frac{1}{2}$ to 3 in. (4 to 7.5 cm) long and wide, broadly rhombic-ovate with truncate base and abruptly acuminate apex, pinnately veined, margin entire, yellow to orange to red-purple fall coloration; *flowers* small in long, slender spikes; capsules 3-valved, splitting and exposing 3, dull white, waxy, persistent seeds. The milky sap is poisonous.
>
> This tree has very rapid growth and has become a major pest of coastal tall grass prairies and a persistent invader of abandoned agricultural lands and bottomland hardwoods in the Coastal Plain from North Carolina to Texas.

Rhamnaceae: The Buckthorn Family

The Rhamnaceae include 49 genera and ca. 900 species of trees and shrubs (sometimes lianas, rarely herbs) widely scattered through the tropics and warmer regions of the world. *Ceanòthus* L., an American genus, consists of ca. 55 shrubby species, many of which are found in the semiarid West. Here they may serve as browse for both sheep and cattle. *Ceanothus* roots contain nitrogen-fixing actinomycetes (see under *Myrica*). The seeds of several *Ceanothus* species are eaten by Native Americans and, to a limited extent, by livestock. The roots, bark, stems, and even leaves of many rhamnaceous plants contain compounds used for pharmaceutical purposes. The fruits of the common jujube, *Zíziphus jujùba* Mill., are edible, and this species is now widely cultivated in many sections of Asia. Although there are a few species that produce timbers of commercial rank, the family is not important for timber.

Ten genera and about 100 species of the Rhamnaceae are found in the United States. *Rhamnus,* however, is the only genus with species of special economic interest. Probably the heaviest (specific gravity 1.3) wood grown in the United States is from leadwood, *Krugiodéndron férreum* (Vahl) Urban, found in southern Florida.

Botanical Features of the Family

Leaves deciduous or persistent, alternate or subopposite, simple, stipulate.

Flowers perfect or polygamous, small, actinomorphic, mostly 5-parted, ovary 2- or 3-carpellate, superior, entomophilous.

Fruit drupe or capsule, sometimes winged.

RHAMNUS L.   buckthorn, cascara

This genus has 125 species of trees and shrubs widely scattered through the temperate and tropical forests of both hemispheres. *Rhámnus cathártica* L., the European buckthorn, was introduced into the United States and Canada for decorative purposes and has become naturalized and a serious weed in many sections of the East. *Rhamnus frángula* L., alder buckthorn, also introduced from Europe, has become naturalized in eastern North America and is also becoming a serious invasive plant. All buckthorns contain toxic glycosides. Several have been used for dyes, medicines (purgatives), and charcoal.

Twelve species of *Rhamnus* are native to the United States, and five of these are arborescent. Two species are native to Canada. *Rhamnus purshiana* is an important tree of the Pacific coast.

Botanical Features of the Genus

Leaves deciduous or persistent, alternate or subopposite, simple, stipulate.
Flowers perfect or polygamous, axillary.
Fruit a drupe, fleshy, several-seeded or sometimes 1-seeded.

Rhámnus purshiàna DC.   cascara buckthorn, bearberry, bitter-bark, chittimwood

Distinguishing Characteristics (Fig. 9-105). *Leaves* deciduous or tardily so, simple, alternate or subopposite, 2 to 6 in. (5 to 15 cm) long, elliptical to oblong-ovate or -obovate, finely serrate or entire, and often remotely revolute; pinnate secondary veins conspicuous, parallel, and with a slight curve upward; fall color yellow. *Flowers* axillary in cymes, yellowish green, 5-parted, appearing in spring and early summer. *Fruit* globose, bluish black, with 2 or 3 obovoid stones. *Twigs* slender, with naked, hoary tomentose buds; leaf scars elevated, narrow lunate, with a few scattered bundle scars; stipular scars conspicuous; pith homogeneous. *Bark* smooth, becoming scaly with age.

Range. Southern and southwestern British Columbia, south to northern California; also in the Rockies of northern Idaho and western Montana. Elevation: sea level to 5000 ft. Disturbed area, roadsides, and understory of mixed forests.

General Description. Cascara buckthorn is a small tree usually not more than 30 to 40 ft tall and 10 to 15 in. dbh (max. 60 by 3 ft). It is moderately gregarious and often forms small groves on moist bottomlands and burns. It is by no means restricted to such sites, however, and may also be found on gravelly or sandy soils, near sea level, or at moderately high elevations. It occurs as an understory species in Douglas-fir forests and is commonly mixed with grand fir, western hemlock, bigleaf and vine maples, and red alder in many sections of the Pacific Northwest.

FIGURE 9-105   *Rhamnus purshiana*, cascara buckthorn. (1) Twig ×1. (2) Leaf ×$\frac{3}{5}$.

Cascara buckthorn is a prolific annual seeder. Moist forest litter and mucky soils make the best seedbeds. Growth is extremely rapid on such sites, and maturity is reached in about 50 years.

When the early Spanish missionaries explored northern California, they found that the Native Americans near the Oregon border were using the intensely bitter bark of this tree as a cathartic, and it was christened "cascara sagrada," or the sacred bark. In modern times, Portland, Oregon, has become the center for collectors to bring in their annual "peel." The peeled tree dies, but vigorous sprouts seem to ensure a continuing supply of the bark.

Sapindaceae: The Soapberry Family

This family of 131 genera and 1450 species is found mostly in tropical and subtropical climates with only a few representatives in the North Temperate Zone. They are trees, shrubs, and vines. Some are favorite ornamentals in North America (for example, Goldenrain tree, *Koelreutéria paniculàta* Laxm.). Some others are important for timbers, or for tropical edible fruits, such as lychee (*Lìtchi*) and akee (*Blìghia*).

The Sapindaceae, Hippocastanaceae, and Aceraceae are very closly related and there is recent evidence indicating that all three should be included in the Sapindaceae s. lat. (Judd et al., 1994, 1999).

Botanical Features of the Family

Leaves persistent or deciduous, alternate (rarely opposite), pinnate or bipinnate, or simple, mostly estipulate, often with toxic saponins.

Flowers small, imperfect (tree monoecious or dioecious), 4- or 5-parted, ovary 3-carpellate, superior, actinomorphic or zygomorphic, entomophilous.

Fruit a berry, drupe, capsule, nut, samara, or schizocarp.

Sapíndus drummóndii Hook. & Arn.   western soapberry

Distinguishing Characteristics. *Leaves* tardily deciduous, alternate, pinnately compound, 5 to 8 in. (13 to 20 cm) long with 11 to 19 leaflets, alternate or subopposite on the rachis, each $1\frac{1}{2}$ to 3 in. (4 to 7.5 cm) long, $\frac{3}{8}$ to $\frac{3}{4}$ in. (1 to 2 cm) wide, lanceolate and falcate, acuminate, entire. *Flowers* small, yellow-white, in erect panicles, in late spring or summer. *Fruit* berrylike, $\frac{3}{8}$ to $\frac{1}{2}$ in. (10 to 12 mm) in diameter, yellow to orange turning black, poisonous. *Twig* yellow-green, puberulent, with triangular to 3-lobed leaf scar and 3 bundle scars, buds with 2 scales, terminal bud lacking, stipular scar lacking, pith solid and homogeneous. *Bark* light gray, becoming rough and furrowed.

Range. Southwestern Missouri, south to Louisiana, west to southern Arizona, northeast to southeastern Colorado. Elevation: 2400 to 6200 ft. Along streams, limestone uplands, in open grasslands in western plains, upper desert, and on mountain slopes in hardwood forests.

General Description. This is a medium-sized tree 20 to 40 ft tall and to 1 ft dbh. The fruits have been used as a substitute for soap for washing clothes; however, these fruits are poisonous if eaten and may cause a skin rash. The wood splits easily and is used for making baskets.

Hippocastanaceae: The Buckeye Family

The Hippocastanaceae include two genera and ca. 15 species of trees and shrubs scattered in the forests of northern South America, Central America, Mexico, eastern and western United States, southeastern Europe, eastern Asia, and India. As a group, they are of little value for the timber that they produce, but many of them, either because of their showy flowers or handsome foliage, are highly prized for ornamental purposes. One genus, *Aesculus,* is represented in North America. The twigs, leaves, and seeds are poisonous if eaten.

Botanical Features of the Family and Genus

Leaves deciduous, opposite, palmately compound with (3)5–9 leaflets, each short-stalked, serrate, petiole long, estipulate.

Flowers perfect or often imperfect in the same panicle; calyx connate and 5-lobed; corolla 4- to 5-parted, zygomorphic; ovary superior; entomophilous; appearing with the leaves in erect, terminal, many-flowered panicles.

Fruit a leathery capsule with 1 to 6 large brown seeds each with a large, light-colored hilum.

AESCULUS L.   buckeye, horsechestnut

The genus *Aesculus* consists of ca. 13 species of trees or large shrubs widely distributed throughout the forests of the Northern Hemisphere. Only two species are important forest trees. Their showy flowers make them useful ornamentals, although litter from the husks and seeds can be quite messy. Horsechestnut and most buckeyes suffer from various leaf diseases and scorch.

Aèsculus flàva Soland. (*A. octándra* Marsh.)   yellow buckeye

Distinguishing Characteristics (Figs. 9-106, 9-107, and 9-108).   *Leaves* palmately 5-folio-late, with nearly elliptical, serrate leaflets, fall color yellow. *Flowers* yellowish, petals 4, with the stamens usually shorter than the 2 lateral petals; pedicels with glandular hairs. *Fruit* smooth. *Twigs* stout, with large nonresinous terminal buds with imbricate scales; leaf scars large, obdeltate with several bundle scars arranged in a V-shaped pattern; pith homogeneous. *Bark* breaking up into large scaly plates.

Range.   Extreme southwestern Pennsylvania, south along the Appalachians to northern Georgia and Alabama, west through southern Ohio to southern Illinois. Elevation: 5000 to 6300 ft. Mesic slopes and coves, and along rivers and streams.

General Description.   The largest of the buckeyes, yellow buckeye is a medium-sized to large tree 60 to 90 ft tall and 2 to 3 ft dbh (max. 152 by 5 ft). Best development is made on deep fertile soils in the mountains of North Carolina and Tennessee, where it occurs in mixture with other hardwoods. It is often a bottomland species, but in the southern Appalachians it is found in rich, high mountain slopes and coves, where it is sometimes one of the codominants with basswood and sugar maple. Growth is fairly rapid, and maturity is reached in 60 to 80 years. It is tolerant and maintains itself well in the climax forests of its range and can live for 430 years. The wood is used for furniture and for turning and carving.

Aèsculus glàbra Willd.   Ohio buckeye, fetid buckeye

Distinguishing Characteristics.   This species differs from *A. flava* in the following features: leaves slightly smaller, the 5 leaflets more nearly lanceolate; bruised foliage and twigs give off a

FIGURE 9-106 *Aesculus flava,* yellow buckeye. (1) Twig $\times 1\frac{1}{4}$. (2) Leaf $\times\frac{1}{2}$. (3) Capsule $\times\frac{3}{4}$; the atypical short prickles seen on the left side are due to introgression from *A. glabra.* (4) Seed $\times\frac{3}{4}$.

FIGURE 9-107 Flowering branch of yellow buckeye. (*Photo by J. W. Hardin.*)

FIGURE 9-108   Bark of yellow buckeye. (*Courtesy of U.S. Forest Service.*)

disagreeable odor; stamens longer than the 4 light yellow petals (Fig. 9-109); pedicels without glandular hairs; fruit somewhat prickly; and bud scales prominently keeled.

Range.   Southern Michigan, east to western Pennsylvania, south to central Alabama, west to Texas, and north to central Iowa. Elevation: 500 to 2000 ft. Valleys and mountain slopes, along stream and riverbanks, flatwoods, and often isolated in open fields. The western populations, with 7 to 11 narrower leaflets, are considered as *A. glabra* var. *arguta* (Buckl.) Robinson (*A. arguta* Buckl.).

General Description.   This is a tolerant, medium-sized tree, 30 to 70 ft tall and 1 to 2 ft dbh with an irregular, rounded crown. It is the state tree of Ohio and used extensively as an ornamental. The wood is used for furniture, for boxes, and locally for fuel.

Several other species, briefly mentioned here, are conspicuous in their native habitats and cultivated as ornamentals.

Aèsculus parviflòra Walt.   bottlebrush buckeye

This is shrubby with small, white flowers in a long, slender, erect inflorescence. A native mainly of Alabama with a disjunct in South Carolina, it is widely cultivated northward and eastward.

Aèsculus califórnica (Spach) Nutt.   California buckeye

A small tree in the mountains of California, it has whitish to pinkish flowers in very showy, long, slender inflorescences.

Aèsculus sylvática Bartr.   painted buckeye

This is a yellow-flowered shrub or small tree in the Piedmont of the southeastern United States.

Aèsculus pàvia L.   red buckeye

This is a red-flowered shrub or small tree of the Coastal Plain of the Southeast but cultivated beyond its native area. A yellow-flowered variety occurs in southwestern Texas.

Aèsculus hippocástanum L.   horsechestnut

Originally a native of the Balkan peninsula, this is now widely planted throughout the world as a street and shade tree. It was introduced into the United States in 1576. It is characterized by seven obovate leaflets, white flowers in a compact panicle, prickly fruit, and dark brown to nearly black, sticky buds. Red horsechestnut, *A.* ×*cárnea* Hayne, readily recognized by its pinkish to scarlet flowers, is a widespread ornamental and of hybrid origin between the horsechestnut and red buckeye.

FIGURE 9-109   *Aesculus glabra,* Ohio buckeye. Flowering branch. (*Courtesy of U.S. Forest Service.*)

Interspecific hybridization and introgression between sympatric species have been the topic of considerable interest. *Aesculus glabra × flava, A. flava × sylvatica,* and *A. pavia × sylvatica* are good examples of both local and dispersed introgression (Hardin, 1957b). See DePamphilis and Wyatt (1989, 1990) for more recent studies.

Aceraceae: The Maple Family

The Maple family includes two genera with 113 species of trees and shrubs. One genus, *Dipterònia* Oliv., includes two small trees, both of central China; the remaining species are in the genus *Acer.*

Botanical Features of the Family

Leaves deciduous (rarely persistent), opposite, simple or compound, mostly estipulate; the simple leaves usually palmately veined and lobed and long-petioled; the compound leaves 1-pinnate.

Flowers regular; imperfect and perfect (most species polygamous or dioecious); borne in (1) racemes, panicles, corymbs, or fascicles, which appear before or with the leaves, or (2) in lateral fascicles from separate flower buds, which appear before the leaves unfold; *calyx* normally 5-parted; *petals* 5 or 0; *stamens* 4 to 12 (mostly 7 or 8); *pistils* 2-lobed, 2-celled, compressed, each lobe winged.

Fruit a double (rarely triple) samara, united at the base, each half long-winged and 1-seeded; *seeds* compressed, lacking endosperm. (In *Dipteronia,* the wing completely surrounds the seed cavity.)

Twigs moderately stout to slender; *pith* homogeneous, terete; *terminal buds* with either imbricate or valvate scales; *lateral buds* similar but smaller and sometimes collateral; *leaf scars* more or less U-shaped; *bundle scars* 3, rarely 5 to 7 or more; *stipular scars* rarely present.

ACER L.   maple

The genus *Acer* consists of ca. 110 species of trees and shrubs widely scattered through the Northern Hemisphere, but most abundant in the eastern Himalayan Mountains and in central China. Old world maples range southward to the mountains of Java and northern Africa; in the new world, they are found from Alaska and Canada to the mountains of Guatemala. Many species and cultivars, both Asiatic and European, are used in North America as popular ornamentals. These include the sycamore maple, *A. pseudoplátanus* L.; plane-tree maple (Shantung maple), *A. truncàtum* Bunge; paperbark maple, *A. gríseum* (Franch.) Pax; Amur maple, *A. ginnàla* Maxim.; Japanese maple, *A. palmàtum* Thunb. ex J.A.Murr.; and fullmoon maple, *A. japónicum* Thunb. ex J.A.Murr. These have many cultivars selected for their attractive foliage and bark. Another very popular street and yard tree in eastern North America is the Norway maple, *A. platanóides* L., with broad 7-lobed leaves and milky sap. It has become naturalized in several areas of the United States and Canada.

Of the 14 maples indigenous to North America, seven (Table 9-25) are important as forest trees and as deer browse throughout the year. Some are a source of honey. Some hybridization and possibly introgression occur among related, sympatric species. Maple is the state tree of Rhode Island.

Maple pollen may be allergenic. Boxelder causes the most problems.

Àcer sáccharum Marsh.   sugar maple

Distinguishing Characteristics (Figs. 9-110, 9-111)

Leaves $3\frac{1}{2}$ to $5\frac{1}{2}$ in. (9 to 14 cm) in diameter; *shape* orbicular, usually palmately 5-lobed (rarely 3-lobed); *margin* of lobes entire or sparingly sinuate-toothed; *apex* acuminate; *base* cordate; *surfaces* glabrous, bright green above, paler below; *fall color* yellow, orange, or red.

TABLE 9-25 COMPARISON OF IMPORTANT MAPLES

Species	Leaves	Flowers	Samaras
A. saccharum sugar maple	$3\frac{1}{2}$–$5\frac{1}{2}$ in. (9–14 cm) in diameter, 5-lobed; glabrous below; margin entire; sinuses rounded	Appearing with the leaves; bright yellow, long-pedicelled, apetalous; polygamous	Autumnal; U-shaped; wings ca. 1 in. (2.5 cm) long, slightly divergent
A. barbatum southern sugar maple	$1\frac{1}{2}$–3 in. (4–7.5 cm) in diameter, 5-lobed; pubescent below; margin entire; sinuses rounded	As above	As above
A. nigrum black maple	4–$5\frac{1}{2}$ in. (10–14 cm) in diameter, mostly 3-lobed; pubescent below; margin entire; sinuses rounded	As above	As above
A. macrophyllum bigleaf maple	6–10 in. (15–25 cm) in diameter, deeply 5-lobed; glabrous below; margin entire; sinuses rounded	As above, but petals present	Autumnal; wing 1–$1\frac{1}{2}$ in. (2.5–4 cm) long, slightly divergent; outside of seed cavity covered with stout, pale brown hairs
A. rubrum red maple	$2\frac{1}{2}$–4 in. (6–10 cm) in diameter, usually 3–5-lobed; sides of the middle lobe mostly converging toward the apex; glaucous below; margin serrate; sinuses acute	Appearing before the leaves; reddish or yellowish, short- to long- pedicelled, petals present; polygamous	Vernal; wings ca. $\frac{3}{4}$ in. (2 cm) long, slightly divergent
A. saccharinum silver maple	4–6 in. (10–15 cm) in diameter, 5-lobed; sides of middle lobe usually divergent; silvery below; margin serrate; sinuses acute	As above, but apetalous	Vernal; wings ca. 2 in. (5 cm) long, extremely divergent
A. negundo boxelder	1-pinnately compound with 3 to 7 leaflets	Appearing with or before the leaves; yellow-green, apetalous; imperfect (tree dioecious)	Autumnal; V-shaped; wings ca. $1\frac{1}{2}$ in. (4 cm) long, slightly convergent at tips

Flowers perfect and staminate (pistil abortive); bright yellow, long-pedicelled, appearing with the leaves in crowded, umbel-like corymbs, apetalous.

Fruit autumnal, borne on slender stems, somewhat horseshoe-shaped with nearly parallel or slightly divergent wings, about 1 to $1\frac{1}{4}$ in. (2.5 to 3 cm) long.

Twigs slender, shiny, and brownish with light-colored lenticels; *pith* white; *terminal buds* $\frac{1}{4}$ to $\frac{3}{8}$ in. (6 to 9 mm) long, acute, sharply pointed, with 4 to 8 pairs of visible scales; *lateral buds* smaller; *leaf scars* V-shaped; *bundle scars* 3.

Bark gray, on older trees deeply furrowed, with long, irregular, thick plates or ridges, sometimes scaly, very variable.

Range

Northeastern (Map 9-36). Local disjuncts in central North Carolina, northwestern South Carolina, and northeastern North Dakota. Elevation: sea level to 2500 ft northward, 3000 to 5500 ft in the southern Appalachians. Mesic forests, rich mountain coves and slopes.

FIGURE 9-110 *Acer saccharum*, sugar maple. (1) Twig ×1$\frac{1}{4}$. (2) Inflorescence ×$\frac{1}{2}$. (3) Staminate flower ×2. (4) Perfect flower ×2. (5) Double samara ×$\frac{3}{4}$. (6) Leaf ×$\frac{1}{2}$. (7) Bark.

FIGURE 9-111   Open-grown sugar maple. (*Courtesy of U.S. Forest Service.*)

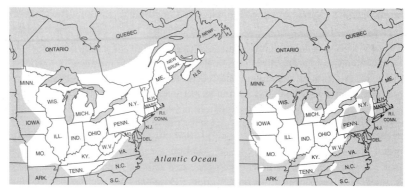

MAP 9-36   **(Left)** *Acer saccharum.* **(Right)** *Acer nigrum.*

General Description

Sugar maple commonly attains a height of 60 to 80 ft and a dbh of 2 ft (max. 150 by 7 ft), and under forest conditions develops a clear, straight, full bole; in the open, the trunk often branches near the ground, and a large dense, rounded, or ovoid crown is produced; the root system is shallow and wide-spreading to deep, depending upon the soil.

Best growth is made on moist, rich, well-drained soils, but the species will persist on more sterile sites. Sugar maple is found in many forest-cover types and is a major component in six of them (Braun, 1950; Burns and Honkala, 1990).

In good seed years (2- to 5-year intervals), just before the leaves expand, nearly every tree is so covered with flowers that at a distance it appears to be enveloped in a yellow haze. By autumn, enormous quantities of the winged fruits have developed and are released. Seed traps indicate a fall of 8 million per acre under some old-growth stands. The following spring countless numbers of seedlings unfold their straplike cotyledons as they emerge from their winter covering of leaves. They prosper even under a heavy forest cover and are extremely tolerant. One of the most tolerant of major species, sugar maple responds to release from extreme and prolonged suppression. It also regenerates by stump sprouts and sometimes root suckers. It may attain an age of 300 to 400 years.

A drastic decline of sugar maples in the New England states and the eastern Canadian provinces was first noticed in the 1980s. A possible cause is acid rain, which affects the availability of soil nutrients, particularly calcium and magnesium. However, mortality may be due to a combination of nutritional stress and insect-defoliation stress.

Besides its primary importance as a timber producer, sugar maple is used as an ornamental tree, is an important source of browse, and also is tapped for valuable syrup and sugar. Thirty-two gallons of the spring sap may be boiled down to a gallon of syrup or 8 lbs of sugar, but the sugar content of the sap varies widely from tree to tree (Leaf and Watterston, 1964). Trees with the highest possible sugar content are being sought to use in tree-breeding programs for the production of high-yielding sugar orchards.

A variant with light gray, almost chalky bark is known as chalk maple, *A. leucoderme* Small or *A. saccharum* var. *leucoderme* (Small) Sarg. It is rare and local from North Carolina and eastern Tennessee west to Texas and Oklahoma.

Sugar maple is the state tree of New York, Vermont, West Virginia, and Wisconsin. The leaf, slightly stylized, is the emblem on the Canadian flag.

Àcer barbàtum Michx.   southern sugar maple, Florida maple

Distinguishing Characteristics. *Leaves* $1\frac{1}{2}$ to 3 in. (4 to 7.5 cm) long and wide, with 3 to 5 lobes and entire margin, pubescent beneath; fall color yellow to red. *Flowers* yellow to red. *Twigs* tan and slender. *Bark* light gray, smooth, becoming furrowed.

Range. Piedmont and Coastal Plain from Virginia to central Florida and west to eastern Texas and Oklahoma. Elevation: sea level to 2000 ft. Valleys, bottomlands, and upland slopes, and growing best on alkaline soils.

General Description. This is a tree to 60 ft tall and 2 ft dbh (max. 125 ft by 3 ft). This southern sugar maple differs from the typical sugar maple in having leaves pubescent and often glaucous beneath, usually smaller, and the lobes round to acute at the apex. The treatment of this species has sometimes been as *A. saccharum* var. *floridanum* (Chapm.) Small & Heller or *A. saccharum* subsp. *floridanum* (Chapm.) Desmarais (see Desmarais [1952] and Kreibel [1957]). It is not tapped for syrup but is becoming important as an ornamental or shade tree.

Àcer nìgrum Michx. f.   black maple

Distinguishing Characteristics. *Leaves* somewhat like those of sugar maple but usually 3-lobed (Fig. 9-112), pubescent in varying degrees, and with a drooping habit; base of petiole with 2, often foliaceous stipules; fall color yellow-orange. *Fruit* often with a slightly larger seed cavity than that of sugar maple. *Twigs* commonly stouter, with conspicuous warty lenticels and larger, more hairy buds. *Bark* more corrugated than that of sugar maple.

Range. Northeastern (Map 9-36). Local disjuncts in western Virginia, western North Carolina, northwestern Tennessee, and northwestern Arkansas. Elevation: sea level to 2500 ft. Moist valleys and upland mixed forests.

General Description. The silvical features of this species are similar to those of sugar maple, except that black maple will grow on the more moist soils of river bottoms, and perhaps is more tolerant of drier and hotter conditions. Where the ranges of sugar and black maple overlap, frequent hybridization occurs. It is similar to sugar maple in wood characteristics and sap quality and production.

There has been considerable debate regarding whether to recognize black maple as a distinct species or as a variety or subspecies of the sugar maple. See Desmarais (1952) and Kreibel (1957) for additional discussions.

FIGURE 9-112   *Acer nigrum,* black maple. (1) Twig ×1$\frac{1}{4}$. (2) Leaf ×$\frac{1}{2}$. (3) Bark.

Àcer macrophýllum Pursh   bigleaf maple

Distinguishing Characteristics (Fig. 9-113)

Leaves 6 to 10 in. (15 to 25 cm) in diameter; *shape* orbicular or nearly so, usually palmately 5-lobed; *margin* of lobes entire to sinuate or sparingly toothed; *apex* acute;

FIGURE 9-113 *Acer macrophyllum,* bigleaf maple. (1) Twig ×$1\frac{1}{4}$. (2) Raceme of staminate flowers ×$\frac{3}{4}$. (3) Raceme of perfect and staminate flowers ×$\frac{3}{4}$. (4) Staminate, pistillate, and perfect flowers ×2. (5) Double samara ×$\frac{3}{4}$. (6) Bark.

base cordate; *surfaces* glabrous, bright green above, pale below; *petioles* with milky sap; *fall color* yellow or orange.

Flowers perfect and staminate; yellow, scented; in slightly puberulent racemes appearing with the leaves.

Fruit 1 to $1\frac{1}{2}$ in. (2.5 to 4 cm) long with slightly divergent wings; the portion covering the seed densely pubescent.

Twigs stout, dark red, or reddish brown to greenish brown, often dotted with rather conspicuous lenticels; *pith* white; *terminal buds* stout, blunt, with 3 or 4 pairs of green- to reddish-colored scales; *lateral buds* small, slightly appressed; *leaf scars* V-shaped to U-shaped, with 5 to 9 bundle scars.

Bark light gray-brown and smooth on young stems but becoming darker and deeply furrowed on old trunks.

Range

Western (Map 9-37). Elevation: sea level to 1000 ft northward, 3000 to 5500 ft southward. Stream banks, rain forests, and moist canyons.

General Description

Bigleaf maple is one of the few commercial hardwoods on the Pacific coast. It grows on a variety of soils throughout its range and is usually a small- to medium-sized tree some 50 ft tall and $1\frac{1}{2}$ ft dbh. Best development is made on rich bottomlands, where it attains a height of more than 80 ft and a dbh of 3 to 4 ft (max. 158 by 9 ft). The trees are usually scattered or in small groves in association with both coniferous and other broad-leaved species common to its range. In some sections of southwestern Oregon, however, this maple is the principal forest species, particularly where it invades logged and burned lands.

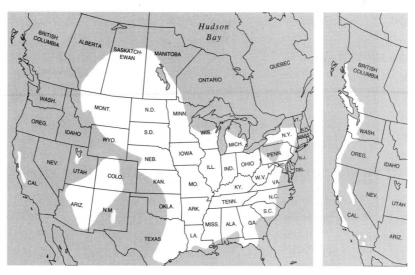

MAP 9-37   **(Left)** *Acer negundo.* **(Right)** *Acer macrophyllum.*

A rather narrow crown is developed under forest conditions, and the bole is often free of limbs for one-half to two-thirds of its length. In more or less open situations, the trunk usually divides a short distance above the ground into several stout, ascending branches, forming a rather compact ovate to subglobose crown. Regardless of habitat, however, the root system is generally shallow and wide-spreading.

An abundance of seed is produced annually, and natural regeneration by this means is excellent; stump sprouts are also quite vigorous. Growth is rapid during the first 40 to 60 years, but decreases considerably later, and maturity is eventually reached between 200 and 300 years. Old trees are commonly defective; large burls occasionally develop along the boles. The demand for western maple burls and fancy grained logs has had much to do with increasing the commercial importance of this species. The wood is used for veneer, furniture, piano frames, woodenware, and locally for fuel. It is an excellent shade tree.

Àcer rùbrum L.   red maple

Distinguishing Characteristics (Fig. 9-114)

Leaves $2\frac{1}{2}$ to 4 in. (6 to 10 cm) in diameter; *shape* orbicular, palmately 3- or 5-lobed (usually 3) with acute sinuses, sides of the terminal lobe usually convergent; *margin* of lobes serrate; *surfaces* light green above, at maturity paler and glaucous below; *fall color* red, orange, or yellow.

Flowers perfect and staminate; appearing in early spring before the leaves, yellow or bright red; short- to long-pedicelled, fascicled; *corolla* present.

Fruit borne in clusters on long slender stems; wings slightly divergent, ca. $\frac{3}{4}$ in. (2 cm) long, maturing in late spring.

Twigs slender, dark red, lustrous, odorless or nearly so, dotted with minute lenticels; *terminal buds* obtuse with 2 to 4 pairs of visible, red scales; *lateral* and *collateral buds* smaller than the terminal, slightly stalked; *leaf scars* V-shaped; *bundle scars* 3.

Bark on young trees smooth and light gray, eventually on the older trunks breaking up into long, narrow, scaly plates separated by shallow fissures.

Range

Eastern (Map 9-38). Elevation: sea level to 6000 ft. This has a very broad ecological range from drier upland slopes to floodplains and swamps.

General Description

Red maple is a medium-sized tree 50 to 70 ft tall and 12 to 24 in. dbh (max. 145 by 6 ft) with a long, fairly clear bole, an irregular or rounded crown, and a shallow root system.

Red maple is one of the first trees to flower in the spring, long before the leaves appear. The fruit matures quickly and is shed in early summer. It germinates immediately or may hold over until the following spring. Growth is rapid, and maturity is reached in 70 to 80 years, although certain individuals may attain an age of 150 years. There are reports of it reaching 300 years (Tyrrell et al., 1998). Many second-growth stands are of sprout origin, but quality is often poor due to butt rot. On a tolerance scale, this tree is intermediate; it often pioneers effectively on disturbed sites.

FIGURE 9-114 *Acer rubrum,* red maple. (1) Perfect flowers $\times\frac{3}{4}$. (2) Staminate flowers $\times\frac{3}{4}$. (3) Double samara $\times\frac{3}{4}$. (4) Leaf $\times\frac{1}{2}$. (5) Bark of old tree. (6) Flower buds $\times1\frac{1}{4}$. (7) Twig $\times1\frac{1}{4}$.

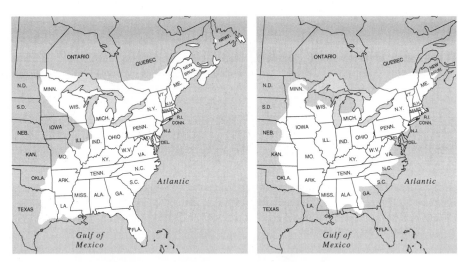

MAP 9-38 (Left) *Acer rubrum.* **(Right)** *Acer saccharinum.*

Red maple is characteristic both of swampy sites and of drier locations of moderately moist, sandy loam soils, or even on rocky uplands. Over its extensive range, it is a most cosmopolitan species, occurring in 54 forest-cover types, and with many associates, including both hardwoods and conifers. It is said to be one of the most abundant and widespread trees of eastern North America. Easily top-killed by fire, it responds with vigorous sprouts. During the last century, red maple has gone from being a minor component of most eastern deciduous forests to now being a significant component or the dominant of many types of forests. Its success is due to the ability to thrive in a wide variety of habitats and ecological conditions created by changes in land use. For a more thorough discussion, see Abrams (1998).

Red maple is used widely as an ornamental and shade tree, it tolerates a great range of soil conditions, provides a spectacular display in early spring with red flowers and fruits, then after the first fall frosts, its leaves often turn a brilliant scarlet or yellow. Twigs and leaves are excellent deer browse, and the fruits are eaten by small mammals. Partially wilted leaves, however, are poisonous to livestock, particularly horses and ponies.

Red maple is the state tree of Rhode Island.

In the pocosins and swamps of the Southeast, the smaller, shallowly 3-lobed leaves have been recognized as var. *trilobum* T.&G. ex K. Koch.

Àcer saccharìnum L.   silver maple

Distinguishing Characteristics (Fig. 9-115)

Leaves 4 to 6 in. (10 to 15 cm) in diameter; *shape* orbicular, deeply palmately 5-lobed, the sides of the terminal lobe divergent, the lobes often again lobed, sinuses often narrowly rounded; *margin* of lobes serrate; *surfaces* pale green above, silvery below at maturity; *fall color* pale yellow to orange.

Flowers perfect and staminate; among the first to appear in the spring and long before the leaves unfold; *corolla* lacking, otherwise similar to those of red maple except greenish yellow.

FIGURE 9-115   *Acer saccharinum,* silver maple. (1) Staminate and perfect flowers ×1. (2) Double samara ×$\frac{3}{4}$. (3) Leaf ×$\frac{1}{2}$. (4) Twig ×$1\frac{1}{4}$. (5) Bark.

Fruit the largest of the eastern maples, with widely divergent wings $1\frac{1}{2}$ to $2\frac{1}{2}$ in. (4 to 6 cm) long; often aborted on one side; maturing in late spring, germinating as soon as released.

Twigs and buds very similar to those of red maple, but often more reddish brown, and with a slightly fetid odor when bruised.

Bark on young trees silvery gray, later breaking up into long, thin, scaly plates that are unattached at the ends.

Range

Eastern (Map 9-38). Disjuncts in eastern Louisiana and upper Michigan. Elevation: sea level to 2000 ft, and to ca. 4500 ft in the southern Appalachians. Floodplains, swamps, moist mountain coves and slopes.

General Description

Silver maple is a medium-sized tree 60 to 80 ft tall and 2 to 3 ft dbh (max. 125 by 7 ft) and usually has a short bole that divides near the ground into several upright branches; the crown is wide-spreading, and the root system is shallow. This maple is a characteristic bottomland species and is not found on dry soils. Young trees recover well even from several weeks of inundation (Hosner, 1960).

Silver maple is a very fast-growing tree, moderately intolerant, reaching maturity in about 125 years. The branches are brittle and often break off during high winds or when loaded with snow or ice; this contributes to early disintegration because wood-destroying fungi readily enter the exposed wood. Silver maple is a common urban tree due to early rapid growth and pleasing appearance. However, some communities do not permit its planting as a street tree because of its many liabilities. Several horticultural varieties are used for ornamental plantings, including a cut-leaf form. Planted trees seem to do well even on dry clay soils. Cultivars from crosses with red maple are generally superior as ornamental shade trees.

Àcer pensylvánicum L.   striped maple, moosewood

A short-lived (to 70 yrs) shrub or small tree to 35 ft tall, it is distinguished by its green bark with conspicuous white, vertical stripes; leaves with 3 shallow and broad terminal lobes and serrulate or doubly serrate margins; yellow fall color; glabrous twigs; and pendulous terminal racemes of yellow-green flowers (Fig. 9-116). It is mainly a tree of southeastern Canada, west to Michigan and northern Wisconsin, and south along the higher elevations (5500 ft) of the Appalachians to northern

FIGURE 9-116   *Acer pensylvanicum, striped maple. Flowering branch with leaf. (Courtesy of U.S. Forest Service.)*

Georgia. Dense understory stands of striped maple present great difficulty in regenerating more valuable overstory species. It is a source of food for wildlife. Individual stems of striped maple can change from producing male flowers to female flowers as the stem matures (Hibbs and Fischer, 1979). Individuals of other maple species have changed gender (Primack and McCall, 1986).

Àcer spicàtum Lam. mountain maple

A shrub or small tree to 25 ft tall, it is distinguished by its brown, nonstriped bark; leaves with 3 shallow and broad terminal lobes and coarsely serrate margins; orange-red fall color; finely pubescent twigs; and erect terminal panicle of pale yellow flowers (Fig. 9-117). It is primarily a northern tree of southeastern Canada, south to Wisconsin, and also in the higher elevations (6000 ft) of the Appalachians to northern Georgia.

FIGURE 9-117 *Acer spicatum,* mountain maple. Flowering branch with leaves. (*Courtesy of U.S. Forest Service.*)

Àcer negúndo L. boxelder, ash-leafed maple

Distinguishing Characteristics (Fig. 9-118)

Leaves pinnately compound with 3 to 7 (rarely 9) leaflets, ca. 6 in. (15 cm) long, $3\frac{1}{2}$ in. (9 cm) wide; *leaflets* short-stalked; *terminal leaflet* often 3-lobed; shape very variable, mostly ovate, oval, obovate, or ovate-lanceolate; *apex* acuminate; *base* cuneate, rounded, or cordate; *margin* coarsely serrate or lobed; *surfaces* light green, glabrous, or

FIGURE 9-118 *Acer negundo,* boxelder. (1) Twig $\times 1\frac{1}{4}$. (2) Leaf $\times\frac{1}{2}$. (3) Double samara $\times\frac{3}{4}$. (4) Bark. (*Photo by R.A. Cockrell.*)

slightly pubescent above, pale green, pubescent along the veins below; *rachis* stout, enlarged at the base; *fall color* pale yellow.

Flowers imperfect (plant dioecious), apetalous, yellow-green, the *staminate* fascicled, the *pistillate* in drooping racemes, appearing with or before the leaves.

Fruit borne on slender stems, V-shaped, with slightly convergent wings 1 to $1\frac{1}{2}$ in. (2.5 to 4 cm) long.

Twigs stout, green to purplish green, lustrous or covered with a glaucous bloom, lenticellate; *pith* terete, white; *terminal buds* ovoid, with about 4 visible, usually bluish white, tomentose scales; *lateral buds* short-stalked, appressed; *leaf scars* V-shaped, *bundle scars* 3 (rarely 5).

Bark thin, light brown, with narrow, rounded, anastomosing ridges separated by shallow fissures; on old trees more deeply furrowed.

Range

Eastern and central (Map 9-37). Disjunct areas in Colorado, New Mexico, Arizona, Utah, and California; naturalized in southeastern Canada, northeastern United States, and northwestern United States. Elevation: sea level to 8000 ft. Stream and riverbanks, floodplains, swamp margins, and moist upland slopes.

General Description

Boxelder is usually a small- or medium-sized tree, 30 to 60 ft tall and to $2\frac{1}{2}$ ft dbh (max. 95 to 6 ft), with an irregular bole, shallow root system, and bushy, spreading crown. It has little if any commercial importance, but it is one of the most common and best known of the maples. However, because of its compound leaves, many people are not aware that it is a maple. It is the most widely distributed maple in North America. Most common on deep, moist soils of bottomlands, it is also found on poorer sites and is perhaps the most aggressive of the maples in unfavorable locations. It has become an urban weed in many localities. The early settlers in the middle west were acquainted with its hardiness, especially in extremes of climate, and planted it widely as a street tree and around their homesteads. Boxelder, however, is not a decorative tree, and although it makes rapid growth, it is short-lived (rarely to 100 yrs) and usually of poor form. It is moderately tolerant. Sprouts usually emerge along the trunk, and when these are removed, others appear in increasing numbers.

Anacardiaceae: The Cashew Family

This family includes ca. 70 genera and 875 species of trees and shrubs found mostly in the warmer regions of the world. Various drugs, dyes, waxes, and tannins are obtained from the juice, which is either milky or clear and acrid, turning black upon drying; this feature is utilized in the production of Chinese lacquer, which is obtained from the sap of *Toxicodéndron vernicifliùum* (Stokes) Barkley. Some species have highly colored woods and are valuable as timber producers, whereas others, such as quebracho (*Schinópsis quebrácho-colorádo* [Schldl.] Bark. & Meyer) of South America, one of the heaviest and hardest woods in the world, yield large amounts of tannin. Pistachio nuts of commerce come from *Pistácia véra* L., and *P. chinénsis* Bunge is becoming very popular as an ornamental lawn, street, or park tree in the southern states. The popular, edible cashew nut comes from the tropical American species, *Anacárdium occidentále* L. The shell of the cashew nut contains a poison that causes severe burns and is used by certain tribes of natives to produce scarred designs on the skin, similar in appearance to tattoos. Another well-known tropical species (*Mangífera índica* L.) produces the mango fruit, which is sold in most markets. Brazilian pepper, *Schìnus terebinthifòlius* Raddi, a shrub or small tree from South America, has become naturalized and is an invasive weed particularly in Florida and Hawaii. *Cotìnus coggýgria* Scop., common smoke tree, is frequently seen in urban plantings for its showy, hairy, inflorescence stalks, which are pink to purple, and for its fall foliage ranging from yellow to purple. *Cotinus obovàtus* Raf., American smoketree, is a relatively rare but beautiful small tree for ornamental plantings. It is native from southeastern Tennessee to central Texas.

Botanical Features of the Family

Leaves deciduous or persistent, usually alternate, simple or compound, estipulate.

Flowers imperfect (plants dioecious or polygamo-dioecious), actinomorphic, mostly 5-parted, ovary superior and 3-carpellate, entomophilous.

Fruit a drupe or nutlike.

RHUS L.   sumac

Autumn foliage colors of many of the sumacs are brilliant red and outstanding. None of these are poisonous, and many are important browse for wildlife throughout the year. They are also a source of honey. There are 11 species of native trees north of Mexico.

Rhùs tỳphina L.   staghorn sumac (Fig. 9-119)

This is a small, deciduous tree or large shrub with stout pubescent branches that bear a resemblance to a stag's horns "in the velvet." *Leaves* large and pinnately compound with 11 to 31 leaflets, coarsely serrate on the margins; fall color purplish red, orange-red, or yellowish. *Flowers* small, in compact terminal panicles, appearing in spring after the leaves. *Fruits* small, in upright, compact, red, cone-shaped panicles. *Twigs* stout and pubescent with milky sap; leaf scars alternate, large, U-shaped and nearly encircling bud, with many bundle scars; terminal bud absent, laterals round, small, and pubescent; pith brown and large. It is found along edges of forests and roadsides and in open fields in southeastern Canada and northeastern quarter of the United States and south in the Appalachians. It hybridizes with *R. glabra*.

The nomenclature of this tree is not fully resolved. Based on the rules dealing with type specimens, this well-known name may have to be changed to *R. hirta* (L.) Sudworth.

Rhùs glàbra L.   smooth sumac

This is a deciduous shrub or small tree similar to the staghorn sumac, except that the branches are glabrous and glaucescent and the fruit is borne in more open clusters. The fall color is red. It is the most common of the sumacs, from across southern Canada southward to northwestern Florida and west to Texas and locally to Nevada and Washington. It is found in open woodlands, roadsides,

FIGURE 9-119   *Rhus typhina*, staghorn sumac. (1) Twig ×1$\frac{1}{4}$. (2) Infructescence ×$\frac{1}{2}$. (3) Leaf ×$\frac{1}{2}$.

grasslands, clearings, and waste places. It hybridizes with *R. typhina*. The fruits and twigs are an important wildlife food.

Rhùs copallìna L.   shining sumac, flameleaf sumac

This is a small, deciduous shrub to small tree (to 26 ft) of the East from southern Ontario to Florida and Texas; leaflets 7 to 27, entire to barely serrate, the rachis of the pinnate leaves is conspicuously winged; fall color dark reddish purple.

Rhùs lanceolàta (Gray) Britton   prairie sumac

This is similar to shining sumac but has narrower leaflets and larger clusters of fruits. It is a large shrub or small tree of Texas and southern New Mexico. It often forms thickets on dry rocky slopes and hills. The pinnate leaves with 13–19 leaflets turn reddish purple in the fall.

Rhùs integrifòlia (Nutt.) Brewer & Wats.   lemonade sumac

> This is an evergreen, aromatic shrub or small tree. The sour, sticky, dark red, hairy fruits can be used to make a refreshing drink. It is found in coastal southern California.

TOXICODENDRON Mill.   poison sumacs

> Some authors continue to treat these species in the genus *Rhus*. Although not forest trees, these species are so **poisonous** to the touch that they should be recognized by all field biologists, hunters, fishermen, and campers generally, and taught to all children at an early age. They should be the best known and among the most dreaded of all plants. Contact dermatitis is an allergic reaction to an oil in the cell sap, and although some people are not sensitive to this, sensitivy may be acquired and it is foolish to handle the plants without some fear and extreme care. Direct or indirect contact with the bruised leaves or other portions of the plant is necessary to produce the irritation, which does not, however, begin until several hours or even a few days after the contact. It is also dangerous to walk through smoke from burning leaves or to handle clothing, pets, tools, or other articles that have bruised the leaves, twigs, or roots (Lampe and McCann, 1985).

Toxicodéndron vérnix (L.) Kuntze   poison-sumac (Fig. 9-120)

> This is a large, deciduous shrub or small tree of the East, from southeastern Canada south to central Florida, west to eastern Texas and southeastern Minnesota. *Leaves* pinnately compound with 7 to 13 entire leaflets, the petiole, rachis, and petiolules usually reddish; fall color orange to scarlet; *fruit* ivory-white in color and borne in open, pendent panicles; *twigs* stout, yellowish brown, and usually mottled; leaf scars shield-shaped with numerous bundle scars; terminal bud present with 2 purplish scales; *bark* light to dark gray. It is found in northern bogs, along creeks, and in southern swamps, pocosins, and moist to wet woods in the open.

Toxicodéndron rádicans (L.) Kuntze   poison-ivy (Fig. 9-121)

> This is a deciduous vine or small shrub with characteristic trifoliolate leaves, which turn red in the fall; fruit yellowish-white; twigs and naked buds brownish, the leaf scars V- or U-shaped, with several bundle scars, the lenticels usually conspicuous. This is a ubiquitous weed in all areas of the East.

Toxicodéndron pubéscens Mill.   eastern poison-oak

> This is a species of the Southeast. It is deciduous, shrublike, and the leaflets are more conspicuously 3- to 7-lobed and pubescent than in the preceding species. It is relatively uncommon and occurs in dry woodlands, thickets, old fields, and sandy habitats.

Toxicodéndron diversilòbum (Torr. and Gray) Greene   western poison-oak
            (Fig. 9-122)

> This is a common, deciduous shrub or small tree of canyons, slopes, and oak woodlands along the Pacific coast from British Columbia to Mexico. It often forms thickets and has increased in abundance due to the disturbance of the soil. It is a diffusely branched, leafy shrub of dry habitats.

FIGURE 9-120   *Toxicodendron vernix,* poison-sumac. (1) Twig ×1$\frac{1}{4}$. (2) Leaf ×$\frac{1}{2}$. (3) Infructescence ×$\frac{1}{2}$.

Simaroubaceae: The Quassia Family

The quassia or bitterwood family with 12 genera and ca. 110 species is chiefly tropical and subtropical and is represented in North America by only five native and one natural-ized tree species. The family is one of trees and shrubs often with very bitter bark and other parts. One of the best-known tropical species is *Quàssia amára* L., bitterwood, of northern South America. The wood contains a water-soluble bitter principle, which is used commercially in the manufacture of insecticides and in medicines. One introduced tree, tree-of-heaven, has become naturalized and increasingly common.

FIGURE 9-121 *Toxicodendron radicans,* poison ivy. (1) Twig ×1 $\frac{1}{4}$. (2) Trifoliolate leaf ×$\frac{1}{2}$. (3) Infructescence ×$\frac{1}{2}$. (*Courtesy of U.S. Department of Agriculture.*)

Botanical Features of the Family

Leaves alternate; pinnately compound, unifoliolate, or simple; stipulate or estipulate.

Flowers small, in large panicles, regular, 3- to 8-parted; pistils 2 to 5 separate, ovaries superior; plants dioecious or polygamous; entomophilous or ornithophilous.

Fruit a capsule, drupe, berry, or samara.

Ailánthus altíssima (Mill.) Swingle   tree-of-heaven

Distinguishing Characteristics (Fig. 9-123). *Leaves* alternate, 1-pinnately compound, 1 to 2 ft (3 to 6 dm) long with 13 to 25, or more, ovate-lanceolate leaflets, each 3 to 5 in. (7.5 to 13 cm) long, 1 or 2 in. (2.5 to 5 cm) wide; apex acuminate; base unequally rounded or subcordate; margin

FIGURE 9-122 *Toxicodendron diversilobum,* western poison-oak. Leafy branch with fruits. (*Courtesy of U.S. Forest Service.*)

entire except for 1 to 5 rounded, basal teeth, each with a prominent dark green gland beneath near the tooth apex; surfaces puberulent and glabrate; fall color yellow. *Flowers* small, yellowish, in large, terminal panicles, appearing in late spring and early summer, the trees usually dioecious. *Fruit* an oblong, twisted samara, $1\frac{1}{4}$ to $1\frac{1}{2}$ in. (3 to 4 cm) long, with the seed cavity in the center, cream or reddish brown, 1 to 5 per flower; borne in large clusters and many remaining on the tree during the winter. *Twig* light brown, very stout, pith homogeneous, light brown; leaf scars heart-shaped with many bundle scars; stipular scars lacking; terminal bud lacking. *Bark* light brown-gray, smooth and becoming rough and fissured.

Range. Naturalized throughout temperate North America in disturbed areas, waste places, roadsides, and around buildings.

General Description. This tree was introduced from eastern Asia in 1784 and has been used considerably for ornamental purposes and to reclaim degraded landscapes. Although mostly disdained for decades, there is increased interest in tree-of-heaven because of its ability to thrive under the worst urban conditions. The name "tree-of-heaven" is rather dubious because the staminate flowers, bruised leaves, and broken twigs have an exceedingly disagreeable odor. This intolerant species grows rapidly and is very hardy, especially in its ability to flourish on poor

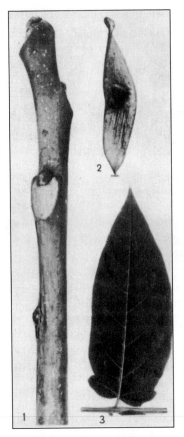

FIGURE 9-123 *Ailanthus altissima*, tree-of-heaven. (1) Twig ×1¼. (2) Samara ×1. (3) One leaflet attached to part of rachis ×½.

hard-packed soils in the smoky atmosphere of industrial cities. The height is 50 to 80 ft and dbh of 1 to 2 ft (max. 90 ft by 3½ ft), with a spreading, rounded, open crown. Seedlings and root suckers are plentiful and aggressive, and when once established, the tree is difficult to eradicate. Stump sprouts are also numerous and sometimes grow in length at the rate of an inch a day. This species is very weedy and a real pest on good sites in prime hardwood forests. It is a source of honey.

Meliaceae: The Mahogany Family

The Mahogany family, which includes some of the finest known cabinet and furniture woods, consists of ca. 50 genera and 565 species of trees and shrubs, mostly tropical. An extremely important member of this family is neem (*Azedeiráchta índica* A. Juss.), of India, where it is revered for its many uses. In fact, various parts of the tree have been used for so many things that it has been called "the village pharmacy." Neem is now widely cultivated in the tropics and subtropics and used for fuel, fodder, food, timber, soaps, toothpaste, lotions, oils, cosmetics, fertilizers, insecticides, and medicines. Having stood the test of modern research, neem stands true to its Sanskrit name, "arishta," meaning "reliever of sickness." In the United States, the evergreen neem trees can be found cultivated in southern Florida, southern California, Arizona, and Washington, D.C.

The family is represented in the United States by the naturalized chinaberry and a single native species, the West Indies mahogany, *Swietènia mahágoni* (L.) Jacq., which reaches its northern limit in extreme southern Florida. It was much exploited for fine cabinet and furniture work, but trade is now zero and wild populations are now of poor form. Bigleaf mahogany, *S. macrophýlla* King, is a highly traded tropical timber species from Central and South America. It is now rare due to five centuries of exploitation. Its cultivation in plantations is seriously hampered by a shoot boring insect.

Botanical Features of the Family

Leaves deciduous or persistent; alternate; simple, unifoliolate, pinnately or bipinnately compound, the leaflets often oblique; estipulate.

Flowers perfect or imperfect, regular, 4 to 5 parted, stamens 8 to 10 monadelphous; pistil 1, superior; entomophilous.

Fruit a capsule or drupe.

Mèlia azédarach L.   chinaberry

Distinguishing Characteristics (Fig. 9-124).   *Leaves* bipinnate, 8 to 18 in. (20 to 46 cm) long, leaflets 1 to 2 in. (2.5 to 5 cm) long and $\frac{3}{8}$ to $\frac{3}{4}$ in. (1 to 1.9 cm) wide, lanceolate or ovate, serrate or lobed. *Flowers* lilac-colored, fragrant and showy in large terminal panicles, appearing in spring after the leaves. *Fruit* a yellow drupe, ca. $\frac{5}{8}$ in. (15 mm) in diameter, wrinkled, remaining attached through the winter; poisonous if eaten. *Twig* stout, leaf scars 3-lobed, bundle scars 9 in 3 C-shaped groups, terminal bud lacking, buds appearing naked, stipular scars lacking, pith homogeneous and white.

Range.   Chinaberry has become naturalized and occurs in dry soils or disturbed areas, pastures, barn lots, and roadside fencerows, usually below 1000 ft elevation, in the southeastern states and also California.

General Description.   This tree, originally a native of the Orient, has been extensively cultivated in the South as an ornamental and shade tree. It is deciduous, or evergreen southward, and reaches 40 ft tall and 1 ft dbh. The fruits are disseminated by birds but are poisonous to livestock and poultry.

Rutaceae: The Rue Family

This is a large family of ca. 150 genera and 1800 species of trees and shrubs distributed over the warmer and temperate regions of the world. All are characterized by a bitter-tasting, aromatic, volatile oil.

FIGURE 9-124   *Melia azedarach,* chinaberry. Branch with drupes persisting in the winter. (*Courtesy of Florida Experiment Station.*)

The family is of importance in horticulture, and the most noteworthy introduced genus is *Cítrus* (orange, tangerine, lemon, lime, grapefruit) with a number of species and varieties with edible fruit. Orange blossom is the state flower of Florida, and orange-blossom honey is popular. Several other tropical genera produce excellent cabinet woods and structural timbers.

There are four native genera in North America. The more common species are Hercules-club (also called toothache tree), *Zanthóxylum clàva-hérculis* L., a small tree of the Southeast; common prickly-ash, *Z. americànum* Mill., a shrub or small tree of eastern United States and eastern Canada; baretta, *Heliétta parvifòlia* (Gray) Benth., found in southern Texas and northeastern Mexico; torchwood (sea amyris), *Ámyris elemífera* L., of southern Florida and the West Indies; and *Ptelea,* described here.

Botanical Features of the Family

Leaves deciduous or persistent, alternate or some opposite, mostly compound or unifoliolate, with glandular-punctate dots, estipulate, aromatic with lemonlike odor when crushed.

Flowers usually regular, white or greenish, bisexual or unisexual, 4 to 5 parted, stamens 8 to 10, pistil 1, superior; entomophilous.

Fruit a drupe, follicle, samara, or hesperidium (the pulp in citrus fruits is derived from enlarged hairs).

Stem with or without thorns.

Ptèlea trifoliàta L.   common hoptree

Distinguishing Characteristics.   *Leaves* deciduous, palmately trifoliolate, 4 to 7 in. (10 to 18 cm) long, leaflets each 2 to 4 in. (5 to 10 cm) long, $\frac{3}{4}$ to 2 in. (2 to 5 cm) wide, ovate or elliptical, acuminate, entire or serrulate, with small punctate dots; fall color yellow. *Flowers* greenish white, ca. $\frac{3}{8}$ in. (10 mm) wide, in terminal panicles, appearing with the leaves, foul-smelling and pollinated by carrion flies. *Fruit* a round, waferlike samara, ca. $\frac{7}{8}$ in. (22 mm) in diameter, yellow-brown, in drooping clusters. *Twigs* slender, brown, pubescent, with a rank odor, leaf scar horseshoe-shaped with 3 bundle scars, stipular scars lacking, terminal bud lacking, lateral buds naked, pith homogeneous, lenticels large and brown. *Bark* gray-brown, thin, smooth or scaly, aromatic.

Range.   Southern Ontario, east to western New York and New Jersey, south to southern Florida, west to Texas, north to southern Wisconsin; also local west to Arizona and southern Utah and into Mexico. Elevation: near sea level to 8500 ft in the southwestern mountains. Rocky slopes and edges of woods.

General Description.   Small aromatic tree to 20 ft tall and 6 in. dbh. This widespread and variable species has been divided into several subspecies and varieties (Bailey, 1962).

Araliaceae: The Ginseng Family

This is a family of 47 genera and 1325 species widely distributed in temperate and tropical regions. They are herbs, shrubs, trees, and woody vines. It is best known for the commonly cultivated English ivy (*Hédera hèlix* L.); the very popular indoor/outdoor ornamental,

Schefflèra; and the commercially important herbal, ginseng (*Pánax quinquefòlius* L.), with root extracts now in cosmetics and medicinal or quasi-medicinal preparations.

Some recent evidence indicates that this family should be included within the Apiaceae s. lat. (Judd et al., 1994, 1999).

Botanical Features of the Family

Leaves alternate (rarely opposite or wholed), simple or pinnately or palmately compound and often very large; with or without stipules.

Flowers perfect or imperfect; in a head, umbel, or panicle; 5-parted, ovary inferior; entomophilous.

Fruit a berry, drupe, or schizocarp.

Aràlia spinòsa L.   devils-walkingstick, prickly ash

Distinguishing Characteristics.   *Leaves* large, alternate, mostly bipinnate, 24 to 60 in. (6 to 15 dm) long and wide, often clustered at the top of an unbranched trunk (or with a few branches); the leaflets many, $1\frac{1}{4}$ to 4 in. (3 to 10 cm) long, ovate or elliptical, serrulate, often with prickles on midrib beneath; fall color yellow. *Flowers* small, white, in large terminal panicles to 48 in. (12 dm) long, in late summer. *Fruit* a purplish black berry, $\frac{1}{4}$ in. (6 mm) in diameter on red pedicels and infructescence branches, persisting into the winter. *Twig* stout, with a circle of prickles at the nodes or scattered; leaf scar nearly encircling twig with one row of many bundle scars; stipular scar lacking; pith solid, large.

Range.   New York and New Jersey, south to central Florida, west to eastern Texas, north to southeastern Missouri; naturalized northward into New England, southern Ontario, and Wisconsin. Elevation: near sea level to 3500 ft northward, to 5000 ft in the southern Appalachians. Moist soils along streams in hardwood forests, edge of woods, and roadsides.

General Description.   Shrub or tree to 30 ft tall (max. 74 ft) and 8 in. dbh, the trunk and branches prickly, the prickles straight, slender, and sharp-pointed. The trunk is unbranched or with few branches, and the large, compound leaves are clustered near the top, forming a unique and characteristic appearance. The fruits are eaten by birds and small mammals.

ASTERIDAE

Oleaceae: The Olive Family

This family includes 24 genera and some 615 species of trees and shrubs distributed mostly through the temperate and tropical forests of the Northern Hemisphere. The genera *Fraxinus* and *Olea* are noted for the many fine timbers that they produce, and *Olèa euròpaea* L. furnishes the olives and olive oil of commerce. Originally a native of the Mediterranean basin, the olive is now widely cultivated throughout the warmer regions of the world, and several varieties are grown in California. Among the shrubby members of this family commonly used for ornamental purposes are *Syrínga* (lilac), *Osmánthus* (tea olive), *Forsýthia* (goldenbell), *Jasmínum* (jasmine), and *Ligústrum* (privet). Some *Ligustrum* species have become naturalized, and particularly *L. sinénse* Lour. is weedy and forms dense thickets in wet areas of the Southeast. *Chionánthus virgínicus* L., the native fringetree, and several ashes, *Fraxinus,* are also used for ornamental purposes.

Four genera are represented by native arborescent species in the United States, but only one, *Fraxinus,* is an important timber producer (Wilson and Wood, 1959).

Botanical Features of the Family

Leaves deciduous or persistent, opposite or rarely alternate, simple or pinnately compound, estipulate.

Flowers perfect and/or imperfect, actinomorphic; *calyx* 4-lobed or lacking; *corolla* 4- or rarely 5- to 16-lobed, or lacking; *stamens* 2 (rarely 3 to 5), adnate to the corolla and alternating with its lobes, rudimentary or lacking in the pistillate flowers; *pistils* 1, superior, the ovary 2- (rarely 3-) celled, with 2 ovules in each cell, with a single style and a 2-lobed stigma, rudimentary or lacking in the staminate flowers; entomophilous or anemophilous.

Fruit a samara, capsule, berry, or drupe.

FRAXINUS L.   ash

The genus *Fraxinus* includes ca. 65 species of trees, rarely shrubs, largely restricted to temperate regions of the Northern Hemisphere but extending into the tropical forests of Java and Cuba. *Fráxinus excélsior* L., European ash, is often planted in North America as a street and shade tree. Flowering ash, *F. órnus* L., also a native of both Europe and Asia, is commonly used ornamentally in the Pacific Northwest; this species produces the "manna" of commerce, a medicinal substance used in the preparation of laxatives. A white wax is obtained from scale insects that feed on *F. chinénsis* Roxb. Numerous other species produce timbers of intrinsic value. A beautiful veneer stock, which is sold in this country under the name of Tamo, is from *F. mandshùrica* Rupr. and *F. sieboldiàna* Blume, the Manchurian and Japanese Siebold ashes, respectively.

The arborescent flora of the United States contains 16 or 17 species of ash; four are native in Canada. Two are considered of primary importance (Table 9-26) and seven

TABLE 9-26   COMPARISON OF IMPORTANT NATIVE ASHES

Species	Leaves	Twigs	Samaras
F. americana white ash	With 5 to 9(13) leaflets; serrate to entire, petioluled, base rounded to acute, glabrous to pubescent and "papillose" below	Terete, glabrous or pubescent; leaf scars notched at top	Lanceolate to oblanceolate; wing extending only to the top of the seed cavity
F. pennsylvanica green ash	With 7 to 9 leaflets; sharply serrate, petioluled, base cuneate, glabrous to pubescent below, not "papillose"	Terete, glabrous to velvety pubescent; leaf scars truncate to shallowly notched at top	Elliptical to narrowly oblanceolate; wing extending along the seed cavity

others are briefly described, four in the East and three in the West. Most are important as deer browse and the fruits as food for birds and small mammals.

Interspecific hybridization does occur, and although hybridization between *F. americana* and *F. pennsylvanica* has been suggested as an origin for the Biltmore ash (Miller, 1955; Santamour, 1962), there is no micromorphological evidence to support this hypothesis (Hardin and Beckmann, 1982).

The pollen is allergenic.

Botanical Features of the Genus

Leaves deciduous, opposite, odd-pinnately compound (rarely reduced to a single leaflet), the leaflets serrate or entire.

Flowers perfect, and/or imperfect, appearing in the early spring before or with the unfolding of the leaves; *calyx* 4-lobed or lacking; *corolla* with 2 to 6 petals, or absent (the latter in all North American species except *F. cuspidata*); *stamens* 2 to 4; *pistils* 1, with a 2-celled ovary; anemophilous.

Fruit a 1-seeded samara with an elongated terminal wing; seed narrow and elongated, with endosperm.

Twigs slender to stout, glabrous or pubescent; *pith* homogeneous, terete between the nodes; *terminal buds* present, with 1 to 3 pairs of scales; *lateral buds* similar to the terminals but smaller; *leaf scars* suborbicular to semicircular, sometimes notched on the upper edge; *bundle scars* numerous, arranged in an open U- or V-shaped line, or sometimes an ellipse; *stipular scars* lacking.

Fráxinus americàna L.   white ash

Distinguishing Characteristics (Fig. 9-125)

Leaves 8 to 12 in. (20 to 30 cm) long, with 5 to 9 (mostly 7, and rarely 11 or 13) leaflets; *leaflets* $2\frac{1}{2}$ to 5 in. (6 to 13 cm) long, $1\frac{1}{4}$ to $2\frac{3}{4}$ in. (3 to 7 cm) wide; *shape* ovate to oblong lanceolate, elliptical, or oval; *apex* acute to acuminate; *base* rounded to acute, the stalks stout and narrowly winged; *margin* serrate, remotely crenate-serrate, or entire; *surfaces* dark green, glabrous above; whitish or pale green, glabrous or pubescent below, "papillose" formed by minute raised cuticlar projections; *rachis* slightly grooved, glabrous; *fall color* yellow, bronze, or purplish.

Flowers imperfect (trees dioecious), apetalous, both sexes appearing in glabrous panicles before or with leaves; *calyx* 4-parted, minute.

Fruit a lanceolate to oblanceolate samara, 1 to 2 in. (2.5 to 5 cm) long, and ca. $\frac{1}{4}$ in. (6 mm) wide; *wing* usually rounded or slightly emarginate (rarely pointed) at the apex, not greatly narrowed at the seed cavity and not extending along side of seed.

Twigs stout, glabrous or pubescent, dark green to gray-green, occasionally purplish, lustrous, lenticellate; *terminal buds* broadly ovoid, obtuse, covered with 4 to 6 brownish scales; *lateral buds* somewhat triangular, the upper pair at the same level as the terminal bud; *superposed buds* occasionally present on vigorous shoots; *leaf scars* semiorbicular, usually notched at the top; *bundle scars* numerous and arranged in a broadly U-shaped line, sometimes nearly closed at the top.

FIGURE 9-125 *Fraxinus americana,* white ash. (1) Twig $\times 1\frac{1}{4}$. (2) Staminate inflorescence $\times\frac{1}{2}$. (3) Staminate flower $\times 4$. (4) Pistillate inflorescences $\times\frac{1}{2}$. (5) Pistillate flower $\times 4$. (6) Samara $\times 1$. (7) Leaf $\times\frac{1}{3}$. (8) Bark.

Bark ashy gray, or on young stems sometimes with an orange tinge, later dark gray, thick, with deep diamond-shaped furrows and narrow forking ridges.

Range

Eastern (Map 9-39). Disjuncts in eastern South Carolina and upper Michigan. Elevation: sea level to 2000 ft northward, to 5000 ft in the southern Appalachians. Moist upland slopes, valleys, coves, and dry to mesic woods; not tolerant of poorly drained soils.

General Description

White ash, the most abundant and important of American ashes, is a tree 70 to 80 ft tall and 2 to 3 ft dbh (max. 160 by 7 ft). The bole is long, straight, clear, and cylindrical and terminates below in a root system that is deep in porous soils but shallow and spreading on rocky sites; the crown is somewhat open. This tree is intermediate in tolerance except in the seedling stage, when it is often found in considerable numbers under forest cover. White ash commonly occurs on deep, moist, fertile upland soils. In the South, it may be found on loamy ridges in the bottoms or elsewhere on well-drained slopes and coves. This ash is a component of some 24 forest-cover types, but in all these it develops well only on the most fertile soils. Rare in original forests, white ash is quite abundant in many places today because it aggressively colonizes abandoned farmland, especially in the Northeast.

Although some seed may be produced each year, good crops develop only at 3- to 5-year intervals. On good sites, growth at first is rapid, and a height of 6 ft may be attained by the end of the third season. Sprouts from young trees are vigorous, often growing 5 ft or more in height the first year.

The Biltmore ash, recognized by some authors, is a form of white ash with twigs, petioles, rachises, and backs of the leaflets densely pubescent. Typical white ash is glabrous.

White ash is used for baseball bats, tennis racquets, hockey sticks, oars, and other playground and sports equipment. It is also an excellent ornamental or shade tree due to its form and fall color. Ash yellows is the most serious disease affecting this species.

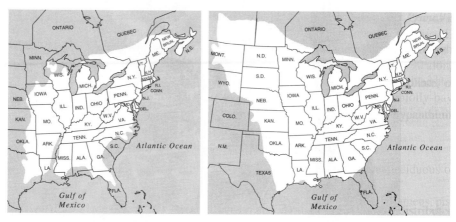

MAP 9-39   **(Left)** *Fraxinus americana.* **(Right)** *Fraxinus pennsylvanica.*

Fráxinus pennsylvánica Marsh.   green ash, red ash

Distinguishing Characteristics (Fig. 9-126)

Leaves 6 to 10 in. (15 to 25 cm) long, with 7 to 9 leaflets; *leaflets* 2 to 5 in. (5 to 13 cm) long, 1 to $1\frac{1}{2}$ in. (2.5 to 4 cm) wide; *shape* lanceolate to ovate-lanceolate, or elliptical; *margin* serrate, sometimes entire below the middle; *apex* acute to acuminate; *base* gradually cuneate to unequally obtuse, the stalks slender; *surfaces* yellow-green above; pale and glabrous to silky pubescent below, not "papillose"; *rachis* moderately stout, grooved, glabrous to pubescent; *fall color* yellow.

Flowers imperfect (trees dioecious), apetalous, both sexes borne in glabrous to tomentose panicles that appear after the leaves have begun to unfold; *calyx* remotely lobed or cup-shaped.

Fruit a narrowly oblanceolate samara, $1\frac{1}{4}$ to $2\frac{1}{4}$ in. (3 to 6 cm) long, $\frac{1}{4}$ in. (6 mm) or less wide; *wing* pointed or slightly emarginate at the apex, abruptly narrowed and extending along the slender seed cavity.

Twigs stout to moderately slender, flattened at the nodes, gray to greenish brown, glabrous to velvety pubescent; *terminal buds* conical to ovate, rusty brown, pubescent; *lateral buds* reniform to triangular, the upper pair at nearly the same level as the terminal; *leaf scars* semicircular, shallowly notched to usually straight along the upper edge; *bundle scars* forming a U- or V-shaped line.

Bark gray-brown, thick, furrowed into narrow, irregular, interlacing, scaly ridges.

FIGURE 9-126   *Fraxinus pennsylvanica,* green ash. (1) Twig $\times\frac{1}{4}$. (2) Leaf $\times\frac{1}{3}$. (3) Samara $\times$1.

Range

Eastern (Map 9-39). Elevation: sea level to 3000 ft in the southern Appalachians. Poorly drained bottomlands, stream and lake borders, swamps, and wet woodlands.

General Description

Green ash is the most widely distributed of American ashes. It is a small- to medium-sized tree 30 to 50 ft tall and 20 in. dbh (max. 145 by 6 ft) with a broad irregular crown, a short, usually poorly formed trunk, and a superficial fibrous root system. It is common, especially through the Midwest, as a scattered tree along stream banks and the borders of swamps. Best development appears to be reached east of the Appalachians. Green ash has been widely planted throughout the plains states and adjacent Canada. It is exceedingly hardy to climatic extremes; although naturally a moist bottomland or stream-bank tree, it will, when once established, persist on dry sterile soils, and it has been used successfully in shelterbelts. It is also a common street tree in the East. Under forest competition, green ash is intolerant to moderately tolerant. Throughout its range, green ash is not a major component of forest stands.

As in white ash, green ash also has both glabrous and densely pubescent forms. The glabrous form is generally called green ash and the pubescent form called red ash; the wood, however, is sold as white ash. These forms have been treated as varieties by some authors.

Fráxinus caroliniàna Mill.   Carolina ash

Distinguishing Characteristics. *Leaves* of 5 to 7 leaflets, each 2 to $4\frac{1}{2}$ in. (5 to 11 cm) long, 1 to 2 in. (2.5 to 5 cm) wide, elliptical to ovate, lanceolate, pale below, with slender stalks, serrate; fall color yellow to bronze. *Fruit* $1\frac{1}{4}$ to 2 in. (3 to 5 cm) long, yellow-brown, broadly elliptical, wing extending to base of seed, and sometimes 3-winged. *Twigs* slender, glabrous, green to brown. *Bark* light gray, thin, and scaly, becoming rough and furrowed.

Range. Coastal Plain from northeastern Virginia to southern Florida and west to southeastern Texas and southern Arkansas. Elevation: sea level to 500 ft. Swamps, stream and riverbanks, flatwoods depressions, and pond margins.

General Description. Tree to 50 ft tall. This is one of the most variable of the ashes, and several varieties have been described. The variation is in the pubescence of twigs and leaves and the shape and size of the samaras.

Fráxinus nìgra Marsh.   black ash

Distinguishing Characteristics (Fig. 9-127). *Leaves* with 7 to 13 oblong to oblong-lanceolate, serrate, sessile leaflets, each 4 to $5\frac{1}{2}$ in. (10 to 14 cm) long, 1 to $1\frac{1}{2}$ in. (2.5 to 4 cm) wide; fall color yellow-brown. *Flowers* perfect and imperfect (polygamous or dioecious), soon naked as the minute calyx falls away, borne in panicles before the leaves. *Fruit* an oblong-elliptical samara, 1 to $1\frac{3}{4}$ in. (2.5 to 4.5 cm) long, with a wide wing extending along the flattened and indistinct seed. *Twigs* stout, grayish; terminal buds ovate-conical, dark brown to nearly black, laterals rounded, the first pair borne some distance below the terminal. *Bark* grayish, relatively smooth, later shallowly furrowed with corky ridges; becoming scaly.

Range. Southeastern Manitoba, east to Newfoundland, south to West Virginia, west to Iowa; disjuncts in northeastern North Dakota and northern Virginia. Swamps, peat bogs, stream and riverbanks.

General Description. Black ash, a typically northern tree, varies from 40 to 50 ft in height and 18 in. dbh (max. 108 by 5 ft). It is slow growing, reaching 60 ft tall and 10 in. dbh in 100 years.

FIGURE 9-127 *Fraxinus nigra,* black ash. (1) Twig ×1$\frac{1}{4}$. (2) Leaf ×$\frac{1}{3}$. (3) Samara ×1. (4) Bark.

Black ash can live up to 260 years. The bole is often poorly shaped, supports a small open crown, and terminates below in a very shallow, fibrous root system. This species occurs as a scattered tree associated with the wet-site species throughout its range. Black ash is intolerant and is not found under heavy forest cover. Its seeds can live up to at least 8 years in the soil. In a few localities, it is abundant and of some importance, although the wood is of a much poorer quality than that of the other northern ashes. From early times, black ash has furnished material for pack baskets fabricated by the Native Americans of the Northeast.

Fráxinus profúnda (Bush) Bush (*F. tomentosa* Michx. f.)   pumpkin ash

Distinguishing Characteristics. *Leaves* of 7 to 9 leaflets, each 3 to 7 in. (7.5 to 18 cm) long, 2 to 3 in. (5 to 7.5 cm) wide, elliptical or lanceolate, entire to slightly serrulate, pubescent below particularly along midrib, and with a slender stalk. *Fruit* 2 to 3 in. (5 to 7.5 cm) long, broadly elliptical, oblanceolate, or spatulate, with wing extending to base. *Twigs* stout, light gray, densely pubescent when young, leaf scar deeply notched at top. *Bark* gray, furrowed into diamond-shaped patterns.

Range. Coastal Plain of southern Maryland and southeastern Virginia, south to northern Florida, west to western Louisiana, north to southeastern Missouri, southern Illinois, and southwestern Ohio. Elevation: near sea level to 500 ft. Swamps and bottomlands.

General Description. Tree to ca. 100 ft tall and 2 ft dbh (max. 130 ft by 5.5 ft). This is an uncommon tree similar to green ash but with larger leaves and larger samaras. It is moderately tolerant and often occurs with baldcypress, water tupelo, and swamp cottonwood. Pumpkin ash produces high-quality lumber.

Fráxinus quadrangulàta Michx. blue ash

Distinguishing Characteristics (Fig. 9-128). *Leaves* with 7 to 11 leaflets, each lanceolate to oblong-ovate, $3\frac{1}{4}$ to 5 in. (8 to 13 cm) long, 1 to $1\frac{1}{2}$ in. (2.5 to 4 cm) wide, serrate, with a slender stalk; fall color yellow. *Flowers* perfect, soon naked as the calyx falls away, borne in panicles as the leaves unfold. *Fruit* oblong to spatulate, 1 to 2 in. (2.5 to 5 cm) long, with wing surrounding the flattened seed. *Twigs* stout, 4-angled and 4-winged between the nodes; terminal buds broadly ovoid, reddish brown. *Bark* on mature trees with long, loose, scaly plates, giving a shaggy appearance to the trunk.

Range. Southern Wisconsin, southern Michigan, and western Ohio, south to northwestern Georgia, west to eastern Oklahoma; disjuncts in southwestern Ontario, eastern Kansas, and northwestern Missouri. Elevation: 400 to 2000 ft. Dry, rocky, limestone slopes and moist river valleys.

General Description. Blue ash is a medium-sized tree to 80 ft tall and 2 ft dbh (max. 120 by 4 ft) that occurs especially on dry limestone uplands through the Ohio and Upper Mississippi Valleys and can live for 400 years. Here its associates include chinkapin and other oaks, hickories, and redbud. This ash occurs as a scattered tree and is relatively rare in comparison with white ash, which often accompanies it on the better soils. The inner bark contains a mucilaginous substance that turns blue upon exposure to the air. A blue dye prepared by macerating the bark in water was used by the pioneers for dyeing cloth. Blue ash has been planted to some extent in the prairie region and the East, where its drought tolerance should make it a suitable street tree. Its silvical features are similar to those of the other upland ashes.

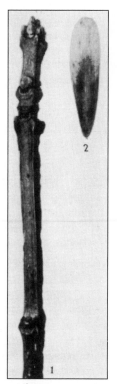

Fráxinus anómala Torr. ex Wats. singleleaf ash

Distinguishing Characteristics. *Leaves* appearing simple (actually unifoliolate) or occasionally pinnate with 2 to 3 leaflets, $1\frac{1}{2}$ to 2 in. (4 to 5 cm) long, 1 to 2 in. (2.5 to 5 cm) wide, ovate and entire. *Fruit* ca. $\frac{3}{4}$ in. (19 mm) long, $\frac{3}{8}$ in. (9 mm) wide, elliptical, with wing extending to base. *Twigs* slender, brown, glabrous, 4-angled or slightly winged. *Bark* brown and furrowed.

Range. Northeastern Utah and western Colorado, south to northwestern New Mexico, west to eastern California. Elevation: usually 2000 to 6500 ft and to 11,000 ft in California. Dry canyons, hillsides, and ponderosa pine forests.

General Description. Shrub or small tree to 25 ft tall and 6 in. dbh. This species of ash is unique because of its unifoliolate leaf.

Fráxinus latifòlia Benth. Oregon ash

Distinguishing Characteristics (Fig. 9-129). *Leaves* 5 to 12 in. (13 to 30 cm) long, with 5 to 9 leaflets, each $1\frac{1}{2}$ to 4 in. (4 to 10 cm) long, 1 to $1\frac{1}{2}$ in. (2.5 to 4 cm) wide, ovate, obovate, or elliptical, sessile or slender-stalked, serrulate or entire, usually tomentose below; fall color yellow. *Flowers* paniculate, appearing with the leaves (plants dioecious). *Fruit* oblong, elliptical, oblanceolate, or spatulate,

FIGURE 9-128 *Fraxinus quadrangulata,* blue ash. (1) Twig $\times 1\frac{1}{4}$. (2) Samara $\times 1$.

FIGURE 9-129 *Fraxinus latifolia,* Oregon ash. (1) Twig ×1$\frac{1}{4}$. (2) Leaf ×$\frac{1}{2}$. (3) Staminate inflorescences ×1. (4) Samara ×1. (5) Bark.

$1\frac{1}{4}$ to 2 in. (3 to 5 cm) long, the wing extending nearly to the base of the slightly flattened seed. *Twigs* stout, densely tomentose, with conical terminal buds and small, ovoid lateral buds; leaf scars suborbicular, the bundle scars arranged in a U-shaped line. *Bark* dark gray to gray-brown, with an interwoven pattern similar to that of white ash.

Range.   Western Washington, south along the coast to San Francisco Bay, and south along the lower western slopes of the Sierra Nevada to southern California. Elevation: sea level to 5500 ft. Stream and riverbanks and bottomlands, or in drier canyons.

General Description.   Oregon ash is a timber tree of secondary importance, but it is the only ash native to the Pacific Northwest. Ordinarily it attains a height of 60 to 80 ft and a dbh of 2 to 4 ft (max. 100 by 7 ft). A clear, symmetrical bole supporting a narrow, compact crown is found in forest-grown trees. The root system is moderately shallow but wide-spreading, and the trees are unusually windfirm. Nearly pure fringes or strips of this ash occur along the banks of water courses and margins of swamps, but it occurs more abundantly in mixture with other species.

Seed is produced after the thirtieth year, and heavy crops are released every 3 to 5 years thereafter. Seedling trees can endure moderate shading when there is an abundance of soil moisture, although they require considerably more light after passing the sapling stage. However, trees that have been suppressed for protracted lengths of time are able to resume a normal rate of growth soon after being released. The growth rate is moderately rapid for the first 60 to 100 years, after which it decreases gradually, and maturity is ultimately reached in about 200 to 250 years. Natural regeneration is also accomplished by sprouts from the root collar. The wood is especially valuable for fuel.

Fráxinus velùtina Torr.   velvet ash

Distinguishing Characteristics.   *Leaves* of 5 to 9 leaflets, each 1 to 3 in. (2.5 to 7.5 cm) long, $\frac{3}{8}$ to $1\frac{1}{4}$ in. (1 to 3 cm) wide, lanceolate to elliptical, with slender stalk, pubescent below; fall color yellow. *Fruit* $\frac{3}{4}$ to $1\frac{1}{4}$ in. (2 to 3 cm) long, narrow, oblanceolate, wing extending to midway of seed. *Twigs* gray or brown, pubescent when young. *Bark* gray, furrowed with broad, scaly ridges.

Range.   Western Texas, southern New Mexico, Arizona, southwestern Utah, southern Nevada, and southern California. Elevation: 2500 to 7000 ft. Stream banks, canyons, mountain slopes, and desert grasslands.

General Description.   Tree to 40 ft tall and 1 ft dbh. This is the common ash of the Southwest and is frequently planted for shade.

OSMANTHUS Lour.   osmanthus

There are ca. 15 species, mostly Asiatic. Several are cultivated as attractive, evergreen ornamentals with very aromatic flowers appearing in the fall. Two popular species are *O. heterophýllus* (G. Don) P. S. Green, holly tea olive, with many cultivars, and *O. fràgrans* Lour., fragrant tea olive. There is only one native species in North America.

Osmánthus americànus (L.) Benth. & Hook. f. ex Gray   devilwood, wild-olive

Distinguishing Characteristics.   *Leaves* opposite, persistent, $3\frac{1}{2}$ to 5 in. (9 to 13 cm) long, $\frac{3}{4}$ to $1\frac{1}{2}$ in. (2 to 4 cm) wide, lanceolate to elliptical, thick and coriaceous, entire, and sometimes revolute. *Flowers* small, fragrant, white or yellowish, bell-shaped with 4 petal lobes, in axillary clusters; appearing in spring. *Fruit* berrylike, $\frac{3}{8}$ to $\frac{3}{4}$ in. (10 to 19 mm) long, elliptical, dark blue. *Bark* gray, scaly.

Range.   Coastal Plain from southeastern Virginia to central Florida, west to southeastern Louisiana. Elevation: sea level to 500 ft. It occurs in moist bay forests, maritime forests, edge of swamps, in river valleys, and on dry sandy ridges and dunes.

General Description.   This is a small, evergreen tree to 30 ft tall and 1 ft dbh. It is easily distinguished from the other bay or maritime trees by its opposite leaves. The fruits are eaten by birds and small mammals.

Scrophulariaceae: The Figwort Family

This family of mostly herbs, some shrubs, and few trees has ca. 269 genera and 5100 species with a cosmopolitan distribution but mainly in the temperate zone. Of economic importance are foxglove, *Digitális purpurèa* L. (source of digitalis glycosides used in medicine), and ornamental herbs such as the common snapdragon (*Antirrhìnum màjus* L.). *Paulownia* is the only tree of this family in the United States.

Botanical Features of the Family

Leaves alternate, opposite or whorled, simple or compound, estipulate, deciduous.

Flowers perfect, irregular, 5-parted with connate calyx and connate and tubular corolla; *stamens* 2 or 4(5), epipetalous; *carpels* 2, fused, ovary superior; entomophilous.

Fruit a capsule with seeds winged or angled.

Paulòwnia tomentòsa (Thunb.) Sieb. & Zucc. ex Steud.   royal paulownia, empress-tree

Distinguishing Characteristics.   *Leaves* opposite, 6 to 16 in. (15 to 40 cm) long and wide, broadly ovate, apex acuminate, base cordate, the lower surface pubescent with erect, branched hairs (seen if scraped off and examined under a microscope); petiole 4 to 8 in. (10 to 20 cm) long. *Flowers* ca. 2 in. (5 cm) long, broadly tubular with 5 unequal lobes, pale violet and fragrant, in erect panicles to 1 ft (30 cm) long, appearing in early spring before the leaves, and from conspicuous, tan buds formed the previous summer. *Fruit* an ovoid capsule, 1 to $1\frac{1}{2}$ in. (2.5 to 4 cm) long, pointed at apex, splitting into 2 valves, with small winged seeds, the dehisced capsules turning black and persisting through the winter. *Twigs* stout, light brown, pubescent, leaf scar nearly elliptical, bundle scars many in a nearly closed ellipse, stipular scars lacking, terminal bud lacking, bud scales ca. 4 and blunt, the pith hollow in the second-year branch. *Bark* gray-brown, with furrows.

Range.   This native of China was introduced into the United States in 1834 as a very attractive ornamental, particularly when in flower. It has now become naturalized in the eastern United States and has become invasive along roadsides, edges of woods, along riverbanks, in waste places, and in pastures. It is planted for mine spoil reclamation.

General Description.   This is a deciduous tree to 50 ft tall and 2 ft dbh. It has a very rapid growth on good sites, with first-year saplings exceeding 7 ft tall and leaves up to 2 ft (60 cm) long.

Paulownia somewhat resembles catalpa vegetatively; however, catalpa differs in its often whorled leaves, lacking the stalked glands and branched hairs on the lower leaf surface, and with continuous pith in the second-year branch. The flowers and fruits are very different. *Paulownia* is still placed in the Bignoniaceae with *Catalpa* by some authors (Duncan and Duncan, 1988). However, it has been shown conclusively that it belongs in the Scrophulariaceae and only superficially resembles *Catalpa.* Royal paulownia wood commands a high price because of its significance in Japan.

Bignoniaceae: The Trumpet-Creeper Family

The Bignoniaceae include 109 genera and 750 species of trees, shrubs, vines, and herbs, the majority of which are tropical. *Cybístax dónnell-smíthii* (Rose) Siebert, a native of Mexico and Central America, has beautiful wood that enters the American markets under the names of "prima vera" or "white mahogany." A number of other very fine timbers come from the Indo-Malayan species of the genus *Stereospérmum.* The sausage-tree, *Kigèlia africàna* (Lam.) Benth., a tropical species, so named because of its peculiar sausage-shaped fruits suspended on long slender stems, is cultivated as an ornamental in southern Florida and southern California.

Four genera with arborescent forms are native to the United States, but only one (*Catalpa*) is of sufficient importance to be described.

Botanical Features of the Family

Leaves mostly deciduous, opposite or whorled (rarely alternate), simple or compound, estipulate.

Flowers perfect, usually large and showy, tubular, zygomorphic, 5-parted, ovary superior; entomophilous and ornithophilous.

Fruit a capsule or rarely berrylike, seeds usually winged.

CATALPA Scop.   catalpa

Catalpa consists of 11 species distributed through the forests of eastern Asia, eastern North America, and the West Indies. Two species are included in the arborescent flora of the United States and Canada. They are native to the southeastern states but have become widely naturalized in the East and are cultivated in the West. They are very similar and often difficult to distinguish. The trees are used ornamentally for their large, showy panicles of white flowers and long, slender, cigar-shaped fruits. They are also well-known for the "catalpa worms" (caterpillars) used for fish bait and for the wood, which is resistant to decay and used for fence posts.

Botanical Features of the Genus

Leaves deciduous, opposite or whorled in threes, simple, long-petioled, the blade heart-shaped, sometimes palmately lobed, pubescent below with unbranched hairs; fall color yellowish green or brownish yellow.

Flowers large and showy in many-flowered panicles or corymbs; corolla tubular, 2-lipped; appearing in late spring after the leaves.

Fruit a long, pendent, semipersistent, slender, terete capsule with many flat, 2-winged seeds, fringed or tufted at the ends of the opposing wings.

Twig stout; terminal bud lacking, lateral buds small, hemispherical, with ca. 6 pointed bud scales; leaf scars nearly circular, raised with a depressed center, the bundle scars many in a closed ellipse; estipulate; pith solid and homogeneous, white.

Catálpa speciòsa Warder ex Engelm.   northern catalpa, western catalpa

Distinguishing Characteristics (Fig. 9-130).   *Leaves* 6 to 12 in. (15 to 30 cm) long, ovate or oval, entire, gradually acuminate at apex, cordate at base, generally inodorous; petiole 4 to 6 in. (10 to 15 cm) long. *Flowers* in few-flowered panicles, each white with 2 yellow-orange lines and thin purplish brown spots and stripes in the throat and on the lower lobe, corolla 2 to $2\frac{1}{2}$ in. (5 to 6.5 cm) across. *Fruit* 8 to 24 in. (20 to 60 cm) long, $\frac{1}{2}$ to $\frac{5}{8}$ in. (12 to 15 mm) in diameter, with a thick wall; seed wings rounded and fringed at the ends. *Bark* reddish brown, thick with thick scales.

Range.   Native in southwestern Indiana and southern Illinois, western Kentucky and western Tennessee, southeastern Missouri, and northeastern Arkansas. Elevation: 200 to 500 ft. Alluvial forests, base of bluffs, riverbanks.

General Description.   Northern catalpa normally attains a height of 60 ft and dbh to 2 ft (maximum 120 by 6 ft). The bole is well formed if grown under proper conditions of soil, moisture, and side shading, but otherwise it may be crooked. The species occurs as scattered trees and is exceedingly reactive to soil conditions. On rich moist soils, a growth of $2\frac{1}{2}$ ft in height and a dbh increase of $\frac{1}{2}$ in. per year may be attained, but on poorer locations the growth is much less. Catalpa is intolerant, but natural pruning does not take place readily, and best results are obtained in rather dense, pure stands. The seasoned timber is very durable when used for posts or ties, although the living trees are subject to heart rot. This species has been widely planted through the Midwest, and it is said that on the proper soil this catalpa will produce more good fence posts in a short time than any other native tree. It is also commonly planted as an ornamental, although its fruits and large leaves are very messy.

Catálpa bignonioìdes Walt.   southern catalpa, common catalpa

Distinguishing Characteristics.   *Leaves* 5 to 10 in. (13 to 25 cm) long with an abruptly acuminate apex, sometimes shallowly 3-lobed, and with an unpleasant odor if crushed. *Flowers* in many-flowered panicles, each white with 2 yellow-orange lines and prominent purple spots and stripes in the throat and on the lower lobe, corolla $1\frac{1}{4}$ to 2 in. (4 to 5 cm) across. *Capsule* 6 to 15 in. (15 to 38 cm) long, $\frac{5}{16}$ to $\frac{3}{8}$ in. (8 to 10 mm) in diameter, with a thin wall; seed wings narrowed and tufted at the ends. *Bark* gray-brown, thin with thin scales.

Range.   Native of southwestern Georgia and northwestern Florida to central Mississippi. Elevation: 100 to 500 ft. Stream banks, low woods, floodplains.

General Description.   This is a small tree 20 to 40 ft tall and dbh to 2 ft. It is widely cultivated as an attractive ornamental. In comparison with *C. speciosa,* this southern catalpa has somewhat smaller leaves that are sometimes 3-lobed, panicle with more flowers, smaller flowers with more prominent purple spotting, blooming about two weeks later, smaller fruits with thinner walls, and bark with thinner scales. These are rather subtle differences, however, and the two are not easily identified unless both are available for direct comparison.

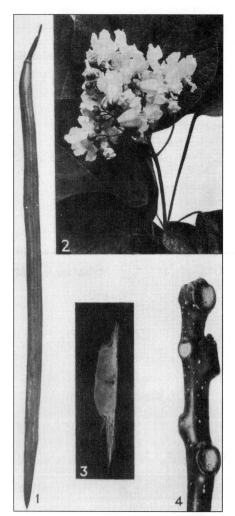

FIGURE 9-130   *Catalpa speciosa,* northern catalpa. (1) Capsule $\times\frac{1}{3}$. (2) Inflorescence $\times\frac{1}{4}$. (3) Seed $\times\frac{3}{4}$. (4) Twig $\times 1$.

LILIOPSIDA—THE MONOCOTS

ARECIDAE

Arecaceae (Palmae): The Palm Family

This family has 203 genera and 2650 species and is pantropical in distribution, with a few in the subtropics and warm-temperate latitudes. Only about 10 genera are native to the United States, one in the Southwest (*Washingtonia*) and the others in the Southeast. They are evergreen trees, shrubs, and woody vines. Palms are unusual in many ways and are one of the most typical tropical plants. The usually unbranched *trunks* are formed only by primary growth with many scattered vascular bundles; therefore, they have no true wood or bark as in dicot trees. Although most are erect, some are prostrate. The *leaves,* although appearing compound when mature, are actually simple before they expand. Weak zones along the folds mechanically split as the leaf enlarges.

The palm family is one of three most important families (with legumes and grasses) for shelter, construction materials, furniture (rattan), food, fiber, wax, and oil. The numerous uses by humans are closely tied into the folklore of the countries where they occur and provide many of the necessities of the peoples and wildlife (Hora, 1981). Well-known members are the date palm (*Phoènix dactylífera* L.), coconut palm (*Còcos nucífera* L.), and African oil palm (*Elaèis quineènsis* Jacq.). Many serve as ornamentals for lawns and streets as well as inside of homes and other buildings. Fine guides to the palms are by Henderson et al. (1995) and Meerow (1992).

Botanical Features of the Family

Leaves persistent, alternate, often very large, forming dense terminal rosettes, long-petiolate with a broad to tubular sheath that often splits and is open at maturity; simple but mechanically splitting into narrow segments, into either pinnate (feather), palmate (fan), a somewhat intermediate costa-palmate, or bipinnate forms and often with free fibers from the margins or tips of the segments; the petiole may or may not be prickly.

Flowers small, fragrant, perfect or imperfect, regular, perianth 3 to 6 parted, stamens 6, ovary 3-carpellate and superior; inflorescence a panicle enclosed at first by a large spathe; entomophilous, rarely anemophilous.

Fruit a fleshy or fibrous drupe, usually 1-seeded with a large endosperm and small, undifferentiated embryo.

Sàbal palmétto (Walt.) Lodd. ex J.A.& J.H.Schult.   cabbage palmetto, palmetto palm

Distinguishing Characteristics (Fig. 9-131).   *Leaves* fan-shaped (costa-palmate) from a short, stout, arching midrib, the segments rigid and erect with threadlike fibers along the edges, blade 4 to 7 ft. (1.2 to 2.1 m) in diameter, petiole 5 to 8 ft. (1.5 to 2.4 m) long, without prickles. *Flowers* small, whitish yellow, in large, drooping panicles initially enclosed by a large, woody spathe; appearing mid to late summer. *Fruit* a globose, fleshy, black drupe, ca. $\frac{3}{8}$ in. (10 mm) in diameter. *Trunk* unbranched, gray-brown, smooth or ridged, often with persistent leaf bases that split in the center into a Y shape.

FIGURE 9-131 *Sabal palmetto,* cabbage palmetto. (*Courtesy of U.S. Forest Service.*)

Range. Outer Coastal Plain from southeastern North Carolina to southern Florida and the Keys and west along the Gulf coast to northwestern Florida. Found along dunes, edges of salt marshes, hammocks, and inland flatwoods.

General Description. This is the common native tree palm to 50 ft tall and $1\frac{1}{2}$ ft dbh in the Southeast and a commonly planted ornamental. It is the state tree of Florida and South Carolina. See Brown (1976) for a description of reproductive biology, ecological life history, ecology, and distribution. Palmetto honey is of economic importance.

Washingtònia filífera (Linden ex André) H. Wendl.   California washingtonia, California fanpalm,
                         desert palm

   Distinguishing Characteristics (Fig. 9-132).   *Leaves* fan-shaped (costa-palmate), 3 to 5 ft
(0.9 to 1.5 m) in diameter, petiole 3 to 5 ft (0.9 to 1.5 m) long with hooked prickles below, the
blade segments are pendant with many thread like filaments from the margins; old brown leaves
hang down against the trunk in a thick mass. *Flowers* fragrant, white, appearing in late summer.
Fruit an ellipsoidal, black drupe, ca. $\frac{3}{8}$ in. (10 mm) in diameter. *Trunk* gray and smooth.

FIGURE 9-132   (Left) *Washingtonia robusta*, Mexican washingtonia. (Right) *Washingtonia filifera*, California washingtonia. (*Photo by J. W. Hardin.*)

Range.   Southwestern Arizona and southeastern California and south into Mexico. Elevation: 490 to 3300 ft. Moist, alkaline soils along streams in narrow mountain canyons and oases.

General Description.   This is the largest native palm of the United States and the only native palm of the Southwest. It was named for George Washington. It reaches to 75 ft tall and 3 ft dbh. It is also cultivated across the southern United States and is one of the most familiar and widely planted of the palms. The fruits have been an important food for Native Americans.

The characteristic, persistent, dried leaves that hang down as a massive "skirt" or "petticoat" provide cover and nesting sites for some birds and small mammals. But, when grown in yards and along streets, these skirts are usually cut because they are fire hazards and breeding sites for rats.

A similar but taller (to 100 ft or more) and thinner-trunked (to $1\frac{1}{2}$ ft dbh) palm commonly cultivated across the southern United States is the Mexican washingtonia or Mexican fan palm, *W. robústa* H. Wendl., introduced from Mexico (Fig. 9-132). The leaves are not as deeply split, the segments have fewer of the threadlike filaments, and the petiole is prickly throughout. It is faster growing than California washingtonia.

COMMELINIDAE

Poaceae (Gramineae): The Grass Family

The Poaceae are a huge family with 668 genera and 9500 species, worldwide in distribution. The grasses are primarily herbs, except for the bamboos.

Bamboos are perennial, of tree size (to 100 ft tall and 1 ft dbh) and "woody" but without secondary growth; *stems* hollow in the internodes and green, yellow, brown, black, or red; *leaves* alternate, sheathing at the base, the blades flat, long, straplike, mostly distichous, and often narrowed and jointed between the blade and sheath, and the blade is deciduous from this joint; *flowers* individually lacking a perianth and inconspicuous, but enclosed in bracts and forming a conspicuous primary inflorescence (spikelet) and these forming a larger secondary inflorescence, anemophilous; *fruit* a grain with a single seed.

There are many genera and species of bamboos, mainly from subtropical and tropical forests. These have numerous uses in papermaking, handicrafts, engineering, pipes, construction (called the "poor man's timber," [Laarman and Sedjo, 1992]), walking sticks, furniture, fishing poles, tools, split for mats, baskets, as edible young shoots, and food and shelter for wildlife.

Nine genera are cultivated in the United States as ornamentals. These are of two types, "clump-forming" and "running" bamboos. Although these seldom flower in North America, the running type can spread rapidly by rhizomes and are difficult to contain. Both types often persist from cultivation and can escape by means of discarded pieces of rhizome.

LILIIDAE

Agavaceae: The Century-Plant Family

This is a small family, closely related to the lilies, with 13 genera and 210 species widespread in warm, arid regions of the Americas but now planted worldwide. They are short-stemmed herbs with a basal rosette or sparsely and irregularly branched trees with terminal rosettes; the leaves are often stiff and succulent. Several genera are important as ornamentals, such as *Yúcca, Agáve, Sansevièria* (mother-in-law's tongue), as a source of fibers (*Agave,* sisal hemp; *Sansevieria,* bowstring hemp), and for a strong drink (pulque, fermented sap from *Agave* stems that is distilled into mescal or tequila).

Botanical Features of the Family

Leaves alternate and forming a close spiral, simple, basal or apical in rosettes, sheathing, usually fibrous, often succulent, stiff and sharp pointed, narrow, tapered and dagger- or swordlike, entire or prickly margined.

Flowers perfect or imperfect (and plants dioecious), mostly regular, 3-parted, ovary superior or inferior; inflorescence usually a panicle; pollinated by insects, birds, or bats.

Fruit a capsule or berry.

AGAVE L.   century plant

This genus, with over 100 species from the southwestern United States to tropical South America, has succulent, spine-tipped leaves in a basal rosette, and an inferior ovary. They form conspicuous plants of the landscape. Agaves flower only once at the end of their lifetime. During vegetative growth, this plant has a dense, basal rosette of large leaves and no obvious stem. Then in 8 to 25 years (not a century as the common name implies), it rapidly sends up a tall stem with a large panicle of yellow to greenish flowers. The plant dies as the seeds mature.

Although not considered as "trees" in the usual sense and generally not included in tree books, there are several species of tree-size agaves in the Southwest, some reaching 25 to 35 ft tall. The one described here serves to show the general habit and size.

The agaves are important for many desert animals.

Agáve párryi Engelm.   Parry agave

Distinguishing Characteristics (Fig. 9-133). *Rosette* 1 to 2 ft (3 to 6 dm) tall and 2 to 4 ft (6 to 12 dm) in diameter. *Leaves* 1 to 2 ft (3 to 6 dm) long, 3 to 4 in. (7.5 to 10 cm) wide, ashy gray, with a terminal spine tip. *Inflorescence* to over 16 ft (4.8 m) tall, the lateral branches horizontal and bearing many terminal flowers. *Flowers* with reddish buds, yellow when open. *Fruit* a dehiscing capsule.

Range.   This is found at 4500 to 8000 ft in the arid Southwest from central Arizona to western Texas and northern Mexico.

YUCCA L.   yucca

This genus with 30 species in usually dry, subtropical America, differs primarily from *Agave* by its superior ovary. Many species have the same habit as that of *Agave* with the basal rosette. However, several reach tree size. The most striking treelike yucca, because of its unusual branching pattern, is the Joshua-tree of the Southwest. Yucca is the state flower of New Mexico.

FIGURE 9-133   *Agave parryi*, Parry agave. (*Courtesy of U.S. Forest Service.*)

Yúcca brevifòlia Engelm.   Joshua-tree

Distinguishing Characteristics (Fig. 9-134).   *Leaves* evergreen, long, stiff, swordlike, the margins sharply serrate, and the apex spine-tipped, 8 to 14 in. (20 to 35 cm) long, $\frac{3}{8}$ to $\frac{5}{8}$ in. (8 to 15 mm) wide, clustered and forming a terminal rosette at the ends of branches. *Flowers* $1\frac{1}{4}$ to $1\frac{1}{2}$ in. (3 to 4 cm) long, bell-shaped with 6 greenish pale yellow tepals, waxy and night-blooming, in erect, terminal panicles 30 to 50 cm long; appearing in early spring. *Fruit* $2\frac{1}{2}$ to $4\frac{3}{4}$ in. (6 to 12 cm) long, 2 in. (5 cm) in diameter, elliptical, becoming a dry but indehiscent capsule. *Trunk* irregularly branched, reddish brown or gray, rough and deeply furrowed into plates; the branches often covered with dead leaves hanging downward.

Range.   Southwestern Utah, southern Nevada, western Arizona, and southern California. Elevation: 2000 to 6500 ft. Dry soils of the Mohave and Sonoran deserts, and close to pinyon pines in the lower conifer forests of the San Bernardino Mountains. It occurs in groves on plains, slopes, and mesas.

General Description.   This is an irregularly branched, picturesque tree from a single trunk and the largest of the yuccas with a height to 32 ft tall and dbh of 1 to 3 ft. It is important as food, for nesting, and as shelter for wildlife in the area.

Yuccas have an interesting interdependence with the Yucca moth—the moth is necessary for pollination, and the developing seeds provide food for the larvae. After gathering pollen from a flower and rolling the pollen into a little ball, the moth lays its eggs in the ovary of another flower and packs the pollen into the stigma, thus both pollinating the flower and providing a safe, food-filled place for the developing larvae, which later burrow out of the fruit.

FIGURE 9-134   *Yucca brevifolia*, Joshua-tree. (*Photo by J. W. Hardin.*)

10

DENDROCHRONOLOGY

Dendrochronology is the determination of the age and growth pattern of trees by the comparative study of annual tree rings (Fig. 10-1, *a* & *b*), as seen on a stump, at the end of a log, or from cores taken by an increment borer. Although this science of tree-ring analysis and its implications are not part of dendrology as it is defined, it seems appropriate to at least briefly mention this fascinating area of study and its contribution to our understanding of tree growth and long-term environmental trends.

The study was developed initially to infer information about archeological remains and various cultures (Douglass, 1920). Detailed prehistoric cultural chronologies, initially in the southwestern United States and more recently in the eastern United States, have been determined through tree-ring analysis (Dean, 1969; Stahle and Wolfman, 1985).

The determination of the *number* of annual rings informs us about the age of an individual or stand. This alone has been very useful in numerous cases such as determining the age of the intermountain bristlecone pine in Nevada (Currey, 1965) and the baldcypress in North Carolina (Stahle et al., 1988). These are the oldest trees, thus far determined, west and east of the Mississippi River. However, the *pattern* of annual rings displayed by hardwood and conifer species of temperate climates is informative in several important ways such as the correlation of extreme drought years with historical events (Fig. 10-1, *c*) (Stahle et al., 1998).

The success of dendrochronological studies depends much on one's understanding of a particular tree species' growth rates on various types of sites and on its longevity and ability to persist on stressful sites. Excellent summaries of dendrochronology and its applications are by Fritts (1976) and Fritts and Swetnam (1989). Recent advances in measuring techniques and statistical analyses and an appreciation for the power of this

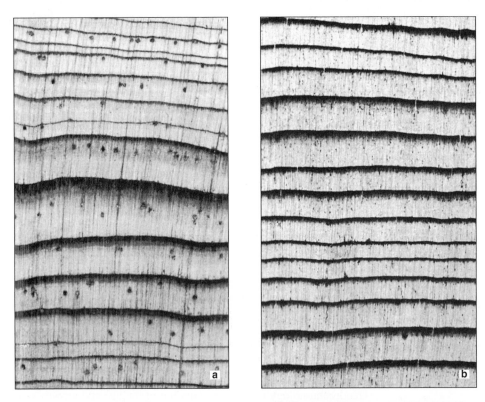

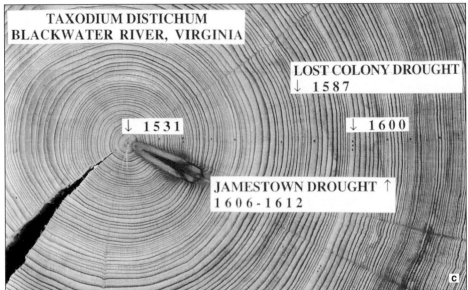

**TAXODIUM DISTICHUM
BLACKWATER RIVER, VIRGINIA**

LOST COLONY DROUGHT
↓ 1587

↓ 1531

↓ 1600

JAMESTOWN DROUGHT ↑
1606-1612

FIGURE 10-1   Cross sections of wood showing annual rings. (*a*) ponderosa pine, sensitive to climatic variation. (*b*) eastern hemlock, growing in an equable climate. (*c*) baldcypress found at Blackwater River, Virginia. The exact calendar dating of each ring was determined with dendrochronology, where these patterns were synchronized with similar patterns in old living trees in the region. Periods of severe drought can be implicated in the fate of the Lost Colony and the appalling death rate in the Jamestown Colony. The gradual decline in ring width after 1612 is not climatic but simply reflects the increasing size and age of the tree. (*Figure c courtesy of David W. Stahle.*)

tool in eastern North America have made dendrochronology more commonly applied to a wide range of questions.

Our understanding of whether climatic "aberrations" of today are indeed unusual can be improved through dendrochronology (Cook and Jacoby, 1977; Stahle and Cleaveland, 1992; Stahle et al., 1985). Tree-ring analyses have explored the historical amount of sunshine for a region (Stahle et al., 1991) and examined historical hydrological changes in rivers (Cleaveland and Stahle, 1989; Jones et al., 1984).

Tree-ring analyses have been especially important since the 1970s as scientists have examined tree growth in relation to atmospheric deposition (Cook and Zedaker, 1992; Fox et al., 1986; LeBlanc, 1992), climate change (Cook et al., 1991), and even to monitor time trends in the deposition of certain pollutants (Ragsdale and Berish, 1988). Tree growth has been correlated with climate and the El Niño–Southern Oscillation (Stahle and Cleaveland, 1992), allowing an historical examination of this highly significant and primary cause of interannual variations in global climate.

Dendrochronology has been used to examine forest disturbance history (e.g., the fire history of giant sequoia groves during the past thousands of years relative to annual fluctuations in precipitation and temperature [Swetnam, 1993] and stand dynamics in relation to fire occurrence in other western conifer forests [Johnson and Fryer, 1989; Pitcher, 1987]). Historical outbreaks of insect attacks on trees have been investigated with tree rings (Kienast and Schweingruber, 1986).

Additionally, dendrochronology has enhanced the value of old, often gnarled trees, especially on stressful sites, and is one reason for the increased interest in finding old-growth sites in eastern North America (Kelley et al., 1992).

So trees, particularly old trees, capture in their annual rings the effect of environmental conditions through many, many centuries. This window into the past enhances our understanding of both environmental and human history and enables scientists to address some of the most urgent environmental issues of today.

11

AFTER DENDROLOGY, WHAT?

After a course in dendrology, one may think that there is little more to learn except just more trees. After all, the trees have been named and classified (Little, 1979), their characteristics described, drawn, or photographed (Elias, 1989; Farrar, 1995; Little, 1980a and b; Young and Young, 1992), habitats identified (Burns and Honkala, 1990), and ranges mapped (Little, 1971, 1976, 1977, 1978; Viereck and Little, 1975).

Literature prior to the 1950s did deal mainly with these basic facts. Since the late 1960s, though, there has been an explosion of research on woody plants. Some studies were floristic for state or regions (Clark, 1971; Godfrey, 1988; Leopold et al., 1998); some revisionary (Burckhalter, 1992; Furlow, 1979); some others dealt with genetics and improvement of commercial tree species (Nienstaedt and Teich, 1971; Zobel and Talbert, 1984); others dealt with physiological ecology (Kozlowski et al., 1990); ecological life history including pollination ecology and reproductive strategies (Brown, 1976); within-species variation and provenance studies (Adams and Turner, 1970; Giannasi, 1978; Howe et al., 1993; Kreibel, 1993); between-species variation in micromorphology (Hardin, 1992); adaptive radiation (Stone, 1973); physiology and autecology (Kramer and Kozlowski, 1979; Kozlowski et al., 1990; Zobel, 1969); hybridization and introgression (Barnes et al., 1974; dePamphilis and Wyatt, 1989, 1990; Flake et al., 1978; Hardin, 1975; Ledig et al., 1969; Rushton, 1993; Tucker, 1993); genetic diversity and structure (Ballal et al., 1994; Berg and Hamrick, 1993; Gibson and Hamrick, 1991; Hamrick et al., 1992; Loveless and Hamrick, 1984; Mosseler et al., 1993); population genetics and relative gene flow correlated with life history (Hamrick et al., 1979; Mosseler et al., 1993); genetic effects of forest fragmentation (Gauthier et al., 1992; Young et al., 1993); concern for conservation of rare species (Geburek, 1997; Newton et al., 1999; Yang and Yeh, 1992); and relationships of species, genera, or families (Hart, 1987; Judd et al., 1994; Malusa, 1992; Murrell, 1993; Price and Lowenstein, 1989; Price et al., 1987; Robson et al., 1993; Strauss and Doerkson, 1990). More and more of these studies since the 1970s have used newer techniques of SEM, chromatography of flavonoids, electophoresis of allozymes, radioimmunoassay of seed proteins, and RNA or DNA analysis.

We are now at a stage where the emphasis will be on inter- and intraspecific genetic variation (using modern techniques such as DNA sequencing); the adaptive significance of anatomical, morphological, cytological, and chemical differences; ecological life histories and reproductive strategies; phenetic, phyletic, and cladistic relationships among families, genera, and species; population biology and the dynamics of woody plant populations; and the mechanisms of speciation and evolution. Call it *modern dendrology.*

Numerous questions need to be asked of the modern dendrologist. For instance, what do we know about the function and importance of the individual or species population in an ecosystem? What do we know of the genetic basis of characters and character states that are used in our classifications? If genes are involved in phenotypic plasticity, what role do they play in the overall genetic architecture of the plant? What do we know about the genetic structure, ecological life history, or autecology of most of our forest trees? What do we know about the extent or dimension of gene exchange in and between populations or the minimum effective size of the breeding populations of trees? And, how are these parameters affected by the increasing fragmentation of our forests? Why are certain tree species rare and endangered or threatened? How can we manage and maintain the critical habitats of rare and endangered tree species or the endangered herbaceous species of a forest floor? What do we know of the details of interspecific competition, plant-plant interactions, speciation, ecophenic vs. ecotypic vs. clinal variation? Has a wide-ranging species undergone ecotypic differentiation, or does it have a highly plastic "general-purpose genotype"? How is the degree of genetic variability correlated with life history and reproductive strategies? What do we really know of the nature of the effective reproductive barriers that maintain the morphological/ecological units that we call species? Does introgression in forest trees actually occur as much as some think, or is the apparent pattern of introgression due to some other cause? Is the introgression that does exist merely "evolutionary noise," or could it be a consistent feature of the adaptiveness of many plant species (a syngameon or compilospecies) and necessary for continued ecological exploitation and future progressive evolution or even survival? Are invasions of exotic species, due to human introductions or global change, going to disrupt the plant communities that we recognize today?

The modern dendrologist faces exciting questions, and the answers to many of these are of utmost significance for forest management and maintenance of genetic diversity in our forest ecosystems throughout the world.

We hope this book provides the user the foundation for a lifetime of learning about trees, whether as a part of his or her professional life, as a rewarding lifetime hobby, or as a well-educated citizen.

There *is indeed* a great deal more—*after dendrology.*

GLOSSARY

abortive   Lacking or defective.

abscission layer   Layer of cells at base of leaf that, when developed, causes the leaf to die and eventually drop.

accessory   An additional part.

achene   A small, dry, 1-seeded, unwinged fruit.

acicular   Needlelike.

actinomorphic   Radially symmetrical; regular flower.

aculeate   "Spiny" or "prickly" leaf margin.

acuminate   Long, tapering; attenuated.

acute   The shape of an acute angle; sharp pointed but not attenuated.

adnate   Fusion of unlike parts of a flower (e.g., perianth and stamens, which form the hypanthium).

agamospermy   The asexual formation of a seed.

aggregate   A cluster of fruits developed from separate pistils of a single flower and remaining together at maturity.

allelopathy   The process in which a plant releases into the environment a chemical compound that inhibits or stimulates the growth of another plant in the same or neighboring habitat. In native plant communities, allelopathy may determine distribution patterns of plants in relation to their associates; in forestry, allelopathy may also affect yields or forest regeneration. Some important allelopathic trees are black walnut, eucalyptus, sugar maple, hackberry, balsam poplar, and sassafras.

allergenic   Causing an allergic reaction such as dermatitis or nasal/bronchial irritation in susceptible individuals.

allogamy   Cross-fertilization between flowers; including geitonogamy and xenogamy.

allopatric   Two taxa with nonoverlapping geographical ranges.

ament   A flexible, often pendent spike bearing apetalous, unisexual flowers.

anastomosing   Interlacing.

androecium   The collective term for all the stamens of a flower.

androgynous   Having both staminate and pistillate flowers in the same inflorescence.

anemochory   Wind dispersal of diaspores.

anemophilous   Wind-pollinated.

aneuploid   A chromosome number that is not an exact multiple of the base haploid number within a group.

angiosperm   The flowering plants; literally "enclosed seeds" in a carpel.

anther   The pollen-bearing portion of the stamen.

anthesis   The time of opening of a flower when pollination can occur.

apetalous   Without petals.

apiculate   Abruptly pointed.

apomixis (n.), **apomictic** (adj.)   Asexual reproduction as a substitute for the normal sexual process.

apophysis   The part of a cone scale that is exposed when the cone is closed.

a posteriori   Derived after observation of the facts.

appressed   Flattened against.

a priori   Derived by reasoning before the facts.

arborescent   Treelike.

arcuate   Pinnate venation in which the entire secondary veins narrowly arch upward toward the apex.

aril   A fleshy appendage growing from the point of attachment of a seed and often covering the seed.

aristate   Bearing a stiff bristle at the apex.

articulate   Joined or jointed.

artificial classification   Based upon a few convenient characters and *a priori* reasoning.

asepalous   Without sepals.

association   A group of species occurring in the same habitat with one or more dominant species as well as associated species; often designated by the names of the dominants (e.g., spruce–fir or pine–oak).

attenuate   Long, tapering; acuminate.

auriculate   With earlike lobes at the base.

autecology   Ecology of an individual plant or species.

autogamy   Self-fertilization; pollination within one bisexual flower.

awl (awn)   Subulate leaf type of conifers that are short, stiff, and tapered to a point.

axil   Upper angle between the stem and attached organ.

axile   Situated on the axis.

axillary   In an axil.

axis   The central line of an organ; the "stem" to which structures are attached.

ballochory (autochory)   Seed dispersal by ballistic means—being shot out of the fruit.

bark   The outer layer of a woody stem; all tissues from the vascular cambium outward.

barochory   Diaspore dispersal by weight (gravity).

basionym   The name-bringing synonym in any change in position or rank of a taxon.

bast   Fibrous constituents of the bark.

berry   A simple fleshy fruit with the entire pericarp fleshy; with one to many seeds.

bi- or bis-   Prefix denoting two, double, or twice.

bifid   Two-cleft; split into two parts.

bilabiate   Two-lipped.

bipinnate   Twice pinnate.

bisexual   Having both sex organs in the same flower; perfect, monoclinous.

blade   Lamina; the flattened, expanded part of a leaf.

B.M.   Board measure.

bole   Trunk of a tree.

boss (n.), **bossed** (adj.)   A raised projection, usually pointed but not prickly.

bract   A modified leaf subtending a flower, branch of an inflorescence, or conifer cone scale.

bracteole   A tertiary bract.

bractlet   A secondary bract.

bud   An embryonic axis with its appendages.

bundle scar   A small scar in a leaf scar left by a vascular bundle.

buttress   Swollen base of tree trunk.

ca. (circa)   About; approximate.

cache   A secure place of storage, as a hoard of nuts.

calcicole (calciphile)   A plant restricted to, or growing best on, alkaline soils.

calyx   Collective term for all the sepals of a flower.

cambium   Lateral meristem responsible for secondary growth.

campanulate   Bell-shaped.

capitate   Shaped like a head.

capsule   A simple, dry, dehiscent fruit, the product of a compound pistil.

carpel   The basic unit of the gynoecium; a simple pistil; the megasporophyll of a flower.

catkin   See "ament."

character   A feature such as leaf shape or bark color.

character state   A particular character expression, such as ovate leaf shape or gray bark color.

ciliate   Descriptive of a margin fringed with hairs.

cladistics   A taxonomic approach to phylogeny emphasizing the branching of evolutionary lines and origin of character divergence.

cladogram   Diagram of cladistic relationships.

cleft   Divided into lobes separated by sinuses that extend more than halfway to the midrib.

climax   The terminal community in ecological succession capable of self-replacement under the prevailing climatic, edaphic, physiographic, biotic, and pyric (fire) conditions.

cline   A gradient of character states correlated with a geographical or ecological gradient.

clone   A population of individuals propagated vegetatively or by apomixis from a single original.

collateral   Accessory buds arranged on either side of the lateral bud.

comose   Seeds with long hairs on the surface, as in cotton or cottonwood.

community   All the organisms inhabiting a common environment and interacting with one another.

compilospecies   A species that incorporates genes, and characters, from other species through hybridization.

complete   A flower having sepals, petals, stamens, and carpels.

compound   Composed of several parts; compound leaf of leaflets, compound pistil of fused carpels.

compressed   Flattened.

cone   An aggregation of sporangia-bearing structures at the tip of a stem; best restricted to the conifers among seed plants.

conifer   A gymnosperm bearing cones.

coning   Development and maturation of a cone; "flowering" in conifers.

connate   Fusion of similar parts of a flower (e.g., sympetalous corolla).

continuum   A gradual change in species composition along an environmental gradient.

convergent evolution   The independent origin of similar features in different evolutionary lines due to similar adaptive pressures.

cordate   Heart-shaped, with the notch at the base.

coriaceous   Thick and leathery.

cork   Outer bark of the cork oak; generally the outer layer of bark.

corolla   Collective term for all the petals of a flower.

corticolous   Growing on the bark of tree trunks or branches.

corymb   Short, broad, more or less flat-topped inflorescence, with branches arising spirally along the stem.

costa-palmate   Leaf of some palms that appear fan-shaped but have a short midrib (costa).

cove   A sheltered valley between opposing slopes.

crenate   Margin with rounded or blunt teeth.

crenulate   Finely crenate.

crown   The upper part of a tree with branches.

cultivar   A horticultural variant, either selected from the wild or produced in cultivation, but maintained under cultivation; reproduced by either sexual or asexual means.

cuneate   Wedge-shaped, tapering to a narrow base.

cupule   A cuplike structure below or surrounding one or more nuts in the Fagaceae, formed by fusion of inflorescence axes bearing bracts that are modified into either scales (*Quercus, Lithocarpus*) or spines (*Fagus, Castanea, Chrysolepis*).

cuspidate   Abruptly tipped with a sharp, rigid point.

cyme   Inflorescence consisting of a central rachis terminated early by a flower, then lower lateral branches developing and either ending with a flower or another cyme.

cymose   Cymelike.

dbh (DBH)   Diameter breast-high (4 1/2 ft or 1.3 m from the ground or 18 in. above butt swell).

deciduous   Not persistent; parts (leaves) dropping in the fall.

decompound   More than once divided or compounded.

decurrent   Said of a leaf base that extends down and is fused to the twig below the point of divergence; tree habit of diffusely branched, spreading form (deliquescent).

decussate   Opposite, 4-ranked leaf arrangement; each pair at right angles to the pairs above and below.

dehiscent   Opening.

deliquescent   Descriptive of a tree form lacking one central axis but rather with diffuse branches.

deltate   Triangular or delta-shaped; "**deltoid**" is the three-dimensional form.

dentate   Margin with sharp teeth pointing outward.

denticulate   Minutely dentate.

determinate short shoot   Spur branch without a functional terminal bud.

diadelphous   Stamens formed into two groups by fusion of the filaments.

diaspore   The dispersed part of a plant; either seed, fruit, or cone.

dichotomous   Y-branching; equal forking into pairs.

diclinous   Flower that is unisexual; literally meaning "two couches."

dicotyledonous   (**dicot**)   Having two cotyledons; members of the Magnoliopsida.

dimorphous   Having two morphological forms.

dioecious   Plant having unisexual flowers or cones; with only one sex per plant; literally meaning "two houses."

disjunct   A species population separated by a considerable distance from the main distributional range.

distichous   Leaves all in one plane (i.e., in two ranks); either opposite or a 1/2 phyllotaxy if alternate.

double serrate   Margin with large teeth, each tooth having smaller teeth.

drupaceous   Drupelike.

drupe   A simple, usually 1-seeded, fleshy fruit with the outer pericarp fleshy and the endocarp bony; a stone fruit.

drupelet   A small drupe.

echinate   Armed with prickles.

ecology   Branch of biology dealing with relationships between organisms and their environment.

ecophene   A direct, nongenetic modification of a plant form by some environmental factor.

ecotone   A transitional zone between two habitats and communities.

ecotype   A heritable, morphological, or physiological form resulting from selection by an ecological condition.

eglandular   Without glands.

ellipsoid   A three-dimensional form having the outline of an ellipse.

elliptical   Plane figure resembling an ellipse, being widest at the center.

emarginate   Apex with a shallow, broad notch.

embryo   A young stage of the sporophyte plant in a seed.

endemic   A species restricted to a very limited, native, geographical area.

endocarp   Inner layer of the pericarp (fruit wall).

endosperm   The food reserve outside the embryo within a seed; a postzygotic structure of Angiosperms.

endozoochory   Diaspore dispersal by being eaten by animals and passing through the digestive tract.

entire   Smooth margin.

entomophilous   Insect-pollinated.

epicormic   A shoot arising from an adventitious or dormant bud on a stem or branch of a woody plant; coppice or "water" sprout.

epigeal (or **epigeous**)   Seed germination in which the cotyledons of the seedling are above the ground surface, as in maple or pine.

epigynous   Perianth appearing to arise from above the ovary; often used to describe a flower with an inferior ovary.

epipetalous   The fusion of filaments to the corolla; appearing as if the stamens were borne on the petals.

erose   Irregularly toothed, eroded or "chewed."

estipulate   Without stipules.

evergreen   Remaining green throughout the year; not losing all the leaves at one time.

excurrent   Tree form with a main axis or trunk extending to the top of the crown and with secondary branches from the trunk.

exocarp   Outer layer of the pericarp (fruit wall).

exozoochory   Diaspore dispersal by being carried on the outside of an animal in fur or feathers, etc.

exserted   Extending out, beyond the surrounding structures.

extant   Living at the present time.

extinct   Eliminated entirely.

extirpated   Eliminated from only a portion of the natural range.

extrorse   Facing or opening outward.

falcate   Sickle- or scythe-shaped.

false whorl   An apparent whorl of secondary branches in pines; considered "false" because the actual arrangement is a tight spiral.

family   A level in the classification hierarchy between genus (below) and order (above); indicated by the suffix *-aceae*, pronounced *-ay-see-ee*.

fascicle   A cluster or bundle.

fastigiate   A very narrow tree form having upright branches close to the trunk.

fertilization   Union of male and female gametes.

filament   The stalk of the stamen that supports the anther.

fimbriate-margined   Appearing fringed.

flabellate   Fan-shaped.

flora   The kinds of plants, taken collectively, that occupy a specified region.

flower   A determinate short shoot usually bearing a perianth, stamens, and carpels.

fluted   Regularly marked by alternating ridges and groovelike depressions.

foliaceous   Leaflike in appearance and texture.

follicle   A single carpellate, dry fruit dehiscing along one line (suture).

forest cover types A forest association, designated by the dominant species of trees.

formation One or more plant communities exhibiting a definite structure or life form and occurring in similar habitats (e.g., grassland, marsh, conifer forest, or deciduous forest).

fossil Remains of life of the geological past.

fruit The seed-bearing organ of a flowering plant; the ripened ovary.

fugacious Soon falling or fading; not permanent.

fusiform Spindle-shaped.

gallery forest A narrow forest along a waterway through a landscape otherwise devoid of trees.

gametophyte Haploid phase of life cycle that produces the gametes.

geitonogamy Cross-fertilization between flowers on a single plant; genetically the same as autogamy.

genotype Genetic constitution of an individual.

genus (sing.), **genera** (pl.) The level in the classification hierarchy between family (above) and species (below).

glabrate Becoming glabrous or nearly so.

glabrous Smooth; lacking hairs.

glandular Having glandular hairs; or the nature of a gland.

glaucescent Slightly glaucous.

glaucous Covered with a white wax in the form of minute flakes, rods, or granules; with a "bloom."

globose Spherical, globular.

glochids Minute, barbed prickles in some cacti.

gregarious Growing in clusters; colonial.

gymnosperm Seed plants bearing "naked seeds" or actually ovules exposed at pollination.

gynoecium Collective term for all the carpels of a flower.

habit The shape or form of a plant.

habitat The locality in which an organism lives.

halophyte A plant restricted to, or growing best in, high salt concentrations (salt flats, dunes, salt marshes).

hammock An isolated area of southern hardwood forests on an elevated, well-drained site, surrounded by extensive pines or marsh; a hardwood forest on deep loamy soils rich in humus.

head A spherical or flat-topped inflorescence of sessile or nearly sessile flowers clustered on a common receptacle.

herbarium A collection of plant specimens, pressed, dried, mounted on sheets, identified, and classified.

hesperidium Type of berry with a thick, leathery rind and many radial sections.

heteroblastic change Transition from juvenile to adult with a more or less abrupt change in morphology.

heteromorphism Having different forms of a structure, such as leaves.

heterophylly Having two or more different forms of leaves on the same plant.

hilum Scar or point of attachment of a seed.

hirsute With stiff or coarse hairs.

hoary With dense, grayish white hairs.

homoecious Plant with perfect flowers (synoecious).

homoplastic similarity Similar character(s) in unrelated groups derived by convergent evolution.

hybrid A cross between two different plants (usually species).

hybridization Process of crossing two unlike plants (usually species).

hybrid swarm Population consisting of hybrids, backcrosses, and successive generations.

hydrochory Diaspore dispersal by water.

hypanthium Structure of a flower usually derived by adnation of perianth and stamens, forming a cup or tube; may be free of ovary (resulting in perigny) or fused with ovary (epigny).

hypogeal (or **hypogeous**) Seed germination in which the cotyledons of the seedling remain at or below the ground surface, as in oak and hickory.

hypogynous Perianth arising below the ovary; often used to describe a flower with a superior ovary.

imbricate Overlapping.

imperfect Flower with only one sex; unisexual, diclinous.

inbreeding Crossing within a single plant; autogamy or geitonogamy.

incised Cut sharply and irregularly, to near the midrib.

incomplete Flower lacking one of the four whorls (calyx, corolla, androecium, gynoecium).

indehiscent Not splitting.

indeterminate short shoot Spur branch having a functional terminal bud.

indigenous Native and original to an area.

inequilateral   Asymmetrical leaf base.

inferior ovary   Ovary appearing below the perianth and stamens; the hypanthium fused with the ovary.

inflated   Bladderlike.

inflorescence   The arrangement of flowers on a branch system.

infraspecific   Classification categories below the species (subspecies, variety, forma).

infructescence   The arrangement of fruits on a branch system.

inserted   Attached to or growing out of.

internode   That portion of a stem between two nodes.

introgression (introgressive hybridization)   A flow of genes from one species to another across a fairly strong barrier to interbreeding by means of hybridization and backcrossing.

introrse   Facing or opening inward.

involucral   Belonging to an involucre.

involucrate   Possessing an involucre.

involucre   A cluster of bracts subtending a flower or inflorescence.

irregular   Zygomorphic; bilateral symmetry in a flower.

keel   A central ridge; the united lower two petals of a papilionaceous flower (Fabaceae).

laciniate   Cut into lobes separated by deep, narrow, irregular incisions.

lactiferous   Production of milky sap in certain species.

lamina   Leaf blade.

lanceolate   Lance-shaped; much longer than wide, widest near the base and tapering to the apex.

lateral bud   Bud in the axil of a leaf.

latex   Milky sap.

layering   Rooting from the lower side of branches.

leaf   Primary lateral appendage of a stem; usually flattened and photosynthetic.

leaflet   A division of a compound leaf.

leaf scar   A scar left on a twig when the leaf drops.

legume   A dry fruit from a single carpel, either dehiscing along two lines (sutures) or indehiscent.

lenticel   A small corky spot or line on the surface of a twig, branch, or trunk.

lenticellate   Having lenticels on the twig.

lenticular   Shaped like a double-convex lens.

lepidote   Covered with minute, flattened, peltate trichomes or glands.

liana   Woody vine.

ligulate   Strap-shaped.

linear   Very long and narrow with nearly parallel sides.

lobe   Any protruding segment of an organ.

lobed   Leaf cut from 1/4 to 1/2 the distance from the margin to the midrib; more generally, any degree of lobing.

long shoot   Stem with normally elongated internodes.

lunate   Crescent-shaped.

lustrous   Glossy, shiny.

marcescence (n.), marcescent (adj.)   Leaves withering and turning brown but persisting on the tree.

membranous   Thin, more or less flexible.

meristem   A localized plant tissue capable of cell division and giving rise to new cells and tissues.

merous   Having a specified number of parts.

mesic   Intermediate moisture regime, between xeric (dry) and hydric (wet).

mesocarp   Middle layer of the pericarp (fruit wall).

midrib   The primary vein of a leaf.

monadelphous   Stamens united into a tube by the union of the filaments.

monoclinous   Flower that is bisexual; literally meaning "one couch."

monocotyledonous (monocot)   Having but one cotyledon; members of the Liliopsida.

monoecious   Plant with both male and female flowers or cones; literally meaning "one house."

monotypic   Having only one representative (e.g., a monotypic genus has only one species).

morphology   Branch of biology dealing with form and structure.

mucro   A short, abrupt tip of a leaf or cone scale.

mucronate   Terminated by a mucro; bristle-tipped.

multiple   Coalesced fruits of an infructescence.

mutation   A change in the amount or structure of the genetic material; an inheritable change of a gene from one form to another.

naked bud   Bud without scales; usually covered with dense hairs.

natural classification   Grouping based upon many characters of the plant and *a posteriori* reasoning.

naturalized   An introduced plant, escaped from cultivation, and reproducing naturally for enough years to have become part of the flora.

naval stores   Resin products used originally for ships, later for medicine and industry; tar and pitch are from burning resinous wood in the absence of air; turpentine and rosin are distillation products from resin collected from living trees.

nitrogen fixation   The assimilation of atmospheric nitgrogen into ammonia and nitrates by true bacteria or actinomycetes associated with the roots of certain plants.

node   The place on a twig that bears one or more leaves.

nomenclature   The naming of an organism.

nut   An indehiscent, dry fruit, usually 1-seeded (though from a compound ovary), with a hard pericarp; sometimes partially or entirely encased in an involucre or husk formed of bracts.

nutlet   Small nut.

oblique   Asymmetrical base of leaf.

oblong   Longer than wide, with margins nearly parallel.

obtuse   Blunt.

orbicular   Circular or nearly so.

ornithophilous   Bird-pollinated.

outcrossing   Crossing between individuals; xenogamy.

oval   Broadly elliptical with the width greater than 1/2 the length.

ovary   The basal portion of a pistil that encloses the ovules.

ovate   Longer than wide, with the widest part below the middle; the outline of an egg.

ovoid   Three-dimensional structure with the shape of an egg.

ovulate   Pertaining to the ovule, or possessing ovules.

ovule   The structure that develops into a seed after fertilization of the enclosed egg.

ovuliferous scale   A structure of a conifer cone in the axil of (and sometimes fused with) a bract and bearing one or more ovules.

palmate   Radiating from a common point; as fingers from the palm of a hand.

palmlike   Tree form with an unbranched trunk and top rosette of leaves.

palynology   A study of pollen and spores; either extinct or extant.

panduriform   Fiddle-shaped; obovate with concave sides below the middle and with two small basal lobes.

panicle   A compound, multibranched inflorescence, usually pyramidal in shape, in which the branches from the primary axis are again branched.

paniculate   In the form of a panicle.

papilionaceous (papinionate)   Butterflylike; descriptive of the flowers of the Fabaceae.

papillose   Covered with minute, rounded projections from the cells.

parallel veins   Any individual, free veins from base to apex of leaf.

parted   Divided by sinuses that extend nearly to the midrib.

patristic similarity   Similarity of characters derived from a common ancestor.

pectinate   Comblike; divided into long, narrow segments.

pedicel   The stalk of an individual flower in an inflorescence.

pedicelled   With a pedicel.

peduncle   Stalk of a single flower as in magnolia; or, the stalk of an inflorescence (from base to first branch).

peg   Short projection on a conifer twig bearing a sessile or petiolate leaf and remaining on a twig following leaf drop.

peltate   Shield-shaped and attached from the center by a supporting stalk; the shape of an umbrella.

pendant, pendulous   Drooping, hanging down.

penniveined   Veined in a pinnate manner.

perennial   Lasting for 3 or more years.

perfect   Flowers with both sexes; bisexual, monoclinous.

perianth   The collective term for both calyx and corolla.

pericarp   Fruit wall, developed from the ovary wall.

periderm   Outer bark; a protecting layer of cork cells; forming first just beneath the epidermis, later deeper in cortex or secondary phloem.

perigynous   Perianth appearing to arise from a hypanthium surrounding but not fused to the superior ovary; often used to describe the flower with this structure.

persistent   "Evergreen" leaves lasting until the next season's leaves appear, or persisting for several years.

petal   One unit of the corolla; inner perianth part.

petaloid   Petallike.

petiolate   Having a petiole.

petiole   The stalk of a leaf.

petiolule   The stalk of a leaflet.

phenetic classification   Grouping based on overall similarity without regard to evolutionary relationships.

phenology   The study of time and sequence of recurring stages in the life history of an organism.

phenotype   The evident characteristics of an organism due to the interaction of genotype and environment.

phenotypic plasticity   Nonheritable variation in characteristics; ecophenic variation.

phloem   Conducting tissue, mainly of sieve-tube elements; inner bark.

phyllotaxy   Leaf arrangement on a stem.

phylogeny (n.), **phylogenetic** (adj.)   Evolutionary history of a group.

physiognomy   The general appearance of a plant community, based on external features (e.g., tall, evergreen trees; low, deciduous trees; pine savannah; or bog).

pinnate   Featherlike, with a central axis and lateral branches.

pinnipalmate   Leaf venation somewhat intermediate between pinnate and palmate in which the lowermost pair of secondary veins are slightly larger and with larger tertiary veins than the other secondaries.

pistil   The evident ovary-style-stigma of a flower; a single separate carpel, or the fusion product of several carpels forming one compound pistil.

pistillate   Flower with pistil(s) and without fertile stamens; female.

pollen   An immature stage of the male gametophyte with a hard, resistant cell wall; carried by some agent in the process of pollination.

pollination   Transfer of pollen from pollen sac (or anther) to ovule (or stigma).

polygamous   Mixture of perfect and unisexual flowers and either mainly monoecious or dioecious (i.e. polygamo-monoecious or polygamo- dioecious).

pome   A fleshy fruit from an inferior ovary with the hypanthium forming most of the fleshy structure.

prickle   Sharp-pointed, slender outgrowth of the epidermis of a twig, leaf, or fruit husk.

provenance   Origin; locality where found or collected; the forester's term for a local race or ecotype.

pseudoterminal bud   The distal lateral bud on a twig that assumes the function of the terminal bud; the twig apex is aborted; recognized by its adjacent leaf scar and aborted tip of the twig.

puberulent   Minutely hairy, scarcely visible to the unaided eye.

pubescent   A general term meaning hairy; or, fine, soft, short hairs.

pulvinus (sing.), **pulvini** (pl.)   The swollen base of a petiole functioning in turgor movements of the leaf.

pyriform   Pear-shaped.

raceme   Inflorescence with a central axis bearing pedicelled flowers.

rachilla   Branch of a rachis.

rachis   Axis of a compound leaf or of an inflorescence.

receptacle   The portion of the axis that bears the floral organs.

recombination   The bringing together of chromosomes (and genes) as a result of fertilization, and the formation of gene combinations in the offspring that differ from those of the parents.

recurved   Curved downward or backward.

reflexed   Bent downward.

regular   Actinomorphic; radial symmetry in a flower.

reniform   Kidney-shaped; wider than long and with a broad cordate base.

repand   Slightly and irregularly wavy.

resin cyst   Cell or cavity (blister) with resin.

reticulate   Forming an interlacing network.

retuse   Apex with a shallow and narrow notch.

revolute   Rolled backward, with the margin rolled toward lower side.

rhombic   Oval, but angular at the sides; diamond-shaped.

rosette   A tight cluster of leaves, either basal or apical on a stem.

rufous   Reddish brown.

rugose   Wrinkled (leaf surface) with sunken veins.

samara   An indehiscent, dry, winged fruit.

samaroid   Samaralike.

savanna   Grassland with widely spaced trees.

scabrous   With short, bristly hairs; rough to the touch.

scale   Small, often appressed leaf; part of the conifer cone that bears ovules.

schizocarp   Dehiscent fruit that splits into individual locules (carpels) each appearing as a simple fruit, as the schizocarpic samaras of maple.

scurfy   Covered with minute scales.

seed   A ripened ovule, following fertilization, and normally with an embryo.

segregation   The separation of chromosomes (and genes) at the time of meiosis.

self-incompatibility   Genetically determined inability for self-fertilization.

sensu   According to; following.

sepal   A unit of the calyx; outer perianth part.

serotinous   Cones that remain closed long after the seeds inside are mature.

serrate   With sharp teeth pointing forward.

serrulate   Minutely serrate.

sessile   Lacking a stalk.

sheath   A leaf base partly or completely surrounding the stem.

short shoot   Short, lateral branch without internode elongation; either determinate or indeterminate.

simple   Of one piece; not compound.

sinuate   Shallowly indented, wavy in a horizontal plane. (See "undulate" for comparison.)

sinus   A recess, cleft, or gap between two lobes.

s. lat. (sensu lato)   In the inclusive sense; broadly defined.

spatulate   Spatula-shaped; elongated with widest point near rounded apex and tapering to the base.

speciation   Formation of a species.

species (sing. and pl.)   The unit of the classification hierarchy below the genus; defined generally as a population of potentially interbreeding individuals set apart from another species by more or less consistent and persistent differences in morphology, ecology, and breeding behavior.

spike   Inflorescence consisting of a central axis bearing sessile flowers.

spine   A sharp, modified leaf or stipule.

spinose   Spinelike or with spines.

spiral   Leaf arrangement on a stem that winds around from base to apex; alternate.

sporophyte   Diploid phase of the life cycle that includes the vascular plant.

spray   Flattened, frondlike branchlets in some conifers (Cupressaceae).

spur   A short, compact branch with no internode elongation; short shoot.

s. str. (sensu stricto)   In the restricted sense; narrowly defined.

stamen   The basic unit of the androecium; the microsporophyll of a flower.

staminate   Flower with stamens and without fertile pistils; male.

stellate   Star-shaped.

stem   Main aerial axis of a plant.

sterigma (sing.), **sterigmata** (pl.)   See "peg."

sterile   Lacking reproductive structures.

stigma   Top part of the pistil that is receptive of pollen.

stipular scar   Scar left on twig after stipules drop.

stipulate   With stipules.

stipule   A leafy appendage attached to the petiole or twig at the base of the petiole; in pairs, one on each side or sometimes fused together.

striate   With fine grooves, ridges, or lines of color.

strigose   Having stiff, straight, appressed hairs, usually pointing the same direction.

strobilus (sing.), **strobili** (pl)   See "cone."

style   Slender part of a pistil between the ovary and stigma; if lacking, then the stigma is sessile.

subulate   Short, narrow, tapered, sharp-pointed leaf; awn or awl.

succession   The sequential process of community change; the sequence of communities that replace one another in a given area culminating in a climax community.

succulent   Juicy, pulpy, fleshy.

superior ovary   Ovary above the origin of the perianth and stamens.

superposed   Bud(s) inserted just above the lateral bud.

suture   Line of dehiscence.

sympatric   Two or more taxa with overlapping geographical ranges.

sympetalous   With petals fused.

syndrome   A number of characteristics occurring together and characterizing a specific function or adaptation.

syngameon   The most inclusive group of introgressing species.

synoecious   Plant with perfect flowers (homoecious).

synonym   A rejected plant name that is replaced by another for some nomenclatural or taxonomic reason.

synzoochory   Diaspore dispersal by animals collecting seeds or fruits for a cache or for burying.

systematics   Study of the diversity of organisms including description, classification, nomenclature, and evolutionary relationships.

taxon (sing.), **taxa** (pl.)   General term for any taxonomic group, at any unspecified level of the classification hierarchy.

taxonomy   See "systematics."

tepal   Undifferentiated perianth part, separate or fused.

terete   Round in cross section.

terminal bud   Bud that terminates the twig; lacking an associated leaf scar.

ternate   In threes.

thorn   A sharp, modified branch in the axil of a leaf or terminating the twig.

tomentose   Densely wooly with soft, curled, matted hairs.

tomentum   The dense, matted hairs of a tomentose condition.

trichome   Any type of hair on the surface.

trichotomous   Three-forked.

trident   Three-toothed.

trifoliolate   With three leaflets.

trigonous   Three-angled.

tripinnate   Three-times pinnate.

truncate   As though cut off at right angle to the midrib.

turbinate   Top-shaped.

twig   The last year's growth of a woody stem.

type specimen   A specimen designated to serve as a reference point for a scientific name (species or infraspecific taxon); it is not necessarily the most typical. The *type* of a genus is a species; the *type* of a family is a genus and the one that forms the root of the family name.

umbel   Inflorescence, often flat-topped, consisting of several pedicelled flowers all attached at the same point at the top of the peduncle.

umbellate   Umbel-like.

umbo   A protuberance, as on the apophysis of a cone scale.

undulate   Wavy in a vertical plane; crisped. (See "sinuate" for comparison.)

unifoliolate   A leaf having only a single leaflet; usually described as "simple."

unisexual   Having only one sex; either diclinous flowers or dioecious plants.

valvate   Meeting at the edges and not overlapping, as in bud scales.

valve   One of the sections into which dehiscent fruits are split.

vascular   With xylem and phloem tissues.

vein   Strand of vascular tissue in a leaf or other laminar structure.

vernal   Appearing in spring.

verticillate   Whorled.

villous   With long, silky, straight hairs.

whorl   Cyclic arrangement, three or more per node.

wood   Secondary xylem.

woolly   With long, matted hairs.

xenogamy   Cross-fertilization between flowers on different plants; outcrossing.

xeric   Dry.

xerophyte   A plant adapted to dry conditions.

xylem   Conducting tissue formed mainly of tracheids or vessels and tracheids.

zygomorphic   Bilaterally symmetrical; irregular flower.

REFERENCES

Abrams, M. D., 1992. Fire and the development of oak forests. *BioScience* **42:**346–353.

———, 1998. The red maple paradox. *BioScience* **48:**355–364.

Adams, R. D. (ed.), 1990. Aspen Symposium '89, Proceedings. *USDA For. Serv. Gen. Tech. Rep. NC-140,* North Central For. Exp. Sta., St. Paul, MN.

Adams, R. P., 1986. Geographic variation in *Juniperus silicicola* and *J. virginiana* of the southeastern United States: Multivariate analyses of morphology and terpenoids. *Taxon* **35:**61–75.

———, 1993. *Juniperus. In* Flora of North America Editorial Committee (eds.), *Flora of North America North of Mexico,* Vol. 2, pp. 412–420. Oxford University Press, New York.

———, and B. L. Turner, 1970. Chemosystematic and numerical studies of natural populations of *Juniperus ashei* Buch. *Taxon* **19:**728–751.

Anderson, E., 1949. *Introgressive Hybridization.* Wiley, New York.

Avery, T. E., and H. E. Burkhart, 1983. *Forest Measurements,* 3d ed. McGraw-Hill, New York.

Bailey, L. H., Hortorium Staff, 1976. *Hortus Third: A Concise Dictionary of Plants Cultivated in the United States and Canada.* Macmillan, New York.

Bailey, V. L., 1962. Revision *of* the genus *Ptelea* (Rutaceae). *Brittonia* **14:**1–45.

Ballal, S. R., S. A. Fore, and S. I. Guttman, 1994. Apparent gene flow and genetic structure of *Acer saccharum* subpopulations in forest fragments. *Can. J. Bot.* **72:**1311–1315.

Bakuzis, E. V., and H. L. Hansen, 1965. *Balsam Fir.* University of Minnesota Press, Minneapolis.

Baranski, M. J., 1975. An analysis of variation within white oak *(Quercus alba* L.). *N.C. Agric. Exp. Sta. Tech. Bull. 236,* Raleigh.

Barnes, B. V., 1975. Phenotypic variation of trembling aspen in western North America. *For. Sci.* **22:**319–328.

———, and B. P. Dancik, 1985. Characteristics and origin of a new birch species, *Betula murrayana,* from southeastern Michigan. *Canad. J. Bot.* **63:**223–226.

———, ———, and T. L. Sharik, 1974. Natural hybridization of yellow birch and paper birch. *For. Sci.* **20:**215–221.

———, and W. H. Wagner, Jr., 1981. *Michigan Trees: A Guide to the Trees of Michigan and the Great Lakes Region.* University of Michigan Press, Ann Arbor.

———, D. R. Zak, S. R. Denton, and S. H. Spurr, 1998. *Forest Ecology,* 4th ed. John Wiley & Sons, New York.

Berg, E. E., and J. H. Hamrick, 1993. Regional genetic variation in turkey oak, *Quercus laevis. Canad. J. Forest Res.* **23:**1270–1274.

Berlin, B., 1992. *Ethnobiological Classification.* Princeton University Press, Princeton, NJ.

Black-Schaefer, C. L., and R. L. Beckmann, 1989. Foliar flavonoids and the determination of ploidy and gender in *Fraxinus americana* and *F. pennsylvanica* (Oleaceae). *Castanea* **54:**115–118.

Bogle, A. L., 1997. Leitneriaceae. *In* Flora of North America Editorial Committee (eds.), *Flora of North America North of Mexico,* Vol. 3, pp. 414–415. Oxford University Press, New York.

Borror, D. J., 1960. *Dictionary of Word Roots and Combining Forms.* Mayfield, Palo Alto, CA.

Braun, E. L., 1950. *Deciduous Forests of Eastern North America.* Blakiston, Philadelphia, PA.

Briggs, D., and S. M. Walters, 1984. *Plant Variation and Evolution,* 2d ed. Cambridge University Press, New York.

Brown, K., 1976. Ecological studies of the cabbage palmetto, *Sabal palmetto. Principes* **20:**3–10, 49–56, 98–115, 148–156.

Brunsfeld, S.J., P.S. and D.S. Soltis, P. A. Gadek, C. J. Quinn, D.D. Strenge, and T. A. Ranker, 1994. Phylogenetic relationships among the genera of Taxodiaceae and Cupressaceae: Evidence from rbcL sequences. *Syst. Bot.* **19:**253–262.

Buchholz, J. T., 1938. Cone formation in *Sequoia gigantea:* I. "The relation of stem size and tissue development to cone formation," II. "The history of the seed cone." *Amer. J. Bot.* **25:**296–305.

———, 1939. The generic segregation of the sequoias. *Amer. J. Bot.* **26:**535–538.

Burckhalter, R. E., 1992. The genus *Nyssa* (Cornaceae) in North America: A revision. *Sida* **15:**323–342.

Burns, R. M., and B. H. Honkala. 1990. Silvics of North America, Vol. 1, Conifers; Vol. 2, Hardwoods. *USDA Handb. 654,* Washington, D.C.

Camp, W. H., 1950. A biogeographic and paragenetic analysis of the American beech. *Yearb. Amer. Phil. Soc.* **1950:**166–169.

Campbell, C. S., and T. A. Dickinson, 1990. Apomixis, patterns of morphological variation, and species concepts in subfam. Maloideae (Rosaceae). *Syst. Bot.* **15:**124–135.

Clark, C. M., 1998. Chloroplast microsatellite differentiation in eastern North American *abies* mill. M.S. thesis, Dept. of Botany, North Carolina State University, Raleigh.

Clark, R. C., 1971. The woody plants of Alabama. *Ann. Missouri Bot. Gard.* **58:**99–242.

Clarkson, R. B., and D. E. Fairbrothers, 1970. A serological and electrophoretic investigation of eastern North American *Abies* (Pinaceae). *Taxon* **19:**720–727.

Cleaveland, M. K., and D. W. Stahle, 1989. Tree ring analysis of surplus and deficit runoff in the White River, Arkansas. *Water Resource Res.* **25:**1391–1401.

Clewell, A. F., and D. B. Ward, 1987. "White cedar in Florida and along the northern Gulf coast," in A. D. Laderman (ed.), *Atlantic White Cedar Wetlands,* pp. 69–82. Westview Press, Boulder and London.

Collier, H. O. J., 1963. Aspirin. *Sci. Amer.* **209:**97–108.

Cook, E. R., T. Bird, M. Peterson, M. Barbetti, B. Buckley, R. D'Arrigo, R. Francey, and P. Tans, 1991. Climatic change in Tasmania inferred from a 1089–year tree-ring chronology of Huon pine. *Science* **253:**1266–1268.

———, and G. Jacoby, Jr., 1977. Tree-ring-drought relationships in the Hudson Valley, New York. *Science* **198:**399–401.

———, and S. M. Zedaker, 1992. "The dendroecology of red spruce decline," in C. Eager and M. B. Adams (eds.), *Ecology and Decline of Red Spruce in the Eastern United States,* pp. 192–231. Springer-Verlag, New York.

Cooke, G. B., 1946. *The Planting and Growing of Cork Oak Trees in the United States.* Crown Cork and Seal Co., Baltimore, MD.

Coombes, A. J., 1985. *Dictionary of Plant Names.* Timber Press, Portland, OR.

Cooper, A. W., and E. P. Mercer, 1977. Morphological variation in *Fagus grandifolia* Ehrh. in North Carolina. *J. Elisha Mitchell Sci. Soc.* **93:**136–149.

Cottam, W. P., J. M. Tucker, and F. S. Santamour, Jr., 1982. Oak hybridization at the University of Utah. *State Arboretum of Utah Publ. 1,* Salt Lake City.

Critchfield, W. B., 1957. Geographic variation in *Pinus contorta. Maria Moors Cabot Found. Publ. 3,* Cambridge, MA.

———, 1984. Impact of the Pleistocene on the genetic structure of North American conifers. *Proc. North Amer. Forest Biol. Work. (Logan, Utah)* **8:**70–118.

———, and E. L. Little, Jr., 1966. Geographic distribution of the pines of the world. *USDA Miscl. Publ. 991,* Washington, D.C.

Cronquist, A., 1988. *The Evolution and Classification of Flowering Plants,* 2d ed. The New York Botanical Garden, Bronx, NY.

Cunningham, F. E., 1957. A seed key for five Northeastern birches. *J. For.* **55:**844.

Currey, D. R., 1965. An ancient Bristlecone pine stand in eastern Nevada. *Ecology* **46:**564.

Dallimore, W., and A. B. Jackson, 1967. A *Handbook of Coniferae and Ginkgoaceae,* 4th ed. by S. G. Harrison. St. Martin's Press, New York.

Daubenmire, R.F, and J. Daubenmire, 1968. Forest vegetation of eastern Washington and northern Idaho. *Wash. Agri. Exp. Sta. Tech. Bull. 60,* Pullman, WA.

Davis, M. B., 1976. Pleistocene biogeography of temperate deciduous forests. *Geosci. Man.* **13:**13–26.

———, (ed.), 1996. *Eastern Old-growth Forests: Prospects for Rediscovery and Recovery.* Island Press, Washington, D.C.

Dean, J. S., 1969. Chronological analysis of Tsegi phase sites in northeastern Arizona. *Papers of the Laboratory of Tree-Ring Research 3,* Tucson, AZ.

DeByle, N. V., and N. P. Winokur (eds.), 1985. Aspen: Ecology and management in the western United States. *USDA For. Serv. Gen. Tech. Rep. RM-119, Rocky Mtn. For. Range Exp. Sta.,* Fort Collins, CO.

Decker, J. P., 1952. Tolerance is a good technical term. *J. For.* **50:**40–41.

———, 1959. Shade-tolerance: A revised semantic analysis. *For. Sci.* **5:**93.

DeHond, P. E., and C. S. Campbell, 1989. Multivariate analyses of hybridization between *Betula cordifolia* and *B. populifolia* (Betulaceae). *Canad. J. Bot.* **67:**2252–2260.

Delcourt, H. R., P. A. Delcourt, and T. Webb, III, 1983. Dynamic plant ecology: The spectrum of vegetational change in space and time. *Quaternary Sci. Rev.* **1:**153–175.

Delcourt, P. A., and H. R. Delcourt, 1981. "Vegetation maps for eastern North America: 40,000 yr B.P. to the present," in R. C. Romans (ed.), *Geobotany II.* Plenum, New York.

Del Tredici, P., 1991. Ginkgos and people—A thousand years of interaction. *Arnoldia* **51**(2):2–15.

Dengler, H. W., 1966. Say it with Holly. *Amer. For.* **72:**1–3.

———, 1967. Bayberries and bayberry candles. *Amer. For.* **73:**5–7.

Den Ouden, P., and B. K. Boom, 1965. *Manual of Cultivated Conifers Hardy in the Cold- and Warm-Temperate Zone.* Martinus Nijhoff, The Hague.

de Oliveira, M. A., and L. de Oliveira, 1994. *The Cork.* ICEP, Lisboa, Portugal.

dePamphilis, C. W., and R. Wyatt, 1989. Hybridization and introgression in buckeyes (*Aesculus:* Hippocastanaceae): A review of the evidence and a hypothesis to explain long-distance gene flow. *Syst. Bot.* **14:**593–611.

———, and ———, 1990. Electrophoretic confirmation of interspecific hybridization in *Aesculus* (Hippocastanaceae) and the genetic structure of a broad hybrid zone. *Evolution* **44:**1295–1317.

Desmarais, Y., 1952. Dynamics of leaf variation in the sugar maples. *Brittonia* **7:**347–387.

Dickinson, T. A., 1985. Biology of Canadian weeds, *Crataegus crus-galli* L. *sensu lato. Canad. J. Pl. Sci.* **65:**641–654.

Dirr, M. A., 1997. *Dirr's Hardy Trees and Shrubs.* Timber Press, Portland, OR.

———, 1998. *Manual of Woody Landscape Plants: Their Identification, Ornamental Characteristics, Culture, Propagation and Uses,* ed. 5. Stipes Publishing Co., Champaign, IL.

Dorr, L. J., and K. C. Nixon, 1985. Typification of the oak (*Quercus*) taxa described by S. B. Buckley (1809–1884). *Taxon* **34:**211–228.

Douglass, A. E., 1920. Evidence of climatic effects in the annual rings of trees. *Ecology* **1:**24–32.

Duncan, W. H., and M. B. Duncan, 1988. *Trees of the Southeastern United States.* University of Georgia Press, Athens.

Eckenwalder, J. E., 1976. Re-evaluation of Cupressaceae and Taxodiaceae: A proposed merger. *Madroño* **23:**237–256.

———, 1980. Foliar heteromorphism in *Populus* (Salicaceae), a source of confusion in the taxonomy of Tertiary leaf remains. *Syst. Bot.* **5:**366–383.

Elias, T. S., 1989. *Field Guide to North American Trees.* Grolier Book Clubs, Danbury, CT.

Elliott, D., 1994. Bee gums, rabbit gums & mountain toothbrushes. *Wildlife in North Carolina* **58**(10):25–27.

Elmore, F. H., 1976. *Shrubs and Trees of the Southwest Uplands.* Southwest Parks and Monuments Assoc., Tucson, AZ.

Eyde, R. H., 1963. Morphological and paleobotanical studies of the Nyssaceae. I. A Survey of the modern species and their fruits. *J. Arnold Arbor.* **44:**1–52.

———,1988. Comprehending *Cornus:* Puzzles and progress in the systematics of the dogwoods. *Bot. Rev.* **54:**233–351.

Eyre, F. H. (ed.), 1980. *Forest Cover Types of the United States and Canada.* Society of American Foresters, Washington, D.C.

Faegri, K., and L. van der Pijl, 1979. *The Principles of Pollination Ecology,* 3d ed. Pergamon Press, New York.

Farjon, A., 1998. *World Checklist and Bibliography of Conifers.* Royal Botanic Garden, Kew, England.

Farrar, J. L., 1995. *Trees in Canada.* Fitzhenry & Whiteside Limited, Markham, Ontario and the Canadian Forest Service, Ottawa.

Farrar, R. M., Jr. (ed.), 1990. Proceedings of the symposium on the management of longleaf pine; 1998 April 4–6, Long Beach, MS. *Gen. Tech. Rep. SO-75, USDA Southern For. Exp. Sta.,* New Orleans, LA.

Fernald, M. L., 1950. *Gray's Manual of Botany,* 8th ed. American Book, New York.

Flake, R. H., E. von Rudloff, and B. L. Turner, 1969. Quantitative study of clinal variation in *Juniperus virginiana* using terpenoid data. *Proc. Nat. Acad. Sci.* **64:**487–494.

———, L. Urbatsch, and B. L. Turner, 1978. Chemical documentation of allopatric introgression in *Juniperus. Syst. Bot.* **3:**129–144.

FNA (Flora of North America Editorial Committee eds.), 1993. *Flora of North America North of Mexico, Vol. 1, Introduction, Vol. 2, Pteridophytes and Gymnosperms.* Oxford University Press, New York.

FNA (Flora of Notrth America Editorial Committee eds.), 1997. *Flora of North America North of Mexico, Vol. 3, Magnoliophyta: Magnoliidae and Hamamelidae.* Oxford University Press, New York.

Forest, H. S., R. J. Cook, and C. N. Bebee, 1990. The American chestnut: A bibliography. *USDA National Agricultural Library, Bibliographies and Literature of Agriculture No. 103*, Nat. Agric. Library, Beltsville, MD.

Fowells, H. A., 1949. Cork oak planting tests in California. *J. For.* **47:**357–365.

Fox, C. A., W. B. Kincaid, T. H. Nash, III, D. L. Young, and H. C. Fritts, 1986. Tree-ring variation in western larch (*Larix occidentalis*) exposed to sulfur dioxide emmisions. *Can. J. For. Res.* **16:**283–292.

Franklin, E. C., 1970. Survey of mutant forms and inbreeding depression in species of the family Pinaceae. *USDA Forest Serv. Res. Pap. SE-61,* Washington, D.C.

Fraver, S., 1992. The insulating value of serotinous cones in protecting pitch pine (*Pinus rigida*) seeds from high temperature. *J. Penn. Acad. Sci.* **65:**112–116.

Fritts, H. C., 1976. *Tree Rings and Climate.* Academic Press, London.

———, and T. W. Swetnam, 1989. Dendroecology: A tool for evaluating variations in past and present forest environments. *Adv. Ecol. Res.* **19:**111–188.

Frodin, D. G., 1984. *Guide to Standard Floras of the World.* Cambridge University Press, New York.

Furlow, J. J., 1979. The systematics of the American species of *Alnus* (Betulaceae). *Rhodora* **81:**1–121, 151–248.

———, 1997. Betulaceae. *In* Flora of North America Editorial Committee (eds.), *Flora of North America North of Mexico*, Vol. 3, pp. 507–538. Oxford University Press, New York.

Fry, W., and J. R. White, 1930. *Big Trees.* Stanford University Press, Stanford, CA.

Garin, G. I., 1958. Longleaf pines can form vigorous sprouts. *J. For.* **56:**430.

Gauthier, J.-P. Simon, and Y. Bergeron, 1992. Genetic structure and variability in jack pine populations: Effects of insularity. *Can. J. Bot.* **22:**1958–1965.

Geburek, T., 1997. Isozymes and DNA markers in gene conservation of forest trees. *Biodiv. & Conserv.* **6:**1639–1654.

Giannasi, D. E., 1978. Generic relationships in the Ulmaceae based on flavonoid chemistry. *Taxon* **27:**331–344.

Gibson, J. P., and J. L. Hamrick, 1991. Heterogeniety in pollen allele frequencies among cones, whorls, and trees of table mountain pine (*Pinus pungens*). *Amer. J. Bot.* **78:**1244–1251.

Gillis, W. T., 1971. Systematics and ecology of poison-ivy and the poison-oaks *(Toxicodendron*, Anacardiaceae). *Rhodora* **73:**72–159, 161–237, 370–443, 465–540.

Gleason, H. A., and A. Cronquist, 1991. *Manual of Vascular Plants of Northeastern United States and Adjacent Canada*, 2d ed. The New York Botanical Garden, Bronx, NY.

Gledhill, D., 1989. *The Names of Plants.* Cambridge University Press, New York.

Godfrey, R. K., 1988. *Trees, Shrubs, and Woody Vines of Northern Florida and Adjacent Georgia and Alabama.* University of Georgia Press, Athens.

Gower, S. T., and J. H. Richards, 1990. Larches: Deciduous conifers in an evergreen world. *BioScience* **40:**818–826.

Graham, S. A., R. P. Harrison, Jr., and C. E. Westell, Jr., 1963. *Aspens: Phoenix Trees of the Great Lakes Region.* University of Michigan Press, Ann Arbor.

Grant, V., 1963. *The Origin of Adaptations.* Columbia University Press, New York.

———, 1981. *Plant Speciation*, 2d ed. Columbia University Press, New York.

Grant, W. F., and B. K. Thompson, 1975. Observations on Canadian birches, *Betula cordifolia, B. neoalaskana, B. populifolia, B. papyrifera*, and *B.* ×*caerulea*. *Canad. J. Bot.* **53:** 1478–1490.

Greuter, W., et al., 1994. International code of botanical nomenclature (Tokyo code). *Regnum Veget.,* vol. 131.

Grewal, H. (ed.), 1987. Bibliography of lodgepole pine literature, *Rep. NOR-X-291, Can. For. Serv. North For. Center,* Edmonton, Alberta.

Griffin, J. R., P. M. McDonald, and P. C. Muick (comp.), 1987. California Oaks: A bibliography. *USDA For. Serv. Gen. Tech. Rep. PSW-96, Pac. Southwest For. Range Exp. Sta.,* Berkeley, CA.

Griffiths, M., 1994. *Index of Garden Plants.* Timber Press, Portland, OR.

Hamrick, J. L., M. J. W. Godt, and S. L. Sherman-Broyles. 1992. Factors influencing levels of genetic diversity in woody plant species. *New Forests* **5:**95–124.

———, Y. B. Linhart, and J. B. Mitton, 1979. Relationships between life history characteristics and electrophoretically detectable genetic variation in plants. *Ann. Rev. Ecol. Syst.* **10:**173–200.

Hardin, J. W., 1957a. A revision of the American Hippocastanaceae. *Brittonia* **9:**145–171, 173–195.

———, 1957b. Studies in the Hippocastanaceae, IV. Hybridization in *Aesculus*. *Rhodora* **59:**185–203.

———, 1974. Studies of the southeastern United States flora IV. Oleaceae. *Sida* **5:**274–285.

———, 1975. Hybridization and introgression in *Quercus alba*. *J. Arnold Arbor.* **56:**336–363.

———, 1979. *Quercus prinus* L.—nomen ambiguum. *Taxon* **28:**355–357.

———, 1990. Variation patterns and recognition of varieties of *Tilia americana* s.l. *Syst. Bot.* **15:**33–48.

———, 1992. Foliar morphology of the common trees of North Carolina and adjacent states. *N.C. Agric. Res. Serv. Tech. Bul. 298,* Raleigh.

———, and R. L. Beckmann, 1982. Atlas of foliar surface features in woody plants, V. *Fraxinus* (Oleaceae) of eastern North America. *Brittonia* **34:**129–140.

———, and G. P. Johnson, 1985. Atlas of foliar surface features in woody plants, VIII. *Fagus* and *Castanea* (Fagaceae) of eastern North America. *Bull. Torrey Bot. Club* **112:**11–20.

———, and D. E. Stone, 1984. Atlas of foliar surface features in woody plants, VI. *Carya* (Juglandaceae) of North America. *Brittonia* **36:**140–153.

Harlow, W. M., W. A. Cote, Jr., and A. C. Day, 1962. The opening mechanism of pine cone scales. *J. For.* **62:**538–540.

Harris, J. G., and M. W. Harris, 1994. *Plant Identification Terminology: An Illustrated Glossary.* Spring Lake Publishing, Payson, UT.

Hart, J. A., 1987. A cladistic analysis of conifers: Preliminary results. *J. Arnold Arbor.* **68:**269–307.

Heiser, C. B., 1973. Introgression re-examined. *Bot. Rev.* **39:**347–366.

Henderson, A., G. Galeano, and R. Bernal, 1995. *Field Guide to the Palms of the Americas.* Princeton University Press, Princeton, NJ.

Hennon, P. E., and A. S. Harris, 1997. Annotated bibliography of *Chamaecyparis nootkatensis. Gen. Tech. Rep. PNW-GTR-413, USDA Pac. Northwest Res. Sta.*, Portland, OR.

———, and C. G. Shaw III, 1997. The enigma of yellow-cedar decline. What is killing these long-lived, defensive trees? *J. For.* **95:**4–10.

Hibbs, D. E., and B. C. Fischer, 1979. Sexual and vegetative reproduction of striped maple (*Acer pensylvanicum* L.). *Bull. Torrey Bot. Club* **106:**222–226.

Hoff, R. J., J. I. Qualls, and D. O. Coffen, 1987. Western white pine: An annotated bibliography. *Gen. Tech. Rep. INT-232, USDA Intermountain Res. Sta.*, Ogden, UT.

Holmgren, P. K., N. H. Holmgren, and L. C. Barnett (eds.), 1990. *Index Herbariorum, pt. 1: The Herbaria of the World.* The New York Botanical Garden, Bronx, NY.

Hora, B. (ed.), 1981. *The Oxford Encyclopedia of Trees of the World.* Oxford University Press, Oxford.

Hori, T., R. W. Ridge, W. Tulecke, P. Del Tredici, J. Tremouillaux-Guiller, and H. Tobe (eds.), 1997. *Ginkgo biloba a Global Treasure from Biology to Medicine.* Springer, Tokyo.

Hosner, J. F., 1960. Relative tolerance to complete inundation of fourteen bottomland tree species. *For. Sci.* **6:**246–251.

Houston, D. R., 1994. Major new tree disease epidemics: Beech bark disease. *Ann. Rev. Phytopath.* **32:**75–87.

Howe, G., W. Hackett, R. Klevorn, and G. Furnier, 1993. Photoperiodic effects on growth and development of black cottonwood. *Amer. J. Bot.* **80**(6):74 (abstract).

Hu, H. H., 1948. How Metasequoia, the "living fossil," was discovered in China. *J. New York Bot. Gard.* **49:**201–207.

Hunt, R. S., 1993. *Abies. In* Flora of North America Editorial Committee (eds.), *Flora of North America North of Mexico,* Vol. 2, pp. 354–362. Oxford University Press, New York.

Iverson, L. R., A. M. Prasad, B. J. Hale, and E. K. Sutherland, 1999. Atlas of current and potential future distributions of common trees of the eastern United States. *USDA Forest Serv. Gen. Tech. Rept. NE-265,* Washington, D.C.

Jacobs, B. F., C. R. Werth, and S. I. Guttman, 1984. Genetic relationships in *Abies* (fir) of eastern United States: An electrophoretic study. *Can. J. Bot.* **62:** 609–616.

Jacobson, G.L, Jr., T. Webb, III, and E. C. Grimm, 1987. "Patterns and rates of vegetation change during the deglaciation of eastern North America," in W. F. Ruddiman and H. E. Wright, Jr. (eds.), *The Geology of North America,* Vol. K-3, *North America and Adjacent Oceans During the Last Deglaciation,* Geol. Soc. Amer., Boulder, CO.

Jeffrey, C., 1973. *Biological Nomenclature.* Edward Arnold Ltd., London.

Jepson, W. L., 1910. *The Silva of California.* University of California Press, Berkeley.

Johnson, E. A., and G. I. Fryer, 1989. Population dynamics in lodgepole pine—Engelmann spruce forests. *Ecology* **70:**1335–1345.

Johnson, G. P., 1988. Revision of *Castanea* sect. *Balanocastanon* (Fagaceae). *J. Arnold Arbor.* **69:**25–49.

Johnston, V. R., 1995. Pinyon pine—juniper woodland. *Fremontia* **23**(2):14–21.

Jones, G, N., 1968. Taxonomy of American species of Linden *(Tilia). Illinois Biol. Monogr. 39,* Urbana.

Jones, P. D., K. R. Briffa, and J. R. Pilcher, 1984. Riverflow reconstruction from tree rings in southern Britain. *J. Climatol.* **4:**461–472.

Jones, S. B., Jr., and A. E. Luchsinger, 1986. *Plant Systematics,* 2d ed. McGraw-Hill, New York.

Judd, W. S., C. S. Campbell, E. A. Kellogg, and P. F. Stevens, 1999. *Plant Systematics: A Phylogenetic Approach.* Sinauer Assoc., Inc., Sunderland, MA.

———, R. W. Sanders, and M. J. Donoghue, 1994. Angiosperm family pairs: Preliminary phylogenetic analyses. *Harvard Papers in Botany* **5:**1–51.

Kapp, R. O., 1969. *How to Know Pollen and Spores.* Wm. C. Brown, Dubuque, IA.

Kelly, P. E., E. R. Cook, and D. W. Larson, 1992. Constrained growth, cambial mortality, and dendrochronology of ancient *Thuja occidentalis* on cliffs of the Niagra Escarpment: An eastern version of bristlecone pine? *Inter. J. Plant Sci.* **153**:117–127.

Kienast, F., and F. H. Schweingruber, 1986. Dendroecological studies in the Front Range, Colorado, U.S.A. *Arct. Alp. Res.* **18**:277–288.

Kologiski, R. L., 1977. The phytosociology of the Green Swamp, North Carolina. *N.C. Agric. Exp. Sta. Tech. Bull. 250,* Raleigh.

Kozlowski, T. T., P. J. Kramer, and S. G. Pallardy, 1990. *The Physiological Ecology of Woody Plants.* Academic Press, San Diego, CA.

Kral, R., 1993. *Pinus. In* Flora of North America Editorial Committee (eds.), *Flora of North America North of Mexico,* Vol. 2, pp. 373–398. Oxford University Press, New York.

Kramer, P. J., and T. T. Kozlowski, 1979. *Physiology of Woody Plants.* Academic Press, New York.

Kreibel, H. B., 1957. Patterns of genetic variation in sugar maple. *Ohio Agric. Exp. Sta. Bull. 791,* Columbus.

———, 1993. Intraspecific variation of growth and adaptive traits in North American oak species. *Ann. Sci. For.* **50**(Suppl. 1):153s-165s.

Kuchler, A.W., 1964. Potential natural vegetation of the conterminous United States. *Amer. Geogr. Soc. Spec. Publ. 36.*

Laarman, J. G., and R. A. Sedjo, 1992. *Global Forests: Issues for Six Billion People.* McGraw-Hill, New York.

Laderman, A. D., 1989. The ecology of the Atlantic white cedar wetlands: A community profile. *U.S. Fish Wildl. Serv. Biol. Rep. 85 (7.21),* Washington, D.C.

Lampe, K. F., and M. A. McCann, 1985. *AMA Handbook of Poisonous and Injurious Plants.* AMA, Chicago, IL.

Landers, J. L., D. H. van Lear, and W. D. Boyer, 1995. The longleaf pine forests of the Southeast: Requiem or renaissance? *J. For.* **93**:39–44.

Langdon, O. G., 1963. Range of South Florida Slash pine. *J. For.* **61**:384–385.

Lanner, R. M., 1975. *Pinyon Pines and Junipers in the Southwestern Woodlands: The Pinyon-Juniper Ecosystem: A Symposium.* Utah State University, Logan.

Leaf, A. L., and K. G. Watterston, 1964. Chemical analysis of sugar maple sap and foliage as related to sap and sugar yields. *For. Sci.* **10**:288–292.

LeBlanc, D. C., 1992. Spatial and temporal variation in the prevalence of growth decline in red spruce populations of the northeastern United States. *Can. J. For. Res.* **22**:1351–1363.

Ledig, F. T., R. W. Wilson, J. W. Duffield, and G. Maxwell, 1969. A descriminant analysis of introgression between *Quercus prinus* L. and *Quercus alba* L. *Bull. Torrey Bot. Club* **96**:156–163.

Leopold, D. J., W. C. McComb, and R. N. Muller, 1998. *Trees of the Central Hardwood Forests of North America.* Timber Press, Portland, OR.

Li, Hui-Lin, 1964. *Metasequoia:* A living fossil. *Amer. Sci.* **54**:93–109.

Linnaeus, C., 1753. *Species Plantarum.* Stockholm, Sweden.

Little, E. L., Jr., 1971. Atlas of United States trees, Vol. 1, Conifers and important hardwoods. *USDA Miscl. Publ. 1146,* Washington, D.C.

———, 1976. Atlas of United States trees, Vol. 3, Minor western hardwoods. *USDA Miscl. Publ. 1314,* Washington, D.C.

———, 1977. Atlas of United States trees, Vol. 4, Minor eastern hardwoods. *USDA Miscl. Publ. 1342,* Washington, D.C.

———, 1978. Atlas of United States trees, Vol. 5, Florida. *USDA Miscl. Publ. 1361,* Washington, D.C.

———, 1979. Checklist of United States Trees (Native and Naturalized). *USDA Handb. 541,* Washington, D.C.

———, 1980a. *The Audubon Society Field Guide to North American Trees, Eastern Region.* Alfred A. Knopf, New York.

———, 1980b. *The Audubon Society Field Guide to North American Trees, Western Region.* Alfred A. Knopf, New York.

———, and W. B. Critchfield, 1969. Sub-divisions of the genus *Pinus* (Pines). *USDA Miscl. Publ. 1144,* Washington, D.C.

———, and K. W. Dorman, 1954. Slash pine *(Pinus elliottii),* including South Florida Slash pine. *USDA Southeast. For. Exp. Sta. Pap. 36,* Washington, D.C.

——— and B. H. Honkala, 1976. Trees and shrubs of the United States: A bibliography for identification. *USDA Miscl. Publ. 1336,* Washington, D.C.

Liu, Z., Y. Yu, and S. B. Carpenter, 1996. Baldcypress and pondcypress. An annotated bibliography 1890–1995. *Louisiana State Univ. Agric. Center Bull. 851, Louisiana Agric. Exp. Sta.,* Baton Rouge.

Loveless, M. D., and J. L. Hamrick, 1984. Ecological determinants of genetic structure in plant populations. *Ann. Rev. Ecol. Syst.* **15**:65–95.

Mabberley, D. J., 1997. *The Plant Book: A Portable Dictionary of the Vascular Plants,* 2d ed. Cambridge University Press, New York.

Malusa, J., 1992. Phylogeny and biogeography of the pinyon pines (*Pinus* subsect. *Cembroides*). *Syst. Bot.* **17:**42–66.

Manos, P. S., 1993. Foliar trichome variation in *Quercus* section *Protobalanus* (Fagaceae). *Sida* **15:**391–403.

McCune, B., 1988. Ecological diversity in North American pines. *Amer. J. Bot.* **75:**353–368.

Meerow, A. W., 1992. *Betrock's Guide to Landscape Palms.* Betrock Information Systems, Cooper City, FL.

Mergen, F., 1963. Ecotypic variation in *Pinus strobus. Ecology* **44:**716–727.

Meyer, F. G., and J. W. Hardin, 1987. Status of the name *Aesculus flava* Solander (Hippocastanaceae). *J. Arnold Arbor.* **68:**335–341.

Meyer, R. E., and S. Meola, 1978. Morphological characteristics of leaves and stems of selected Texas woody plants. *USDA Agric. Res. Serv. Tech. Bull. 1564.* Hyattsville, MD.

Millar, C. I., S. H. Strauss, M. T. Conkle, and R. D. Westfall, 1988. Allozyme differentiation and biosystematics of the California closed-cone pines. *Syst. Bot.* **13:**351–370.

Miller, G. N., 1955. The genus *Fraxinus,* the ashes, in North America, north of Mexico. *Cornell University Agric. Exp. Sta. Mem.* **335:**1–64.

Miller, R. F., and P. E. Wigand, 1994. Holocene changes in semiarid pinyon-juniper woodlands. *BioScience* **44:**465–474.

Minore, D., 1983. Western redcedar: A literature review. *USDA For. Serv. Gen. Tech. Rep. PNW-150, Pac. Northwest For. Range Exp. Sta.,* Portland, OR.

Mirov, N. T., 1967. *The Genus Pinus.* Ronald Press, New York.

Mitton, J. B., and M. C. Grant, 1996. Genetic variation and the natural history of quaking aspen. *BioScience* **46:**25–31.

Mosseler, A. J., K. N. Egger, and D. J. Innes, 1993. Life history and the loss of genetic diversity in red pines. *Amer. J. Bot.* **80**(6):81 (abstract).

Murrell, Z. E. 1993. Phylogenetic relationships in *Cornus* (Cornaceae). *Syst. Bot.* **18:**469–495.

Newton, A. C., T. R. Allnutt, A. C. M. Gillies, A. J. Lowe, and R. A. Ennos, 1999. Molecular phylogeography, intraspecific variation and the conservation of tree species. *TREE* **14:**140–145.

Nienstaedt, H., and A. Teich, 1971. The genetics of white spruce. *USDA For. Serv. Res. Pap. WO-15,* Washington, D.C.

Nixon, K. C., 1993. Infrageneric classification of *Quercus* (Fagaceae) and typification of sectional names. *Ann. Sci. For.* **50**(Suppl. 1):25s-34s.

———, 1997. Fagaceae. *In* Flora of North America Editorial Committee (eds.), *Flora of North America North of Mexico,* Vol. 3, pp. 436–506. Oxford University Press, New York.

Norstog, K. J., and T. J. Nichols, 1998. *The Biology of the Cycads.* Cornell University Press, Ithaca, NY.

O'Connell, M. M., M. D. Bentley, C. S. Campbell, and B. J. W. Cole, 1988. Betulin and lupeol in bark from four white-barked birches. *Phytochemistry* **27:**2175–2176.

Palmer, E. J., 1948. Hybrid oaks of North America. *J. Arnold Arbor.* **29:**1–48.

Parks, C. R., J. F. Wendel, M. M. Sewell, and Y.-L. Qiu, 1994. The significance of allozyme variation and introgression in the *Liriodendron tulipifera* complex (Magnoliaceae). *Amer J. Bot.* **81:**878–889.

Pitcher, D. C., 1987. Fire history and age structure in red fir forests of Sequoia National Park, California. *Can. J. For. Res.* **17:**582–587.

Price, R. A., and J. M. Lowenstein, 1989. An immunological comparison of the Sciadopityaceae, Taxodiaceae, and Cupressaceae. *Syst. Bot.* **14:**141–149.

———, J. Olsen-Stojkovich, and J. M. Lowenstein, 1987. Relationships among the genera of Pinaceae: An immunological comparison. *Syst. Bot.* **12:**91– 97.

Primack, R. B., and C. McCall, 1986. Gender variation in a red maple population (*Acer rubrum:* Aceraceae): A seven-year study of a "polygamodioecious" species. *Amer. J. Bot.* **73:**1239–1248.

Proctor, M., P. Yeo, and A. Lack, 1996. *The Natural History of Pollination.* Timber Press, Portland, OR.

Radford, A. E., 1986. *Fundamentals of Plant Systematics.* Harper & Row, New York.

———, W. C. Dickison, J. R. Massey, and C. R. Bell, 1974. *Vascular Plant Systematics.* Harper & Row, New York.

Ragsdale, H. L., and C. W. Berish, 1988. The decline of lead in tree rings of *Carya* spp. in urban Atlanta, GA, USA. *Biogeochemistry* **6:**21–29.

Reed, P. B., Jr., 1988. National list of plant species that occur in wetlands: National summary. *U.S. Fish Wildl. Serv. Biol. Rept.* **88**(24).

Reveal, J. L., and M. L. Seldin, 1976. On the identity of *Halesia carolina* L. (Styracaceae). *Taxon* **25:**123–140.

Robson, K. A., J. Maze, R. K. Scagel, and S. Banerjee, 1993. Ontogeny, phylogeny, and intraspecific variation in North American *Abies* Mill. (Pinaceae): An emperical approach to organization and evolution. *Taxon* **42:**17–34.

Rushton, B. S., 1993. Natural hybridization within the genus *Quercus* L. *Ann. Sci. For.* **50**(Suppl. 1):73s-90s.

Santamour, F. S., Jr., 1962. The relation between polyploidy and morphology in white and biltmore ashes. *Bull. Torrey Bot. Club* **89**:228–232.

Schmidt, T. L., and R. J. Piva, 1995. An annotated bibliography of eastern redcedar, *Resource Bull. NC-166, USDA North Central For. Exp. Sta.,* St. Paul, MN.

Schmidt, W. C., and K. J. McDonald, 1990. Proceedings of the symposium on whitebark pine ecosystems: Ecology and management of a high-mountain resource; 1989 March 29–31; Bozeman, MT. *Gen. Tech. Rep. INT-270, USDA Intermountain Res. Sta.,* Ogden, UT.

———, and ———, 1995. Ecology and management of Larix forests: A look ahead. Proceedings of an international symposium; 1992 October 5–9; Whitefish, MT. *Gen. Tech. Rep. INT-GTR-319, USDA Intermountain Res. Sta.,* Ogden, UT.

Schreiner, E. J., 1949. Poplars can be bred to order. *USDA Yearbook Separate 2119*, Washington, DC.

———, 1950. Genetics in relation to forestry. *J. For.* **48**:33–38.

Schulman, E. 1958. Bristlecone pine, oldest known living thing. *Natl. Geogr.* **113**:355–372.

Schultz, R. P., 1997. Loblolly pine. The ecology and culture of loblolly pine (*Pinus taeda* L.). *USDA Agri. Hand. 713,* Washington, DC.

Scott, C. W., 1960. *Pinus radiata,* Food and Agric. *Organization of the United Nations,* Rome.

Sharik, T. L., and B. V. Barnes, 1976. Phenology of shoot growth among diverse populations of yellow birch *(Betula alleghaniensis)* and sweet birch *(B. lenta).* *Canad. J. Bot.* **54**:2122–2129.

Smith, J. P., Jr., 1977. *Vascular Plant Families.* Mad River Press, Eureka, CA.

Smouse, P. E., and L. C. Saylor, 1973a. Studies of the *Pinus rigida-serotina* complex: I. A Study of geographic variation. *Ann. Missouri Bot. Gard.* **60**:174–191.

———, and ———, 1973b. Studies of the *Pinus rigida-serotina* complex: II. Natural hybridization among *Pinus rigida-serotina* complex, *P. taeda* and *P. echinata*. *Ann. Missouri Bot. Gard.* **60**:192–203.

Spongberg, S. A., 1990. *A Reunion of Trees.* Harvard University Press, Cambridge, MA.

———, and J. Ma, 1997. *Cladrastis* (Leguminosae subfamily Faboideae tribe Sophoreae): A historic and taxonomic overview. *Intern. Dendrology Soc.Yearbook* **1996**:27–35.

Squillace, A. E., 1966. Geographic variation in Slash pine. *For. Sci. Monogr.* 10.

Stace, C. A., 1989. *Plant Taxonomy and Biosystematics,* 2d ed. Edward Arnold, London.

Stahle, D. W., and M. K. Cleaveland, 1992. Reconstruction and analysis of spring rainfall over southeastern U.S. for the past 1000 years. *Bull. Amer. Meteorol. Soc.* **73**:1947–1961.

———, ———, D. B. Blanton, M. D. Therrell, and D. A. Gray, 1998. The Lost Colony and Jamestown Droughts. *Science* **280**:564–567.

———, ———, and R. S. Cerveny, 1991. Tree-ring reconstructed sunshine duration over central USA. *Inter. J. Climat.* **11**:285–295.

———, ———, and J. G. Hehr, 1985. A 450–year drought reconstruction for Arkansas, United States. *Nature* **316**:530–532.

———, ———, and ———, 1988. North Carolina climate changes reconstructed from tree rings: A.D. 372 to 1985. *Science* **240**:1517–1519.

———, and D. Wolfman, 1985. The potential for archeological tree-ring dating in eastern North America. *Advances in Archaeological Method and Theory* **8**:279–302.

Stearn, W. T., 1992. *Botanical Latin.* David and Charles, London.

Stebbins, G. L., Jr., 1950. *Variation and Evolution in Plants.* Columbia University Press, New York.

Steiner, K. C., 1979. Patterns of variation in bud-burst timing among populations in several *Pinus* species. *Silvae Genet.* **28**:185–194.

Steinhoff, R . J ., and J . W. Andresen, 1971. Geographic variation in *Pinus flexilis* and *P. strobiformis* and its bearing on the taxonomic status. *Silvae Genet.* **20**:159–167.

Stettler, R. F., H. D. Bradshaw, Jr., P. E. Heilman, and T. M. Hinkley (eds.), 1996. *Biology of Populus and Its Implications for Management and Conservation.* NRC Research Press, Ottawa, Ontario.

Stone, D. E., 1973. Patterns in the evolution of amentiferous fruits. *Brittonia* **25**:371–384.

———, 1997. *Carya. In* Flora of North America Editorial Committee (eds.), *Flora of North America North of Mexico*, Vol. 3, pp. 417–425. Oxford University Press, New York.

Stone, E. L., Jr., and M. H. Stone, 1943. "Dormant" versus adventitious buds. *Science* **98**:62.

Strauss, S. H, and A. H. Doerksen, 1990. Restriction fragment analysis of pine phylogeny. *Evolution* **44**:1081–1096.

Stuessy, T. F., 1990. *Plant Taxonomy.* Columbia University Press, New York.

Sudworth, G. B., 1967. *Forest Trees of the Pacific Slope.* Dover, New York.

Swetnam, T. W., 1993. Fire history and climate change in giant sequoia groves. *Science* **262**:885–889.

Taylor, R. J., 1993. *Picea. In* Flora of North America Editorial Committee (eds.), *Flora of North America North of Mexico*, Vol. 2, pp. 369–373. Oxford University Press, New York.

Terrell, E. E., S. R. Hill, J. H. Wiersema, and W. E. Rice, 1986. A checklist of names for 3000 vascular plants of economic importance. *USDA Handb. 505*, Washington, D.C.

Thor, E., and P. E. Barnett, 1974. Taxonomy of *Abies* in the southern Appalachians: Variation in balsam monoterpenes and wood properties. *For. Sci.* **20:**32–40.

Tiner, R. W., 1991. The concept of a hydrophyte for wetland identification. *BioScience* **41:**236–247.

Trehane, P., C. D. Brickell, B. R. Baam, W. L. A. Hetterscheid, A. C. Leslie, J. McNeill, S. A. Spongberg, and F. Vrugtman (eds.), 1995. *The International Code of Nomenclature for Cultivated Plants.* Quarterjack Publishing, Wimborne, UK.

Tucker, J. M., 1993. Hybridization in California oaks. *J. Internatl. Oak Soc.* **3:** 4–13.

———, and H. S. Haskell, 1960. *Quercus dunnii* and *Q. chrysolepis* in Arizona. *Brittonia* **12:**196–219.

Tyrrell, L. E., G. J. Nowacki, T. R. Crow, D. S. Buckley, E. A. Nauertz, J. N. Niese, J. L. Rollinger, and J. C. Zazada, 1998. Information about old growth for selected forest type groups in the eastern United States. *USDA For. Serv. Gen. Tech. Rep. NC-197, North Central For. Exp. Sta.*, St. Paul, MN.

van der Pijl, L., 1982. *Principles of Dispersal in Higher Plants,* 3d ed. Springer-Verlag, Berlin.

Viereck, L. A., and E. L. Little, Jr., 1975. Atlas of United States trees: Vol. 2, Alaska trees and common shrubs. *USDA Miscl. Publ. 1293,* Washington, D.C.

Walker, L. C., 1998. *The North American Forests: The Geography, Ecology, and Silviculture.* CRC Press, Boca Raton, FL.

Waller, G.R. (ed.), 1987. *Allelochemicals: Role in Agriculture and Forestry.* ACS Symposium Series 30. Amer. Chem. Society, Washington, D.C.

Walters, D. R., and D. J. Keil, 1996. *Vascular Plant Taxonomy,* 4th ed. Kendall/Hunt, Dubuque, IA.

Ward, D. B., 1963. Contributions to the flora of Florida—2, Pines (Pinaceae). *Castanea* **28:**1–10.

———, and A. F. Clewell, 1989. Atlantic white cedar *(Chamaecyparis thyoides)* in the southern states. *Florida Scientist* **52:**8–47.

Watson, F. D., 1983. A taxonomic study of pondcypress and baldcypress *(Taxodium* Richard). Ph.D. dissertation, Dept. of Botany, North Carolina State University, Raleigh.

———, 1993. *Taxodium. In* Flora of North America Editorial Committee (eds.), *Flora of North America North of Mexico,* Vol. 2, pp. 403–404. Oxford University Press, New York.

Weatherspoon, C. P., Y. R. Iwamoto, and D. D. Piirto, 1986. Proceedings of the workshop on management of giant sequoia; 1985 May 24–25, Reedley, CA. *Gen. Tech. Rep. PSW-95, Pacific Southwest For. and Range Exp. Sta.,* Berkeley, CA.

Westerkamp, C., and H. Demmelmeyer, 1997. *Leaf Surfaces of Central European Woody Plants.* Gebrüder Borntraeger, Berlin.

Whittaker, R. H., 1956. Vegetation of the Great Smoky Mountains. *Ecol. Monogr.* **26:**1–80.

Wilson, K. A., and C. E. Wood, Jr., 1959. The genera of Oleaceae in the Southeastern United States. *J. Arnold Arbor.* **40:**369–384.

Wilson, K. L., 1997. Casuarinaceae. *In* Flora of North America Editorial Committee (eds.), *Flora of North America North of Mexico,* Vol. 3, pp. 539–541. Oxford University Press, New York.

Woodland, D. W., 1997. *Contemporary Plant Systematics,* 2d ed. Andrews University Press, Berrien Springs, MI.

Wright, H. E., 1983. *Late Quaternary Environments of the United States:* Vol. 2, *The Holocene.* University of Minnesota Press, Minneapolis.

Wright, J. W., 1955. Species crossability in spruce in relation to distribution and taxonomy. *For. Sci.* **1:**319–349.

Yang, R.-C., and F. C. Yeh, 1992. Genetic consequences of in situ and ex situ conservation of forest trees. *Forest. Chron.* **68:**720–729.

Yazvenko, S. B., and D. J. Rapport, 1997. The history of ponderosa pine pathology. *J. For.* **95:**16–20.

Young, A. G., H. G. Merriam, and S. I. Warwick, 1993. The effects of forest fragmentation on genetic variation in *Acer saccharum* Marsh. (sugar maple) populations. *Heredity* **71:**277–289.

Young, J. A., and C. G. Young, 1992. *Seeds of Woody Plants.* Timber Press, Portland, OR.

Zimmerman, M. H., and C. L. Brown, 1971. *Trees: Structure and Function.* Springer-Verlag, New York.

Zobel, B. J., 1951. The natural hybrid between Coulter and Jeffrey pines. *Evolution* **5:**405–413.

———, and J. Talbert, 1984. *Applied Forest Tree Improvement.* Wiley, New York.

———, G. Van Wyk, and P. Stahl, 1987. *Growing Exotic Forests.* Wiley, New York.

Zobel, D. B., 1969. Factors affecting the distribution of *Pinus pungens,* an Appalachian endemic. *Ecol. Monogr.* **39:**303–333.

———, L. F. Roth, and G. M. Hawk, 1985. Ecology, pathology, and management of Port-Orford-cedar *(Chamaecyparis lawsoniana). Gen. Tech. Rep. PNW-184, USDA Pacific Northwest For. and Range Exp. Sta.,* Portland. OR.

Zomlefer, W. B., 1994. *Guide to Flowering Plant Families.* University of North Carolina Press, Chapel Hill.

Zon, R., 1907. A new explanation of the tolerance and intolerance of trees. *Proc. Soc. Amer. For.* **2:**79–94.

INDEX

Page numbers in **boldface** indicate illustrations; names in *italics* indicate synonyms.

Abies, 188
  alba, 190
  amabilis, 195, **196**
  arizonica, 208
  balsamea, 191, **192, 194**
    var. *fraseri,* 194
    var. phanerolepis, 193
  bifolia, 208
  bracteata, 209
  cephalonica, 190
  concolor, 204, **205**
    var. concolor, 206
    var. lowiana, 206
  firma, 190
  fraseri, 194
  grandis, 201, **202, 203**
  homolepis, 190
  lasiocarpa, 206, **207, 208**
    var. arizonica, 208
  lowiana, 206
  magnifica, 197, **198**
    var. shastensis, 197
  nordmanniana, 190
  pindrow, 190
  pinsapo, 190
  procera, 199, **200**
  religiosa, 190
  sibirica, 190
Abietaceae, 103

Acacia, 416
Acer, 444
  barbatum, 448
  ginnala, 444
  griseum, 444
  japonicum, 444
  leucoderme, 448
  macrophyllum, 449, **450**
  negundo, 458
  nigrum, **449**
  palmatum, 444
  pensylvanicum, 456
  platanoides, 444
  pseudoplatanus, 444
  rubrum, 452, **453**
    var. trilobum, 454
  saccharinum, 454, **455**
  saccharum, 444, **446, 447**
    var. or subsp.
      floridanum, 448
    var. leucoderme, 448
  spicatum, **457**
  truncatum, 444
ACERACEAE, 443
Aesculus, 440
  arguta, 442
  californica, 442
  ×carnea, 442
  flava, 440, **441, 442**

  glabra, 440, **443**
    var. arguta, 442
  hippocastanum, 442
  octandra, 440
  parviflora, 442
  pavia, 442
  sylvatica, 442
African oil palm, 483
Agamospermy, 51
Agathis australis, 102
AGAVACEAE, 486
Agave, 487
  Parry, 487, **488**
  parryi, 487, **488**
Ailanthus altissima, 464, **466**
Akee, 439
Alaska-cedar, 235, **236**
Alaska-cypress, 235
Albizia julibrissin, 417
Alder, 370
  black, 370
  European, 370
  hazel, 371
  red, 371, **372**
  Sierra, 373
  speckled, 371
  white, 370, 373
Alder buckthorn, 438
Aleurites fordii, 437

517